CASE

SHOP MANUAL C-201

Models ■ C ■ D ■ L ■ LA ■ R ■ S ■ V ■ VA

Models ■ 200B ■ 300 ■ 300B ■ 350 ■ 400B ■ 500B ■ 600B

Models ■ 400 ■ 700B ■ 800B

I&T

SHOP SERVICE

Information and Instructions

This shop manual contains several sections each covering a specific group of wheel type tractors. The Tab Index on the preceding page can be used to locate the section pertaining to each group of tractors. Each section contains the necessary specifications and the brief but terse procedural data needed by a mechanic when repairing a tractor on which he has had no previous actual experience.

Within each section, the material is arranged in a systematic order beginning with an index which is followed immediately by a Table of Condensed Service Specifications. These specifications include dimensions, fits, clearances and timing instructions. Next in order of arrangement is the procedures paragraphs.

In the procedures paragraphs, the order of presentation starts with the front axle system and steering and proceeding toward the rear axle. The last paragraphs are devoted to the power take-off and power lift systems. Interspersed where needed are additional tabular specifications pertaining to wear limits, torquing, etc.

HOW TO USE THE INDEX

Suppose you want to know the procedure for R&R (remove and reinstall) of the engine camshaft. Your first step is to look in the index under the main heading of ENGINE until you find the entry "Camshaft." Now read to the right where under the column covering the tractor you are repairing, you will find a number which indicates the beginning paragraph pertaining to the camshaft. To locate this wanted paragraph in the manual, turn the pages until the running index appearing on the top outside corner of each page contains the number you are seeking. In this paragraph you will find the information concerning the removal of the camshaft.

More information available at Clymer.com
Phone: 805-498-6703

Haynes Publishing Group
Sparkford Nr Yeovil
Somerset BA22 7JJ England

Haynes North America, Inc
859 Lawrence Drive
Newbury Park
California 91320 USA

ISBN-10: 0-87288-373-6
ISBN-13: 978-0-87288-373-4

CASE

Models ■C ■D ■L ■LA ■R ■S ■V ■VA

Previously contained in I & T Shop Service Manual No. C-2

SHOP MANUAL
J. I. CASE

SERIES	PRODUCTION	SERIES	PRODUCTION
C	1937-1939	R	1937-1940
D	1939-Current	S	1941-Current
L	1937-1940	V	1940-1942
LA	1940-Current	VA	1942-Current

Serial number located on instrument panel name plate. On all series except V and VA, engine serial number is stamped on left rear side of engine block in vicinity of air cleaner. On series V and VA, engine serial number is stamped on right side of engine block.

BUILT IN THESE VERSIONS

	MODEL DESIGNATIONS Tricycle		MODEL DESIGNATIONS Axle Type	
SERIES	Single Wheel	Double Wheel	Non-Adjustable	Adjustable
C	CC	CC	C-CC-CO	—
D	DC-DC3	DC-DC3	D-DC-DC4-DCS-DH-DO-DS-DV	DC-DC3
L	—	—	L	—
LA	—	—	LA	—
R	RC	RC	R	RC
S	SC	SC	S-SO	SC
V	VC	VC	V-VO	VC
VA	VAC-VAC11	VAC-VAC12	VA-VAO	VAC-VAC13-VAC14-VAH

(For tractor cross-sectional views, refer to pages 45 and 59.)

INDEX (By Starting Paragraph)

CONDENSED SERVICE DATA

Tractor Model	C	D	L	LA	R	S	V	VA
GENERAL								
Engine Make	Own	Own	Own	Own	Own	Own	Own	Own
Engine Model	C	D	L	LA	FC	S	F124	VA
Cylinders	4	4	4	4	4	4	4	4
Bore—Inches	3 7/8 or 4.0	3 7/8 or 4.0	4 5/8	4 5/8	3 1/4	3 1/2 or 3 5/8	3	3 1/4
Stroke—Inches	5 1/4	5 1/4	6	6	4	4	4 3/8	3 3/4
Displacement—Cubic Inches	260 or 276.4	260 or 276.4	404	404	133	154 or 165	124	124
Compression Ratio—Gas., 3 7/8 Inch Bore		5.58		4.83				
Compression Ratio—Dist., 3 7/8 Inch Bore		4.88		4.09				
Pistons Removed From:	A&B	A&B	A&B	A&B	A&B	A&B	Above	Above
Main Brgs., Number of:	3	3	3	3	3	3	3	3
Main Brgs., Adjustable?	Yes	Yes	Yes	Yes	(12)	No	No	No
Rod Brgs., Adjustable?	Yes	Yes	Yes	Yes	(12)	No	No	No
Cylinder Sleeves	Wet	Wet	Wet	Wet	Dry	Wet	Dry	Wet
Forward Speeds	3	4	3	4	3	4	4	4
Generator Make				←——Auto-Lite——→				
Starter Make				←——Auto-Lite——→				
TUNE-UP								
Firing Order	1-3-4-2	1-3-4-2	1-3-4-2	1-3-4-2	1-3-4-2	1-3-4-2	1-3-4-2	1-3-4-2
Valve Tappet Gap—Inlet	18C	18C	18C	18C	10C	10C	14C	14C
Valve Tappet Gap—Exhaust	18C	18C	18C	18C	12C	10C	14C	14C
Inlet Valve Face Angle	45	45	45	45	45	45	30	30
Exhaust Valve Face Angle	45	45	45	45	45	45	45	45
Inlet Seat Angle	45	45	45	45	45	45	30	30
Exhaust Seat Angle	45	45	45	45	45	45	45	45
Ignition Distributor Make							←——Auto-Lite——→	
							See paragraph 133	
Ignition Magneto Make	Case	Case	Bosch	Case	Case	Case	F-M	Case
Ignition Magneto Model	4JMA	41	MJA-4B112	4CMA	4CMA	41	FMJ-4B	41
Magneto Model Before Tractor 4607033		4CMA						
Distributor Breaker Gap							.020	.020
Magneto Breaker Gap	.015	.010	.015	.015	.015	.010	.020	.010
Distributor Timing Retard							TC	TC
Distributor Timing—Full Advance								
Magneto Impulse Trip Point	TC	TC	TC	TC	TC	1/2" BTC	TC	3/16" ATC
Magneto Lag Angle—Degrees	25	25	30	30	35	25	15	25
Magneto Lag Angle Prior Tractor 4607033		30						
Magneto Running Timing	25°		30°	30°	35°		15°	
Flywheel Mark Indicating:								
Magneto Impulse Trip Point	D	D	D	D	DCI	3/16" Hole	DC	
Distributor Retard Timing							DC	DC
Spark Plug Make	AC	AC	AC	AC	AC	AC	AC	AC
Model for Gasoline & LPG	75	85S Com	AC75	85S Com	84	45 Com	85	87 Com
Model for Low Octane	75	85S Com	AC75	85S Com	84	45 Com	85	87 Com
Electrode Gap	.025	.025	.025	.030	.030	.025	.025	.025
Carburetor Make	Zenith	Zenith	Zenith	Zenith	Zenith	Zenith	M-S	M-S Zenith
Model & Assembly No.	Refer to Carburetor Application Table, paragraph 98.							
Float Setting (Marvel Schebler)							9/32	9/32
Float Setting (Zenith)	1 39/64	1 39/64	1 49/64	1 49/64	1 5/32	1 5/32		1 5/32
Calibration	Refer to Model Number in Tractor Components Manual.							
Engine Low Idle RPM	500	500	500	500	400	500	500	500
Engine High Idle RPM	1175	1390	1200	1200	1555	1760	1600	*
Engine Loaded RPM	1100	1200	1100	1100	1425	1550	1425	*
Belt Pulley Loaded RPM	C-973 CC-846	818	780	779	991	1078	969	*
Belt Pulley No Load RPM		948	860	850	1080	1224	1088	*
PTO Loaded RPM	536	540	550	550	543	541	559	*
PTO No Load RPM	573	622	600	600	595	614	628	*
SIZES-CAPACITIES-CLEARANCES								
(Clearances in Thousandths)								
Crankshaft Journal Diameter	2.250	2.205	2.750	2.750	2.1245	2.500	2.250	2.249
Crankpin Diameter	2.2495	2.2495	2.7495	2.7495	1.7495	2.3745	1.937	1.937
Rod C to C Length	10 3/4	10 3/4	12	12	7 1/4	8 1/2	7	7
Camshaft Journal Dia., Front	1.8075	1.8075	2.2665	2.2665	2.1235	1.9025	1.872	1.7495
Camshaft Journal Dia., Center	1.7775	1.7775	2.2465	2.2465	1.8735	1.8715	1.746	1.7495
Camshaft Journal Dia., Rear	1.7475	1.7475	2.2265	2.2265	1.4985	1.8405	1.247	1.7495
Piston Pin Diameter	1.2333	1.2333	1.4848	1.4848	.875	.9993	.8592	.8592
Valve Steam Diameter	7/16	7/16	7/16	7/16	5/16	5/16	11/32	11/32
Compression Ring Width	1/8	1/8	5/32	5/32	1/8	1/8	1/8	1/8
Oil Ring—Width	1/4	1/4	5/32	5/32	3/16	3/16	1/4	1/4

*Refer to paragraph 105A

Tractor Model	C	D	L	LA	R	S	V	VA
SIZES—CAPACITIES—CLEARANCES continued ...								
Main Brgs., Diam. Clearance..................	1.5-3	1.5-3	1.5-3	1.5-3	.5-2.5	1.5-2.5	1.5-3	1.5-3
Rod Brgs., Diam. Clearance...................	1.5-3	1.5-3	1.5-3	1.5-3	.5-2.5	1-3	1.5-2	1.5-2
Piston Skirt Clearance.......................	4-5	4-5	6-8	6-8	3-4	3.5-5	3-4	3-4
Crankshaft End Play..........................	4-10	4-10	4-10	4-10	5-7	4-8	4-6	4-6
Camshaft Bearing Clearance..................	2-4	2-4	2-4	2-4	1.5-2	2-4	2-4	2-4
Cooling System—Gallons......................	5.0	6.75	12.5	15.25	4.25	4.0	3.0	3.25
Crankcase Oil—Quarts........................	7.0	7.0	12.0	12.0	7.0	5.0	4.0	4.0
Trans. & Diff.—Quarts.......................	36.0	42.0	68.0	68.0	43.0	43.0	20.0	28.0
(Early Production)........................						36.0		
Final Drive, Each—Quarts....................	0.0	6.0	0.0	0.0	0.0	0.0	0.0	1.5
Torque Tube—Quarts..........................	0.0	0.0	0.0	0.0	0.0	0.0	8.0	8.0
(Refill)...................................								6.0
Hydraulic System—Quarts....................	0.0	10.0	0.0	13.0	0.0	9.5	(7)	(6)
(If 2 work cylinders)......................		11.5				11.0		
Add for BP—Quarts...........................	0.0	0.0	0.0	0.0	0.0	0.0	0.0	0.0
Add for PTO Continuous type—Quarts.......	0.0	10.0	0.0	0.0	0.0	10.0	0.0	0.0

(6) Torque tube is reservoir for hydraulic system. (7) Models with Kingston pump mounted on belt pulley unit use the torque tube lubricant. All other models use the transmission oil. (12) Bearings have shims which only control crush of shell insert.

FRONT SYSTEM—TRICYCLE AND AXLE TYPE

ADJUSTMENT OF PEDESTAL (Vertical Spindle) & FRONT STEERING GEAR

This section applies to the complete front system on all models except that it does not cover the steering gear units which are mounted on the torque tube or transmission. It does cover the steering mechanism mounted in the front support. For information on servicing the steering gears which are mounted on torque tube or transmission refer to MAIN STEERING GEAR beginning with paragraph 22.

Single Wheel Models CC-DC-DC3-RC-SC-VC-VAC prior 5570000 (Refer to paragraph 3 for dual wheel type)

The front axle pedestal (support) on these models houses the integral vertical spindle and wheel fork. On models VAC prior 5570000 and VC, the vertical spindle rotates in a bushing (3—Fig. C1) at the upper end, and in a taper roller bearing (5) at the lower end. On all other models, the vertical spindle rotates in two bushings. Refer to Fig. C2.

An all models, except VAC and VC, spindle end thrust is carried by either a roller or ball thrust type bearing (20). On VAC and VC, the end thrust is carried by the vertical spindle taper roller bearing (5—Fig. C1).

1. To adjust vertical spindle end play on all except models VC and VAC, vary the number of shims (18—Fig. C2) which are located between lower side of steering spindle arm (8) and top of pedestal (22), until all end play

is removed and yet spindle rotates freely. To gain access to shims (18), remove the steering spindle arm.

On models VC and VAC, vertical spindle end play adjustment is con-trolled by the spindle retaining nut (1—Fig. C1) which should be tightened to remove all end play and yet permit same to rotate freely. To gain access to the nut, remove tractor grille.

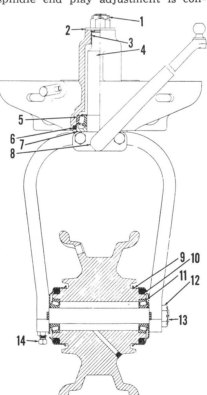

Fig. C1—Model VAC prior 5570000 and VC single wheel tricycle version front support assembly.

1. Spindle retaining nut
2. Washer
3. Bushing
4. Wheel fork & vertical spindle
5. Bearing cup & cone
6. Felt seal
7. Felt seal retainer
8. Steering arm
9. Felt retainer
10. Felt
11. Bearing cup & cone
12. Axle retaining nut
13. Wheel axle shaft
14. Set screw

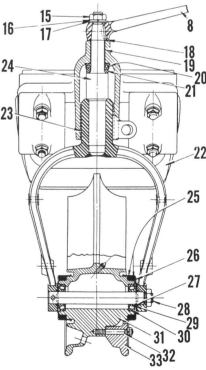

Fig. C2—Model SC single wheel tricycle version front support assembly. Models CC-DC-DC3-RC are similar.

8. Steering arm
15. Spindle retaining nut
16. Lock washer
17. Washer
18. Shims
19. Bushing
20. Thrust bearing
21. Bearing collar
22. Pedestal (support)
23. Bushing
24. Wheel fork & vertical spindle
25. Felt (2 used)
26. Retainer
27. Seal
28. Wheel axle shaft
29. ...
30. Bearing cup & cone
31. Hub (serviced with brg. cups)
32. Rim retaining stud

Single Wheel Model VAC after 5570000
(Refer also to paragraph 4)

The front support Fig. C3 on this model carries the main steering gear unit housing and integral vertical spindle and fork assembly.

2. To adjust vertical spindle bearings (two taper roller bearings) first remove tractor grille. Remove steering gear housing cap (1) and cotter pin from spindle nut. Tighten spindle nut (3) to remove all end play and yet permit same to rotate freely.

2A. To reduce play in steering wormshaft bearings, first remove tractor grille. Then, disconnect lower shaft (38—Fig. C4) from wormshaft universal joint (37). Move wormshaft (30) to determine amount of end play which is present. Remove 3 cap screws retaining worm housing sleeve cap (28 —Fig. C4) to steering gear housing, and remove the sleeve cap to gain access to shims (29). Remove shims until shaft has zero end play. The plastic shims can be identified for thickness by their color. Green colored shims are .003; blue, .005 and brown, .010. After adjusting worm shaft bearings, check the gear mesh.

2B. To adjust gear mesh, first adjust wormshaft bearings as outlined in paragraph 2A. Locate mid-position of the gear by turning steering gear from one extreme position to the other; then, back half way. With gear in mid-position, remove 3 cap screws which retain worm housing sleeve and cap (28) to gear housing. Slowly rotate steering worm sleeve (34) which is eccentric, while checking amount of gear backlash by feeling for play in vertical spindle. Continue to rotate steering worm sleeve until zero play in vertical spindle is obtained, when steering gear is in mid-position.

With mesh adjustment completed, temporarily reinstall the three retaining cap screws and check the mesh adjustment by rotating the steering gear from full left to full right. Wormshaft should have a slight amount of backlash when gear is moved either way from the mid-position or high point.

Dual Wheel Models CC-DC-DC3-RC-SC-VC-VAC prior 5570000
(Single wheel is paragraph 1)

The front pedestal on these models houses the vertical spindle. On early production CC models, refer to Fig.

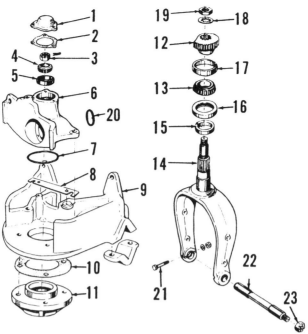

1. Gear housing cap
2. Gasket
3. Spindle nut
4. Bearing cone
5. Bearing cup
6. Steering gear housing
7. "O" ring
8. Radiator bracket
9. Front support
10. Gasket
11. Spindle bearing carrier
12. Sector
13. Bearing cone
14. Wheel fork & vertical spindle
15. Thrust collar
16. Oil seal
17. Bearing cup
18. Washer
19. Sector retaining nut
20. "O" ring
21. Retaining bolt
22. Wheel axle shaft
23. Retaining nut

Fig. C3—Model VAC after 5570000 single wheel tricycle version front support and main steering gear. Refer to Fig. C4 for steering gear worm and shaft assembly.

26. Cap screw
27. Cap screw
28. Sleeve worm housing cap
29. Shims
30. Steering worm
31. Ball bearing race
32. Ball bearing & cage
33. Key
34. Steering worm sleeve
36. Bushing
37. Universal joint yoke
38. Lower steering shaft
39. Bracket
40. Upper steering shaft
41. Upper bearing & extension
42. Bushing
43. Dust seal

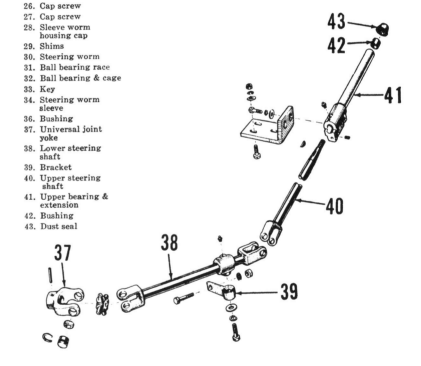

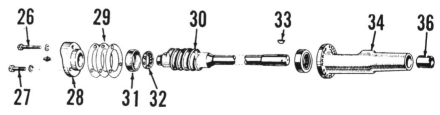

Fig. C4—Model VAC after 5570000 steering gear worm and shaft.

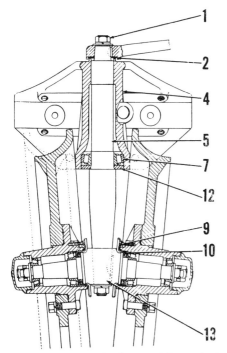

Fig. C5—Early production model CC dual wheel tricycle version front support assembly. Refer to Fig. C6 for legend.

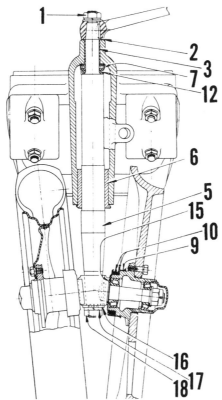

Fig. C7—Models RC and SC dual wheel tricycle version front support assembly. Refer to Fig. C6 for legend.

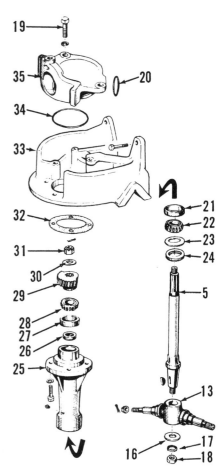

Fig. C9—Model VAC after 5570000 dual wheel tricycle version front support, and main steering gear. Refer to Fig. C4 for steering gear worm and shaft assembly.

5. Vertical spindle	25. Pedestal
13. Horizontal spindles	26. Oil seal
	27. Bearing cup
16. Washer	28. Bearing cone
17. Lock washer	29. Sector
18. Retaining nut	30. Washer
20. "O" ring	31. Retaining nut
21. Bearing cup	32. Gasket
22. Bearing cone	33. Front support
23. Thrust washer	34. "O" ring
24. Oil seal	35. Gear housing

spindle rotates in two bushings. Spindle end thrust on these models is carried by a roller type thrust bearing (7).

3. On all models adjust vertical spindle end play by varying shims, (2 —Fig. C5, C6, C7 or C8) located between lower side of steering arm and top of pedestal until all end play is removed and yet spindle rotates without binding.

Dual Wheel Model VAC after 5570000
(Single wheel is paragraph 2)

The front support, Fig. C9, on this model carries the main steering gear unit housing, front pedestal, and vertical and horizontal axle spindle assembly.

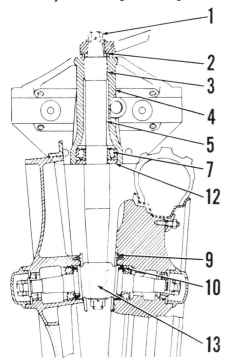

Fig. C6—Models DC, DC3, late production CC, dual wheel front support assembly. Model VAC serial 5261528-5570000 is similar.

1. Retaining nut	9. Retainer
2. Shims	10. Felt seal
3. Bushing	12. Bearing collar
4. Pedestal	13. Horizontal spindles
5. Vertical spindle	15. Seal retainer
6. Bushing	16. Washer
7. Bearing	17. Lock washer
8. Felt seal	18. Retaining nut

C5, the vertical spindle rotates directly in the front support (4) at the upper end and a taper roller thrust bearing (7) at the lower end.

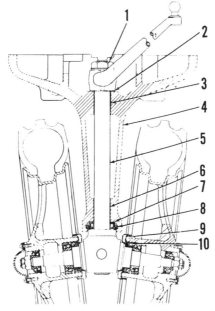

Fig. C8—Models VAC prior 5261528 and VC dual wheel tricycle version front support assembly. Refer to Fig. C6 for legend.

On DC-DC3, and also VAC 5261528 to 5570000, and later production CC models, refer to Fig. C6, the spindle rotates in a bushing (3) at the upper end, and at the lower end in a taper roller bearing (7) which also carries the spindle end thrust.

On RC-SC-VC, and also VAC prior to 5261528, refer to Fig. C7 or C8 the

4. To adjust bearings for vertical spindle (5) (two taper roller bearings) first remove tractor grille. Raise and block-up front of tractor. Remove four cap screws retaining vertical spindle and pedestal assembly (25) to front support and remove assembly. Tighten vertical spindle retaining nut (31) to remove all end play and yet permit same to rotate freely.

4A. To reduce play in steering wormshaft bearings, Fig. C4, follow same procedure as outlined for single wheel models in paragraph 2A.

4B. To adjust gear mesh, follow same procedure as outlined for single wheel models in paragraph 2B.

Adustable Axle Models DC and DC3 after 4511448-RC-SC-VC-VAC prior 5570000-VAH

5. The front support on these models contains a steering spindle to which is attached a center steering arm at the lower end, and a steering arm at the upper end for the drag link which connects with the steering gear.

On late production tractors except model RC, shims (13—Fig. C10 or C11) located between upper steering arm and front support control steering spindle (5) end play. Vary the number of shims to remove all end play. To gain access to the shims on models DC and DC3 after 5600000 and SC, remove upper steering arm (1). On VC, and VAC prior 5570000, remove tractor grille, and upper steering arm to gain access to the shims. On

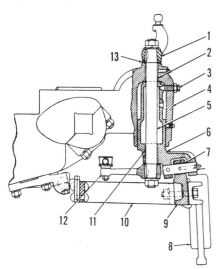

Fig. C10—Models DC and DC3 after 5600000, and SC adjustable front axle version front support assembly. Model VAC prior 5570000 is similar.

2. Bushing
3. Set screw
4. Pedestal (support)
5. Steering spindle
6. Front axle pedestal
7. Bushing
8. Axle extension
9. Main axle member
10. Radius rod
11. Bushing
12. Center steering arm
13. Shims

all model RC tractors and early production units of the other models no shims (13) were used. Shims can and should be installed in these tractors the first time the gears are serviced.

Adjustable Axle Model DC prior 4511449

6. The axle main member support, is similar to that as used on the non-adjustable axle models. There is no vertical spindle.

Adjustable Axle Models VAC and VAH after 5570000

7. The front support, Fig. C12, on these models carries the main steering gear unit housing (35), pivot support, and steering spindle shaft (37). To adjust end play in steering spindle (37) first remove cap screws retaining pivot support (36) to front support and withdraw pivot support. Tighten nut at top end of steering spindle to obtain zero end play and free rotation.

7A. To reduce play in steering wormshaft bearings, follow same procedure as outlined for single wheel models in paragraph 2A.

7B. To adjust gear mesh, follow same procedure as outlined for single wheel models in paragraph 2B.

OVERHAUL PEDESTAL AND FRONT STEERING GEAR

Single Wheel CC-DC-DC3-RC-SC-VC-VAC prior 5570000

On VAC prior 5570000, and VC the vertical spindle rotates in a bushing at the upper end, and a taper roller bearing at the lower end, Fig. C1. On all other models, the vertical spindle rotates in two bushings, Fig. C2.

On all models except VAC prior 5570000, vertical spindle end thrust is carried by either a roller or ball thrust type bearing. On VAC prior 5570000, the end thrust is carried by the vertical spindle taper roller bearing.

8. To disassemble the front support unit, first block-up and support front of tractor. On all models except models VC and VAC, remove steering spindle arm retaining nut and arm and withdraw vertical spindle, wheel fork, and wheel assembly downward and out of front support.

On models VC and VAC prior 5570000, remove tractor grille, and vertical spindle retaining nut. Vertical spindle, wheel fork, and wheel assembly can now be withdrawn downward and out of front support.

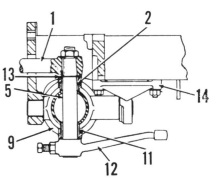

Fig. C11—Model VC adjustable front axle version front support assembly.

1. Steering arm
2. Bushing
5. Steering spindle
9. Main axle member
11. Bushing
12. Center steering arm
13. Shims
14. Support bracket

Vertical spindle bushings require final sizing after installation to provide a diametral clearance of .002-.004 for the lower bushing, and .003-.006 for the upper bushing.

8A. A felt type dust seal (6—Fig. C1) for the vertical spindle is used only on models VAC prior 5570000 and VC. Seal can be renewed after removing the vertical spindle.

8B. When reassembling, adjust vertical spindle end play on all models except VC and VAC by varying the number of shims (18—Fig. C2) until all end play is removed and yet spindle rotates freely. On models VC and VAC, vertical spindle end play adjustment is controlled by the spindle retaining nut (1—Fig. C1), which should be tightened to remove all end play.

Single Wheel Model VAC after 5570000

9. To disassemble spindle support and steering gear unit, first remove tractor grille, and block-up and support front portion of tractor. Remove steering gear housing cap (1—Fig. C3), and cotter pin from spindle nut. Remove spindle retaining nut (3), and spindle bearing carrier (11) cap screws. Bump spindle and fork assembly down and out of upper bearing cone (4) and steering gear housing. The need and procedure for further disassembly of the vertical spindle will be determined by an examination of the parts and with reference to Figs. C3 & C4. Before removing sector (12) from vertical spindle, mark both pieces to insure correct relationship when reassembling.

With the spindle and fork assembly removed from the front support, disconnect wormshaft yoke from lower steering shaft yoke (37—Fig. C4). Remove two cap screws retaining gear

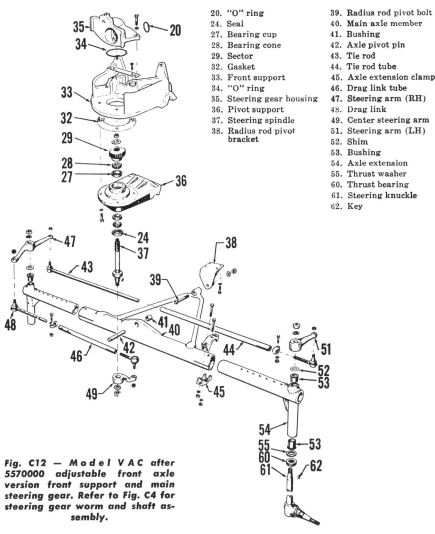

20. "O" ring	39. Radius rod pivot bolt
24. Seal	40. Main axle member
27. Bearing cup	41. Bushing
28. Bearing cone	42. Axle pivot pin
29. Sector	43. Tie rod
32. Gasket	44. Tie rod tube
33. Front support	45. Axle extension clamp
34. "O" ring	46. Drag link tube
35. Steering gear housing	47. Steering arm (RH)
36. Pivot support	48. Drag link
37. Steering spindle	49. Center steering arm
38. Radius rod pivot	51. Steering arm (LH)
bracket	52. Shim
	53. Bushing
	54. Axle extension
	55. Thrust washer
	60. Thrust bearing
	61. Steering knuckle
	62. Key

Fig. C12 — Model VAC after 5570000 adjustable front axle version front support and main steering gear. Refer to Fig. C4 for steering gear worm and shaft assembly.

5570000, *the vertical spindle is equipped with two felt type seals, one at each end of the spindle.*

10. To disassemble front support unit, first block-up and support front of tractor and remove front wheel units. On models VAC prior 5570000 and VC, remove tractor grille. On all models, remove steering spindle arm retaining nut and arm, and withdraw vertical spindle downward and out of front support.

10A. Vertical spindle bushings require final sizing after installation to provide a diametral clearance of .002-.004 for the lower bushing, and .003-.006 for the upper bushing.

10B. When reassembling, adjust vertical spindle end play on all models by varying the number of shims (2—Fig. C5, C6, C7 or C8) until all end play is removed and yet spindle rotates freely.

Dual Wheel Model VAC after 5570000

11. To disassemble spindle support and steering gear unit, first remove tractor grille, and block-up and support front portion of tractor. Remove four cap screws retaining vertical spindle (5—Fig. C9) and pedestal assembly to front support, and remove assembly. The need and the procedure for further disassembly will be determined by an examination of the parts and with reference to Fig. C9. Before removing sector (29) from vertical spindle (5), mark both pieces to insure correct relationship when reassembling. On this model, the vertical steering spindle can be removed from the wheel spindles (13) after removing the retaining nut (18) and by the use of a suitable puller.

With vertical spindle and pedestal assembly removed from the front support, disconnect wormshaft universal joint yoke (37—Fig. C4) from lower steering shaft yoke. Remove two cap screws (19—Fig. C9) retaining gear housing (35) to front support (33), and remove the gear housing wormshaft and worm housing sleeve as an assembly.

To remove wormshaft and worm housing sleeve from gear housing (35), remove three cap screws retaining sleeve (34—Fig. C4) and cap (28) to gear housing. Bump wormshaft and sleeve assembly forward and out of gear housing. The wormshaft bearings and the bushing (36) can be renewed after removing the worm from the housing sleeve. Bushing (36) is pre-

housing (6—Fig. C3) to front support, and remove gear housing, wormshaft and worm sleeve as an assembly.

To remove wormshaft and worm housing sleeve assembly from gear housing, remove three cap screws retaining worm sleeve (34—Fig. C4) and cap (28) to gear housing. Bump wormshaft and sleeve assembly forward and out of gear housing. Wormshaft bearings and the bushing (36) can be renewed after removing the wormshaft from the worm sleeve. Bushing (36) is presized, and if carefully installed requires no final sizing after installation.

9A. When reassembling the gear unit, renew worm housing sleeve "O" ring (20—Fig. C3) and gear housing to front support "O" ring (7).

Adjust vertical spindle bearings, wormshaft bearings, and gear mesh as outlined in paragraphs 2, 2A and 2B.

Dual Wheel Models CC-DC-DC3-RC-SC-VC and VAC prior 5570000

On early production CC models, the vertical spindle rotates directly in the

front support at the upper end, and in a roller thrust bearing at the lower end, refer to Fig. C5. On models DC-DC3 and VAC 5261528 to 5570000, the vertical spindle rotates in a bushing at the upper end, and at the lower end in a taper roller bearing which also carries the spindle end thrust, refer to Fig. C6. On RC, SC, VC and VAC prior 5261528, the spindle rotates in two bushings, refer to Fig. C7 or C8. Spindle end thrust on this group of models is carried by a roller type thrust bearing.

On all dual wheel models except models VAC prior 5261528 and VC, the vertical steering spindle is separate from the wheel spindles. On these models, the wheel spindle block is keyed and retained to the vertical spindle by means of a nut. On models VAC prior 5261528 and VC, the vertical steering spindle is integral with the wheel spindles.

On models VAC prior 4829409 and V, vertical spindle is equipped with one felt seal which is located at the lower end. On model VAC 4829408-

sized and if carefully installed requires no final sizing after installation.

11A. When reassembling the gear unit, renew worm housing sleeve "O" ring (20—Fig. C9) and gear housing to front support "O" ring (34).

Before reinstalling vertical spindle and pedestal to front support, adjust spindle bearings by tightening spindle nut (31) to remove all end play and yet permit same to rotate freely. Adjust worm shaft bearings as outlined for single wheel models in paragraph 2A. Adjust gear mesh by following same procedure as outlined for single wheel models in paragraph 2B.

Adjustable Axle Model DC prior 4511449

11B. The axle main member support is similar to that as used on the non-adjustable axle models.

Adjustable Axle Models DC3 after 4511448-RC-SC

12. Refer to Fig. C10. To renew the steering spindle (5) and/or spindle bushings (2 & 11), first remove nut retaining center steering arm (12) to spindle. Disconnect drag link from upper steering arm. Working from below, bump spindle up and out of center steering arm, and support. Upper steering arm can be removed **before** or **after** removing the steering spindle by using a suitable puller.

Spindle bushings (2 & 11) located in support, and pedestal can be renewed at this time. Bushings require final sizing after installation to provide a .001-.003 diametral clearance.

12A. When reassembling, adjust steering spindle end play by varying the number of shims (13) which are located between upper steering arm and front support so as to remove all end play. Refer to paragraph 5.

Adjustable Axle Models VAC prior 5570000-VAH-VC

On model VC, steering spindle rotates in two bushings (2 & 11—Fig. C11) which are located as shown in Fig. C11. On all other models, steering spindle rotates in two bushings which are located in front axle support.

13. To renew the steering spindle and/or bushings, first remove tractor grille and on VC only, remove radiator. Remove set screw on model VC, and nut on all others which retains the center steering arm to the spindle. Remove upper steering arm, and withdraw the spindle. Steering spindle bushings which require final sizing after installation to provide a .002-.004

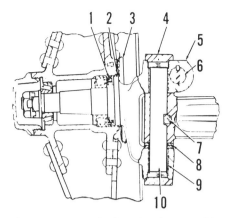

Fig. C13—Front wheel and axle assembly as used on some adjustable and non-adjustable axle models employing the reverse Elliott type spindles (knuckles).

1. Felt seal	6. Spindle arm
2. Retainer	7. Retaining pin
3. Dust shield	8. Thrust washers
4. Spindle cap	9. Bushing
5. Spindle (knuckle)	10. Spindle (king) pin

diameteral clearance, can be renewed at this time.

13A. When reassembling, adjust steering spindle end play by varying the number of shims which are located on the under side of the upper steering arm so as to remove all end play. Refer to paragraph 5.

Adjustable Axle Model VAC after 5570000

Except that the vertical (steering) spindle (37—Fig. C12) in the front support terminates in a center steering arm instead of wheel axle shafts, the construction, overhaul and adjustment procedures of the steering gear are similar to the dual wheel model as outlined in paragraphs 11 and 11A. To remove the gear housing, the preliminary work described in paragraph 14 must first be performed.

14. To remove the steering gear unit first block-up and support front portion of tractor. Disconnect tie rod (drag link) from center steering arm (49). Remove radius rod pivot bolt (39), and axle main member pivot pin (42); then, roll axle with wheels as an assembly away from tractor. Steering unit may now be removed and disassembled as in paragraph 11.

14A. Before reinstalling steering spindle (37) and pivot support assembly (36) to front support (33), adjust spindle bearings by tightening spindle sector nut to remove all end play and yet permit spindle to rotate freely. Adjust worm shaft bearings as outlined for single wheel models in paragraph 2A. Adjust gear mesh by following the procedure as outlined for single wheel models in paragraph 2B.

FRONT AXLE MEMBER
All Models

Models C, CC, D, DC4, DCS, DH, DO, L, LA, R, S, SO, V, VO, VA, VAO are equipped with non-adjustable type axles which have the reverse Elliot type spindles (knuckles) as shown in Fig. C13. Models DC and DC3 after 4511448, RC, SC, VC, VAC, VAH are equipped with adjustable type axles which have the Lemoine type spindles (knuckles) as shown in Fig. C14. Models DC and DC3 prior 4511449 are equipped with an adjustable type axle which has the reverse Elliot type spindles (knuckles) as shown in Fig. C13.

15. On all models, the axle main member complete with wheel spindles (knuckles), and wheels can be removed from the tractor as a single unit. Exact procedure varies with the various models, but is self-evident after observing the actual tractor.

SPINDLE (KNUCKLE) & BUSHINGS
All Models

On adjustable axle models employing the Lemoine type spindles, except model VAC after 5255705, model DC and DC3 prior 4511449, and model VC, the wheel axle shaft (knuckle) is keyed and retained to the steering spindle by means of a nut. On all other models equipped with Lemoine type spindles, the steering spindle is integral with the wheel axle shaft.

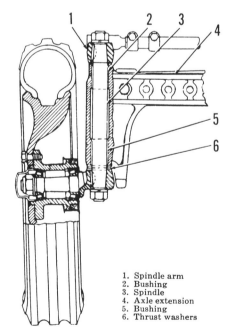

1. Spindle arm	
2. Bushing	
3. Spindle	
4. Axle extension	
5. Bushing	
6. Thrust washers	

Fig. C14—Front wheel and axle assembly as used on some adjustable type axle models employing the Lemoine type spindles (knuckles).

16. Steering spindle (knuckle) bushings should be renewed if the diametral clearance exceeds .020. Bushings require final sizing after installation and should be reamed to provide a diametral clearance of .002-.003.

To remove all wear in systems employing the Lemoine type spindles, it may be necessary to also install new steering spindles.

16A. On series C, D, L, LA, R, both thrust washers for one knuckle are installed between lower side of axle main member and steering spindle. On models VAC prior 5255705, V and VO, one thrust washer is installed at the top and one at the lower end of the steering spindle. On models VAC after 5255705, VAH, VA, VAO, a thrust bearing is installed on the lower end of the axle main member.

Recommended front wheel toe-in of ¼-⅜ inch is adjusted by varying the length of the tie rod or rods.

RADIUS ROD

All Models

17. On all except models VAC after 5570000, V, VA, the radius rod is bolted to the axle main member. On models VAC after 5570000, V, VA, the radius rod is welded to the axle main member. No radius rod is used on the model VC adjustable type axle.

17A. On models VAC prior 5570000, VAH, S, SO, L after 300889, the radius rod pivots at the rear in a ball and socket type pivot. On some of this group, only the socket is renewable as the ball is riveted to the radius rod. On others of this group, both the ball and socket are renewable.

On models L prior 300890, LA, D, DC and DC3 after 4511448, DC4, DH, DO and DV, the radius rod pivots at the rear in a renewable bushing and pin which requires final sizing after installation to provide a diametral clearance of .002-.005.

On models VAC after 5570000, VA, V, R, RC, DC prior 4511449, C, and CC, the radius rod pivots at the rear in a renewable pin which is not bushed.

AXLE PIVOT PIN

All Models

18. On models V and VA non-adjustable type axles, the main axle member pivot pin is welded to the axle support. On all other models, the axle main member pivots on a renewable pivot pin.

18A. On series C, D prior 4511449, L and R, the axle main member pivot is not bushed. On all other models, the axle main member is equipped with a pivot bushing which can be renewed after removing the axle member from the front support. Pivot bushings require final sizing after installation to provide a .002-.005 diametral clearance.

MAIN STEERING GEAR

Information in this section applies ONLY to those steering gear units mounted on the torque tube or the transmission. For service information covering all of the front system including the steering mechanism mounted in the front support, refer to paragraphs 1 through 18A.

**Series C-R-S-V and
Models D & DC & DO
prior 4607033
Models DH & DV prior 4511449
Model L prior 4402720
Models VA-VAO-VAH
Models VAC prior 5570000**

22. WORM SHAFT END PLAY. Refer to Figs. C16, C17, C18, C19, C20,

C21, C22 and C23. First step in making this adjustment is to disconnect throttle linkage from worm shaft column (12), and remove battery and carrier from column on models so equipped. Remove cap screws retaining worm shaft column to transmission housing cover. Remove column and steering worm and shaft assembly.

To reduce end play, remove cotter pin and tighten worm retaining nut (7) sufficiently until worm shaft has zero end play, but rotates freely.

Relock nut, and reinstall column and steering worm assembly, using a new gasket of same thickness as original gasket between column and transmission (gear housing) cover.

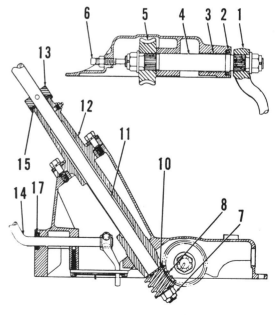

Fig. C17—Series R steering gear assembly.

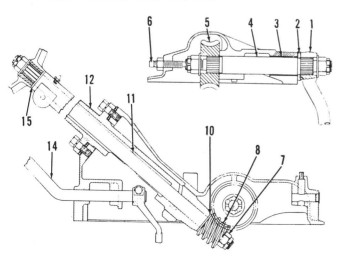

Fig. C16—Series C steering gear assembly.

1. Pitman (drop) arm	7. Adjusting nut	13. Thrust washer collar
2. Oil seal	8. Worm	14. Transmission gear
3. Bushing	10. Thrust washer	shifter lever
4. Worm wheel shaft	11. Worm shaft	15. Thrust washer
5. Worm wheel	12. Worm shaft	17. Oil seal
6. Adjusting screw	column	

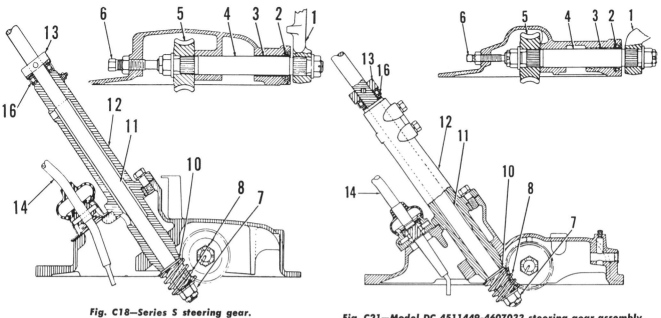

Fig. C18—Series S steering gear.

Fig. C21—Model DC 4511449-4607033 steering gear assembly.

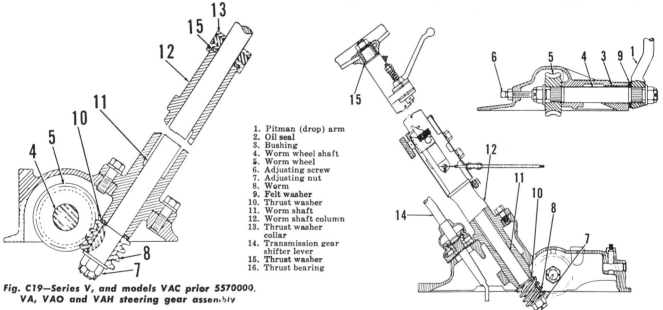

1. Pitman (drop) arm
2. Oil seal
3. Bushing
4. Worm wheel shaft
5. Worm wheel
6. Adjusting screw
7. Adjusting nut
8. Worm
9. Felt washer
10. Thrust washer
11. Worm shaft
12. Worm shaft column
13. Thrust washer collar
14. Transmission gear shifter lever
15. Thrust washer
16. Thrust bearing

Fig. C19—Series V, and models VAC prior 5570000, VA, VAO and VAH steering gear assembly

Fig. C22—Models DC and DH prior 4511449 steering gear.

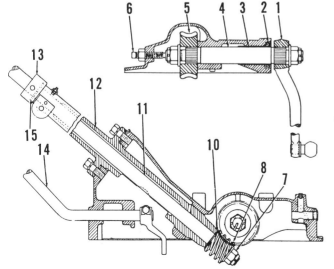

Fig. C20—Models D and DO prior 4607033, and model DV prior 4511449 steering gear assembly.

Fig. C23—Model L prior 4402720 steering gear assembly.

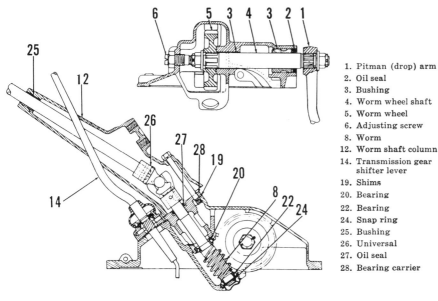

1. Pitman (drop) arm
2. Oil seal
3. Bushing
4. Worm wheel shaft
5. Worm wheel
6. Adjusting screw
8. Worm
12. Worm shaft column
14. Transmission gear shifter lever
19. Shims
20. Bearing
22. Bearing
24. Snap ring
25. Bushing
26. Universal
27. Oil seal
28. Bearing carrier

Fig. C24—Models D 4607033-5600000, DO and DV after 4607033, and DCS and DS steering gear assembly.

ing worm shaft rotates in unbushed machined bores of the steering shaft column but is sometimes provided with a thrust bearing at the steering wheel end. Inspect this end of shaft and renew any worn parts. Adjust gear unit as outlined in preceding paragraphs 22, 23, and 24.

Models D & DC & DV after 4607033-DCS Model LA after 5208165

26. **WORM SHAFT END PLAY.** Refer to Figs. C24, C25, and C26. Steering gear worm and shaft (8) on these models rotates in two taper roller bearings. To reduce play in worm shaft bearings, remove steering wheel, and steering shaft column to transmission cover retaining cap screws, and slide column away from transmission cover. Add or remove shims (19), which are located between column and gear housing on some models, and between worm shaft bearing carrier and gear housing on other models, until worm shaft has zero end play, but rotates freely. Shims are available in .003, .005 and .0125 thickness for this adjustment.

23. **WORM WHEEL (GEAR) SHAFT END PLAY.** End play is controlled by adjusting screw (6) which is located on left side of transmission (gear housing) cover. To adjust, loosen adjusting screw locknut, and turn adjusting screw clockwise to reduce end play. All end play in worm wheel shaft should be eliminated, but gear should not bind. Relock set screw after making the adjustment.

24. **GEAR MESH.** Excessive backlash between worm (8) and worm wheel (5), may be corrected by meshing an unused portion of the worm wheel with the worm. To make this

adjustment, remove steering pitman (drop) arm (1) from worm wheel shaft (4). Rotate steering wheel until worm wheel shaft has moved approximately ⅓ revolution. With front wheels in straight ahead position, reinstall steering pitman (drop) arm to worm wheel shaft.

25. **OVERHAUL.** Disassembly of the gear unit is readily evident after an examination, and removal of the transmission top cover which is also the steering gear housing. Worm wheel shaft bushings require final sizing after installation to provide a diametral clearance of .001-.004. Steer-

27. **WORM WHEEL (GEAR) SHAFT END PLAY.** End play is controlled by adjusting screw (6) which is located on left side of transmission (gear housing) cover. To adjust, loosen adjusting screw locknut, and turn adjusting screw clockwise to reduce end play. All end play in worm wheel shaft should be eliminated but gear

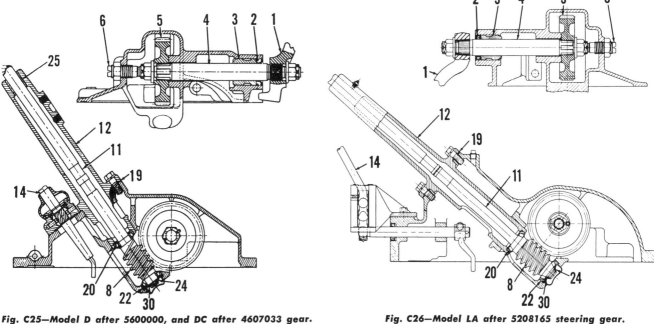

Fig. C25—Model D after 5600000, and DC after 4607033 gear.

1. Pitman (drop) arm	4. Worm wheel shaft	8. Worm	14. Transmission gear shifter lever	20. Bearing	25. Bushing
2. Oil seal	5. Worm wheel	11. Worm shaft	19. Shims	22. Bearing	30. Washer
3. Bushing	6. Adjusting screw	12. Worm shaft column		24. Snap ring	

Fig. C26—Model LA after 5208165 steering gear.

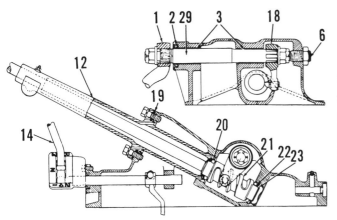

1. Pitman (drop) arm
2. Oil seal
3. Bushing
6. Adjusting screw
12. Worm Shaft column
14. Transmission gear shifter lever
18. Cam lever
19. Shims
20. Bearing
21. Worm and shaft
22. Bearing
23. Snap ring
29. Lever shaft

Fig. C27—Models L after 4402720, and LA prior 5208165 steering gear assembly.

should not bind. Relock set screw after making the adjustment.

28. GEAR MESH. Excessive backlash between worm (8) and worm wheel (5) may be corrected by meshing an unused portion of the worm wheel with the worm. Remove steering pitman (drop) arm from worm wheel shaft. Rotate steering wheel until worm shaft has moved approximately ⅓ revolution. With front wheels in straight ahead position, reinstall steering arm to worm wheel shaft.

29. OVERHAUL. Procedure for dissembly is readily evident after an examination of the unit and removal of the transmission (gear housing) top cover. Worm wheel shaft bushings require final sizing after installation to provide a diametral clearance of .001-.004. Worm shaft rotates in two taper roller bearings.

When reassembling, install worm and shaft assembly, and adjust worm shaft bearings before assembling balance of steering gear.

Adjust gear unit as outlined in preceding paragraphs 26, 27 and 28.

Models L after 4402720— LA prior 5208165

Before adjusting the gear, disconnect drag link or remove steering pitman (drop) arm to remove load from gear and permit locating mid-position of gear unit.

30. WORM SHAFT (CAM) END PLAY. Worm shaft (21—Fig. C27) rotates in two ball bearings the inner races for which are formed by the ground ends of the worm (cam) itself. To reduce play in worm shaft bearings, remove steering shaft column to transmission (gear housing) cover retaining cap screws, and slide column away from transmission cover. Remove shims (19) until worm shaft has zero end play but rotates freely. Shims are available in .003, .005, and .0125 thicknesses for this adjustment.

31. STUD (LEVER SHAFT) MESH. Locate mid-position of steering gear by turning steering wheel from one extreme position to the other, then back half-way. With gear in mid-position, loosen locknut on adjusting screw (6) and turn adjusting screw clockwise until end play in lever shaft is barely perceptible. Shaft should have an increased amount of end play when gear is moved either way off the mid or high point. Steering pitman (drop) arm should be positioned on lever shaft so that steering gear is in mid-position when wheels are straight ahead.

32. OVERHAUL. Procedure for dissembly is readily evident after an examination of the unit, and removal of the transmission (gear housing) top cover. Lever shaft bushings require final sizing after installation to provide a diametral clearance of .001-.004.

When reassembling, install worm and shaft assembly, and adjust worm shaft bearings before assembling balance of steering gear.

Adjust gear unit as outlined in preceding paragraphs 30 and 31.

Model VAC after 5570000

33. OVERHAUL. Steering gear for the single and dual wheel tricycle, and adjustable axle versions for model VAC after 5570000 is located in the front support. For adjustment and overhaul procedures, refer to paragraphs 2 and 9 for single wheel tricycle models; paragraphs 4 and 11 for dual wheel tricycle models; paragraphs 7 and 14 for adjustable axle models.

ENGINE AND COMPONETS

R & R ENGINE WITH CLUTCH
Series C-D-L-LA-R-S

40. To remove the engine with clutch from the tractor, it is necessary to perform a tractor split (detach engine from transmission housing), and detach front axle and radiator support from front face of engine.

40A. TRACTOR SPLIT. To detach engine from transmission as shown in Fig. C29 first drain engine and transmission housing clutch compartment. Place transmission gear shift lever in neutral and block the rear wheels. Disengage clutch. (Clutch is over-center type.) Remove tractor hood. On

adjustable and non-adjustable axle models, disconnect radius rod ball or remove pivot pin from forward end of transmission housing. Disconnect engine controls, fuel line, oil line, and heat indicator from radiator elbow. Disconnect drag line from steering gear pitman (drop) arm. (On S series only, remove air cleaner oil cup.) Remove hand hole plates from both sides of transmission housing clutch compartment. Block-up and support forward end of transmission housing, and support engine in a hoist. Remove bolts, and cap screws attaching the transmission to engine, and separate the two assemblies.

40B. FRONT ASSEMBLY. To remove front support and axle as an assembly from front face of engine, first drain and remove radiator assembly. Remove bolts and cap screws attaching front support to front of engine, and roll front support, front axle and wheels as an assembly away from the engine.

Series V-VA

41. To remove the engine with clutch from the tractor, it is necessary to perform a tractor split (detach engine from torque tube) and on VA series, to perform the additional work of detaching front axle and radiator support from front face of engine. On

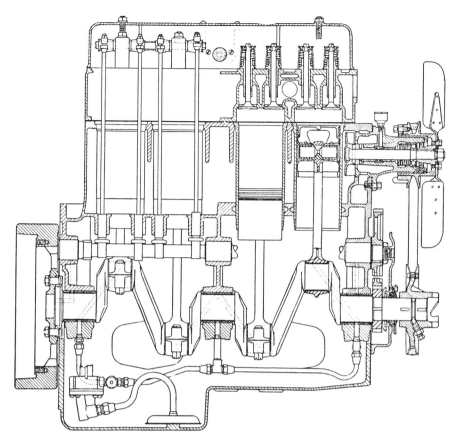

Fig. C28—Series C tractor engine.

support forward end of torque tube, and support engine in a hoist. Remove bolts and nuts attaching torque tube to engine on VA series, or engine to clutch housing on V series, and separate the two assemblies. On V series refer to paragraph 41C, on VA to paragraph 41B.

41B. On VA series, to remove front support and axle as an assembly from front face of engine, first drain cooling system. With radiator bolted to front support, remove bolts and cap screws attaching front axle and radiator support to front face of engine, and roll assembly away from the engine.

41C. On V series support engine, and frame with front axle assembly separately; then, remove engine front support bolt from lower side of timing gear cover, and remove bolts attaching frame to clutch housing. Lift engine out of frame.

CYLINDER HEAD
Series C-D-L-LA-S-VA (Valve-in-Head)

42. To remove the cylinder head on these models, first drain the cooling system. Raise the hood on VA series only. Remove the hood on all other series. On VA series only, remove radiator grille for convenience. Disconnect cylinder head water outlet elbow from front of cylinder head. On all series, except VA, loosen bolts attaching radiator side members to front support. (One bolt is located at the lower outside corners of each radiator side members.) Tilt radiator assembly toward front of tractor. Remove mani-

V series, remove engine from frame assembly after making split.

41A. TRACTOR SPLIT. To detach engine from torque tube, first remove the hood, fuel line to carburetor, and disconnect engine controls. (On V series only, remove fuel tank.) Disconnect drag link from steering gear pitman arm on models so equipped. On

model VAC after 5570000, disconnect and remove lower steering shaft from worm shaft and from upper steering shaft. On adjustable and non-adjustable axle models, disconnect radius rod at the rear pivot. On models equipped with a hydraulic pump which is mounted on the engine, disconnect the pump lines at the pump. Block-up and

Fig. C29—Detaching engine from transmission housing on a model DC tractor. Use a similar method on series C, L, LA, R, and S.

fold and carburetor as a unit. On L and LA series, disconnect rocker arms oiling system. Remove valve cover, rocker arms assembly, and push rods. Remove cylinder head retaining nuts, and lift off head.

When reinstalling cylinder head on S series, install rocker arm shaft with rocker lubricating holes facing the push rods; on D series install the rocker arm shaft with lubricating holes facing downward; and on VA series install rocker arm shaft with lubricating holes facing the valve springs. Tighten head nuts progressively and from the center outward. Retighten after the engine has reached operating temperature. Torque tighten cylinder head retaining nuts to the following values:

Series	Torque Ft. Lbs.
C-D	75
L-LA	125
S	60
VA (½ inch studs)	65
(⅜ inch studs)	45

Adjust both inlet and exhaust valve tappet (rocker arms) gap as follows:

Series	Tappet Gap
C-D	.018 Cold
S	.010 Cold
VA	.014 Cold

Series R-V (L-Head)

43. To remove the cylinder head on these models, first drain the cooling system. Remove hood on R series only, and raise the hood on V series. On V series, remove heat indicator bulb. Remove air cleaner and support bracket, and cylinder head water outlet elbow. Remove cylinder head retaining nuts and lift off head.

When reinstalling cylinder head, torque tighten retaining nuts series R to 70 ft. lbs., and on series V to 60 ft. lbs. Tighten head nuts progressively and from the center outward. Retighten after engine has reached operating temperature. Adjust tappets gaps as follows:

Series	Tappet Gap
R Inlet	.010 Cold
R Exhaust	.012 Cold
V	.014 Cold

VALVES AND VALVE SEATS
Series C-D-L-LA-S-VA

44. Both inlet and exhaust valve tappets should be set cold to .018 on series C, D, L and LA; .010 cold on series S; .014 cold on series VA. Inlet and exhaust valves on series C, D, L, and LA are interchangeable. On series S,

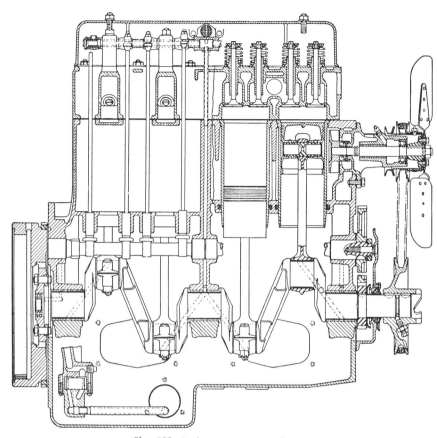

Fig. C30—Series D tractor engine.

and VA, the inlet valves have a slightly larger head than the exhaust valves. On series VA only, the inlet valve stems are provided with a neoprene oil guard.

All inlet valves seat directly in the cylinder head. The exhaust valves on all except series C are provided with seat inserts which are installed in the cylinder head with a .0015-.003 shrink fit.

On series C, D, L, LA, and S, both inlet and exhaust valves have a face and seat angle of 45 degrees. On series VA, the inlet valve face and seat angle is 30 degrees, and the exhaust valve face and seat angle is 45 degrees. Desired seat width is 1/16 inch. Seats can be narrowed, using 30 and 60 degree cutters or stones on all series except VA. On VA series, seats can be narrowed, using 15 and 75 degree cutters.

Series R-V

45. Tappets should be set cold to .010 for inlet, and .012 for exhaust on series R tractors; and .014 cold for both inlet and exhaust on V series. Inlet and exhaust valves are not interchangeable.

Inlet valves on series R and V seat directly in the cylinder head, as do the exhaust valves on series R. The exhaust valves on series V are provided with seat inserts which are installed in the cylinder head with a .0015-.003 shrink fit.

On series R, both inlet and exhaust valves have a face and seat angle of 45 degrees. On series V, the inlet valve face and seat angle is 30 degrees, and

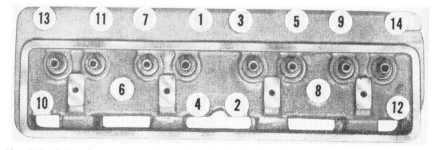

Fig. C31—Tightening sequence for series S cylinder head retaining nuts. Series C, D, L, LA and VA are similar.

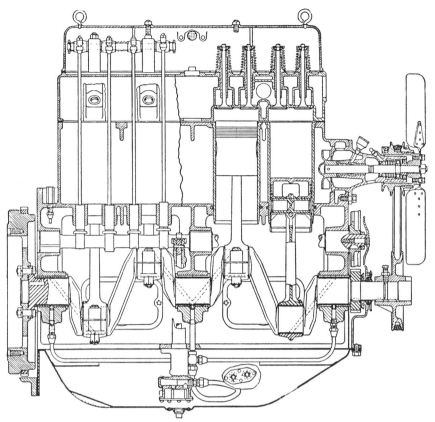

Fig. C32—Series LA tractor engine. Series L is similar.

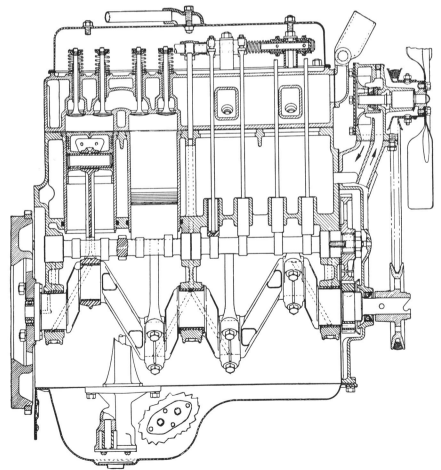

Fig. C33—Series S engine. Later engines have mushroom type tappets.

the exhaust valve face and seat angle is 45 degrees. Desired seat width is 1/16 inch. Seats having a 45 degree angle can be narrowed by using 30 and 60 degree cutters, and 30 degree angle seats can be narrowed by using 15 and 75 degree cutters.

VALVE GUIDES

All Series

46. Valve guides used in series C, L, R, and early production series D and LA engines which are of the shoulder type are interchangeable in any one series. On series S and late production D and LA engines, valve guides are of the shoulderless type and are not interchangeable. On series V and VA, the shoulderless type guides are interchangeable. The late production shoulderless valve guides can be used for servicing series C, D, L and LA engines originally equipped with the shoulder type.

The shoulderless type valve guides should be pressed into the cylinder head on valve-in-head engines until the distance from the valve spring guide surface on cylinder head to top of guide measures as follows:

Series	Distance
D	1.0
LA	$1\frac{5}{8}$
S	$\frac{7}{8}$
VA	31/32

On series V engines, distance from top of guide to gasket surface of cylinder block should measure 1 15/32 inches.

Valve guides should be reamed after installation to the diameters listed so as to provide the correct diametral clearance.

Series	Guide Inside Diameter
C-D-L-LA	.4375–.4385
R	.312
S	.3125–.3135
V-VA	.3422–.3432

Desired diametral clearance of valve stems in guides is as follows:

Series	Valve Stem Diametral Clearance
C-D-L-LA	.0035–.0055
R (Inlet)	.0015–.002
(Exhaust)	.0025–.003
S	.0025–.0045
V-VA (Inlet)	.0015–.0025
(Exhaust)	.004 –.005

VALVE SPRINGS

All Series

47. Inlet and exhaust valve springs are interchangeable in any one series. Springs having damper coils (closely wound coils at one end) should be in-

stalled with damper coils nearest to the cylinder head on all valve-in-head engines, or nearest to the valve head on all L-head engines.

On valve-in-head engines, springs can be renewed after removing the rocker arms and shaft as an assembly. On series R and V L-head engines, springs can be renewed after removing the cylinder head and valve.

An alternate method for removing the valve spring without removing the head on series R and V L-head engines is as follows: Turn adjusting screw down against lock and holding the valve up with a wire inserted through spark plug hole, remove the spring.

Springs which are rusted, distorted or do not meet the pressure specifications shown in the table, should be renewed.

CAM FOLLOWERS (Tappets)
Series C-D-L-LA-S-VA

48. The mushroom type cam followers (tappets) operate directly in machined bores of the cylinder block. (Series S tractors prior to serial 4806466 were originally equipped with cylindrical (barrel) type cam followers. However, mushroom type are being supplied for servicing engines originally equipped with cylindrical type followers.) Cam followers are supplied in the standard size as well as .010 oversize for all series except VA. In addition, cam followers of .020 oversize are available for the series LA. On series VA, only the standard size followers are available. Tappets should have a suggested diametral clearance of .0005-.002 with a maximum allowable clearance of .007 in block bores.

48A. Mushroom type cam followers can be removed after removing the oil pan, rocker arms and shaft assembly, oil pump on all except VA series, battery ignition unit on models so equipped, timing gear cover and camshaft.

On series S equipped with cylindrical (barrel) type cam followers, same should be replaced with the later mushroom type. Cylindrical type followers can be extracted after rocker arms and shaft assembly and push rods have been removed by using a long wooden dowel tapered at one end. Insert dowel into the follower bore and lift the follower out through the push rod opening.

Adjust both inlet and exhaust valve tappet (rocker arm) gaps as follows:

VALVE SPRING SPECIFICATIONS

Series	Free Length	Tension
C-D	2 3/8	39-44 Lbs. @ 1 29/32 or 99-108 Lbs. @ 1 15/32
L-LA	3 1/8	88-96 Lbs. @ 2 1/4
R	2 1/4	32-40 Lbs. @ 1 29/32 or 60-71 Lbs. @ 1 5/8
S	2 1/32	25-28 Lbs. @ 1 3/4 or 61-69 Lbs. @ 1 13/32
V	2	47-53 Lbs. @ 1 45/64 or 96- 104 Lbs. @ 1 27/64
VA	2 3/8	53-59 Lbs. @ 1 7/8 or 108-114 Lbs. @ 1.524

Series	Tappet Gap
C-D-L-LA	.018 Cold
S	.010 Cold
VA	.014 Cold

Series R-V

49. Mushroom type cam followers as used in the series R engines, and cylindrical (barrel) type cam followers as used in the series V engines operate directly in machined bores of the cylinder block. Oversize cam followers are not supplied. Tappets should have a suggested diametral clearance of .0005-.002 with a maximum allowable clearance of .007.

49A. On series R engines, cam followers can be removed after removing the camshaft. On series V engines, any cam follower can be removed by first removing the valve spring which is accomplished by turning the adjusting screw down against the lock and holding the valve up with a wire inserted through the spark plug hole. Now remove the tappet screw and lock element from the tappet, and withdraw the cam follower.

Adjust tappets cold to .010 for inlet and .012 for exhaust on series R engines, and .014 cold for both inlet and exhaust on series V engines.

VALVE ROCKER ARMS
Series C-D-L-LA-S-VA

50. Rocker arms and shaft assembly can be removed after removing the hood (raise hood on VA series), valve cover, oil line on L and LA series, and rocker shafts supports retaining bolts and/or nuts.

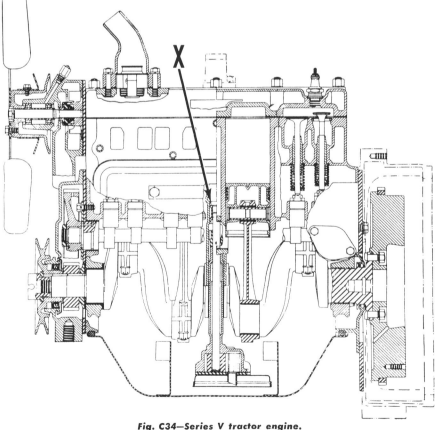

Fig. C34—Series V tractor engine.

Fig. C35—Adjusting rocker arms (tappet gap) on series D tractor engine. Note offset rocker arms. Series C engines are similar.

Fig. C36—Adjusting rocker arms (tappet gap) on series L and LA tractor engines. Note offset rocker arms.

1. Valve cover
3. Crankcase breather
4. Rocker arm shaft supports
5. Rocker arm shaft

Fig. C37—Series S engine rocker arms and shaft assembly. Note offset rocker arms.

Fig. C38—Adjusting rocker arms (tappet gap) on series VA tractor engines. Note offset rocker arms.

Disassembly of the rocker arms and shafts assembly is self-evident after an examination. Bushings and/or shafts should be renewed if the diametral clearance exceeds a suggested value of .008. Rocker arm bushings, which are provided for service, require final sizing after installation to the values as listed so as to provide a diametral clearance of .002-.003.

Series	Bushings I. D.
C-D	.6265-.6275
L-LA-VA	.7515-.7525
S	.624 -.625

Rocker arms are right and left and when reassembling same to the rocker shaft, refer to Figs. C35, C36, C37 and C38.

When reassembling rocker arms and shafts, on series S, install rocker arm shafts with the lubricating hole facing the push rods; on D series install the shaft with the lubricating holes facing downward; and on VA series install the shaft with the lubricating holes facing the valve springs.

Adjust both inlet and exhaust valve tappet (rocker arm) gaps as follows:

Series	Tappet Gap
C-D-L-LA	.018 Cold
S	.010 Cold
VA	.014 Cold

VALVE TIMING
Series C-D-L-LA-S

51. Valves are correctly timed when the single punch mark on the crankshaft gear meshes with an identical double punch mark on the camshaft gear, as shown in Fig. C39. The single punch mark on the camshaft gear should mesh with an identical double punch mark on governor and magneto gear.

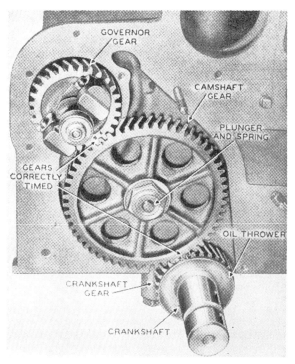

Fig. C39—On series C, D, L, LA and S tractor engines, mesh single punch mark on crankshaft gear with double punch mark on camshaft gear, and mesh single punch mark on camshaft gear with a double punch mark on magneto and governor gear.

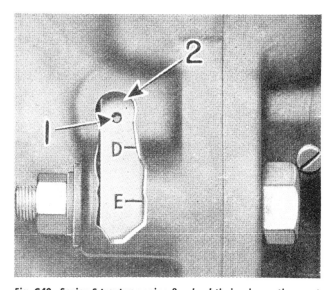

Fig. C40—Series S tractor engine flywheel timing inspection port is located on right side of transmission housing clutch compartment. Magneto impulse trip point (1) is a 3/16 inch hole. Exhaust valve closed mark "E" is similar on series C, D, L, and LA flywheels.

Fig. C41—Series VA tractor engine timing gear train and marks. Mesh single punch mark on crankshaft gear with an identical double punch mark on camshaft gear.

A. Valve timing marks
B. Idler gear timing marks
C. Magneto timing marks
1. Camshaft gear
2. Idler gear
3. Magneto gear
4. Crankshaft gear

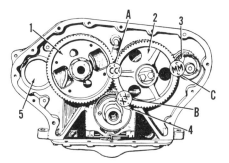

Fig. C42—Series R tractor engine timing gear train and marks.

For checking valve timing without removing the timing gear cover, use the following data: With tappets adjusted to normal operating gap, number one exhaust valve should be just closed when the flywheel mark "E" is visible at the inspection port in right side of clutch housing.

Series R

52. Valves are correctly timed when the idler gear which has three separate marks on its forward face is installed so that the corresponding marks on the camshaft gear, crankshaft gear and magneto gear are in register. In other words, the camshaft gear "C" mark should register with the "C" mark on the idler gear; the crankshaft gear mark "X" should register with the mark "X" on the idler gear; and the magneto gear mark "M" with the mark "M" on the idler gear. Refer to Fig. C42.

To check valve timing when engine is assembled, crank engine until inlet valve of No. 1 cylinder is just starting to open. At this time, the flywheel mark "IN-O" should be in register within ¼ inch with the index pointer over the flywheel.

Series V-VA

53. Valves are correctly timed when the single punch mark on the crankshaft gear is registered with an identical double punch mark on the camshaft gear as shown in Fig. C41. On series V engines equipped with magneto ignition, the timing marks for the governor and magneto gear are located on the rear face of the gear. These marks can be seen after removing the governor.

On series VA, refer to paragraph 134A for procedure for installing magneto adapter and timing magneto gear to the camshaft gear.

To check timing when engine is assembled, crank engine until No. 1 exhaust valve has just started to close. At this time the flywheel mark "DC1 & 4/EX. CL" should be in register within ¼ inch with the timing inspection hole pointer located on the left side of the flywheel housing.

TIMING GEAR COVER & OIL SEAL

Series C-D-L-LA

54. It is possible to renew the crankshaft front oil seal of felt without removing the timing gear cover. The felt seal can be renewed after removing the hood, radiator, crankshaft pulley and seal retainer.

54A. To remove timing gear cover, place support under engine oil pan. Remove hood, and drain oil pan and cooling system. Disconnect drag link from steering gear pitman (drop) arm, water inlet elbow from cylinder head, and radius rod pivot from adjustable and non-adjustable axle models. Disconnect radiator hoses. Remove radiator and front axle support (front frame extension) to cylinder block retaining bolts and cap screws and roll radiator, bracket and axle assembly as a single unit forward and away from the engine. Remove fan and generator belts, and water pump assembly. Remove fan drive pulley from crankshaft (pulley is retained in position by a set screw). Remove crankshaft front seal retainer and felt seal from timing gear cover. On series L and LA, remove three cap screws retaining oil pan to timing gear cover, and loosen all other oil pan retaining cap screws. Remove timing gear cover retaining cap screws and remove the cover.

Series R-V-VA

55. On series R engines, it is possible to renew the crankshaft front oil seal of felt without removing the timing gear cover. The oil seal can be renewed after removing the hood, radiator, and crankshaft pulley.

On series V and VA, although it is possible to renew the oil seals of felt and treated leather without removing the timing gear cover, our suggestion is to remove the cover which entails little additional work.

55A. To remove timing gear cover, first drain engine cooling system, and remove tractor hood on R series only. Raise hood on V and VA series. On series R and V and VA prior to 4829408 remove the radiator. On series VA after 4829408, remove radiator, front axle and front support as a

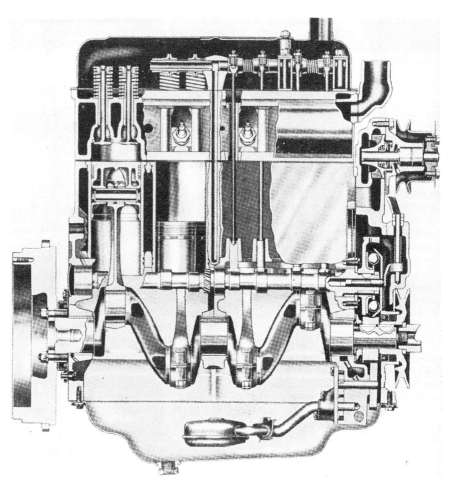

Fig. C43—Series VA tractor engine.

single unit by disconnecting drag link from steering shaft arm, and removing cap screws and bolts which retain front support to front face of engine. On series V, support the engine and remove engine front support bolt from lower side of timing gear cover. On VA series only, remove three oil pan to timing gear cover retaining cap screws and loosen all other pan cap screws.

On series V and VA remove water pump. On series R remove fan assembly. On all series, remove crankshaft pulley. Remove timing gear cover retaining cap screws and withdraw cover.

On series V and VA, install oil seal of treated leather with lip of seal facing the timing gears.

Series S

56. To renew the crankshaft front oil seal, it will be necessary to remove the timing gear cover.

To remove the cover, first drain the cooling system and remove the hood. Support engine and disconnect front axle support (frame extension) with radiator and front axle attached from front face of engine and roll assembly

away from engine. Remove fan belt, and crankshaft pulley. Remove two cap screws attaching oil pan to timing gear cover, and loosen all remaining oil pan retaining cap screws. Remove timing gear cover to cylinder block cap screws and lift off cover. Crankshaft front oil seal of treated leather can be renewed at this time. Seal is installed with lip facing the timing gears.

TIMING GEARS

Series C-D-L-LA-S-V

57. Refer to Fig. C39. Timing gear train on these models consists of three helical gears which include a governor gear that drives both governor and magneto. Cam gear and crank gear can be removed from their respective shafts after removing the timing gear cover and by using a suitable puller. Both gears can be reinstalled without heating. When installing the cam gear, remove the oil pan on series V engines or crankcase inspection plates on all other models and using a heavy bar buck-up the camshaft at one of the lobes near front end of shaft. Mesh the single punch mark on the crank

gear with a double punch mark on the cam gear. On all, but the V series, mesh the double punch mark on the governor gear with the single punch mark on the front face of the cam gear.

57A. On V series engines equipped with magneto ignition, the cam gear has a timing mark on both faces and only the rear face of the governor gear is marked. To bring these marks into register, it is necessary to remove the magneto.

Series R

58. Timing gear train, as shown in Fig. C42, consists of five helical gears. In addition to the cam and crank gears, there are a large idler gear, magneto drive gear and a governor drive gear. The idler gear which meshes with the crank gear drives both cam gear and magneto gear. The governor gear is driven by the cam gear.

Cam gear and crank gear can be removed from their respective shafts after removing the timing gear cover and using a suitable puller. When installing the cam gear, remove crankcase inspection plates and buck-up the camshaft at one of the lobes near front end of shaft with a heavy bar.

The idler gear (2) which has three separate marks on its front face should be installed so that the corresponding marks on the camshaft, crankshaft and magneto gear are registered. In other words, the mark "C" on the cam gear should align with the mark "C" on the idler gear; the mark "X" on the crank gear should align with the mark "X" on the idler gear, and the mark "M" on magneto gear should align with the mark "M" on the idler gear. Timing of the governor gear is not required.

The idler gear rotates on a bushing which is supplied for service. Bushing requires final sizing after installation to provide a diametral clearance of .0015-.0025.

Series VA

59. Timing gear train consists of a cam gear, crank gear, oil pump drive gear, and on models equipped with magneto ignition, a magneto drive gear. Cam gear and crank gear can be removed from their respective shafts after removing the timing gear cover by using a suitable puller. Both gears can be reinstalled without heating. When installing the cam gear, remove the oil pan and buck-up the camshaft gear at one of the lobes near front end of shaft with a heavy bar.

On VA series after 5365686, the flyweight type governor weight assem-

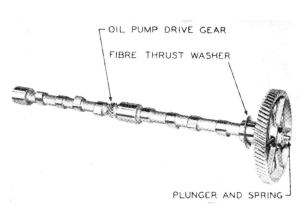

Fig. C46—Series LA tractor engine camshaft gear and assembly. Series C, D, L and S are similar.

bly is riveted to the cam gear. Cam gears on these models are available for service without the weight unit, or with the weight unit already attached.

Mesh the single punch mark on the crank gear with an identical double punch mark on the cam gear as shown in Fig. C41. No timing is required for the oil pump drive gear which meshes with the crank gear. To time the magneto drive gear to the cam gear on models which have the hydraulic pump mounted on the lower side of the belt pulley unit, refer to Magneto Adapter, paragraph 134A. On magneto ignition models which have the hydraulic pump mounted on the engine, mesh single punch mark on cam gear with the double punch mark on magneto and pump drive gear and retime magneto as outlined in paragraph 132.

CAMSHAFT AND BUSHINGS

Series C-D-L-LA-S

60. To remove the camshaft, first remove the timing gear cover as outlined in paragraph 54A or 56, and the oil pan as outlined in paragraph 87 or 88, and the engine oil pump. Remove the rocker arms and shafts assembly, and push rods. Block up the cam followers (tappets) so as to clear the cams and withdraw camshaft and gear from front of engine.

Camshaft journals ride directly in 3 machined bores of the cylinder block. Recommended clearance of journals in bores is .002-.004. The suggested maximum permissible clearance is .007, and when it exceeds this amount it will be necessary to renew the camshaft and/or the cylinder block, or to make up and install bushings. Camshaft journal diameters are listed in the table.

Camshaft end play is automatically controlled by a spring loaded plunger located in forward end of camshaft.

When reinstalling the camshaft and gear unit, make sure all tappets are

in place and that the fibre thrust washer is installed between the rear face of the cam gear and the cylinder block, as shown on Fig. C46.

Series R

61. To remove the camshaft, first remove the timing gear cover, paragraph 55A; oil pan, paragraph 89; and the engine oil pump. Remove cylinder head, valves and valve springs. Block up cam followers with clothes pins or heavy grease. Remove cap screws retaining camshaft thrust plate to cylinder block and withdraw camshaft and gear from front of engine.

Camshaft journals ride in three renewable bushings. Service bushings are presized, and if carefully installed require no final sizing after installation. In renewing the camshaft bushings, it will be necessary to detach engine from clutch housing as outlined in paragraph 40A, and remove the flywheel, and Welch plug which is located at camshaft rear journal. Recommended clearance of journals in bushings is .001-.0015, with a suggested maximum permissible clearance of .007. Camshaft journal diameters are as listed in the table.

Camshaft end play of .003-.005 is controlled by the thickness of the thrust plate which is located behind the camshaft gear. Renew the thrust plate if the end play exceeds .007.

Welch plug located in the cylinder block at the rear camshaft bearing can be renewed after detaching the engine from the clutch housing and removing the flywheel.

Series V

62. To remove the camshaft, first remove the timing gear cover as outlined in paragraph 55A, oil pan as outlined in paragraph 88 and the oil pump. Remove battery ignition unit on models so equipped. Remove the valves, cam followers, and camshaft thrust plate retaining cap screws. Withdraw camshaft and gear from front of engine.

CAMSHAFT JOURNAL DIAMETERS

Series	C&D	L&LA	R	S	V	VA
No. 1 Journal Diameter......	1.807-1.808	2.266-2.267	2.123-2.124	1.902-1.903	1.8715-1.8725	1.749-1.750
No. 2 Journal Diameter......	1.777-1.778	2.246-2.247	1.873-1.874	1.871-1.872	1.7457-1.7465	1.749-1.750
No. 3 Journal Diameter......	1.747-1.748	2.226-2.227	1.498-1.499	1.840-1.841	1.2465-1.2475	1.749-1.750

Camshaft journals ride in three renewable bushings. Service bushings require final sizing after installation to provide a diametral clearance of .002-.004. The suggested maximum permissible clearance is .007. In renewing the camshaft bushings, it will be necessary to detach engine clutch housing from the torque tube as outlined in paragraph 41A, and remove the flywheel, and Welch plug which is located at camshaft rear journal. Camshaft journal diameters are listed in the table.

Camshaft end play of .003-.005 is controlled by the thickness of the thrust plate which is located behind the camshaft gear. Renew thrust plate if the end play exceeds .007.

Welch plug located in the cylinder block at the rear camshaft bearings can be renewed after detaching the engine clutch housing from the torque tube, and removing the flywheel.

Series VA

63. To remove camshaft, first remove the timing gear cover as per paragraph 55A and the oil pan as per paragraph 90. Remove valve cover, rocker arms assembly and push rods. On models prior to 5365686, withdraw governor front race and the nine steel balls from front of camshaft. Remove governor rear race by removing camshaft gear retaining nut. On models after 5365686, the weight unit of the flyweight type governor is riveted to the cam gear.

Remove battery ignition unit on models so equipped. Remove cap screws which retain the camshaft thrust plate to the cylinder block. Block up cam followers and withdraw the camshaft and gear.

Camshaft journals ride in three renewable bushings. Service bushings require final sizing after installation to provide a diametral clearance of .002-.004. The suggested maximum permissible clearance is .007. In renewing the camshaft bushings, it will be necessary to detach engine from torque tube as outlined in paragraph 41A, and remove the flywheel and Welch plug which is located at camshaft rear journal. Camshaft journal diameters are listed in the table.

Camshaft end play of .002-.004 is controlled by the thickness of the thrust plate which is located behind the camshaft gear. Renew thrust plate if the end play exceeds .007.

Welch plug located in the cylinder block at the rear camshaft bearing can be renewed after detaching the engine from the torque tube, and removing the flywheel.

ROD AND PISTON UNITS

Series C-D-L-LA-R-S

64. Connecting rod and piston assemblies are removable from either **above** or **below**. To remove the assemblies from **above**, remove cylinder head, and hand hole plates from sides of oil pan on series C, D and R, or hand hole plates from sides of cylinder block on series L and LA, or oil pan on series S. On series C, L and LA, it will be necessary to remove the oil pump to main bearing oil lines. On series R, it will be necessary to remove the oil pump.

64A. To remove the assemblies from **below** as shown in Fig. C47, remove the oil pan or crankcase hand hole plates. However, it is suggested that the oil pan be removed on series C, D, L, LA and R. On series C, L, and LA, remove the oil pump to main bearing oil lines. On series R, remove the oil pump. When reinstalling the piston and ring assembly, use a ring compressor to enter the rings in the cylinder bore.

64B. Piston and rod units are installed with the rod and cap correlation marks facing away from the camshaft. On series C, D, L and LA, further identification for correct installation will be to install the rods so that the piston pin clamp screw faces opposite to camshaft side of engine. On series S, install the assembly so that the connecting rod cap on the diagonally cut rod is on the camshaft side of the engine, as shown in Fig. C48.

For bearing adjustment procedure on series C, D, L and LA, refer to paragraph 71; on series R, to 72A.

Torque tighten the connecting rod bolts to the following values:

Series	Torque
C-D	70 Ft. Lbs.
L-LA	150 Ft. Lbs.
R	60 Ft. Lbs.
S	65 Ft. Lbs.

Series V-VA

65. Connecting rod and piston assemblies are removed from **above** after removing the oil pan and cylinder head. On series V, the connecting rods are offset. Numbers 1 and 3 have the long part of the bearing toward the flywheel, and Numbers 2 and 4 have the long part of the bearing toward the timing gears.

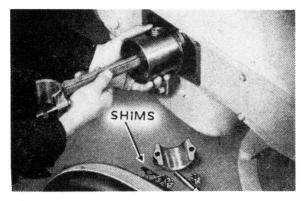

SHIMS

Fig. C47—Removing piston and connecting rod assembly from below on series D engine. Assemblies can also be removed from above after removing cylinder head. Series C, L, LA and R are similar.

Pistons and rods are installed with the rod and cap correlation marks facing toward the camshaft. Replacement rods are not marked and should be installed with the oil spurt hole at lower end of rod facing the camshaft.

Torque tighten the connecting rod bolts to 60 ft. lbs. on series V, and 45-50 ft. lbs. on series VA.

PISTONS, SLEEVES AND RINGS
All Series

66. **PISTONS.** Series C and D engines of 4 inch bore size and series S engines of 3⅝ inch bore size (late production or field converted) are equipped with solid skirt aluminum alloy pistons. Solid skirt iron pistons are used in all other engines. Pistons are available in standard size only for series C, D, L, LA, S and VA engines, and in standard and semi-finished sizes for the series R and V engines.

The recommended piston skirt clearance is .004-.006 on series C and D; .006-.008 on series L and LA; .003-.004 on series R, V and VA; and .0035-.005 on series S. Skirt clearance should be measured at right angles to the piston pin bosses.

67. **SLEEVES.** Series C, D, L, LA, S and VA engines are equipped with wet type sleeves. On series R and V engines, dry type sleeves are used.

With the piston and connecting rod assembly removed from the cylinder block, use a suitable puller to remove the sleeve. Before installing a new sleeve, clean all mating and sealing surfaces in block. All wet type sleeves should be tried in block bores to full depth without packing rings to be sure there is no obstruction. After making trial installation, remove the sleeves and install packing rings dry in grooves in block or sleeves. On installations where the packing ring is installed in the block, lubricate only the sleeve contacting face of the ring. On installations where the packing ring is installed on the sleeve,

lubricate only the outer face of the ring. Use palm oil or vaseline as a lubricant on the neoprene sealing rings.

On wet type sleeve installations only, the top of the sleeve should extend .001-.003 above the top of the cylinder block on all series except VA. On series VA, this standout should be .002-.004. If this standout is in excess of the indicated high limit, check for foreign material under sleeve flange.

To test, reassemble all but the hand hole plates on the C, D, L and LA series or all but the oil pan on the S and VA series and fill the cooling system with cold water and check for leaks around bottoms of sleeves.

On series R and V, coat the outer surface of the dry type sleeve with a mixture of powdered aluminum and penetrating oil. Press sleeves into block bores until top surface of sleeve is flush with top surface of block. After installing the sleeve, check piston skirt clearance and sleeve bore for distortion. Top of sleeve should not be below top surface of block. On series V only, install sleeves so that notches or slots at lower end of sleeve are at

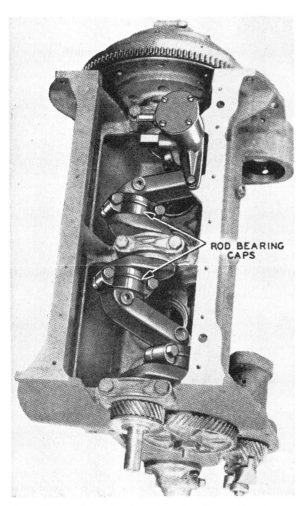

Fig. C48—On the series S engine, the diagonally cut connecting rod at the crankpin end should be installed so that the rod cap is on the camshaft side of the engine.

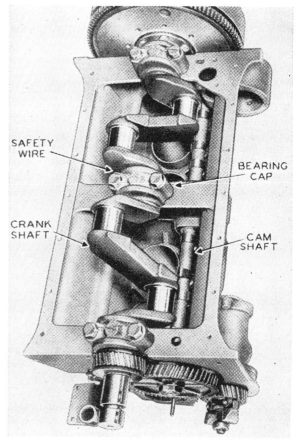

Fig. C49—Crankshaft installation in series D engine.

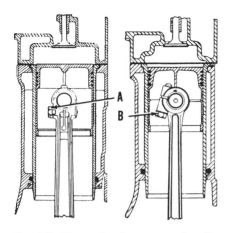

Fig. C50—Piston pin clamp screw installation. (A) Series L and LA. (B) Series C and D.

right angles (crosswise) to crankshaft center-line.

Series C and D engines (with head No. A2873AA, casting 5576A or head No. 2399A, casting 5505A) were designed for operation on low cost distillate type fuel. When such engines are converted to 4 inch bore by installation of Case 4865AA sleeve set, the resulting increase in compression ratio requires the use of medium octane gasoline fuel. It is recommended that the original hot manifold be replaced by a cooler gasoline type manifold.

68. PISTON RINGS. Pistons used in the series C of 3⅞ inch bore and R engines are equipped with 3 rings. On all other series, the pistons are equipped with 4 rings. Compression rings, which have a dot or the word "TOP" marked on one side, should be installed on the piston with the marked side toward the top.

Recommended piston ring end gap is as follows:

Series	Comp. Rings	Oil Ring
C-D	.015-.025	.010-.020
L-LA	.020-.030	.015-.025
R-S	.010-.020	.010-.020
V-VA	.010-.015	.010-.015

Recommended piston ring side clearance in thousandths is as follows:

Series	1st. Comp. Ring	2nd & 3rd Comp. Ring	Oil Ring
C-D	1.5-3	1 -2.5	1 -2.5
L-LA	1.5-2.5	1.5-2.5	1.5-2.5
R	1.5-3	1.5-3	1.5-3
S	1.5-3	1 -2.5	1.5-2.5
V	1.5-2	1.5-2	1.5-2
VA	2 -3.5	1 -2.5	1 -2

PISTON PINS

Series C-D-L-LA

69. Pin is locked in connecting rod, as shown in Fig. C50. Pin should be fitted to a finger push fit (.0005

clearance) in the unbushed piston. The piston pins are supplied only in the standard size.

To remove the piston pin on the C and D series remove the clamp screw (B) from the rod. On L and LA series loosen the screw (A) but do not remove it. Piston can be installed either way to the rod. When reinstalling, line up the notch in piston pin with clamp screw hole so that screw will enter threads easily. Tighten screw and lock same with tabs of washer.

Series R-S-V-VA

70. The floating type piston pins which are available only in the standard size are retained in the piston bosses by snap rings. Piston can be installed either way to the rod.

CONNECTING RODS AND BEARINGS

Series C-D-L-LA

71. Connecting rod bearings are of the cast-in babbitt type, and are provided with shims for adjustment. Rod bearings can be adjusted through hand holes provided in sides of oil pan or cylinder block. When removing shims, remove the same amount (thickness) from each side.

Re-babbitted connecting rod exchange service is available through the factory. Connecting rods with cast-in babbitt bearings are supplied in undersizes of 0.010 and 0.020 as well as standard.

Replacement rods are not marked but should be installed with the piston pin clamp screw facing away from camshaft side of engine.

Series	C-D	L-LA
Crankpin Dia. (mean)	2.2495	2.7495
Running Clearance	1.5- 3	1.5- 3
Side Clearance	5-12	5-12
Bolt Torque (ft. lbs.)	70	150

Series R

72. Connecting rod bearings are of the slip-in type, provided with shims for a limited range of adjustment. Refer to paragraph 72A for bearing adjustment procedure. Rod bearings can be adjusted through hand holes provided in sides of oil pan.

Bearings are available in undersizes of 0.020 and 0.040, as well as standard.

Series	R
Crankpin Dia. (mean)	1.7495
Running Clearance	.5- 2.5
Side Clearance	4-7
Bolt Torque (ft. lbs.)	60

72A. BEARING ADJUSTMENT. Connecting rod bearing shims, which are provided for a limited range of

adjustment, do not fit between the bearing shell parting surfaces, but only between the bearing cap and its mating surface. These shims control the amount of crush or pinch placed on the shell and only indirectly control the bearing running clearance of .0005-.0025. Individual shim thickness of the factory supplied shim pack of three shims is .002.

It is considered a better practice to install new bearings rather than attempt to adjust them. In an emergency where new bearings are not available, and it is necessary to remove one .002 thick shim from each side to reduce the running clearance, it is also necessary to decrease the height of a pair of bearing shells by .002 so as to retain the correct bearing shell crush. Removal of metal from the parting surface of shell (reducing shell height) can be accomplished with fine emery paper and a flat surface, making an occasional check on the bearing shell height to prevent removal of too much metal.

For new bearing shells and crankshaft installation, use the standard shim pack of three .002 shims which will automatically provide the correct running clearance and bearing crush.

Series S-V and VA

73. Connecting rod bearings are of the shimless, non-adjustable, slip-in, precision shell type. When installing new shells be sure nib or projection on shell engages the milled notch in both the rod and cap.

Bearing shells are available in undersizes of .002, .010 and .020 as well as standard for series S. On series V and VA bearing shells are available in undersizes of .002, .020, .022, .040 and standard as well as semi-finished.

Replacement rods are not marked and should be installed on series VA engines with the oil spurt hole in rod facing camshaft side of engine.

On series V, the connecting rods are offset. Numbers 1 and 3 have the long part of the bearing toward the flywheel, and Numbers 2 and 4 have the long part of the bearing toward the timing gears.

On series S, the diagonally cut connecting rod at the crankpin end should be installed so that the rod cap is on the camshaft side of the engine, Fig. C48, when the piston is at either top or bottom dead center positions.

Series	S	V-VA
Crankpin Dia. (mean)	2.3745	1.937
Running Clearance	1-3	1.5-2
Side Clearance	5-11	8-12
Bolt Torque (ft. lbs.)	65	60

CRANKSHAFT AND BEARINGS

Series C-D-L-LA

74. Shaft is supported on three slip-in type bearings provided with shims for adjustment. These shims fit between the bearing shell parting surfaces. Crankshaft end play of .004-.010 is controlled by flanged ends of center main bearing. Only unfinished main bearing shells which require line-boring are available for service.

To renew one or more main bearings, first remove the engine, and crankshaft. Install new shells with one standard factory shim pack on each side of the shell parting surfaces; then torque tighten bearing cap retaining bolts to 150 ft. lbs. Line-boring can now be accomplished to provide the correct diametral running clearance of .0015-.003.

To remove crankshaft, it is necessary to remove the engine as in paragraph 40A and 40B, clutch, flywheel, timing gear cover, and oil pan. On C, L, and LA, also remove the oil pump to main bearing cap oil lines.

Check the crankshaft journals for wear, scoring and out-of-round condition against the values as listed:

Series	C-D	L-LA
Journal Dia. (mean)	2.2495	2.7495
Running Clearance	1.5-3	1.5-3
Crankpin Dia. (mean)	2.2495	2.7495
Renew or regrind if out-of-round more than	.0025	.0025
Shaft End Play	4-10	4-10
Bolt Torque (ft. lbs.)	150	150

Series R

75. Shaft is supported on three slip-in type bearings provided with shims for a limited range of adjustment. Refer to paragraph 75A. Crankshaft end play of .005-.007 is controlled by flanged ends of center main bearing. Bearings, which are available in undersizes of .020 and .040 as well as standard, are renewable from below without removing the crankshaft.

To remove crankshaft, it is necessary to remove the engine as in paragraph 40A and 40B, clutch, flywheel, timing gear cover, and oil pan.

Check the crankshaft journals for wear, scoring and out-of-round condition against the values as listed:

Series	R
Journal Dia. (mean)	2.1245
Running Clearance	.5-2.5
Crankpin Dia. (mean)	1.7495
Renew or regrind if out-of-round more than	.0025
Shaft End Play	5-7
Bolt Torque (ft. lbs.)	90

75A. **BEARING ADJUSTMENT.** Crankshaft main bearing shims, which are provided for a very limited range of adjustment, do not fit between the bearing shell parting surfaces, but only between the bearing cap and its mating surface. Refer to paragraph 72A for further information.

Series S-V-VA

76. Shaft is supported on three shimless, non-adjustable, slip-in, precision type main bearings, renewable from below without removing the crankshaft.

Bearings are supplied for service in undersizes of .002, .005, .010 and .020, as well as standard on series S engines, and on series V and VA engines, in undersizes of .002, .020, .022, .040, also standard and semi-finished. Use of the semi-finished bearing shells requires the removal of the engine and crankshaft, and line-boring equipment.

On series S engines, crankshaft end play of .004-.008 is controlled by flanged ends of center main bearing.

On series V engines, crankshaft end play is controlled by the flange at the rear of the front bearing shell, and a thrust washer just forward of the front bearing shell. End play of .004-.006 is adjusted by means of shims which are located between a thrust washer and shoulder of crankshaft number 1 main journal. To gain access to shims, remove crankshaft gear.

On series VA engines, crankshaft end play of .004-.006 is controlled in a manner similar to that as used on the series V, except that a thrust washer is used in lieu of the number 1 main bearing shell flange. To gain access to shims, remove crankshaft gear.

On V series, the front and rear main bearing caps can be removed after removal of oil pan front and rear filler blocks. On VA series, to remove the combination rear main bearing cap and filler block, first remove the two cap screws which retain rear oil seal retainer. When reinstalling rear main bearing cap and filler block, renew the packing which is located in grooves on sides of filler block.

To remove crankshaft, it is necessary to remove the engine as outlined in paragraphs 40A and 40B for S series, or paragraphs 41A and 41B or 41C for V and VA series, clutch, flywheel, timing gear cover, and oil pan.

Check the crankshaft journals for wear, scoring and out-of-round condition against the values as listed.

Series	S	V-VA
Journal Dia. (mean)	2.4995	2.2495
Running Clearance	1.5-2.5	1.5-3
Crankpin Dia. (mean)	2.3745	1.937
Renew or regrind if out-of-round more than	.0025	.0025
Shaft End Play	4-8	4-6
Bolt Torque (ft. lbs.)	100	95

CRANKSHAFT REAR OIL SEAL

Series C-D-L-LA-S

77. A crankshaft rear oil seal is not used in these engines.

Series R

78. Crankshaft rear oil seal is contained in the one-piece main bearing closure plate which is bolted to the rear of the cylinder block. Oil seal can be renewed after detaching engine from transmission housing as outlined in paragraph 40A and removing clutch and flywheel.

Series V

Upper half of the cork oil seal is held in a rear oil guard which is located in a groove in the rear face of the cylinder block. The lower half of the cork seal is retained in the oil pan rear filler block.

79. To renew the lower oil seal, remove oil pan and rear filler block. For upper seal renewal, rotate rear oil guard, which contains upper half of seal, out of its groove in cylinder block. To do this, press on one end of guard with a punch or screw driver and rotate crankshaft. The oil guard should turn with the crankshaft. When reinstalling, lubricate the surface of the crankshaft where cork seal rides and install oil guard in same manner as was removed. Do not soak the cork in oil prior to installing as it will swell and retard installation. Renewal procedure for lower half of cork seal is self-evident after removing the oil pan and rear filler block.

Series VA

Shaft rear oil seal of treated leather is contained in the one piece retainer which is bolted to the rear face of the cylinder block, and rear main bearing cap.

80. To renew the seal, first detach engine from torque tube as outlined in paragraph 41A, and remove clutch, flywheel and oil pan. Remove three cap screws attaching upper half of seal retainer to rear face of cylinder block. Working through oil pan opening, remove two cap screws attaching oil seal retainer to rear main bearing cap, and remove oil seal. Install new

seal in retainer, so that lip of seal faces toward front of engine.

CRANKSHAFT FRONT OIL SEAL

All Series

81. It is possible to renew the crankshaft front oil seal of felt as used on series C, D, L, LA, and R without removing the timing gear cover. On series S, to renew the front oil seal of treated leather, it will be necessary to remove the timing gear cover.

On series V and VA, although it is possible to renew the oil seals of felt and treated leather without removing the timing gear cover, our suggestion is to remove the cover which entails little additional work.

When renewing the oil seal on series V and VA without removing the timing gear cover, it is advisable to loosen all of the cover retaining cap screws when removing the old seal, and to run the engine with them loosened so as to facilitate centering of the new seal to the crankshaft.

On engines using the treated leather seal, install the seal with the lip facing the timing gears.

FLYWHEEL & GEAR

All Series

82. On series C, D, L, LA, R, and S, flywheel removal requires detaching engine from transmission housing as in paragraph 40A, and removal of engine clutch. On series V and VA, flywheel removal requires detaching engine from torque tube on VA series, or clutch housing from torque tube on V series as in paragraph 41A, and removal of engine clutch.

The starter ring gear can be renewed on series V and VA without removing the flywheel. Install ring gear with bevel end of teeth facing the transmission.

On all other series, including series C and L which are equipped with starters, renewal of the starter ring gear requires removal of the flywheel. The bevel end of ring teeth should face toward the engine timing gears.

OIL PAN

Series C-D

87. Method of removing the oil pan is as follows: First drain transmission housing clutch compartment, and oil pan. Remove radius rod and tie rod from models so equipped. On series D only, remove oil pump screen (2—Fig. C51) from left side of oil

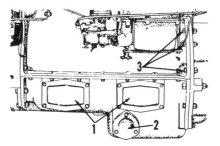

Fig. C51—Oil pan installation on series C and D engines.

1. Hand hole plates   2. Oil pump screen

pan. Remove bolts and cap screws attaching oil pan to transmission housing, and oil pan to front axle support (front frame extension). Loosen bolts attaching cylinder block to transmission housing, and cylinder block to front axle support. Carefully separate pan from front and rear gaskets. By following this procedure you will prevent damaging the timing gear cover gasket and the cylinder block to transmission housing gasket. Next, remove all cap screws attaching pan to bottom of cylinder block and remove the oil pan.

Series L-LA-S-V

88. Method of removing the oil pan is as follows: First drain oil. Remove radius rod, and oil pump screen from left side of oil pan on models so equipped. Remove oil pan attaching cap screws and lower the pan.

88A. On V series, to renew the oil pan gaskets when pan is off, it will be necessary to remove the front and rear filler blocks. These aluminum castings contain cork strips which seal the front and rear arches of the oil pan. When removing the front filler block, carefully separate it from the crankcase front end plate gasket as renewal of end plate gasket will require removal of timing gear cover and front end plate. The rear filler block also contains the lower half of crankshaft rear oil seal.

Series R

89. Method of removing the oil pan is as follows: First support engine in a hoist, and remove the radiator and front axle bracket (front frame extension). Remove the starting motor.

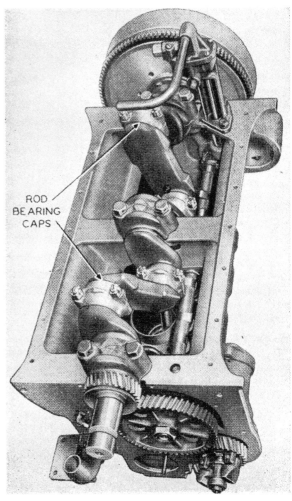

Fig. C52—Oil pump installation on series D engines. Series C is similar.

ROD BEARING CAPS

Fig. C53—Oil pump installation on series LA engines. Series L is similar.

1. Pinion key
2. Retaining pin
3. Pump drive pinion
4. Thrust washer
5. Pump body
6. Drive shaft
7. Relief valve
8. Relief valve spring
9. Driving gear key
10. Body driving gear
11. Cover
12. Body idler gear
13. Idler gear shaft
14. Inlet tube
15. Outlet elbow

Fig. C54—Oil pump installation on series S engines.

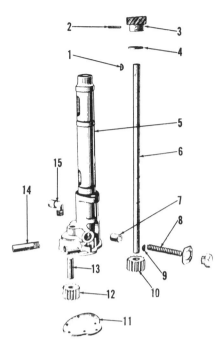

Fig. C55—Oil pump as used on series C, D, L and LA engines. Series S is similar.

Drain oil pan and remove oil pan attaching cap screws and bolts, and lower the oil pan.

Series VA

90. Method of removing the oil pan is as follows: First drain the oil pan, and remove pan to cylinder block attaching cap screws. On models VAC and VAH, oil pan may now be lowered. On VA and VAO, lower the pan far enough to reach inside and hold up the floating oil pump screen; then rotate the pan in a crosswise direction until pan is clear of steering tie rod.

OIL PUMP AND RELIEF VALVE

Series C-D-L-LA-S

91. The gear type pump, Fig. C55, which is driven by the camshaft, is mounted on the underside of the cylinder block as shown in Figs. C52, C53 and C54. Pump removal requires removal of oil pan.

Check the pump internal gears for backlash which should not exceed .006. The recommended diametral clearance between gears and pump body should not exceed .006. Renew pump body and/or shaft or both if clearance exceeds .007.

On series L prior to serial 300835, select gaskets between pump body and cover to obtain .003-.005 clearance between pump gears and cover. On all other series, no gaskets are used between pump body and cover. On these models, if pump gear to cover clearance exceeds .010, renew gears, cover, and/or pump body to obtain a desired clearance of .003-.005.

91A. The adjustable ball type relief valve as used on series C engines prior 4209666, and series L prior 300835 is located in the pump body. Relief valve adjustment for a pressure of 30-35 psi on C series, and 35-40 psi on L series can be made by working through the oil pan inspection hand hole plates.

Series C engines after 4209665, L series after 300834, and all other series have a non-adjustable plunger type relief valve which is located in the pump body. Normal oil pressure on C and D series is 30-35 psi; 35-40 psi on L and LA series; and 25-30 psi on S series.

Series R

92. The gear type pump, which is driven by the camshaft, is mounted on the underside of the cylinder block. To remove the oil pump, first remove hand hole cover from right rear side of oil pan, and cover from lower side of oil pan. Working through side hand hole cover opening, remove pump to cylinder block attaching cap screws. Pump can now be removed through opening in lower side of oil pan.

Check the pump internal gears for backlash which should not exceed .006 The recommended diametral clearance between gears and pump body should not exceed .006. Renew pump body and/or shaft or both if shaft to pump body clearance exceeds .007. Select gaskets between pump body and cover to obtain .003-.005 clearance.

92A. The adjustable ball type relief valve is located externally on the right side of the engine just below the valve chamber. Oil pressure can be adjusted to 15 psi at governed speed by rotating the slotted relief valve adjusting screw.

Series V

93. The gear type pump, which is driven by the camshaft, is mounted on the underside of the cylinder block. Pump removal requires removal of the oil pan.

Check the pump internal gears for backlash which should not exceed .006. The recommended diametral clearance between gears and pump body should not exceed .006. Renew pump shaft bushing which is located at drive gear end of pump body or pump body if diametral clearance exceeds .007. Se-

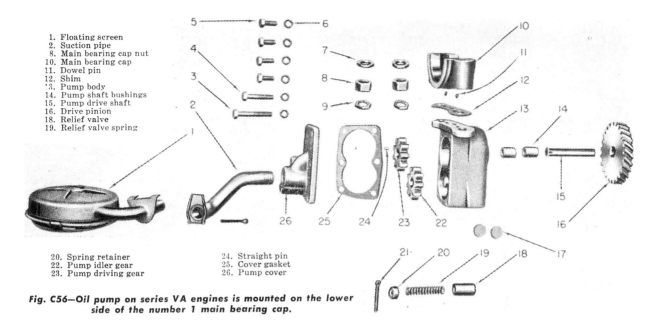

1. Floating screen
2. Suction pipe
8. Main bearing cap nut
10. Main bearing cap
11. Dowel pin
12. Shim
13. Pump body
14. Pump shaft bushings
15. Pump drive shaft
16. Drive pinion
18. Relief valve
19. Relief valve spring

20. Spring retainer
22. Pump idler gear
23. Pump driving gear

24. Straight pin
25. Cover gasket
26. Pump cover

Fig. C56—Oil pump on series VA engines is mounted on the lower side of the number 1 main bearing cap.

lect gaskets between pump body and cover to obtain .003-.005 end clearance.

A spool shaped, bronze sleeve (X— Fig. C34) is pressed into the oil pump body bore of the cylinder block. This sleeve provides a bearing surface for the upper end of pump drive shaft which extends above the pump drive pinion. Renewal of this sleeve may be accomplished when distributor drive shaft or cylinder head, valve chamber cover and oil pump are removed.

CAUTION: Lack of oil pressure may result if this sleeve is loosely seated or improperly positioned as it seals the distributor drive shaft and oil pump shaft bores from the main oil gallery.

93A. The piston type oil relief valve is located externally on right side of engine in the vicinity of the carburetor. Oil pressure can be varied by adding or removing shim washers between spring and retainer plug. Normal oil pressure is 20-24 psi at 1650 rpm.

Series VA

94. The gear type pump shown in Fig. C56 is bolted to the bottom of number one main bearing cap. The pump drive pinion (16) meshes with the crankshaft gear. Pump is accessible after removing the oil pan. To obtain desired .006-.008 backlash of drive gear (16) vary the shims (12).

Check the pump internal gears for backlash which should not exceed .005. Gear diametral clearance in pump body should not exceed .004. A .007 lead gasket is used between pump body and cover. If body (13), pinion (16) or gear (23) is worn, install VTA3408 body and drive gear assembly as components are not available.

Pump shaft bushings (14) are re-

newable only on early production pumps which do not have the drive gear and pinion attached to the shaft by Rollpins. Bushings should be reamed to .501-.503 to allow a running clearance of .002-.0045.

94A. The non-adjustable plunger type valve relief valve is located in the oil pump body. Plunger (18) should fit in the pump body bore with a .003-.005 clearance. Relief valve spring, which has a free length of 1 15/16 inches should test 14¾ pounds at 1⅜ inches.

CARBURETOR

All Series (Gasoline Type)

98. The carburetor application table applies to Case tractors which were built since 1937.

Turning the idle mixture screw towards its seat richens the mixture. Turning the power adjusting needle towards its seat leans the mixture. Idle speed stop screw should be adjusted to permit the engine to idle at 475-500 rpm, on all series except R. On R series, adjust engine idle speed to 400 rpm. Refer to Standard Units manual for calibration and overhaul.

LP-GAS CARBURETOR

Series D and LA tractors are available with LP-gas carburetors designed and built by Ensign. Like other LP-gas systems, these are designed to operate with the fuel supply tank not more than 80% filled. It is important when starting the D and LA series to

CARBURETOR APPLICATION

Series	Make	Model	Assy. No.	Float Setting Inches
C prior 4300000	Zen.	124½TOXP	08672	1 1/2
C prior 4300000	Zen.	124½TOXP	07124	1 1/2
C after 4300000	Zen.	62AXJ9	08964	1 39/64
D prior 4301201	Zen.	124½TOXP	08672	1 1/2
D after 4301201	Zen.	62AJX9	08964	1 39/64
D4511449-5600000	M-S	TSX11		9/32
L prior 4210939	Zen.	K6A	06799	1 49/64
L after 4210939	Zen.	K6A	08714	1 49/64
LA	Zen.	K6A	08714	1 49/64
R	Zen.	193½P	07206	1/4
R	Zen.	161-7	010136	1 5/32
S prior 8027115	Zen.	161JX7	09667	1 5/32
S after 8027115	Zen.	267-8	11535	1 5/32
V Gas prior 4518012......	M-S	TSX42		9/32
V Gas after 4518011......	M-S	TSX43		9/32
V Distillate	M-S	TSX74		9/32
VA	M-S	TSX114		9/32
VA	Zen.	161JX7	11106	1 5/32
VA	Zen.	161JX7	11506	1 5/32
VA Latest	M-S	TSX		9/32

Fig. C57—Ensign LPG model W regulator and model Xg carburetor installation on series D.

open the vapor valve on the supply tank *SLOWLY; if opened too fast, it will shut off the fuel supply to the regulator. Too rapid opening of vapor or liquid valves may cause freezing.*

Series D

The Ensign model Xg carburetor and model W regulator as used on this series have 3 points of mixture adjustment, plus an idle stop screw. Refer to Fig. C57.

99. **IDLE STOP SCREW.** Idle speed stop screw on the carburetor throttle should be adjusted to provide a slow idle speed of 500 crankshaft rpm or 341 belt pulley rpm.

99A. **STARTING SCREW.** Immediately after the engine is started, bring the throttle to the **fully open** position and with the choke **closed**, rotate starting screw until the highest engine speed is obtained. A slightly richer adjustment (counter-clockwise until speed drops slightly) may be desirable for a particular fuel or operating condition. Average adjustment is one turn open.

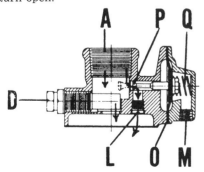

Fig. C58—Sectional view of economizer used on Ensign model Xg carburetor.

A. Fuel inlet
D. Load adjusting screw
L. Orifice
M. Vacuum connection
O. Diaphragm
P. Fuel passage
Q. Spring

99B. **IDLE MIXTURE SCREW.** With the choke **open**, engine warm and idle stop screw set, adjust idle mixture screw, located on regulator, until best idle is obtained. An average adjustment is ¾ of a turn open.

99C. **LOAD SCREW.** In the Xg Ensign carburetor, the so-called load screw (D—Fig. C58) primarily controls the partial load mixture. The richer mixture needed for full power is supplied by a by-pass jet (L) which is opened and closed by the economizer diaphragm spring (Q) and the manifold vacuum. When the manifold vacuum drops below 4-6 inches Hg which will occur at full load (wide-open carburetor throttle) the diaphragm spring opens the non-adjustable power jet. When the vacuum is higher than 10-13 inches Hg which will occur at no load or light load, regardless of rpm, the diaphragm spring is overridden by the higher vacuum and the jet is closed off at which time the mixture is controlled by the load screw and idle screw to provide economizer action. In the Xg carburetor, the vacuum controlled valve attached to the economizer diaphragm enriches the mixture whereas in the Ensign Kgl carburetor (as used on series LA tractors), the plunger attached to the economizer diaphragm leans the mixture.

99D. ADJUSTMENT WITHOUT LOAD. Average adjustment of the load screw, Fig. C57, as recommended by J. I. Case is 3½ turns open. However, to accurately set the load screw and regulator for the fuel and operating conditions as found in a certain locality, the J. I. Case recommendations are as follows:

The engine should be thoroughly

warmed up to operating temperature. The carburetor throttle valve and choke should be **fully open.** With the engine operating at its no load speed of 1390 rpm, rotate the load screw counter-clockwise (rich) until the engine speed decreases; then rotate the load screw clockwise (lean) until the engine runs smoothly. Continue to rotate the load screw until the engine begins to run unevenly; then, rotate the screw counter-clockwise until the engine runs smoothly. Recheck the adjustment and lock the load screw in position. Recheck the idle mixture screw adjustment as per paragraph 99B.

99F. ENSIGN METHOD WITHOUT LOAD. First carefully adjust the idle mixture screw as in paragraph 99B. Disconnect the economizer to carburetor flange vacuum line, and plug the carburetor flange connection. Run engine at high idle speed with hand throttle in position where governor does not regulate the rpm. Adjust load screw to obtain maximum rpm, note its position; then carefully rotate it in a lean direction (clockwise) until the rpm just begins to fall. Rotate the screw to the mid-point of these two positions and tighten the locknut. Unplug the carburetor flange connection and reconnect the economizer vacuum line. The power valve and jet are fully open when using this method.

99G. ANALYZER AND VACUUM GAGE METHOD. In this method, the engine is operated with the carburetor throttle wide open and with sufficient load on the engine to hold the rpm to maximum operating speed (1200) or 300 to 500 rpm slower than maximum operating rpm. One method of loading the engine is to disconnect or short out two or more spark plug wires. Do not disconnect any lines. Set the load screw to give a reading of 12.8 on the analyzer gasoline scale or 14.3 on an analyzer with a LPG scale.

99H. Check the part throttle (partial load) mixture by reducing the opening of the throttle valve and the load on the engine until a manifold vacuum of 10-13 inches is obtained at the same rpm as used in paragraph 99G. The power jet should be closed at this time and the analyzer should now read 13.8-14.5 on the gasoline scale or 14.9-15.5 on the LPG scale. If readings are lower than specified, fuel may be leaking past the power valve; if higher than specified, the power jet orifice may be too small.

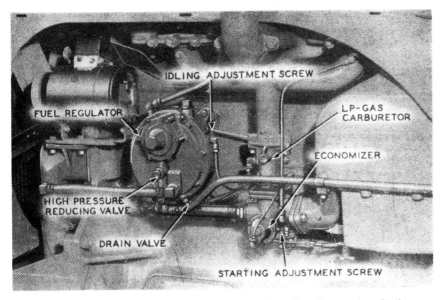

Fig. C59—Ensign LPG model R regulator and model Kgl carburetor installation on series LA.

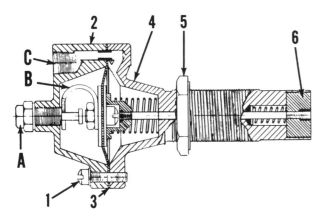

A. Adjusting screw (economizer)
B. Adjusting screw stop
C. Vacuum connection
1. Cover attaching screws
2. Cover
3. Diaphragm
4. Body assy.
5. Locknut
6. Plunger

Fig. C60—Ensign dry gas economizer as usually installed on the model Kgl carburetor.

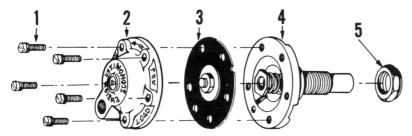

Fig C60A—Special simplified version of the Ensign dry gas economizer as used with the Ensign Kgl carburetor on the series LA tractor. Refer to Fig. C60 for legend.

Series LA

The Ensign model Kgl carburetor and model R regulator as used on this series have 3 points of mixture adjustment, plus an idle stop screw. Refer to Fig. C59.

99J. **IDLE STOP SCREW.** Idle speed stop screw on the carburetor throttle should be adjusted to provide a slow idle speed of 500 crankshaft rpm or 335 belt pulley rpm.

99K. **STARTING SCREW.** Follow the procedure as outlined for the series D in paragraph 99A. Average adjustment is 1¼ turns open.

99L. **IDLE MIXTURE SCREW.** Follow the procedure as outlined for the series D in paragraph 99B. Average adjustment is ½ turn open.

99M. **LOAD SCREW OR ECONOMIZER.** The model Kgl Ensign carburetor used on the LA model tractors is equipped with a special simplified version of the Ensign Dry Gas Economizer, Fig. C60A. The regular version of this economizer is illustrated in Fig. C60 but as used on the LA the adjusting screw (A) and stop (B) have been eliminated. The effect of this change is to provide automatic rather than adjustable control of the partial or light load mixture. In this set-up, the richer mixture needed for maximum power is adjusted by varying the depth to which the entire economizer body is screwed into the fuel passage of the carburetor. The economizer body thus becomes the load screw. When the manifold vacuum is higher than approximately 9 inches Hg, it overcomes the large diaphragm spring and moves the restricting plunger into the fuel passage thus leaning the mixture and providing economizer action. It will be seen therefore that the vacuum controlled plunger in this type of economizer leans the mixture whereas the vacuum controlled valve, Fig. C58, in the economizer of the Xg carburetor richens the mixture and is actually a full power jet. Average adjustment is 4 turns open.

99N. **ADJUSTMENT WITHOUT LOAD.** Carefully adjust the idle mixture as described in paragraph 99B. Remove the suction line from the economizer and plug the line or the line connection at the inlet manifold. Disconnect rod from carburetor throttle lever and rotate idle stop screw to produce high idle rpm of 1275. Loosen locknut (5) on economizer and rotate the entire unit until maximum rpm is obtained. Note this position; then rotate in opposite (counter-clockwise) direction until rpm begins to fall. Rotate the entire unit to the mid-point of these two positions and tighten the locknut. Unplug the manifold connection and reconnect the suction line.

99P. **ANALYZER AND VACUUM GAGE METHOD.** Follow the procedure as outlined for the series D in paragraph 99G and use 1100 rpm as the operating speed.

LP-GAS FILTER

100. Filters used on these systems are subjected to pressures as high as 150 psi and should be able to stand

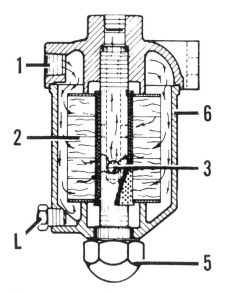

Fig. C61—LP-Gas filter used on series D
and LA.

L. Drain plug
1. Fuel inlet
2. Filter cartridge
3. Outlet passage
5. Stud nut
6. Filter bowl

this pressure without leakage. Unit should be drained periodically at the blow off cock (L—Fig. C61). When major engine work is being performed, it is advisable to remove the lower part of the filter, thoroughly clean the interior and renew the felt cartridge of same if not in good condition.

LP-GAS REGULATOR

Series D (Ensign Model W)

100A. **HOW IT OPERATES.** Fuel from the supply tank enters the regulating unit inlet (A—Fig. C62) at a tank pressure of 25 to 80 psi and is reduced from tank pressure to about 4 psi at the high pressure reducing valve (C) after passing through the strainer (B). Flow through high pressure reducing valve is controlled by the adjacent spring and diaphragm. When the liquid fuel enters the vaporizing chamber (D) via the valve (C) it expands rapidly and is converted from a liquid to a gas by heat from the water jacket (E) which is connected to the coolant system of the engine. The vaporized gas then passes at a pressure slightly below atmospheric pressure via the low-pressure reducing valve (F) into the low pressure chamber (G) where it is drawn off to the carburetor via outlet (H). The low pressure reducing valve is controlled by the larger diaphragm and small spring.

Fuel for the idling range of the engine is supplied from a separate outlet (J) which is connected by tubing to a separate idle fuel connection on the carburetor. Adjustment of the

carburetor idle mixture is controlled by the idle fuel screw (K) and the calibrated orifice (L) in the regulator. The balance line (M) is connected to the air inlet horn of the carburetor so as to reduce the flow of fuel and thus prevent over-richening of the mixture which would otherwise result when the air cleaner or air inlet system becomes restricted.

Series LA (Ensign Model R)

100B. **HOW IT OPERATES.** Liquid LP-Gas from the supply tank enters the model R regulator at (1—Fig. C63) and is reduced from tank pressure to about 4 psi at high pressure reducing valve (3) after passing through the strainer. Flow through high pressure valve is controlled by diaphragm (4), lever (5) and spring (6). When the liquid fuel enters the vaporizer coil (7) it expands rapidly and is converted from a liquid to a gas by heat from the engine coolant system water which surrounds the vaporizing coil. The vaporized gas then passes (at a pressure slightly below atmospheric pressure) via the main valve (8) to the carburetor via the outlet (9). Control of the outlet pressure is obtained by operation of the

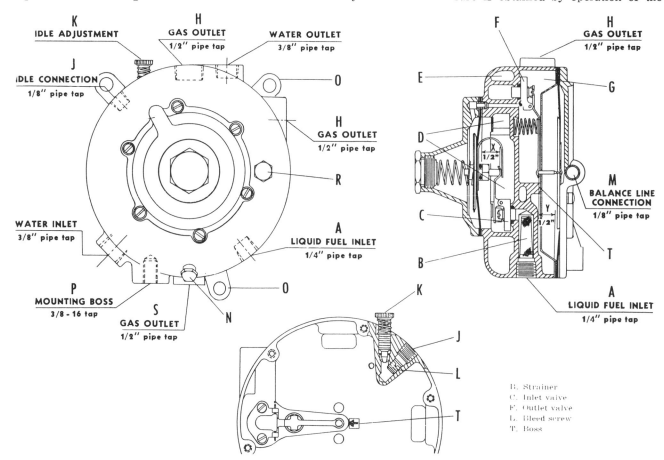

Fig. C62—Model W Ensign LPG regulator used on series D engines—assembled view.

large diaphragm (10), pin (11) and spring (12).

Fuel for the idling range of the engine is supplied by a separate outlet (15) via tube (16) from vapor reserve chamber (17) which is supplied by port (19). Outlet (15) is connected by tubing to a separate connection on the carburetor. Adjustment of the carburetor idle mixture is controlled by the idle fuel screw (20). A tube connects the atmospheric vent (21) to a pitot tube in the carburetor air horn so as to reduce the flow of fuel vapor and thus prevent over-richening of the mixture which would otherwise result when the air cleaner or air inlet system becomes restricted.

TROUBLE-SHOOTING

These procedures apply to all model W Ensign regulators and to model R regulators produced after regulator serial 148372.

100C. **SYMPTOM.** Engine will not idle with Idle Mixture Adjustment Screw in any position.

CAUSE AND CORRECTION. A leaking valve or gasket is the cause of the trouble. Look for leaking low pressure valve caused by deposits on valve or seat. To correct the trouble wash the valve and seat in gasoline or other petroleum solvent.

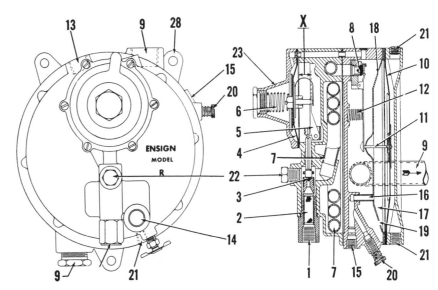

Fig. C63—Model R Ensign LPG regulator used on series LA engines—assembled view.

1. Fuel inlet	8. Main low pressure valve
2. Filter screen	9. Outlet to carburetor
3. High pressure valve	10. Outlet pressure diaphragm
4. Inlet pressure diaphragm	11. Push pin
5. Inlet diaphragm lever	12. Outlet diaphragm spring
6. Inlet diaphragm spring	13. Water inlet
7. Fuel vaporizer coil	14. Water outlet
	15. Idling fuel outlet

17. Vapor reserve chamber
18. Partition plate
19. Orifice
20. Idle fuel adjustment
21. Atmospheric vent
23. Regulator cover

If foregoing remedy does not correct the trouble check for leak at high pressure valve by connecting a low reading (0 to 20 psi) pressure gauge at point (22) on model R regulator or at point (R) on the model W regulator. If the pressure increases **after** a warm engine is stopped, it proves a leak in the high pressure valve. Normal pressure is 3½-5 psi on model W, 4-5 psi on model R except where this model is connected to the engine oil pressure circuit. On model R with oil pressure control the normal pressure should be 1-2 psi with engine stopped; 8.5-10 with engine running.

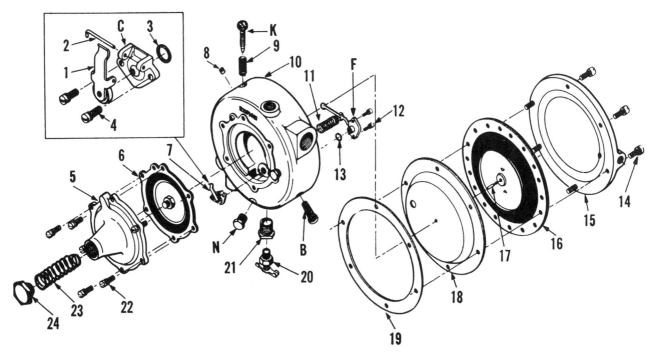

Fig. C62A—Model W Ensign LPG regulator used on series D engines—exploded view.

B. Fuel inlet strainer	2. Pivot pin	8. Bleed screw
C. Valve seat	3. "O" ring	9. Idle screw spring
F. Outlet valve assy.	5. Regulator cover	10. Regulator body
K. Idle adjusting screw	6. Inlet pressure diaphragm	11. Outlet diaphragm spring
1. Inlet diaphragm lever	7. Inlet valve assembly	13. "O" ring

15. Back cover plate	19. Partition plate gasket
16. Outlet pressure diaphragm	20. Drain cock
17. Push pin	21. Reducing bushing
18. Partition plate	23. Inlet diaphragm spring
	24. Spring retainer

Fig. C63A—Using Ensign gauge 8276 to set the fuel inlet valve lever to the dimension as indicated at (X) in Fig. C62.

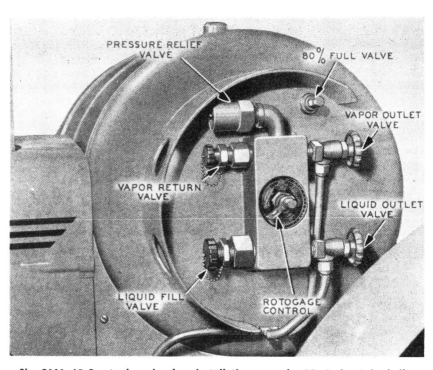

Fig. C64A—LP-Gas tank and valves installation on series LA. Series D is similar.

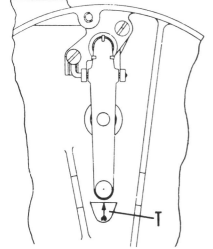

Fig. C63B—Location of post or boss with stamped arrow for the purpose of setting the fuel inlet valve lever.

100D. **SYMPTOM.** Cold regulator shows moisture and frost after standing.

CAUSE AND CORRECTION. Trouble is due either to leaking valves as per paragraph 100C or the valve lev-ers are not properly set. For information on setting of valve lever refer to paragraph 100E.

REGULATOR OVERHAUL

If an approved station is not available the model R (without oil pressure attachment) regulator and the model W can be overhauled as outlined in paragraph 100E.

100E. Remove the unit from the engine and completely disassemble, using Figs. C62A and C64 as references.

Thoroughly wash all parts and blow out all passages with compressed air. Inspect each part carefully and discard any that are worn.

Before reassembling the unit, note dimension (X—Figs. C62 and C63) which is measured from the face on the high pressure side of the casting to the inside of the groove in the valve lever when valve is held firmly shut as shown in Fig. C63A. If dimension (X) which can be measured with Ensign gauge No. 8276 or with a depth

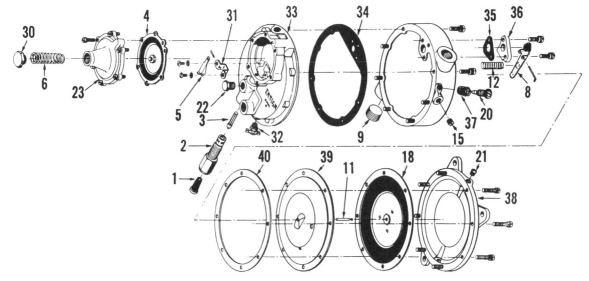

Fig. C64—Model R Ensign LPG regulator used on series LA engines—exploded view.

1. Fuel inlet strainer
2. **Fuel inlet valve**
3. Needle valve
4. Inlet pressure diaphragm
5. Inlet diaphragm lever
6. Inlet diaphragm spring
8. Outlet valve lever
9. Plug (¾)
11. Push pin
12. Outlet valve
15. Bleed screw
18. Outlet pressure diaphragm
20. Idle fuel adjustment
21. Pipe plug (⅛)
23. Regulator cover
30. Spring retainer
31. Pivot support
32. Drain cock
35. Outlet valve gasket
36. Outlet valve seat
37. Idle screw spring
38. Support plate
39. Partition plate
40. Partition plate gasket

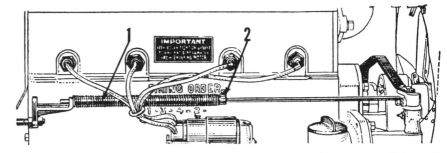

Fig. C66—Series C and L tractor engine governor hand control to governor linkage. To increase engine speed, loosen collar (2) and stretch the spring (1) until desired speed is obtained.

rule is more or less than ½ inch, bend the lever until this setting is obtained. A boss or post (T—Fig. C63B) is machined and marked with an arrow to assist in setting the lever. Be sure to center the lever on the arrow before tightening the screws holding the valve block. The top of the lever should be flush with the top of the boss or post (T).

GOVERNOR
ADJUSTMENT
Series C-L

101. First check governor linkage for free operation. Next, place governor hand control lever and carburetor throttle valve in the wide open position, and adjust length of carburetor throttle shaft to governor arm link rod. Adjust the length of the link rod so that it will be necessary to move the governor arm ¼ inch forward to connect the link rod.

With the linkage properly adjusted and hand control lever in wide open position, adjust governed no load speed of 1175 rpm on series C or 1200 rpm on series L by increasing or decreasing the tension of the governor control spring. To increase the engine speed, loosen the collar (2—Fig. C66) and stretch the spring until desired speed is obtained. To reduce the en-

gine speed, loosen collar (2) and shorten the spring.

If engine has a tendency to surge or hunt, shorten the carburetor throttle shaft to governor arm link rod ½ turn at a time until engine surge is eliminated.

To obtain engine rpm, multiply belt pulley rpm by 1.13 on model C; by 1.31 on model CC; and by 1.39 on series L.

Series D-LA-S

102. First check governor linkage for free operation. (Refer to Fig. C65.) Next, remove the link rod which connects the throttle shaft arm to the governor lever. Pull the hand throttle rod fully rearward (wide open position). Check the distance from the center of the governor link rod ball joint hole in the governor lever to the

front face of the cylinder block.

On series D, this distance should be 7/8 inch as shown in Fig. C67. On series LA, this distance should be 2 inches as shown in Fig. C68. On series S, the governor lever should be parallel to the front face of the cylinder block as shown in Fig. C70.

To obtain the desired dimension, remove the governor lever and bend the lever in the area as shown. Do not try to bend the arm while same is installed as there is a possibility of damaging the governor fork or operating shaft. Maintaining this dimension will prevent the governor lever from striking the generator brace in the wide open position or the generator belt in the idle position on the D series; on LA series from striking the water pump cap screw or grease cup in the wide

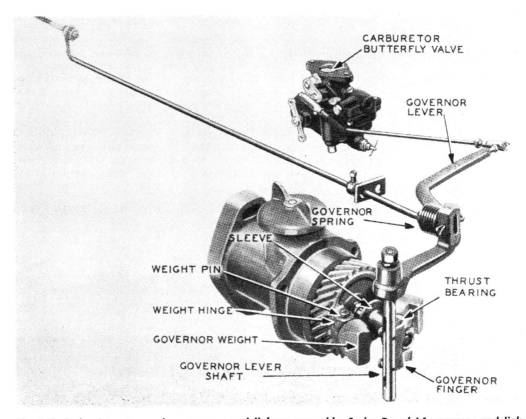

Fig. C65—Series S tractor engine governor and linkage assembly. Series D and LA governor and linkage installation is similar except for construction of governor lever.

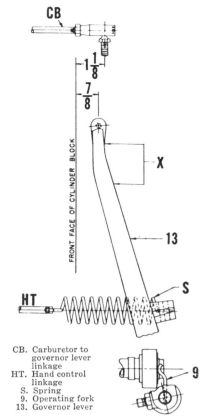

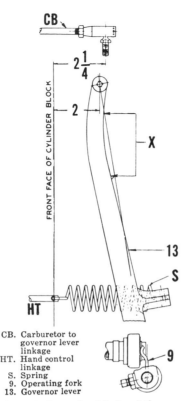

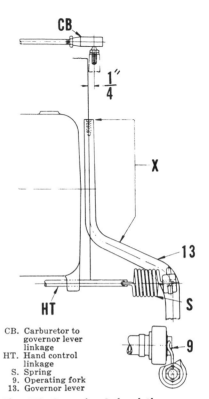

Fig. C67—On series D, bend the governor lever in the area indicated by "X" to obtain the 7/8 inch measurement as shown.

CB. Carburetor to governor lever linkage
HT. Hand control linkage
S. Spring
9. Operating fork
13. Governor lever

Fig. C68—On series LA, bend the governor lever in the area indicated by "X" to obtain the 2 inch measurement as shown.

CB. Carburetor to governor lever linkage
HT. Hand control linkage
S. Spring
9. Operating fork
13. Governor lever

Fig. C70—On series S, bend the governor lever in the area indicated by "X" until the lever is parallel to the front face of the cylinder block.

CB. Carburetor to governor lever linkage
HT. Hand control linkage
S. Spring
9. Operating fork
13. Governor lever

open position or generator belt in the idle position; or on series S from striking cylinder block in the wide open position or the water pump body in the idle position.

With the carburetor throttle valve, and governor lever in the wide open position, adjust the length of the link rod to the dimensions as shown in Fig. C67, C68 or C70, or so that the link rod is approximately 1/4 inch longer than is necessary to connect the throttle valve lever to the governor lever.

102A. Adjust governed no load speed, of 1390 rpm on series D; 1200 rpm on series LA; or 1760 rpm on series S by increasing or decreasing the tension of the governor control spring. To increase the engine speed, shorten the length of the rod which connects the governor lever to the hand control rod as shown in Fig. C69. To decrease engine speed, lengthen the rod.

To obtain engine rpm, multiply belt pulley rpm by 1.47 on series D; by 1.41 on series LA; and by 1.44 on series S.

If governor has a tendency to surge or hunt at wide open throttle, shorten the link rod, which connects the throttle valve to the governor lever, one half turn at a time until surge is eliminated.

Series R

103. No load engine governed speed of 1555 rpm (1080 rpm when checked at the belt pulley) is adjusted by increasing or decreasing the tension on the governor spring (2—Fig. C71).

To increase engine high idle speed, pull governor lever back to full speed position; then, loosen collar (3) and stretch spring until desired speed is obtained. Relock collar in this position. To reduce engine high idle speed loosen collar and shorten the length of spring. Surging of the governor at no load speed can be eliminated by turning the stabilizing screw (10—Fig. C83) in or out until surge is eliminated.

Stabilizing screw is located on back face of governor unit. Do not turn the screw in too far as it will affect idling speed.

103A. To adjust linkage when governor has been overhauled or control parts renewed, first retract the stabilizing screw about 1/2 inch. Push governor lever to the closed position and set spring collar (3) approximately 9 inches from forward end of spring anchor (1) as shown in Fig. C71. Pull governor lever back to wide open position and adjust throttle rod (5) so that there is 1/8 inch clearance between stop pin (9) on carburetor body and full open stop on throttle valve arm (6). Start engine and adjust idling stop

Fig. C69—On series D, LA and S, to increase engine speed, shorten the length of the rod which connects the governor lever to the hand control rod.

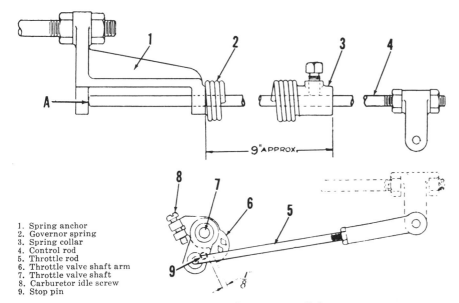

1. Spring anchor
2. Governor spring
3. Spring collar
4. Control rod
5. Throttle rod
6. Throttle valve shaft arm
7. Throttle valve shaft
8. Carburetor idle screw
9. Stop pin

Fig. C71—Series R engine governor linkage.

screw (8) on throttle valve arm to obtain an idling speed of 400 rpm. In some cases the ⅛ inch setting of throttle rod will not permit a satisfactory idle speed and it may be necessary to shorten the rod one turn at a time until proper idle speed is obtained. The ⅛ inch setting may then increase to as much as 3/16 inch.

With the governor lever in idling position, adjust governor anchor (1) position so that the anchor stop just touches the rear end of the control lever rod as shown at point A. At this time, the carburetor throttle valve should be in the idling position if the adjustment is correct.

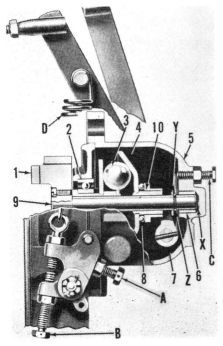

Fig. C72—Series V engine governor.

Series V

104. First check governor linkage for free operation. Next, place governor hand control lever and carburetor throttle valve in wide open (full speed) position, and adjust length of carburetor link rod. Adjust the length of the link rod so that it will be necessary to move the link rod 1/16 inch forward to hook it into the carburetor throttle shaft arm.

Start engine, and move governor hand control lever to wide open position. Engine no load governed speed of 1600 rpm (1088 belt pulley rpm) is adjusted by increasing or decreasing tension on the governor spring (D—Fig. C72). To increase governed speed, rotate screw (A) counter-clockwise and rotate screw (B) clockwise until desired speed is obtained. To decrease engine speed, rotate screw (A) clock-

A. Speed adjusting screw
B. Speed adjusting screw
C. Bumper spring adjusting screw
D. Governor spring
X. Thrust washer
Y. Washer & shims
Z. Snap ring
1. Governor & magneto drive gear
2. Bearing
3. Governor balls
4. Upper race
5. Governor housing
6. Bushing
7. Operating fork & shaft
8. Thrust bearing
9. Governor shaft
10. Fork base

wise, and rotate screw (B) counter-clockwise. Eliminate surge by rotating the bumper screw (C) in a clockwise direction slightly.

Series VA

The VA series engines prior to tractor serial 5365687 are equipped with a flyball type governor unit. Later engines have a flyweight type with the weight unit riveted to the camshaft gear.

105. First check governor linkage for free operation. Next, place governor hand control lever and carburetor throttle valve in wide open (full speed) position, and adjust length of carburetor link rod (B—Fig. C74) so as to slightly pre-load the governor lever. This is done by having the carburetor link rod 1/16 inch longer than is necessary to connect the carburetor throttle valve arm to the governor arm.

105A. No load speed on engines with one set of mounting holes in governor control rod guide is 1625 engine, 1106 belt rpm. If engine has M-S carburetor Case part VTA4714 and governor control rod guide with two sets of mounting holes, the no load speed is 1880 (1650 loaded) engine, 1280 belt pulley rpm. To adust no load speed, start engine and pull governor control lever to full speed position. On VA series engines prior A4700000 rotate the adjusting screw (2—Fig. C73), which is located in the governor control stop, in a clockwise direction to increase engine speed. Rotating the screw in a counter-clockwise direction will decrease the engine speed.

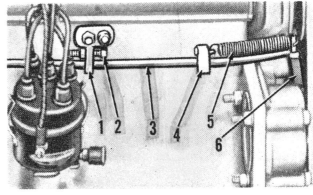

Fig. C73—Governor linkage on series VA engines prior to engine serial A4700000. To increase engine speed, rotate adjusting screw (2) in a clockwise direction.

1. Governor control stop
2. Adjusting screw
3. Control rod
4. Spring anchor
5. Governor spring
6. Governor lever

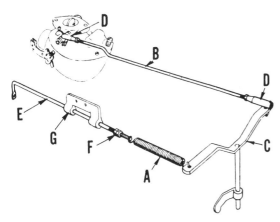

Fig. C74—Governor linkage on series VA engines after engine serial A4700000. To increase engine speed, rotate nuts (F) in a counterclockwise direction.

neto drive gear. Remove cap screws which attach magneto bracket to the engine and withdraw bracket and governor assembly rearward and out of engine.

106A. To remove the governor arm shaft (10—Figs. C76 & C77) from the timing gear cover, insert a screw driver through governor assembly opening in engine and remove the two round head screws attaching governor fork (9) to lever shaft; then, pull shaft out of bushing (11) in timing gear cover.

106B. To reinstall, reverse removal procedure placing governor gear in mesh with camshaft gear so that magneto drive coupling slots are in same position as when removed. Refer to Fig. C75. Retime magneto as outlined in paragraphs 131 or 132.

107. **DISASSEMBLE AND OVERHAUL.** On series D after serial 460-7032 the governor can be disassembled when out of engine by removing the weights (8—Fig. C76) and sleeve (18). Remove gear retaining nut (6) and gear (5) from shaft (24). Remove Woodruff key and withdraw the shaft from the magneto base. Thrust bearing (17) is held in place by the snap ring (16).

107A. Disassembly procedure on D series prior to serial 4607033 and on C series is substantially the same as outlined in paragraph 107 the main difference being in the details of the governor shaft rear seal (2—Fig. C77) and the collar (27).

107B. Shaft bushings (23) should be sized after installation to provide a running clearance of .0015-.0025. Lever shaft bushings (11) should be sized to .499-.500 to provide approxi-

Fig. C75—Magneto drive coupling position when timing the magneto to the engine on series D after serial 4607032 and S. On series D prior serial 4607033, the coupling slots are 30 degrees from the horizontal.

On VA series engines after A4700000, vary the tension on the governor spring by means of the nuts (F—Fig. C74) located on the hand control rod (E).

R & R AND OVERHAUL
Series C-D

106. **R&R GOVERNOR.** Crank engine until No. 1 (front) piston is on compression stroke and the flywheel mark "D" is visible through timing inspection hole which is located on right hand side of clutch compartment, then remove the magneto. On D after 460-7032 and C series, the slots in the magneto drive coupling should be 25 degrees past the horizontal as shown in Fig. C75. On D series tractors prior to 4607033, the drive coupling slots will be 30 degrees past the horizontal. Mark the location of the drive coupling slots on the magneto mounting flange to insure proper timing of the camshaft gear to the governor and mag-

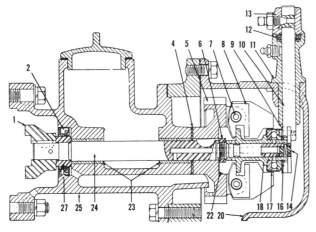

Fig. C76—Series D after tractor serial 4607032 governor unit and magneto drive assembly.

1. Magneto coupling	10. Governor lever shaft	20. Timing gear cover
2. Oil seal	11. Bushing	22. Gear retaining nut lock
4. Thrust washer	12. Felt washer	23. Bushings
5. Governor & magneto drive gear	13. Governor lever	24. Governor shaft
6. Gear retaining nut	14. Shaft plug	25. Magneto bracket
7. Weight pivot pin	16. Snap ring	27. Thrust collar
8. Weight	17. Thrust bearing	28. Collar pin
9. Operating fork	18. Governor sleeve	

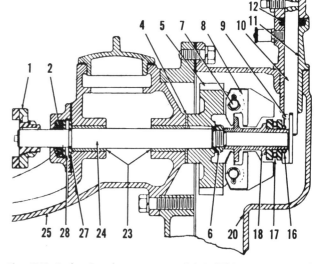

Fig. C77—Series D prior tractor serial 4607033 governor unit and magneto drive assembly.

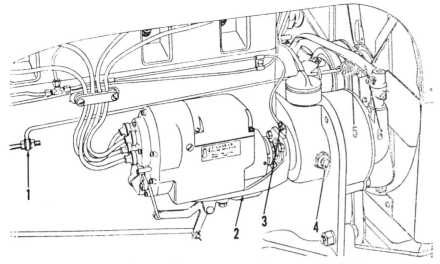

Fig. C78—Series L and LA governor and magneto installation.

1. Speed adjustment	3. Governor oil line	5. Governor spring
2. Magneto base	4. Carrier screw	6. Governor lever

mately .001-.0015 clearance on lever shaft. On series D tractors after serial 5600000 the governor sleeve rotates on two bronze bushings which should be sized to provide .0015-.002 clearance.

Series L-LA

108. **R&R GOVERNOR.** To remove the governor assembly, first remove the radiator, magneto, magneto drive coupling, magneto base (2—Fig. C78) and disconnect governor oil line (3). Remove governor carrier retaining cap screw (4), governor lever (6), and disconnect governor control rod. Remove governor housing cover from front of timing gear cover and withdraw the governor assembly. It may

be necessary to lightly tap rear of assembly with a soft hammer to withdraw governor assembly.

When reinstalling governor assembly mesh governor gear double punch mark with an identical single punch mark on the camshaft gear. Reinstall and retime magneto as outlined in paragraph 131.

109. **DISASSEMBLE AND OVER-HAUL.** Disassembly procedure is same as outlined for series C and D in paragraphs 107 and 107A. Refer to Figs. C78, C79, C80 and C81.

On series L prior 4208797, the governor shaft rotates in two taper roller bearings which should be adjusted to

zero pre-load with governor gear retaining nut. On series L after 4208796, and LA, the governor shaft rotates in two renewable bushings (23—Fig. C79). Service bushings require final sizing after installation to provide a diametral clearance of .0015-.0025, or to an inside diameter of .736-.737.

On series LA prior 4904771, and L, governor lever shaft rotates in either

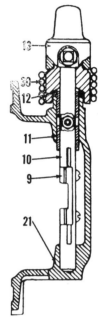

Fig. C80—Governor housing cover, and governor lever shaft and fork assembly as used on series LA prior 4904771. Refer to Fig. C79 for legend.

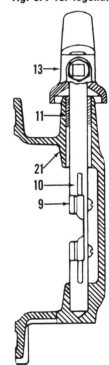

1. Magneto coupling
2. Oil seal
3. Magneto base
4. Thrust washer
5. Governor & magneto drive gear
6. **Gear retaining nut**
7. Weight pivot pin
8. Weight
9. Operating fork
10. Governor lever shaft
11. Bushing
12. Felt washer
13. Governor lever
14. Shaft plug
16. **Snap ring**
17. Thrust bearing
18. Governor sleeve
21. Housing cover
22. Gear retaining nut lock
23. Bushings
24. Governor shaft
26. **Governor carrier**
30. Torsion spring

Fig. C79—Series L after 4208796 and LA after 4904771 governor unit and magneto drive assembly. Series L prior 4208797 is similar except that shaft (24) rotates on taper roller bearings instead of bushings (23). Refer to Figs. C80 and C81 for variations in the governor lever shaft and cover assembly.

Fig. C81—Governor housing cover, and governor lever shaft and fork assembly as used on series L engines. Refer to Fig. C79 for legend.

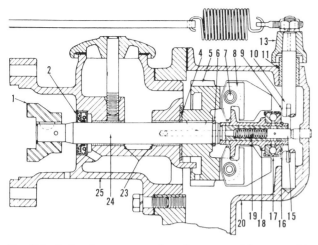

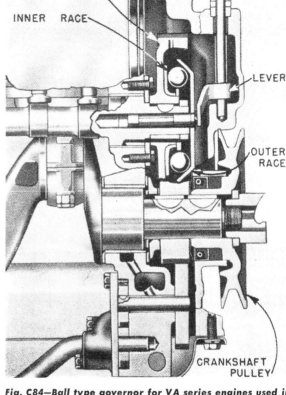

Fig. C82—Series S governor unit and magneto drive assembly.

1. Magneto coupling	9. Operating fork	17. Thrust bearing
4. Thrust washer	10. Governor lever	18. Governor sleeve
5. Governor & mag-	shaft	19. Thrust spring
neto drive gear	11. Bushing	20. Timing gear cover
6. Snap ring	13. Governor lever	23. Bushings
7. Weight pivot pin	15. Thrust plunger	24. Governor shaft
8. Weight	16. Snap ring	25. Magneto bracket

Fig. C84—Ball type governor for VA series engines used in tractors prior to tractor serial 5365687.

one or two renewable thin wall bronze bushings (11—Fig. C80 or C81) which require final sizing after installation to provide a .001-.002 diametral clearance. A similar clearance should be provided for the lever shaft bushing (11—Fig. C79) on LA after 4904771.

Series R

110. R&R GOVERNOR. Governor is mounted on rear face of timing gear cover at right side of engine. To remove unit, first disconnect throttle rod, and governor control rod. Remove carburetor, and governor attaching cap screws. Remove set screw (5—Fig. C83) which locks governor housing to crankcase and withdraw governor unit.

110A. DISASSEMBLE AND OVERHAUL. After removing governor assembly from the engine, it can be dis-

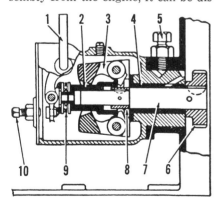

Fig. C83—Series R engine governor is similar to the one as shown except that the governor shaft (7) rotates on a ball bearing instead of a bushing (4) as illustrated.

1. Governor lever	6. Drive gear
2. Governor sleeve	8. Weights support
3. Governor weights	9. Thrust bearing
5. Housing lock screw	10. Stabilizer screw

assembled in the following manner: Remove the pin retaining governor drive gear (6) to shaft, and pull the gear from the shaft. Remove the governor gear Woodruff key. Remove cover and governor lever; then withdraw shaft with weights and thrust sleeve as an assembly. Sleeve (2) can be removed from shaft (7) after removing the weights. The thrust bearing (9) is a press fit on the sleeve. Although Fig. C83 shows the shaft (7) supported in bushing (4) a ball bearing is used in the production governor.

Series S

111. R&R GOVERNOR. To remove the governor, crank engine until No. 1 piston is on compression stroke, and the 3/16 inch diameter hole in the flywheel, Fig. C40, is visible through the clutch compartment timing inspection port. The engine must be cranked to this position for governor removal as governor weights will interfere with the cam gear if engine is rotated to a different position. Remove the magneto. The slots in the magneto drive coupling should be 25 degrees past the horizontal as shown in Fig. C75. Remove the magneto bracket (governor carrier) and governor unit.

111A. When reinstalling the governor, be sure governor shaft thrust spring (19—Fig. C82) and plunger (15) are inserted in governor shaft

bore. Install governor and magneto bracket assembly to engine so that the driving slot of the magneto coupling is 25 degrees (or two of the governor gear teeth) past the horizontal position as shown in Fig. C75. Install bracket attaching cap screws. Retime magneto, as per paragraph 132.

112. DISASSEMBLE AND OVERHAUL. The S series tractor governor, Fig. C82, can be disassembled by following the procedure as outlined for C and D series in paragraph 107. The governor drive gear (5) on this series is retained to the governor shaft by snap ring (6) instead of a nut as used on the other series. A spring loaded plunger (15) in the drive shaft acts on a thrust button in the timing gear cover to take up shaft end play.

The governor shaft rotates in two renewable bushings (23) which require final sizing after installation to an inside diameter of .7355-.7365. On later series S engines, sleeve (18) rotates on two renewable bushings which require final sizing after installation to provide a diametral clearance of .0015-.002. Governor lever shaft rotates in one renewable bushing (11) which is located in timing gear cover. This bushing which requires final sizing to an inside diameter of .499-.500, can be renewed without removing the timing gear cover.

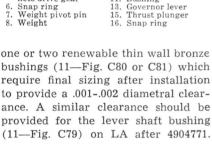

Series V

113. **R&R GOVERNOR.** The governor assembly can be removed from the front of the timing gear cover after removing the tractor grille and radiator.

When reinstalling the governor on tractors equipped with magneto ignition, it will be necessary to remove the magneto so that the governor gear can be correctly timed to the cam gear. Remove magneto, and rotate engine crankshaft until punch mark on rear face of cam gear can be seen through magneto mounting opening. Install governor so that a similar punch mark on rear face of governor gear registers with the punch mark on the cam gear. Reinstall and retime magneto as outlined in paragraph 132.

113A. **DISASSEMBLY AND OVERHAUL.** Disassembly of the governor unit is self-evident after an examination of the unit and with reference to Fig. C72. Drive shaft bushing (6) and operating lever shaft bushing (not shown) require final sizing after installation to provide a .002-.003 diametral clearance.

A thrust washer (X) is installed in the governor body (5) prior to the installation of the drive shaft bushing (6). This washer controls the thrust of the helical drive gear. Suggested drive shaft end play is .004-.020.

Travel of the upper race (4), thrust bearing (8), and fork base (10) on the drive shaft is limited by the thrust washer (Y). Travel can be reduced by the addition of shims in front of this washer (between washer & retaining snap ring (Z)). Full travel should be adjusted to .220-.230.

Series VA

114. **R&R GOVERNOR.** To remove the governor assembly on VA series engines prior to engine serial A470-0000 first remove tractor grille, radiator, crankshaft pulley, and the governor control cover from the front face of the timing gear cover.

On VA series engines after engine serial A4700000, to remove the governor, it will be necessary to remove the timing gear cover as outlined in paragraph 55A.

114A. **DISASSEMBLE AND OVERHAUL.** On engines equipped with the ball type governor unit (tractors prior to tractor serial 5365687) the front race (26—Fig. C85) and steel balls (8) can now be removed. Remove nut (27) which retains the ball driver, inner race and camshaft gear. Remove ball driver (28) and rear race (6). Should the governor balls drop into timing gear case during removal it will be necessary to remove the timing gear cover on engines prior A4700000. When

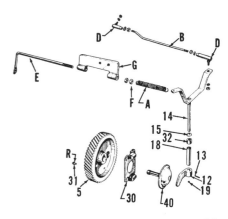

Fig. C86—Flyweight type governor used in VA series engines after tractor serial 5365686. Weight unit (30) is riveted to the cam gear.

5. Cam gear	19. Ball
12. Rocker arm	30. Weight & carrier
13. Taper pin	31. Rivet
14. Control lever	32. Needle bearing
15. Felt seal	40. Upper race
18. Thrust spacer	& shaft

reinstalling be sure pin (29) is in camshaft gear as this pin locks the ball driver (28) to the gear.

114B. On engines (after tractor serial 5365686) equipped with the flyweight type governor unit, the weight unit (30—Fig. C86) is riveted to the cam gear. To remove the weight unit, first remove timing gear cover and cam gear, and derivet the weight unit from the cam gear. The weights are not serviced separately, and must be renewed as an assembly with the carrier. The 3/16 inch diameter by ½ inch long round head rivets which attach the weight unit to the cam gear should be installed with the rivet head contacting the weight unit plate. When reinstalling the cam gear, remove the oil pan and using a heavy bar, buckup the camshaft at one of the lobes near front end of shaft.

114C. On all VA series, the procedure for removal of the governor operating shaft and arm for needle bearing or bushing renewal, is self-evident after removing governor inspection cover on models so equipped, or the timing gear cover on all other models.

COOLING SYSTEM

Series R engines have a thermosiphon type cooling system; a coolant pump is used on all other series. Radiators are not equipped with a pressure cap.

THERMOSTAT

Series C-D-L-LA-S-V-VA

116. Early production tractors of this group are not equipped with a thermostat. On series C, D, L and LA

5. Cam gear	15. Felt seal	24. Retaining ring
6. Inner race	16. Shaft bushing	25. Balls
7. Plain washer	17. Rocker shaft	26. Outer race &
9. Pin	18. Thrust spacer	shaft
10. Ball	19. Ball	27. Cam gear retaining nut
12. Rocker arm	23. Governor cover	28. Ball driver
13. Taper pin	(prior engine	29. Straight pin
14. Control lever	A4700000)	

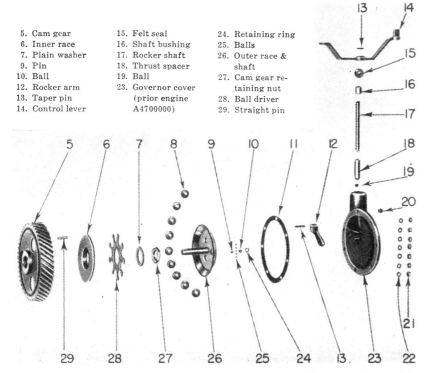

Fig. C85—Ball type governor used on VA series engines prior to engine serial A4700000. On VA series engines after serial A4700000, the use of a separate governor cover (23) is discontinued. Governor cover on these models is formed by the timing gear cover.

which are equipped with a thermostat, the thermostat is located in the radiator upper tank inlet elbow. On series S, and VA, the thermostat is located in the upper hose. On series V, the thermostat is located in the cylinder head water outlet elbow.

Series	Start To Open	Fully Open
C-D-L-LA	175°F	200°F
S Early	160°F	190°F
S Late-V-VA Dist.	175°F	200°F
V-VA Gas	150°F	175°F

FAN BELT

All Series

117. To renew the fan belt on series C, and L, and all early production series D, LA, V, and VA, it will be necessary to remove the radiator assembly. On late production series D and LA, it will be necessary to loosen lower fan shroud, and remove the fan blades retaining cap screws and drop fan to bottom of radiator shell. Also, remove starting crank. On LA series only, unscrew forward part of pulley hub.

117A. On series R, the fan belt can be renewed after removing cap screws retaining fan bracket to cylinder block and moving fan assembly out of place far enough to allow removal and re-installation of the belt.

117B. On series S, and late production V, and VA, fan belt can be renewed without removing the radiator.

FAN

All Series

118. To remove fan blades on all except series R and S, it will be necessary to remove the radiator. On series R and S, the fan blades can be renewed without removing the radiator. Disassembly and overhaul of the R series fan unit is self-evident after an examination of the unit.

RADIATOR & HOSES

All Series

120. Method of removing the radiator is self-evident after examining the tractor. Series C, D, L, LA and R radiators are made in three sections comprising the core unit, and upper and lower tank units. The upper and lower tank units are bolted to the core unit. Series S, V, and VA radiators have the upper and lower tank units welded to the core unit.

120A. All cooling system hoses can be renewed without removing any other part of the system. On series C, D, L, LA and R, the radiator to cylinder block elbow can be removed after removing the elbow retaining cap screws.

WATER PUMP

Series C-D-L-LA-S-V-VA

121. **REMOVE & REINSTALL.** To remove the water pump on series C, D, L, LA, V, and VA, it will be necessary to remove the radiator.

121A. On series S, the water pump can be removed without removing the radiator by first removing the fan blades. Remove fan belt from pump pulley and pump by-pass tube. Remove three cap screws attaching the water pump upper body to the lower body, and remove the pump unit.

Series D prior 4511450-C-L-LA

122. **OVERHAUL.** Pump seal (3) can be renewed by removing the impeller (1). To disassemble pump, Fig. C87 or C88 remove fan blades, pulley and pulley hub. Bump shaft (5) and impeller (1) as an assembly from pump body (20). On all but LA series, remove felt seal retainer (14—Fig. C87), felt seal (13) and washer (9). On LA series, remove shaft spacer (11—Fig. C88) and outer grease seal (12). The ball bearing (8—Fig. C87 or C88)and thrust spring (7) can now be removed. On LA series, the inner grease seal (6—Fig. C88) can be renewed at this time.

The impeller can be removed from the shaft after removing pin (2—Fig. C87 or C88). Impeller is a loose fit on shaft so as to permit its hub to

1. Impeller
2. Retaining pin
3. Seal thrust washer
4. Pump shaft bushing
5. Pump shaft
6. Seal
7. Thrust spring
8. Ball bearing
9. Seal retainer
10. Snap ring
11. Shaft spacer
12. Oil seal
13. Felt seal
14. Seal retainer
15. Fan hub
16. Pump shaft nut lock
17. Pump shaft nut
20. Pump body
22. Seal assembly

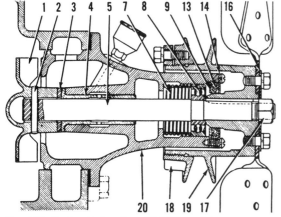

Fig. C87—Series C, D prior 4511450 and L water pump.

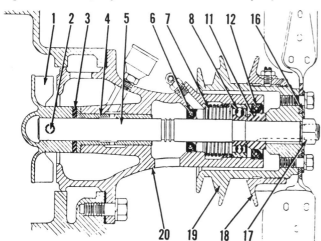

Fig. C88—Series LA water pump.

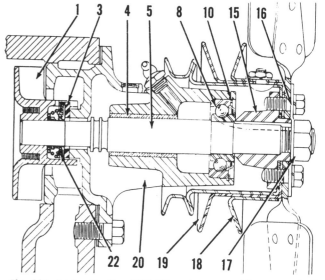

Fig. C89—Series D after 4511449 water pump. The pump shaft should not project beyond the hub of the impeller.

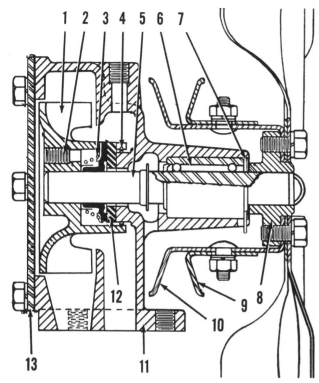

1. Impeller
3. Seal assembly
4. Seal retaining ring
5. Pump shaft
6. Pump shaft bearing
7. Snap ring
8. Fan hub
11. Pump body
12. Thrust washer
13. Pump body cover

Fig. C90—Series S water pump. The pump shaft should not project beyond the hub of the impeller.

align with thrust washer (3) which seals the pump. Surfaces of the impeller, washer and bushing must be smooth and true to effect a proper seal. If impeller is removed from the shaft, repack the impeller cavity with grease.

Impeller shaft bushing (4) requires final sizing after installation to an inside diameter of .874-.875 on series D and LA; and on series C and L so as to provide a suggested .002-.003 diametral clearance.

122A. On LA series, press the inner grease seal (6—Fig. C88) into pump body, with lip facing inward, to a depth of 2 11/16 inches when measured from front face of pump body to seal. When reassembling pump unit, insert shaft with thrust washer and impeller in pump body. Pack the space around the shaft, spring (7) and shaft bearing (8) with water pump grease. Install thrust spring (7—Fig. C87 or C88). Install bearing with shielded side facing the fan blades and make certain that bearing is pressed against shoulder on shaft. The bearing must be a free fit in pump body bore as the thrust spring (7) reacts against the bearing so as to exert pressure on the thrust washer. Thrust washer (3) serves the purpose of a seal between the impeller and flanged surface of shaft bushing (4). On all but LA series, reinstall washer (9—Fig. C87), felt seal (13) and seal re-

tainer (14). On LA series reinstall outer grease seal (12—Fig. C88) with lip facing inward and spacer (11). On all pumps install hub and fan blades.

Series D after 4511449

123. **OVERHAUL.** Pump seal (3—Fig. C89) can be renewed after removing the impeller. The rear face of the impeller (1) is drilled and tapped for the use of two 5/16 inch N.C. bolts and a suitable puller for removing the impeller. The pump body surface which is contacted by thrust washer (3) of seal assembly (22) should be square with shaft bore and smooth. Reface seal seat with a piloted cutter or renew pump body if seal seat is scored or worn.

To renew pump shaft bushing (4) which requires final sizing after installation to an inside diameter of .8735-.875 and/or ball bearing (8), it will be necessary to remove the fan blades, pulley hub and pump impeller. Remove ball bearing retaining snap ring (10), and withdraw shaft and bearing assembly from front of pump. Ball bearing should be installed with the shielded side facing the fan blades. Repack the ball bearing with grease.

Series S

124. **OVERHAUL.** To renew pump seal, first remove pump body cover (13—Fig. C90) and impeller. The rear face of the impeller is drilled and tapped for the use of two 5/16 inch

N.C. bolts and a suitable puller for removing the impeller.

The pump body surface which is contacted by thrust washer (12) of seal assembly (3) should be square with shaft bore and smooth. Reface seal seat with a piloted cutter or renew pump body if seal seat is scored or worn.

The shaft and prelubricated bearing assembly is renewed as a unit. To remove shaft and bearing assembly from pump body when impeller is off, first remove bearing snap ring (7) and press shaft toward front and out of pump body.

Series V

125. **OVERHAUL.** To renew pump seal, first remove pump cover (13—Fig. C91), and impeller (1). On early production engines, the pump impeller is retained to the shaft by means of a set screw. Carbon thrust washer and seal assembly (2) can be removed from impeller after removing snap ring (3). Seal seat (B) in pump body which is contacted by carbon thrust washer must be square with shaft bore and smooth. Reface seal seat with a piloted cutter or renew pump body if seal seat, which is not supplied for service, is scored or worn.

Shaft (11) and/or ball bearings (5) can be renewed after removing fan pulley and impeller, and pressing the shaft forward and out of pump body. Reinstall bearings with the shielded side facing outward or away from each other. Repack bearings with grease.

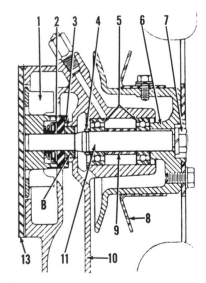

Fig. C91—Series V water pump. Pump seal seat (B) is not supplied for service. The pump shaft should not project beyond the impeller.

1. Impeller
2. Seal assembly
3. Snap ring
4. Snap ring
5. Pump shaft bearings
7. Fan pulley hub retaining nut
9. Bearing spacer
10. Pump body
11. Pump shaft
13. Pump body cover

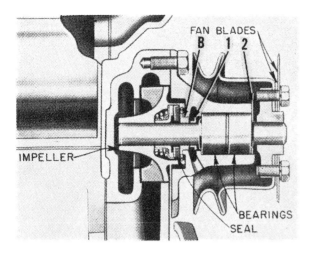

FAN BLADES
B 1 2

IMPELLER

BEARINGS
SEAL

Fig. C92—Series VA water pump. The pump shaft should not project beyond the impeller. Pump seal seat (B) is not supplied for service.

1. Slinger
2. Snap ring

Series VA

Water pumps used on VA series early production engines prior engine serial A4700000 use taper roller bearings for the pump shaft. Later production VA series engines prior to engine serial A4700000 use two ball bearings which can be renewed separately from the pump shaft. Water pumps, as shown in Fig. C92, used on engines after engine serial A4700000 use sealed prelubricated bearings which are serviced only as an assembly with the pump shaft.

126. **OVERHAUL.** To renew pump seal, first remove impeller. Carbon thrust washer and seal assembly can be removed from impeller after removing seal assembly retaining snap ring. Seal seat (B) in pump body which is contacted by carbon thrust washer must be square with shaft bore and smooth. Reface seal seat with a piloted cutter or renew pump body if seal seat, which is not supplied for service, is scored or worn.

Shaft and/or bearings can be renewed after removing pulley hub and impeller, and removing the bearings set screws on early production engines, or snap ring on pumps where the shaft and bearings are serviced as an assembly. Adjust taper roller bearings on pumps so equipped to provide shaft with zero end play. When renewing the bearings in pumps which are equipped with taper roller bearings, these should be replaced with ball bearings.

ELECTRICAL SYSTEM

Various makes and models of magnetos, battery ignition units, generators and regulators, and starting motors have been used on these engines. Refer to Standard Units manual for detailed service and overhaul information by using the make and model of the unit being serviced as a guide.

IGNITION TIMING

Series D after 4607032, C, S, and VA are equipped with magnetos which have a 25 degree lag angle impulse coupling. Series D prior 4607033, L, and LA are equipped with a magneto which has a 30 degree lag angle impulse coupling. Series R is equipped with a magneto which has a 35 degree lag angle impulse coupling. Series V is equipped with a magneto which has a 15 degree lag angle impulse coupling.

Series D prior 4607033-C-L-LA-R

131. **MAGNETO.** To time the base type magneto, which has a manual advance control, first adjust breaker gap to values as listed below:

Magneto	Breaker Gap
Bosch	.015
Case 4CMA	.015
Fairbanks Morse	.020

Remove No. 1 cylinder spark plug and rotate engine crankshaft until air is felt escaping from the spark plug

hole; then continue to rotate the crankshaft until flywheel mark "D" or "DC1" is centered in inspection port.

Set magneto advance and retard lever on the magneto breaker box cover in the full advance position. Advance lever will be down when magneto is in full advance.

With magneto in an upright position, insert a spark plug wire in number 1 terminal (upper right) on the distributor block. Holding the other end of the spark plug wire ⅛ inch from the magneto frame, rotate the impulse coupling in direction of normal rotation until a spark jumps from the spark plug wire.

Without disturbing the settings of either the magneto or flywheel, install the magneto on the mounting bracket. Should the lugs on the driving member of the governor shaft not match (within 5 degrees) the slots in the coupling floating member proceed as follows: First check magneto for correct model and impulse coupling. On base mounted magnetos, loosen the governor shaft driving member lock nut, and rotate the driving member until same registers with the coupling slots.

After installing magneto, temporarily bolt same to magneto mounting base. Insert a piece of .015 shim stock between impulse coupling and driv-

ing member. Tighten magneto mounting bolts and withdraw the shim stock. The use of the .015 shim stock will provide end clearance for the magneto coupling floating member.

Recheck timing by rotating engine crankshaft in a counter-clockwise direction for a fraction of a revolution; then, slowly rotate crankshaft in a clockwise direction until impulse coupling trips. At this time, the flywheel timing mark "D" or "DC1" should be in the center of the timing inspection port.

Firing order is 1-3-4-2.

Series D after 4607032-S-V-VA

132. **MAGNETO.** To time the flange type magneto as used on these engines, first adjust breaker gap to the values as listed below:

Magneto	Breaker Gap
Bosch	.015
Case 4JMA	.015
Case 41	.010
Fairbanks Morse	.020

Remove No. 1 cylinder spark plug and rotate engine crankshaft until air is felt escaping from the spark plug hole; then continue to rotate the crankshaft until flywheel mark "D" on D series, "3/16 hole" on S series, "DC 1&4/EX. CL." on V series, or 3/16 inch after "DC" mark on VA series is centered in inspection port.

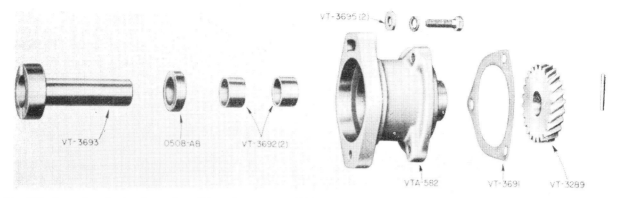

Fig. C93—Magneto adapter is used on VA series tractors which have the hydraulic lift pump mounted on the belt pulley.

With the flywheel timing mark positioned in center of inspection port, the magneto coupling groove in governor shaft or accessory shaft should be approximately 25 degrees past the horizontal position as shown in Fig. C75 on series D and S engines, or just past the horizontal position on series V and VA engines. If coupling groove is not in this position, the governor shaft or accessory shaft gear will have to be retimed to the camshaft gear.

With magneto in an upright position, insert a spark plug wire in number 1 terminal (upper right) on distributor block. Holding the other end of the spark plug wire ⅛ inch from the magneto frame, rotate the impulse coupling in direction of normal rotation until a spark jumps from the spark plug wire.

Without disturbing the settings of either the magneto or flywheel, install magneto to the engine and install the magneto retaining cap screws.

Recheck the timing by rotating engine crankshaft in a counter-clockwise direction for a fraction of a revolution; then slowly rotate crankshaft in a clockwise direction until impulse coupling trips. At this time, the flywheel retard timing mark should be centered in the timing inspection port. Final adjustment can be obtained by rotating the magneto in its mounting flange. Rotating the magneto toward the engine will retard the spark, and moving it away from the engine advances the spark.

Series V-VA

133. **BATTERY IGNITION.** An Auto-Lite model IGW4156B distributor is used on the V series engines, and models IGZ4009A or IGC4803 (year 1942), IAD410IA (years 1943-1947), IAD60032B (year 1948), IAD-60032F (years 1948 & up) are used on the VA series engines. Set breaker contacts to a .020 gap.

To time the distributor to the engine, crank the engine until No. 1 piston is on its compression stroke and the flywheel mark "DC1&4/EX. CL." for VA series, or "DC" for V series is centered in the timing inspection port which is located on the left side of the clutch housing on V series, or on the left side of the torque tube on VA series.

Install the distributor unit to the cylinder block opening with the rotor (finger) under the number one terminal in the cap. Rotate distributor housing in a counter-clockwise direction until breaker contacts have just opened; then, tighten distributor housing clamp. Firing order is 1-3-4-2.

MAGNETO ADAPTER

Series VA

134. The magneto adapter, Fig. C93, as used on VA series tractors, which have the hydraulic lift pump mounted on the belt pulley unit, is mounted behind the timing gear cover on the right side of engine. To remove the adapter, remove magneto, and cap screws retaining the adapter to the engine.

To disassemble, remove the tapered pin retaining the drive gear (VT-3289) to the shaft, and bump shaft rearward out of gear and adapter. The adapter shaft rotates in two bushings (VT-3692) which require final sizing after installation to provide a .002-.004 diametral clearance. Install drive shaft

MODEL L

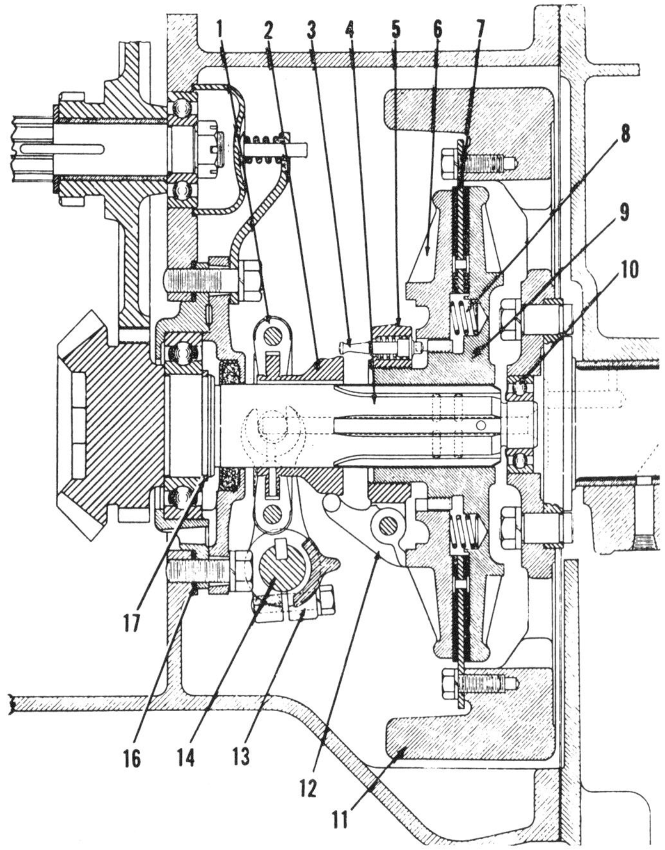

CLUTCH ADJUSTMENT

Series C-D prior 4818100-L-LA

140. Remove inspection plates from left and right hand sides of transmission housing clutch compartment. Place transmission gear shift lever in neutral, and disengage the clutch. Rotate belt pulley, until clutch lockpin (3—Fig. C94), is accessible through inspection hole. Place transmission in gear to prevent clutch shaft from turning. Pull out lock pin (3) and using the pronged tool furnished with the tractor, rotate adjusting yoke (5) **up** or in a clockwise direction (viewed from rear of tractor) until the operating lever requires a distinct pressure to engage the clutch. Release the lockpin, making certain that it engages the floating plate (6).

After adjusting engine clutch, adjust belt pulley brake.

1. Cone collar
2. Sliding cone
3. Lock pin
4. Clutch shaft
5. Adjusting yoke
6. Floating plate
7. Driving disc
8. Release spring
9. Back plate
10. Pilot bearing
11. Flywheel
12. Clutch finger lever
13. Release fork
14. Release fork shaft
16. Shims
17. Snap ring

Fig. C94—Series C clutch installation. Series D Prior 4818100, L and LA are similar except for the use of a double row type ball bearing and treated leather oil seal (instead of the felt type).

oil seal (0508-AB) with lip facing drive gear.

134A. When reinstalling adapter, crank engine until No. 1 piston is coming up on compression stroke; then, continue to crank engine until the flywheel mark "DC1&4/EX. CL." is 3/16 inch past the pointer in timing inspection hole. Install and mesh drive gear with camshaft gear so that the slots in the drive shaft are in a horizontal or slightly past the horizontal position. The magneto drive gear is now correctly timed to the camshaft.

1. Sleeve collar
2. Sliding sleeve
3. Clutch shaft
4. Cover plate
5. Release spring
6. Lock pin
7. Pressure plate
8. Driven disc
9. Pilot bearing
10. Flywheel
11. Adjusting ring
12. Lever roller
13. Clutch lever
14. Release fork
15. Sleeve bushings
16. Shims
17. Snap ring

ENGINE CLUTCH

For adjustment and overhaul data of the power take-off unit clutch, refer to paragraphs 280 through 282. Engine clutch is an over-center type on all series C, L, LA, R and on D and S prior to 1953. Later series D and S except Orchard versions and all series V and VA tractors have spring loaded, single dry plate type.

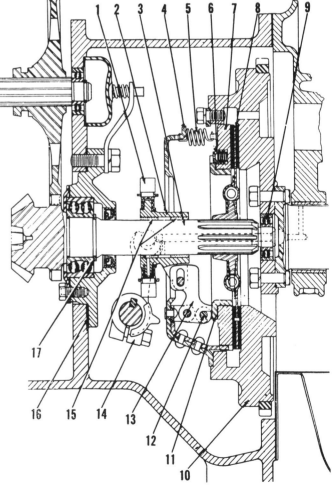

Fig. C95—Series D after 4818099 over-center type clutch installation. Series S over-center type engine clutch is similar although clutches in later S tractors have a cast iron cover.

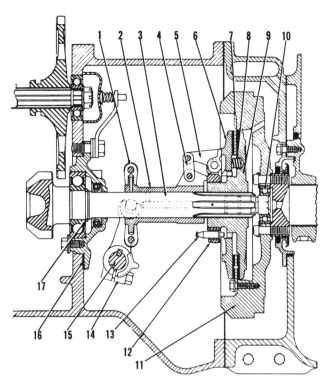

1. Sleeve collar
2. Sliding sleeve
3. Clutch shaft
4. Lever link
5. Clutch finger
6. Floating plate
7. Lined plate (driving disc)
8. Release spring
9. Back plate
10. Pilot bearing
11. Flywheel
12. Adjusting yoke
13. Lock pin
14. Release fork
15. Release fork shaft
16. Shims
17. Snap ring

Fig. C96—Series R clutch installation.

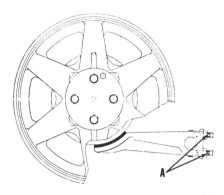

Fig. C98—To adjust pulley brake on series D and LA, loosen both set screws (A), then rotate upper set in to decrease the clearance between the brake and pulley hub.

140A. BELT PULLEY BRAKE ADJUSTMENT. After adjusting the engine clutch, check the belt pulley brake to be sure that it contacts the pulley on series C, L and R, or the pulley hub on series D and LA when the clutch lever is pulled all the way back or when the clutch lever is in the disengaged position. To adjust the brake on series C, L and R, loosen both lock nut (B—Fig. C97) and set screw lock nut. Rotating set screw (A) will move the pulley brake in or out. Tighten nut (B) and set screw lock nut.

On series D and LA, loosen lock nuts on both set screws (A—Fig. C98). Loosening the lower set screw and tightening the upper set screw will ap-

ply the brake earlier during rearward movement of the clutch hand lever. Lock adjustment by tightening the set screw lock nuts.

Series D after 4818099-S Over-Center Type
(For Spring Loaded Type, Refer to Page 120.)

141. Remove inspection plate from left side of transmission housing compartment. Place transmission gear shift lever in neutral, and disengage the clutch. Crank engine until the spring loaded lockpin (6—Fig. C95) in clutch collar adjusting ring (11) is visible through inspection plate opening. A raised boss on the pressure plate (7) or a white paint mark on the cover plate (4) will indicate exact location of the lockpin. With a piece of 1/4 inch or 5/16 inch diameter rod which is bent to form a 110 degree angle, depress the lockpin. With the lockpin depressed, rotate adjusting collar down or in a counter-clockwise direction (viewed from rear of tractor) to tighten the clutch (decrease clearance between lined plate and pressure plate). Rotate the adjusting ring until a noticeable drag is felt as the belt pulley is rotated by hand, indicating that the clutch facings are dragging slightly. Then, rotate the adjusting ring upward (clockwise) until the drag disappears. Clutch should be adjusted to a point where no slippage will occur under load, and yet not so

tight that difficulty is encountered when engaging the operating lever. Release the lockpin, making certain that it engages the adjusting ring.

After adjusting engine clutch, readjust belt pulley brake as outlined in paragraph 140A.

Series R

142. Follow same procedure as outlined in paragraph 140 for the series D prior 4818100, C, L, and LA. Refer to Fig. C96.

Series V-VA

143. With clutch fully engaged, the clutch pedal should have 1½-2 inches of free travel. This amount of free travel produces the desired 1/8-3/16 inch clearance between release bearing face and face of the release levers. This adjustment is made by turning the yoke clevis at the pedal end of the operating rod (9—Fig. C102).

R & R CLUTCH UNIT
Series C-D-L-LA-R-S
(For Spring Loaded Type, Refer to Page 120.)

144. SPLITTING TRACTOR. To remove the clutch assembly, it is necessary to split the tractor (detach engine from transmission). To perform a tractor split proceed as follows: Block the rear wheels, as shown in Fig. C99, disengage clutch, and place transmission in neutral. Drain oil from clutch housing. Remove hood and disconnect engine controls, fuel line, oil pressure gauge line, and heat indicator. Disconnect drag link at steering drop arm, and if tractor is equipped with radius rod disconnect same from transmission housing. On S series only, remove air cleaner oil cup. On all series, remove hand hole covers from sides of transmission housing clutch compartment. Support front end of transmission housing and remove the ¾ or 1 inch bolts from each side of transmission case flange and insert in their place

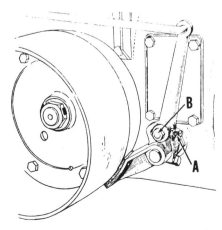

Fig. C97—To adjust pulley brake on series C, L and R, loosen nut (B) and turn set screw (A) in or out.

two special splitting pins. On the C, D, R, and S series, these pins (Case Part 139AA) are ¾ x 13 13/16 inches.

On the L and LA series, pins (Case Part 61AA) are 1 x 14½ inches. A rolling floor jack placed under the engine or A-frame attached to the engine can be used in lieu of the splitting pins. Remove the bolts and cap screws attaching transmission housing to engine, and roll front half of tractor forward as shown.

144A. **R&R CLUTCH.** To remove clutch after separating the transmission housing from the engine proceed as follows: Remove belt pulley brake and clutch lever. Remove cap screws attaching clutch assembly to flywheel and withdraw clutch and sliding cone or sleeve with collar.

Pilot bearing in flywheel can be renewed at this time. Clutch release fork and shaft can be removed at this time.

144B. When reinstalling, place over-center type clutch and cone or sleeve with collar, on the clutch shaft making sure release fork engages arms on collar. On series D prior to 4818100, C, L and LA be certain that the key which prevents sliding cone from rotating on clutch shaft is installed on shaft. Slide the transmission housing up to the engine, guiding the clutch shaft into the pilot bearing. Bolt the transmission housing to engine. Working through the hand holes, bolt clutch assembly to flywheel. Install belt pulley brake and clutch and adjust over-center type clutch as outlined in paragraphs 140, 141, or 142.

Series V-VA

145. **SPLITTING TRACTOR.** To remove the clutch unit, it will be necessary to detach the engine from the torque tube on series VA, and clutch housing from torque tube on series V. To detach engine, proceed as follows: Remove the hood, fuel line to carburetor and disconnect engine controls. On series V, remove gas tank. On series VA which have the hydraulic lift pump mounted on the engine, disconnect pump to torque tube reservoir lines. On all other series, disconnect drag link at steering arm and on tractors equipped with radius rod, disconnect same from torque tube. Support the engine with a hoist and install a jack or support under front end of torque tube. Remove the torque tube to engine or clutch housing bolts and roll front half of tractor forward as shown in Fig. C100.

145A. **R&R CLUTCH.** Remove cap screws retaining clutch assembly to

flywheel and remove clutch assembly. To remove the clutch release bearing and carrier assembly (1—Fig. C102)

Fig. C99—Method of splitting tractor (detaching engine from transmission) on series S tractors. A similar procedure is used on series C, D, L, LA and R tractors.

without removing clutch shaft from torque tube proceed as follows: Disconnect the clutch release rod at the

Fig. C100—Method of splitting tractor (detaching engine from torque tube) on series VA tractors. Series V is similar, except that the split is made between the engine clutch housing and torque tube.

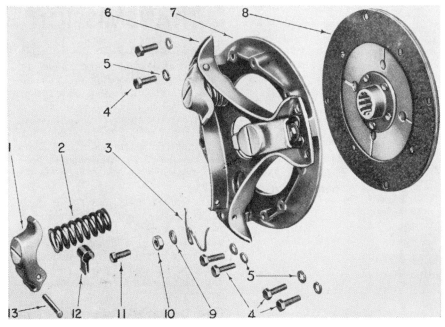

Fig. C101—Series V and VA Auburn clutch as used prior to 1954 production. In latest, pressure plate assembly Case part VTA4706, return clips Case part V14870 take the place of items (3) and (12) shown herewith.

1. Release lever	7. Pressure plate
2. Pressure spring	8. Lined (driven) disc
3. Return spring	9. Adjusting screw washer
6. Bracket	10. Lock nut
	11. Adjusting screw
	12. Cushion spring
	13. Pivot pin

pedal and withdraw release rod forward and out of torque tube and clutch release fork. Remove cap screws attaching the clutch shaft bearing carrier (14) to forward wall of torque tube. Rotate carrier 180 degrees and unscrew release fork pivot pin (23) from carrier. Withdraw release bearing carrier and fork as an assembly.

Oilite bronze pilot bushing (20) in flywheel can be renewed after unbolting clutch assembly from flywheel.

Install driven disc with baffle toward engine flywheel.

OVERHAUL CLUTCH UNIT
Series C-D-L-LA-R-S
(For Spring Loaded Type, Refer to Page 120.)

145B. Disassembly of the clutch unit is self-evident after an examination

and with reference to Figs. C94, C95, and C96. Check toggle pins and toggles, and cone and collar for wear.

The cone or sleeve collar (1) can be removed without detaching the transmission housing from engine. Working through the clutch compartment inspection cover openings, remove the two bolts which attach the collar halves together. Shims located between the parting surfaces of the collar (1) control its running clearance on the cone or sleeve (2). Vary the number of shims to provide sufficient running clearance without permitting the collar to rattle.

Series V-VA

146. For detailed data on clutch overhaul refer to Auburn Clutches in Standard Units manual. Also refer to Fig. C101 and to paragraphs 148 & 148A.

CLUTCH SHAFT & SEAL
Series C-D-L-LA-R-S

147. The clutch shaft (4—Fig. C94 or 3—Figs. C95 & C96) as used in these series, is integral with the transmission main drive bevel pinion. For clutch shaft removal and overhaul procedure, refer to paragraph 165.

147A. The clutch shaft felt type oil seal, Fig. C94, as used in early production series C and L, can be renewed after removing the clutch unit, and unbolting the felt seal retainer from the forward end of the clutch shaft bearing carrier. Later production C and L series, and all other series, use a treated leather type oil seal, Figs. C95 & C96, which is located in the clutch shaft bearing carrier. The oil seal can be renewed after removing the clutch shaft and bearing carrier as outlined in paragraphs 165, 165A and 165B. Install new seal with the lip facing the transmission shafts.

Series V-VA

148. Clutch shaft oil seal (2—Fig. C102) which is located in bearing carrier (14) is accessible for renewal after splitting the tractor as in paragraph 145. After performing a tractor split, remove clutch actuating rod, and cap screws attaching bearing carrier (14) to forward wall in torque tube. Rotate carrier 180 degrees and unscrew release fork pivot pin (23) from carrier. Withdraw bearing carrier and remove seal. Install new seal with the lip facing the transmission.

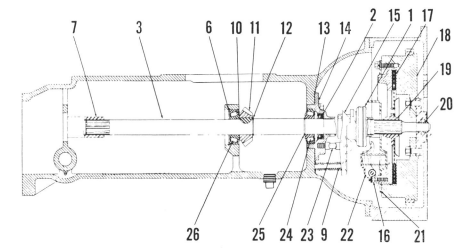

Fig. C102—Series V tractor torque tube and clutch shaft assembly. Shims (13) control clutch shaft bearings adjustment. Series VA torque tube and clutch shaft are similar, except that the clutch housing is integral with the torque tube.

1. Release bearing	9. Clutch rod	15. Clutch release fork	21. Clutch cover
2. Oil seal	10. Shims	16. Adjusting screw	22. Release lever
3. Clutch shaft	11. Belt pulley drive gear	17. Clutch plate facings	23. Release fork pivot pin
6. Snap ring	12. Snap ring	18. Flywheel	24. Bearing
7. Clutch to transmission input shaft coupling	13. Shims	19. Lined plate	25. Snap ring
	14. Bearing carrier	20. Pilot bushing	26. Bearing

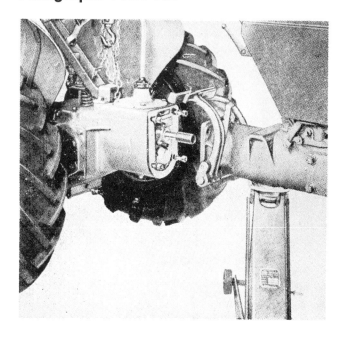

Fig. C103—Detaching transmission housing from torque tube on series VA tractors. Series V is similar.

TRANSMISSION

Series C-D-L-LA-R-S

The transmission gearset, bevel pinion which is located on the transmission end of the engine clutch shaft, bevel ring gear which is located on the transmission first reduction shaft, differential unit, differential spur gear, and final drive sprockets and chains are carried in the transmission housing as shown for the L series in Fig. C104.

On models DCS and DS of the D series tractors, the final drive sprockets and chains apply to those in the transmission housing and not to the chains and sprockets which are contained in separate housings (final drive housings bolted to the transmission housing).

160. **BASIC PROCEDURES.** Transmission shafts are considered to be the bevel pinion which is located on the transmission end of the clutch shaft, the first reduction (belt pulley) shaft, second reduction (sliding gear) shaft, and reverse gear shaft. For differential shaft, differential unit, and differential ring (spur) gear overhaul data, refer to paragraphs 200 through 202.

Shifter Rails and Forks. Shifter rails and forks can be removed after removing transmission top cover. On some models, it will also be necessary to remove the left rear wheel and tire unit to provide clearance for complete removal of shifter rails.

Second Reduction Shaft. The second reduction (sliding gear) shaft can be removed after removing the transmission top cover, shifter rails and forks and power take-off shaft. Also, remove the belt pulley and brake drum from models so equipped.

Bevel Pinion (Clutch Shaft). The bevel pinion which is integral with the clutch shaft can be removed after performing a tractor split (detach transmission housing from engine). Also, remove the power take-off shaft on all series except series R.

First Reduction Shaft. On all series, the first reduction (belt pulley) shaft can be removed after removing the transmission top cover, shifter rails and forks, power take-off shaft, second reduction shaft, and performing a tractor split (detach transmission housing from engine) and removing the bevel pinion (clutch shaft).

Reverse Idler and Shaft. The reverse idler gear can be removed after removing the transmission top cover, shifter rails and forks, power take-off

148A. To remove clutch shaft, perform work as outlined in paragraph 148 for renewing the oil seal; then, remove belt pulley unit from models so equipped. On VA series with hydraulic pump mounted on underside of belt pulley housing, it will be necessary to remove the hydraulic pump to cylinder hoses.

Withdraw clutch shaft and belt pulley drive bevel gear as an assembly. Bearing cones (24 & 26—Fig. C102) will remain on shaft. Shims (10) located between bearing cone (26) and bevel gear (11) are used to position the gear for correct mesh with the belt pulley driven gear. Reinstall clutch shaft with same number (thickness) of belt pulley drive gear shims as were removed. To check bevel gears for correct mesh, it will be necessary to remove the steering gear housing, or steering shaft support from models so equipped. Belt pulley driven gear to drive gear backlash is .004-.008 (equal to a movement of .003-.007 when measured at rivet heads in belt pulley hub). Refer also to paragraph 270.

CLUTCH HOUSING

Series V

149. **REMOVE & REINSTALL.** To remove clutch housing proceed as follows: First remove engine clutch as outlined in paragraphs 145 and 145A. Remove nuts retaining flywheel to crankshaft. Support engine, and frame with front axle assembly separately. Remove bolts attaching frame to clutch housing, and cap screws attaching clutch housing to rear face of cylinder block. Clutch housing can now be removed from cylinder block.

TORQUE TUBE

Series V-VA

153. To remove torque tube, Fig. C102, from tractor, first support engine and torque tube separately. Remove tractor hood, and fuel tank. Disconnect drag link from steering gear drop (pitman) arm on models so equipped. On model VAC after 5570000, disconnect steering shaft universal joint from worm shaft and upper steering shaft and remove lower steering shaft. On adjustable and non-adjustable axle models, disconnect radius rod at the rear pivot. On models equipped with an engine mounted hydraulic pump, disconnect the lines at the pump, and at the reservoir. Remove bolts attaching torque tube to engine on VA series, or torque tube to clutch housing on V series. Refer to Figs. C100 and C102.

153A. Support transmission housing and torque tube separately. Disconnect brake actuating rods from brake pedals. Remove four nuts attaching torque tube to the transmission housing. Remove torque tube by pulling same forward until clutch shaft is free of transmission main drive shaft coupling. Refer to Fig. C103.

153B. If the torque tube is to be renewed, remove the steering gear, or steering shaft support from models so equipped, belt pulley unit, clutch shaft, brake pedals and clutch pedal pivot shafts, hydraulic lift yoke, and other parts.

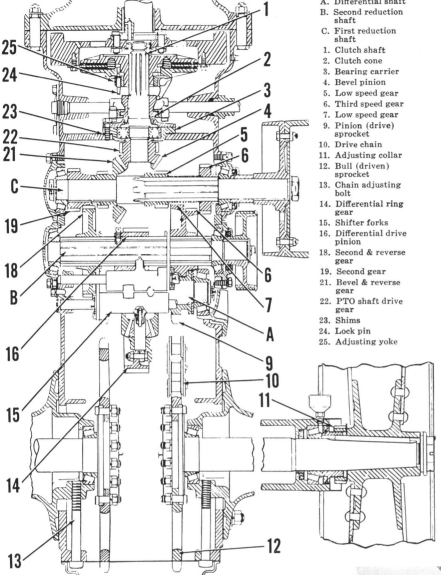

A. Differential shaft
B. Second reduction shaft
C. First reduction shaft
1. Clutch shaft
2. Clutch cone
3. Bearing carrier
4. Bevel pinion
5. Low speed gear
6. Third speed gear
7. Low speed gear
8. Pinion (drive) sprocket
9. Pinion (drive) sprocket
10. Drive chain
11. Adjusting collar
12. Bull (driven) sprocket
13. Chain adjusting bolt
14. Differential ring gear
15. Shifter forks
16. Differential drive pinion
18. Second & reverse gear
19. Second gear
21. Bevel & reverse gear
22. PTO shaft drive gear
23. Shims
24. Lock pin
25. Adjusting yoke

covers from sides of transmission housing clutch compartment. Support front end of transmission housing and remove the ¾ or 1 inch shoulder bolts from each side of transmission case flange and insert in their place two special splitting pins. On the C, D, R, and S series, these pins (Case Part 139AA) are ¾ x 13 13/16 inches. On the L and LA series, pins (Case Part 61AA) are 1 x 14½ inches. A rolling floor jack placed under the engine, or an A-frame attached to the engine can be used in lieu of the splitting pins. Remove the bolts and cap screws attaching transmission housing to engine, and roll front half of tractor forward as shown in Fig. C99.

161A. To facilitate reconnecting transmission to engine on series D after 4818099, and S tractors equipped with an over-center type clutch remove clutch assembly from flywheel. Install clutch including cone or sleeve with collar on clutch shaft and make sure release fork engages collar arms. On all series slide the transmission housing up to the engine, guiding the clutch shaft into the pilot bearing. Bolt the transmission housing to engine. Working through clutch compartment hand holes, bolt clutch assembly to flywheel. Adjust clutch as outlined in paragraph 140, 141 or 142 for over-center types, or page 120 for spring loaded types.

162. **R & R TRANSMISSION TOP COVER.** Refer to Fig. C105. Remove engine hood, fuel tank, and carburetor choke rod. Disconnect drag link at

Fig. C104—Series L tractor engine clutch, transmission gearset, differential, bull and pinion sprockets, drive chains and wheel axle shafts assembly. General arrangement of the aforementioned components except for brakes is similar on series C, D, LA, R and S tractors.

shaft, and second reduction shaft.

161. **TRACTOR SPLIT.** To perform a tractor split (detach transmission housing from engine) proceed as follows: Block the rear wheels, as shown in Fig. C99, disengage clutch, and place transmission in neutral. Drain lubricant from clutch housing. Remove hood and disconnect engine controls, fuel line, oil pressure gauge line, and heat indicator. Disconnect drag link at steering gear pitman (drop) arm, and if tractor is equipped with radius rod, disconnect same from transmission housing.

On series S only remove air cleaner oil cup. On all series remove hand hole

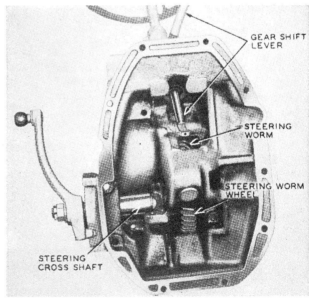

Fig. C105—Series D transmission top cover which contains the steering gear unit and transmission shifter lever. Basic construction of the series C, L, LA, R and S tractors transmission top cover is similar.

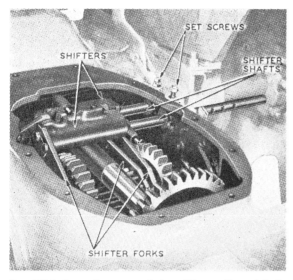

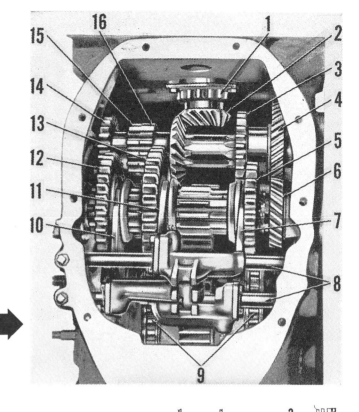

Fig. C106—Removing transmission shifter rails and forks on series D tractors. Series C, L, LA, R and S are similar.

Fig. C107A—Transmission gearset on series S tractors. Series C, D, L, LA and R tractors are similar.

1. PTO drive gear	7. 3rd & 4th shifter fork	12. 2nd & reverse gear
2. Clutch (bevel pinion) shaft	8. Shifter rails	13. 1st gear
3. 3rd gear	9. Drive chains	14. 2nd gear
4. 4th gear	10. 2nd & reverse shifter fork	15. Reverse gear
5. 3rd gear	11. 1st shifter fork	16. 1st gear
6. 4th gear		

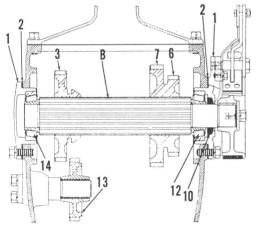

B. Second reduction shaft
1. Bearing carrier
2. Shims
3. 2nd & reverse gears
4. 4th gear
6. 3rd gear
7. 1st gear
10. Oil seal (felt)
11. Oil seal (treated leather)
12. Bearing cup & cone
13. Reverse idler
14. Bearing cup & cone

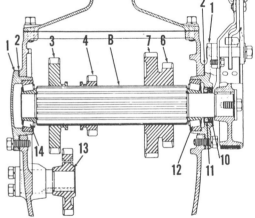

Fig. C107—Series C transmission 2nd reduction (sliding gear) shaft. Shims (2) control bearings adjustment.

Fig. C108 — Series D prior 4805353 transmission 2nd reduction (sliding gear) shaft. Transmission hand brake is used only on models prior 4511449.

steering pitman (drop) arm, oil line at the oil pressure gauge, and the tail light wire at the light switch. On series D, remove battery, an on all other series, disconnect battery ground cable at the transmission cover. Remove center and rear portions of the throttle control rod. On series S, remove four cap screws attaching the instrument panel to transmission housing cover; then, lift panel up and place same on top of battery shield. On all other series, remove two cap screws attaching instrument panel to fuel tank and place the panel with its attached wires on top of engine. Place transmission gear shift lever in neutral. Remove top cover to housing at-

taching cap screws, and remove the cover.

OVERHAUL. Data on overhauling the various components which make up the transmission are outlined in the following paragraphs.

163. SHIFTER RAILS AND FORKS. Refer to Fig. C106. Removal of shifter rails (shafts), and forks requires removal of transmission top cover as per paragraph 162. From outer sides of transmission housing, remove shifter rails cover plates, or machine screws and washers from outer ends of rails. Loosen shifter rail set screws located on gasket surface of top cover opening. Bump rails toward left side

of tractor and remove forks through top cover opening. On some models, complete removal of shifter rails necessitates removal of left rear wheel and tire unit.

164. SECOND REDUCTION SHAFT. Refer to Fig. C107, C108, C109, C110, C111, C112, or C113 for model concerned. It is not necessary to perform a tractor split (detach transmission from engine) for removal of 2nd reduction shaft. To remove shaft and gears, first remove transmission top cover paragraph 162, shifter rails and forks paragraph 163, and pto shaft as outlined in the POWER TAKE-OFF section. Remove belt pulley from right end of 1st reduction shaft and brake

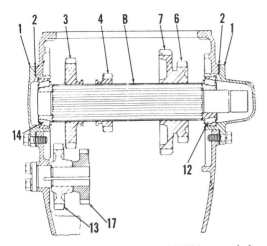

Fig. C109—Series D 4805353-5600000 transmission 2nd reduction (sliding gear) shaft. Shims (2) control bearings adjustment.

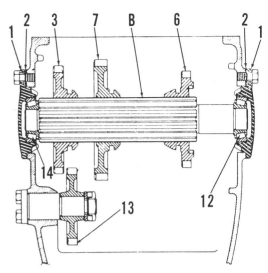

Fig. C112—Series R transmission 2nd reduction (sliding gear) shaft. Shims (2) control bearings adjustment.

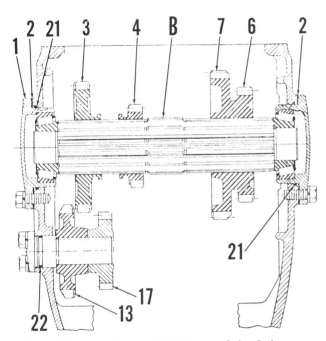

Fig. C110—Series D after 5600000 transmission 2nd reduction (sliding gear) shaft. Shims (2) control bearings adjustment.

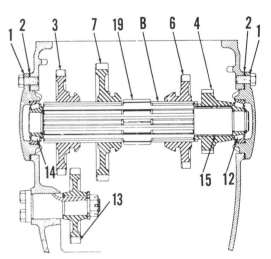

Fig. C113—Series S transmission 2nd reduction (sliding gear) shaft. Shims (2) control bearings adjustment.

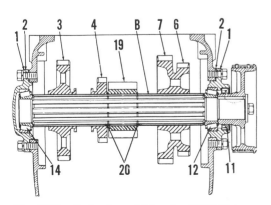

Fig. C111—Series L, and LA prior 5208165 transmission 2nd reduction (sliding gear) shaft. Series LA after 5208164 is similar except that differential drive pinion (19) is integral with the second reduction shaft. Shims (2) control bearings adjustment.

B. Second reduction shaft
1. Bearing carrier
2. Shims
3. 2nd & reverse gear
4. 4th gear
6. 3rd gear
7. 1st gear
11. Seal (treated leather)
12. Bearing cup & cone
13. Reverse idler
14. Bearing cup & cone
15. Bushing
17. Oiler gear
19. Differential drive pinion
20. Snap rings
21. "O" ring
22. "O" ring

drum from right end of 2nd reduction shaft on models so equipped.

On series L, and LA prior 5208165, the differential drive spur pinion (19) is positioned on the 2nd reduction shaft with either snap rings or a set screw. On other series, including LA after 5208164, the spur pinion is integral with the second reduction shaft. On series where the spur pinion is not integral with the shaft, remove snap ring from left side of pinion or loosen spur pinion set screw and move the pinion about 2 inches to the left of its normal position.

On series R and S, remove 2nd reduction shaft bearing carriers from both sides. On all other series, remove the right side bearing carrier and only loosen the left side bearing carrier. Slide the 2nd reduction shaft (B) toward right side of housing. Raise left end of shaft and withdraw same

Fig. C114—Removing the second reduction shaft by raising the left end and withdrawing same out through top cover opening on series LA tractors.

Fig. C115—Engine clutch shaft installation when viewed from transmission housing clutch compartment. Model shown is equipped with a continuous type pto.

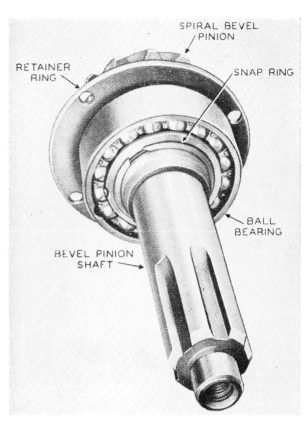

Fig. C116—Engine clutch shaft with integral bevel pinion as used in series C, D, L, LA, R and S tractors equipped with non continuous pto.

through top cover opening as shown in Fig. C114.

Bearings on the 2nd reduction shaft are adjusted by adding or removing shims (2) which are located under the bearing carriers (1). Adjust bearings to provide zero end play to shaft, and yet permit same to rotate without perceptible drag.

165. R & R CLUTCH (BEVEL PINION) SHAFT. Refer to Fig. C94, C95, C96, C104, C115, C116, C117, C118, or C119 for model concerned. To remove the engine clutch shaft, first perform a tractor split as outlined in paragraph 161. On all except series R, remove power take-off shaft (continuous and non-continuous type) as outlined in POWER TAKE-OFF section. The clutch shaft ball type pilot bearing which is located in the flywheel can be renewed at this time after unbolting the engine clutch from the flywheel.

Working through clutch compartment opening, remove engine clutch release fork and shaft, Fig. C115. Remove four cap screws attaching clutch shaft bearing carrier to forward wall in transmission housing and withdraw clutch shaft and bearing carrier assembly.

The bevel pinion mesh position, which is adjusted by means of a special gauge, is controlled with shims (16—Fig. C94, C95, C96 or C117) located between clutch shaft bearing carrier and transmission housing forward wall. For method of adjusting the mesh position and backlash of the main drive bevel pinion (clutch shaft) and ring gear (located on first reduction shaft), refer to paragraphs 186, 186A and 186B.

On series C, L, and R, the clutch shaft with integral bevel pinion can be purchased separately from the bevel ring gear. On series D, LA, and S, the clutch shaft with integral bevel pinion and bevel ring gear (located on first

reduction shaft) are available only as a matched set.

165A. *Overhaul—Models Without Continuous PTO.* With the clutch shaft and bearing carrier assembly removed, Fig. C116, the clutch shaft, bearing, and/or oil seal of treated leather can be renewed after disassembling the bearing carrier.

On models equipped with a treated leather oil seal install same with the lip facing the transmission shafts. The clutch shaft ball bearing is retained to the clutch shaft by a snap ring which is available in three different thicknesses of .096, .0975, and .099. When reassembling the bearing to the clutch shaft, press the bearing on the shaft until same contacts the clutch shaft shoulder; then, install the thickest bearing retaining snap ring so as to eliminate any endwise movement of the bearing inner race on the clutch shaft. Refer to paragraphs 186, 186A and 186B for method of adjusting the bevel pinion and ring gear mesh and backlash.

165B. *Overhaul—Models With Continuous PTO.* Refer to Figs. C117, C118 and C119. After removing the clutch shaft, pto drive sleeve and gear and

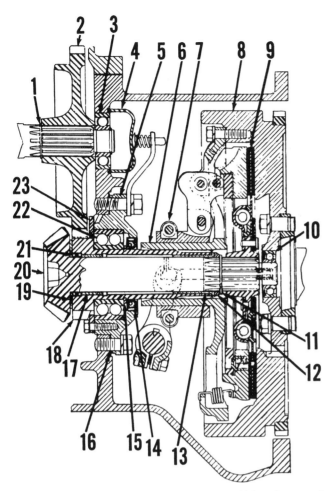

1. Snap ring
2. PTO shaft gear
5. Bearing carrier
6. Engine clutch cover
7. Cone collar
8. Flywheel
9. Driven disc
10. Pilot bearing
11. Snap ring
12. Thrust washer
13. Bushing
14. Oil seal
15. Snap ring
16. Shims
17. Bushing
18. PTO drive sleeve and gear
19. Thrust washer
20. Clutch shaft
21. "O" rings
23. Retainer ring

Fig. C117—Engine clutch, clutch shaft, and pto drive sleeve and gear installation on series D tractors equipped with an over-center type clutch and a continuous type power take-off. Series S tractors equipped with continuous pto and over-center clutch are similar except for differences in the engine clutch. Refer to page 120 for spring loaded clutch used on latest D and S series.

bearing carrier assembly as outlined in paragraph 165, the individual components which make up this assembly can be renewed.

The pto drive sleeve and gear and/or "O" ring oil seals can be renewed after removing clutch shaft snap ring, clutch shaft throw-out collar (thrust washer) and withdrawing clutch shaft from pto drive sleeve and gear. Bushings located in drive sleeve are not supplied for service. Worn bushings require renewal of drive sleeve and bushings as an assembly.

To renew pto drive spur gear sleeve ball bearing, remove cap screws attaching retainer ring to bearing carrier. Press pto drive sleeve and gear assembly and bearing out of the bearing carrier. The treated leather oil seal can be renewed at this time, and should be installed with the lip facing the transmission shafts. The pto drive sleeve gear bearing is retained to the sleeve by a snap ring which is available in three different thicknesses (on D series .0960, .0975 and .0990; on S series .0800, .0815 and .0830). When reassembling the bearing to the sleeve, press the bearing on the sleeve until same contacts the spur gear, then install the thickest bearing retaining snap ring so as to eliminate any end-wise movement of the bearing inner race on the sleeve.

The clutch shaft is retained in the pto drive sleeve and gear by the clutch shaft snap ring which is available in three different thicknesses of .0775, .0805, and .0835 for both the D and S series.

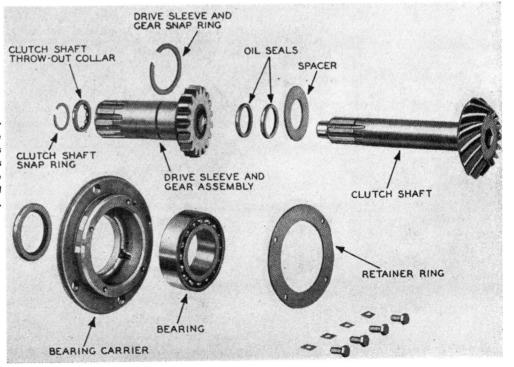

Fig. C118—Exploded view of engine clutch shaft and pto drive sleeve and gear assembly as used on series D tractors equipped with continuous type pto. Series S tractors equipped with continuous pto are similar.

CLUTCH SHAFT THROW-OUT COLLAR

DRIVE SLEEVE AND GEAR SNAP RING

OIL SEALS

SPACER

CLUTCH SHAFT SNAP RING

DRIVE SLEEVE AND GEAR ASSEMBLY

CLUTCH SHAFT

RETAINER RING

BEARING

BEARING CARRIER

When reassembling the clutch shaft to the drive sleeve and gear, install a snap ring which will provide the clutch shaft with .003 end play. If the end play is less than .003, it will be necessary to reduce the thickness of the clutch shaft steel thrust collar. If the end play is greater than .003, the use of a new clutch shaft thrust collar and/or bronze spacer will help decrease the end play.

Refer to paragraphs 186, 186A and

186B for method of adjusting the bevel pinion and ring gear mesh and backlash.

166. FIRST REDUCTION (BELT PULLEY) SHAFT. Refer to Figs. C120 through C128. Removal procedure for this shaft (C) varies with the individual models and same will be outlined in paragraphs 166A through 166E. On all series except D after 4805353, the first reduction shaft is supported on taper roller type bearings. On series D after 4805353, the first reduction shaft is supported by ball type bearings.

On series C, L and R, the bevel ring gear mounted on the first reduction shaft can be purchased separately. On series D, LA and S, the similarly located bevel ring gear is not available separately but only as a matched set with the bevel pinion.

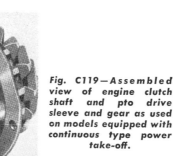

Fig. C119—Assembled view of engine clutch shaft and pto drive sleeve and gear as used on models equipped with continuous type power take-off.

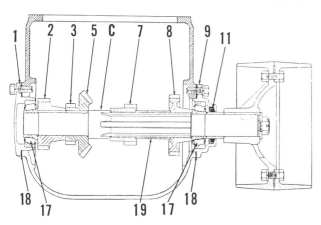

Fig. C120—Series C transmission first reduction (belt pulley) shaft. Shims (18) control bearings adjustment and ring gear (5) to bevel pinion (clutch shaft) backlash. Refer to Fig. C123 for legend.

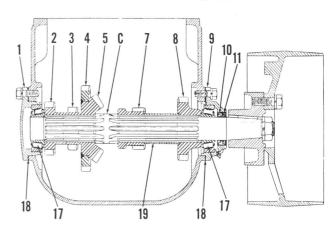

Fig. C122—Series D prior 4805354 transmission first reduction (belt pulley) shaft. Shims (18) control bearings adjustment, and ring gear (5) to bevel pinion (clutch shaft) backlash. Refer to Fig. C123 for legend.

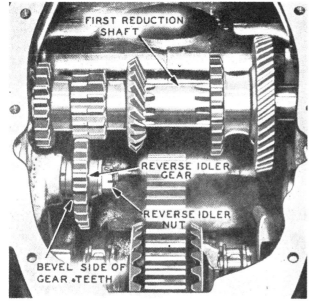

Fig. C121—Series S transmission first reduction (belt pulley) shaft installation. Series D and R are similar.

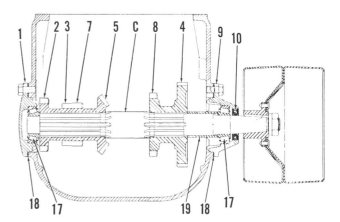

Fig. C123—Series R and S transmission first reduction (belt pulley) shaft. Shims (18) control bearings adjustment, and ring gear (5) to bevel pinion (clutch shaft) backlash.

C. First reduction shaft	5. Bevel ring gear	11. Oil seal (felt)
1. Bearing carrier	7. 1st gear	17. Bearing cup & cone
2. 2nd gear	8. 3rd gear	18. Shims
3. Reverse gear	9. Bearing carrier	19. Split sleeve spacer
4. 4th gear	10. Oil seal (treated leather)	

Fig. C124—Removing the first reduction shaft by raising the left end and withdrawing same out through top cover opening on series D prior 4805354, R and S.

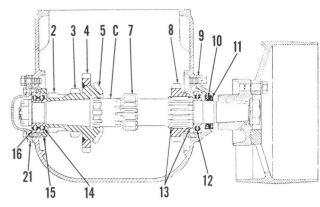

Fig. C125—Series D after 4805353 first reduction (belt pulley) shaft. Shims (16) control ring gear (5) to bevel pinion (clutch shaft) backlash adjustment.

C. First reduction shaft	7. First gear
2. 2nd gear	8. 3rd gear
3. Reverse gear	9. Bearing carrier
4. Fourth speed gear	10. Oil seal (treated leather)
5. Bevel ring gear	11. Oil seal (felt)

12. Bearing
13. Snap ring
14. Bearing
15. Bearing carrier
16. Shims
21. Carrier cap

For mesh and backlash adjustment of the bevel pinion (clutch) shaft, and bevel ring gear which is keyed or splined to the first reduction shaft, refer to paragraphs 186, 186A and 186B.

166A. *Series C.* The first reduction shaft (C—Fig. C120) can be removed after removing the clutch (bevel pinion) shaft as per paragraph 165, and 2nd reduction shaft as per paragraph 164.

After removing the 2nd reduction shaft, remove first reduction shaft bearing carriers (1 & 9) from sides of transmission housing, and keep the shims with their respective carriers. From left side of transmission housing, remove two cap screws retaining reverse idler shaft to housing and withdraw idler shaft. Gear will remain in bottom of transmission housing.

Buck-up the first reduction shaft and bump third gear (8) to the right just far enough to remove the split sleeve spacer (19). After removing spacer, bump third gear (8) to the left until same contacts the first gear (7). Move shaft toward right side of transmission housing and lift left end of shaft up and withdraw same out through top cover opening. Gears can now be removed from the shaft.

Adjust the taper roller bearings with shims (18) to provide zero end play to the shaft and yet permit same to rotate without perceptible drag. These shims, also control the backlash of the main drive bevel gear. Refer to paragraphs 186, 186A and 186B for procedure in adjusting the mesh and

backlash of the bevel pinion (clutch shaft) and ring gear.

166B. *Series D Prior 4805354-R-S.* Refer to Figs. C121, C122, C123 and C124. The first reduction shaft (C) can be removed after first removing the engine clutch (bevel pinion) shaft as per paragraph 165 and 2nd reduction shaft as per paragraph 164.

Remove first reduction shaft bearing carriers (1 & 9) from sides of transmission housing and keep the shims with their respective carriers. On series R and S only, working through top cover opening, remove cotter pin and castellated nut retaining reverse idler gear to shaft. On series D, R and S remove two cap screws retaining reverse idler shaft to left side of transmission housing and withdraw idler shaft. Gear will remain in bottom of transmission housing.

On series D prior 4805354, buck up the first reduction shaft and bump the third gear (8) to the right just far enough to remove the split sleeve spacer (19). After removing the spacer, bump third gear (8) to the left until same contacts the first gear (7).

Move first reduction shaft toward right side of housing, and lift left end of shaft up and withdraw same out through top cover opening as shown in Fig. C124. Gears can now be removed from the shaft.

Adjust the taper roller bearings with shims (18) to provide zero end play to the shaft and yet permit same to ro-

tate without perceptible drag. These shims also control the backlash of the main drive bevel gears. Refer to paragraphs 186, 186A and 186B for procedure in adjusting the mesh and backlash of the bevel pinion (clutch shaft) and ring gear.

166C. *Series D After 4805353.* Refer to Fig. C125. The first reduction shaft (C) can be removed after first removing the clutch (bevel pinion) shaft as per paragraph 165, and 2nd reduction shaft as per paragraph 164.

Remove snap ring (13) from inner side of third gear (8), and slide third gear toward the first gear (7). From left outer side of transmission housing, remove two cap screws retaining reverse idler shaft to the housing and withdraw idler shaft. Gear will remain in bottom of transmission housing.

From left side of transmission housing, remove the eight 5/16 inch cap screws and three ½ inch cap screws from first reduction shaft bearing carrier (15) and bearing cap (21), and remove cap (21) and shims (16). From right side of transmission housing, remove bearing carrier (9).

Slide first reduction shaft toward right side of housing, and lift left end up and withdraw same out through top cover opening as shown in Fig. C126. Gears can now be removed from the shaft. Install treated leather oil seal (10) in right side bearing carrier (9) with lip of seal facing the bevel gear.

Shims (16) inserted between carrier cap (21) and bearing carrier (15) control the backlash of the main drive bevel gears. Refer to paragraphs 186, 186A and 186B for procedure in adjusting the mesh and backlash of the bevel pinion (clutch shaft) and ring gear.

166D. *Series L.* Refer to Fig. C127. The first reduction shaft (C) can be removed after first removing the bevel pinion (clutch) shaft, and 2nd reduction shaft as per paragraph 164.

Remove first reduction shaft bearing carriers (1 & 9) from sides of transmission housing and keep the shims (18) with their respective car-

riers. Buck-up the first reduction shaft, and bump third gear (8) to the right and just far enough to remove the split sleeve spacer (19). After removing the spacer, bump third gear (8) to the left until same contacts the first gear (7). From left end of shaft remove two cap screws and washers retaining bearing cone to shaft. Remove bearing cone by bucking-up the second gear (2), and bumping shaft toward right side of housing. Move first reduction shaft (C) toward right side of transmission housing and lift left end of shaft up and withdraw same out through top cover opening. Gears can now be removed from the shaft.

Adjust the taper roller bearings with shims (18) to provide zero end play to the shaft and yet permit same to rotate without perceptible drag. These shims also control the backlash of the main drive bevel gears. Refer to paragraphs 186, 186A and 186B for procedure in adjusting the mesh and backlash of the bevel pinion (clutch shaft) and ring gear.

166E. *Series LA.* Refer to Fig. C128. The first reduction shaft (C) can be removed after first removing the 2nd reduction shaft as per paragraph 164, and clutch (bevel pinion) shaft as per paragraph 165.

Fig. C126—Removing the first reduction shaft by raising the left end and withdrawing same out through top cover opening on series D after 4805353.

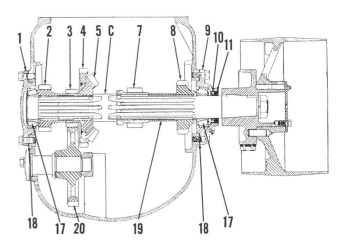

Fig. C128—Series LA first reduction (belt pulley) shaft. Shims (18) control bearings adjustment and ring gear (5) to bevel pinion (clutch shaft) backlash. Refer to Fig. C127 for legend.

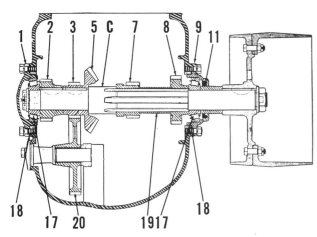

Fig. C127—Series L first reduction (belt pulley) shaft. Shims (18) control bearings adjustment and ring gear (5) to bevel pinion (clutch shaft) backlash.

C. First reduction shaft	5. Bevel ring gear	11. Oil seal (felt)
1. Bearing carrier	7. 1st gear	17. Bearing cup & cone
2. 2nd gear	8. 3rd gear	18. Shims
3. Reverse gear	9. Bearing carrier	19. Split sleeve spacer
4. 4th gear	10. Oil seal (treated leather)	20. Reverse idler

Fig. C128A—Transmission gearset on series S tractors. Series C, D, LA and R tractors are similar.

Loosen cap screws retaining first reduction shaft left side bearing carrier (1), and remove right side bearing carrier (9). Move first reduction shaft toward right side of transmission housing and lift left end of shaft up and withdraw same out through top cover opening. Gears can now be removed from the shaft. Install new oil seal (10) in right side bearing carrier (9) with lip of seal facing the bevel gear.

Adjust taper roller bearings with shims (18) to provide zero end play to the shaft and yet permit same to rotate without perceptible drag. These shims also control the backlash of the main

drive bevel gears. Refer to paragraphs 186, 186A and 186B for procedure in adjusting the mesh and backlash of the bevel pinion (clutch shaft) and ring gear.

167. REVERSE IDLER & SHAFT. To remove reverse idler gear & shaft, first remove the transmission top cover, power take-off shaft, shifter rails and forks, and second reduction shaft as outlined in paragraphs 162, 163 and 164.

On R and S series only, remove cotter pin and nut which retains idler to shaft. On all series, from left side of transmission housing, remove two cap screws retaining reverse idler shaft to transmission. Withdraw reverse idler

shaft, and remove gear through top cover opening. On D series after 480-5353, an oiler gear is also installed on the reverse idler shaft. On C series only, the reverse idler rotates on a renewable bronze bushing. On all other series, the idler gear (including oiler gear on D series) is not equipped with a bushing.

Reinstall reverse idler in the following manner: On series C, R and S with bevel edge of teeth toward left side of transmission housing; on series D prior 4805354; L and LA with long hub of idler toward left side of transmission housing; and on series D after 4805353 with long hubs of both the idler and oiler gear facing each other.

MODEL LA

MODEL VAC (Prior 5570000)

Series V

168. BASIC PROCEDURES. The transmission gearset, main drive bevel pinion and ring gear, and differential unit are carried in the transmission housing as shown in Figs. C129 & C129A. The mesh position of the bevel pinion is fixed and non-adjustable. There are infrequent instances where the failed or worn transmission part is so located that the repair can be completed safely without complete disassembly of the transmission. In effecting such localized repairs, time will be saved by observing the following as a general guide:

Shifter Rails and Forks. Shifter rails and forks which are located on the transmission cover are accessible for overhaul after removing the cover.

Main (Input) Drivegear. Main drivegear bearings (3 & 5) can be adjusted with adjusting nut (1) located on forward end of shaft. Adjustment can be made after detaching transmission housing from torque tube as in paragraph 169A. Removal of main drivegear (6) requires the additional work of removing transmission top cover so as to prevent the loss of the pocket bearing (37). Oil seal (2) can be renewed after detaching torque tube from transmission housing.

Main (Sliding) Gearshaft. Main shaft bearings (16 & 21) can be adjusted with adjusting nut (22) located on rear end of shaft. Adjustment can be made after removing the transmission top cover and differential unit as in paragraphs 169B or 204. Removal of main (sliding gear) shaft (36) requires the additional work of removing the main drivegear.

Countershaft. Countershaft bearings (42 & 54) can be adjusted with adjusting nut (41) located on forward end of shaft. Adjustment can be made after detaching transmission housing from torque tube as in paragraph 169A. Removal of countershaft (39) requires additional work of removing main drivegear and main (sliding gear) shaft.

Reverse Idler. Reverse idler (50) and shaft (47) can be removed after removing the transmission top cover, main drivegear, main (sliding gear) shaft, and countershaft.

Main Drive Bevel Pinion. Mesh position of bevel pinion is non-adjustable. Bevel pinion and shaft can be purchased separately from the bevel ring gear.

Bevel pinion shaft bearings (25 & 33) can be adjusted after detaching

transmission from torque tube, and removing transmission top cover, main drivegear, main (sliding gear) shaft, and countershaft. Removal of bevel pinion shaft requires the additional work of removing the main drive bevel ring gear and differential unit as per paragraph 169B or 204.

169. PREPARATION FOR OVERHAUL. To completely overhaul the transmission, it is necessary to first detach (split) the transmission housing from the torque tube as in paragraph 169A, and to remove the differential from the transmission housing as in paragraphs 169B or 204.

169A. R&R TRANSMISSION FROM TORQUE TUBE. Block up and support transmission housing and torque tube separately. Disconnect brake actuating rods from brake pedals. Remove four nuts attaching torque tube to the transmission housing. Roll transmission assembly rearward until clutch shaft is free of transmission main drive gear and shaft coupling.

169B. R&R DIFFERENTIAL. Remove fenders, platforms, seat assembly, and transmission rear cover or power take-off unit. Drain transmission and differential compartments. Support transmission housing and remove both rear axle and sleeve assemblies. Remove cap screws retaining both differential bearing carriers to transmission housing. Support differential unit, and remove bearing carriers. Remove the bevel ring gear and differential unit.

OVERHAUL. Data on overhauling the various components which make up the transmission are outlined in the following paragraphs. Particular care and attention should be given to the reinstallation and adjustment procedures.

170. SHIFTER RAILS AND FORKS. Transmission top cover contains the shifter rails and forks. Remove both platforms, and cap screws retaining top cover to transmission. To remove shifter forks and rails, first remove shifter rail hole Welch plugs, and set screws from forks and selectors. Bump rails rearward being careful not to lose detent balls, springs, and interlock pins.

When reassembling, install detent balls, springs, and interlock pins in cover and slide rails through cover, forks, and selectors.

171. MAIN (INPUT) DRIVEGEAR. To remove main drivegear (6) first

disconnect torque tube from transmission as in paragraph 169A, and remove both platforms, and transmission top cover. Next, remove bearing carrier (59), and bump the shaft and bearing carrier forward and out of transmission. As shaft assembly is withdrawn from front of transmission, be sure that the pilot bearing (37) does not drop to bottom of housing.

Overhaul is self-evident. Install new oil seal (2) with lip facing the transmission. After gear shaft is installed adjust bearings with nut (1) to permit shaft to revolve freely with no end play. Lock bearing adjustment by staking a portion of the nut into splines of shaft.

172. MAIN (SLIDING) GEARSHAFT. To remove main (sliding) gearshaft (36), and gears (7, 13, 15 and 19), first remove the main (input) drivegear as in paragraph 171, and differential unit as in paragraph 169B. Working through differential compartment, unstake and remove bearings adjusting nut (22). While bucking-up front bearing cone (16) bump shaft forward and out of transmission housing being careful not to lose spacers (18 & 58), and spring washer (20) or narrow spacer and shims.

On some early models, the fore and aft movement of the sliding gearshaft reduction gear (19) is controlled by a spring washer (20). On models equipped with a spring washer, discard the spring washer and use a narrow spacer (Case No. VT3995) and shims (Case Nos. VT3996 or VT3997).

Reassembly is reverse of disassembly. Gears (7, 13 & 15) should be installed as shown. Buck-up shaft and install bearing cone (16). Install spacer (18), reduction gear (19), narrow spacer (20), shims (same thickness as removed or if shims and spacer are being used for the first time, install the remaining components and then check to determine the amount of shims to be installed so as to provide reduction gear (19) with a .005 movement), wide spacer (58), bearing cone (21), and bearing adjustment nut (22).

Adjust main (sliding) gearshaft bearings with nut (22) to permit shaft to revolve freely with no end play. If the zero end play adjustment cannot be obtained after tightening the adjusting nut (22), it is an indication that too many shims are inserted between the narrow and wide spacers, or one of the spacers (18 or 58) is too wide. To correct, remove some

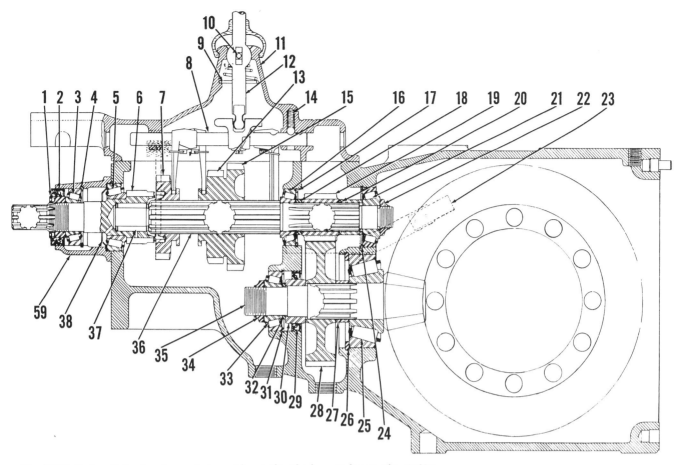

Fig. C129—Series V tractor transmission—side sectional view. Refer to Fig. C129A.

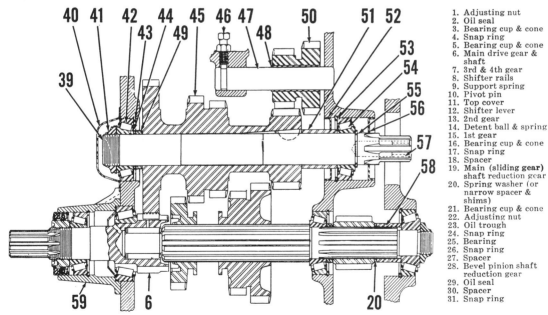

Fig. C129A—Series V tractor transmission—top sectional view. Refer to Fig. C129.

1. Adjusting nut
2. Oil seal
3. Bearing cup & cone
4. Snap ring
5. Bearing cup & cone
6. Main drive gear & shaft
7. 3rd & 4th gear
8. Shifter rails
9. Support spring
10. Pivot pin
11. Top cover
12. Shifter lever
13. 2nd gear
14. Detent ball & spring
15. 1st gear
16. Bearing cup & cone
17. Snap ring
18. Spacer
19. Main (sliding gear) shaft reduction gear
20. Spring washer (or narrow spacer & shims)
21. Bearing cup & cone
22. Adjusting nut
23. Oil trough
24. Snap ring
25. Bearing
26. Snap ring
27. Spacer
28. Bevel pinion shaft reduction gear
29. Oil seal
30. Spacer
31. Snap ring
32. Spring washer (or narrow spacer & shims)
33. Bearing cup & cone
34. Adjusting nut
35. Main drive bevel pinion & shaft
36. Main (sliding gear) shaft
37. Main shaft pilot bearing
38. Oil shield
39. Countershaft
40. Bearing cover
41. Adjusting nut
42. Bearing cup & cone
43. Snap ring
44. Spring washer (or narrow spacer & shims)
45. Countershaft gear cluster
46. Set screw
47. Reverse idler shaft
48. Bushing
49. Spacer
50. Reverse idler
51. Spacer
52. Oil shield
53. Snap ring
54. Bearing cup & cone
55. Snap ring
56. Oil shield
57. PTO shaft bushing
58. Spacer
59. Bearing carrier

shims, or reduce width of spacer by using emery cloth and a flat surface. After obtaining correct bearing adjustment check fore and aft movement of reduction gear (19). Gear movement of .005 is controlled with shims inserted between narrow spacer (20) and wide spacer (58).

173. COUNTERSHAFT. To remove countershaft (39) and gear cluster (45), first remove main (input) drive-gear and main (sliding) gearshaft as in paragraphs 171 and 172. From front face of transmission housing, remove bearing cover (40) and unstake and remove bearings adjusting nut (41).

While bucking-up the gear cluster bump countershaft rearward until cluster gear woodruff key is exposed and remove key. Continue bumping shaft rearward and out of transmission, being careful not to lose spacer (49) and spring washer (44).

On some early models, the fore and

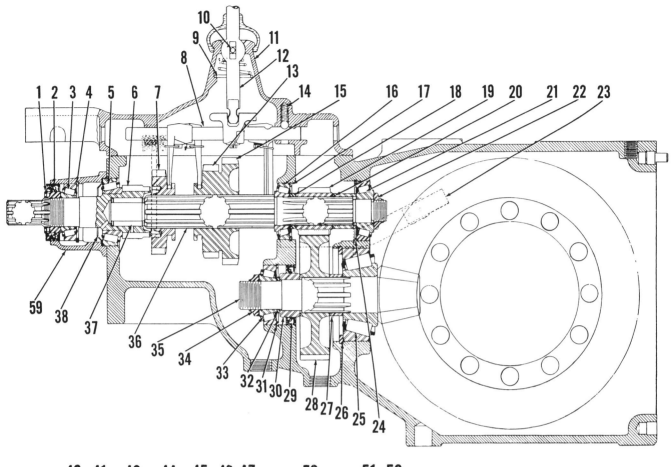

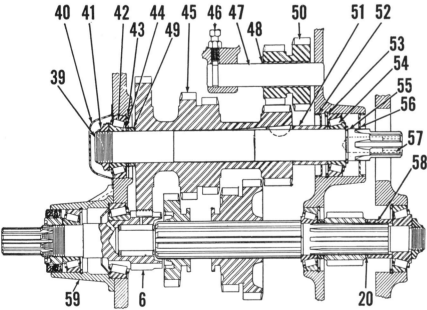

Fig. C129B—Series V tractor transmission—side and top sectional views. Refer to Fig. C129 for legend.

it is an indication that too many shims are inserted between wide spacer (49) and narrow spacer, or one of the spacers is too wide. To correct, remove some shims or reduce width of spacer by using emery paper and a flat surface. Check fore and aft movement of gear cluster. Gear movement of .005 is controlled with shims inserted between the narrow and wide spacers. After bearings are adjusted, lock adjusting nut by staking.

174. REVERSE IDLER. To remove reverse idler (50) and shaft (47), first remove main drive (input) gear, sliding gear shaft, and countershaft as in paragraphs 171, 172 and 173. From outer lower side of transmission housing, remove idler shaft retaining set screw (46). Bump idler shaft rearward and remove idler (50) out through transmission top cover opening.

Idler rotates on a steel-backed, bronze bushing (48) which is pressed in place and sized to provide a .004-.008 diametral clearance. When installing a new bushing, install the bushing so that the lead of the spiral oil groove is in the same direction as the idler rotates (which is clockwise when viewed from rear of transmission).

aft movement of the gear cluster is controlled by a spring washer (44). On models so equipped, discard the spring washer and use a narrow spacer (Case No. VT3995) and shims (Case Nos. VT3996 or VT3997).

The power take-off shaft bronze pilot bushing (57) can be renewed at

this time. New bushing requires final sizing to provide a .002-.003 diametral clearance.

Reassembly is reverse of disassembly. Adjust countershaft bearings with nut (41) to permit shaft to revolve freely with no end play. If the zero end play adjustment cannot be obtained

Notched end of bushing should project $\frac{1}{16}$ inch beyond rear face of idler gear.

175. MAIN DRIVE BEVEL PINION. The main drive bevel pinion & shaft (35) and lower reduction gear (28) can be removed after removing the main (input) drivegear, main (sliding) gearshaft, and countershaft as in paragraphs 171, 172 and 173. Working through transmission top cover opening, unstake bevel pinion shaft bearings adjusting nut (34) and remove nut. Bump bevel pinion shaft rearward and out of transmission, and remove reduction gear (28), spring washer (32) or narrow spacer and shims, and spacers (27 & 30) out through top cover opening.

Oil seal (29) which is installed with lip facing reduction gear (28) can be renewed at this time. Reassembly is reverse of disassembly.

Bevel pinion and shaft can be purchased separately from bevel ring gear. Bevel pinion mesh (fore and aft position) is fixed and non-adjustable.

On some early models, the fore and aft movement of reduction gear (28) is controlled by a spring washer (32). On models so equipped, discard the spring washer and use a narrow spacer (Case No. VT3995) and shims (Case Nos. VT3996 or VT3997).

175A. BEARING ADJUSTMENT. Adjust bevel pinion shaft bearings with nut (34) to provide a slight preload (8 pounds when measured with a spring scale engaged in one of the reduction gear teeth when only this shaft is in the transmission housing). If the bearing preload adjustment cannot be obtained after tightening the adjusting nut, it is an indication that too many shims are inserted between oil seal sleeve (30) and narrow spacer or, one of the spacers is too wide. To correct, remove some shims, or reduce width of spacer by using emery cloth and a flat surface. Check fore and aft movement of reduction gear. Reduction gear movement of .005 is controlled with shims inserted between the narrow spacer and oil seal sleeve (30). After bearings are adjusted, lock the adjustment by staking the nut.

175B. Backlash Adjustment. After installing differential unit, adjust carrier bearings by means of shims (2—Fig. C139A) so that the unit rotates freely, yet without perceptible end play. After adjusting the differential carrier bearings, adjust bevel pinion-to-ring gear backlash to .006-.008 by removing shoms from under one carrier and installing them under the op-

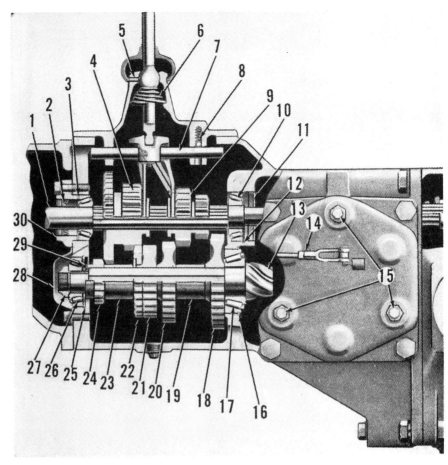

Fig. C130—Models VA, VAC and VAO tractor transmission prior to serial 6013844. Disc brake installation, as shown, is used on models prior 5455000. Model VAH gearset is similar, except that it is equipped with final drive (drop) housings bolted to the sides of the transmission housing. Refer to Fig. C131C for details of latest bevel pinion (countershaft) shaft.

1. Main drive shaft
2. Oil seal
3. Shims
4. 2nd & 4th gear
5. Shifter lever pivot pin
6. Support spring
7. Shifter rail
8. Detent ball & spring
9. 1st & 3rd gear
10. Bearing cup & cone
11. Oil shield
12. Snap ring
13. Main drive bevel pinion & shaft
14. Brake rod
15. Disc brake adjusting screw
16. Snap ring
17. Bearing cup & cone
18. 1st gear
19. Sleeve spacer
20. 3rd gear
21. 2nd gear
22. Oil slinger
23. Sleeve spacer
24. 4th gear
25. Snap ring
26. Bearing cup & cone
27. Adjusting nut
28. Bearing cap
29. Spring washer
30. Bearing retainer

posite bearing carrier. To increase backlash, remove shims from ring gear side and install the same number on the opposite side. This adjustment should be made **after**, never **before**, the bearings have been adjusted.

Series VA

176. BASIC PROCEDURES. Transmission gearset, main drive bevel pinion and ring gear, and differential unit are carried in the transmission housing as shown in Figs. C130 & C139. The mesh position of the bevel pinion is fixed and non-adjustable. In effecting localized repairs, time will be saved by observing the following as a general guide:

Shifter Rails and Forks. Shifter rails and forks which are located on the transmission cover are accessible for

overhaul after removing the cover.

Main Drive (Sliding Gear) Shaft. The oil seal (2) for main drive shaft (1) can be renewed or bearings adjusted with shims (3) after detaching torque tube from transmission and removing bearing retainer (30). Removal of main drive shaft (1) requires the additional work of removing transmission top and rear (or power take-off assembly) covers.

Bevel Pinion Shaft. Bearings (17 & 26) can be adjusted by nut (27) after detaching transmission from torque tube, and removing housing top cover. Removal of bevel pinion shaft requires the additional work of detaching rear axle housing from transmission housing on models VA, VAC, and VAO. On model VAH, remove both final drive (drop) housings. Also, on all VA models, remove brake assemblies and bevel ring gear and differential unit.

Reverse Idler. Reverse idler can be removed after removing the main drive (sliding gear) shaft, and bevel pinion shaft.

177. **R&R TRANSMISSION.** Overhaul of the complete transmission requires the following: Block up and support transmission housing and torque tube separately. Remove four nuts attaching torque tube to the transmission housing. Roll transmission and final drive assembly rearward until clutch shaft is free of transmission main drive (sliding gear) shaft coupling as shown in Fig. C131A.

177A. Drain transmission and differential compartments. Remove fenders, platforms and seat assembly. Support transmission housing as shown in Fig. C131, and on models VA, VAC and VAO only, remove bolts attaching rear axle housing to transmission housing. On models prior 4828245, remove brake housing covers and remove nut (located on inside of brake housing) from stud attaching rear axle housing to transmission. Roll rear axle housing away from transmission housing.

NOTE: The upper right-hand and lower left-hand bolts attaching transmission housing to rear axle housing are precision ground so as to serve as dowels in aligning the two housings.

On models VAH, remove both final drive (drop) assemblies. On all mod-

Fig. C131A—Detaching transmission housing from torque tube on series VA tractors. Series V is similar.

els, remove brake assemblies, and cap screws from differential bearing carriers. Support differential unit and remove bearing carriers. Remove the bevel ring gear and differential unit.

OVERHAUL. Data on overhauling the various components which make up the transmission are outlined in the following paragraphs. Particular care should be given to the reinstallation and adjustment procedures.

178. SHIFTER RAILS AND FORKS. Transmission top cover contains the shifter rails and forks. Remove four

cap screws retaining cover to transmission housing. To remove shifter forks and rails, first remove set screws from forks. Then, bump rails forward and out of cover being careful not to lose detent balls, springs, and interlock pin.

When reassembling, install detent balls, spring and interlock pin in cover, and slide rails through cover and forks.

179. MAIN DRIVE (SLIDING GEAR) SHAFT. To remove this shaft (1—Fig. C131B) and gears (4 & 9), first disconnect torque tube from transmission as in paragraph 177, and remove transmission top cover and axle housing rear cover or transmission rear cover, or power take-off assembly. Remove main drive shaft bearing retainer (30) and bearing adjusting shims (3). Spread the two sliding gears (4 & 9) apart and against walls of transmission housing and place Case Tool No. T122 (a 1½ inches diameter pipe sawed in half, 2¼ inches long) between the two gears, as shown in Fig. C132 so as to permit removal of rear bearing cone (10). If special tool or spacer is not available, slide gear (9) against bearing cone (10) and buck-up gear with a bar while bumping shaft forward. Bump main drive shaft forward which will remove rear bearing cone from shaft and front bearing cup from the housing. Withdraw shaft and remove gears through top cover opening. Install the new oil seal (2) with lip facing the transmission gearset.

With front bearing cone installed on shaft, and rear bearing cup installed

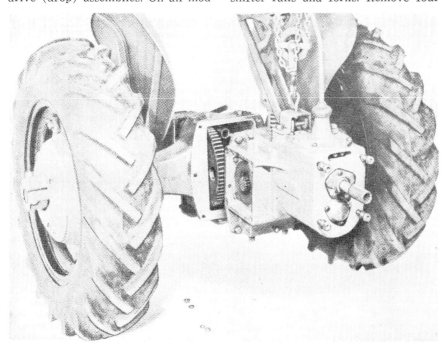

Fig. C131—Removing transmission housing from rear axle housing on models VA, VAC and VAO.

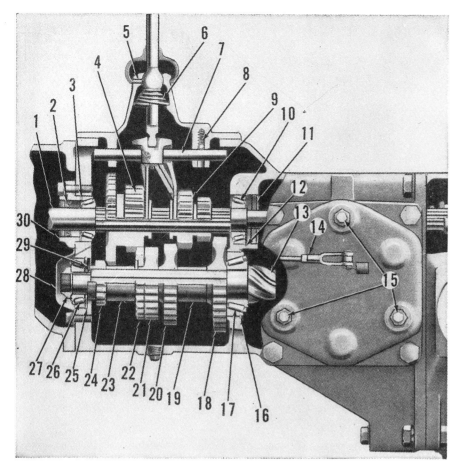

Fig. C131B—Models VA, VAC and VAO tractor transmission prior to tractor serial 6013843. It is similar to the gearset used in VAH except the latter is equipped with final drive (drop) housings bolted to the sides of the transmission housing. Refer to Fig. C131C and paragraph 180 for latest changes in the (countershaft) bevel pinion shaft.

Reassembly is reverse of disassembly. When VA tractors prior to serial 6013843 are being overhauled always discard the spring washer Case part VT3414 (shown at 29 in Fig. C131B) also the 1 21/32 inch long spacer Case VT3425 shown as (19) which is the rear one of two such spacers used on the shaft. Retain the other parts if unworn, and referring to Fig. C131C assemble new shaft spacer V21042, heavy duty spring washer V20217 and large thrust washer V20203 in place of the original spacer. Also at front end of shaft between the bearing cone and gear, assemble the new small thrust washer V20218 as shown.

180A. *Bearing Adjustment.* Adjust bevel pinion shaft bearings with nut (27) to a slight preload of 8-10 pounds when measured with a spring scale engaged in the teeth of the second gear (21) when only this shaft is installed. After adjusting the bearings, lock the adjustment by staking the nut.

Backlash Adjustment. After installing the differential unit, adjust carrier bearings and backlash as outlined for the V series "Backlash Adjustment" in paragraph 175B.

181. REVERSE IDLER. To remove the reverse idler and shaft, first remove the main drive (sliding gear) shaft, and bevel pinion shaft, as in paragraphs 179 and 180. Working through transmission top cover opening, remove idler shaft retaining set screw from idler shaft support. Bump shaft rearward and remove gear.

Idler rotates on a steel-backed bronze bushing which is pressed in place and sized after installation to .862-.863. When installing a new bushing, the lead of the spiral oil groove should be in same direction as the idler rotates (which is clockwise when viewed from the rear of transmission). The notched end of bushing should project 1/16 inch beyond rear face of idler.

in transmission housing, insert sliding gear shaft through front of transmission housing and install gears (4 & 9) as shown. While bucking-up front end of shaft bump cone (10) on to rear end of shaft. Adjust main drive shaft bearings with shims (3) inserted between front bearing cup, and bearing retainer (30) to permit shaft to revolve freely with no end play.

180. BEVEL PINION & SHAFT. The main drive bevel pinion and shaft (13) can be removed after removing the transmission, and main drive (sliding gear) shaft as in paragraphs 177 and 179. Remove bevel ring gear and differential assembly. Remove bearing cap (28) and using a pointed tool, unstake and remove bevel pinion shaft bearings adjusting nut (27). While bucking-up gear (18), bump shaft rearward and out of transmission housing and remove gears, and spacers out through top cover opening.

Bevel pinion and shaft can be purchased separately from bevel ring gear. Bevel pinion mesh (fore and aft position) is non-adjustable.

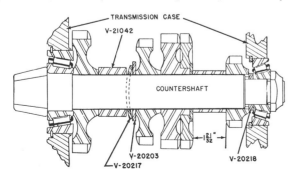

Fig. C131C—When overhauling VA series tractors prior to serial 6013844 always install above numbered parts as per paragraph 180.

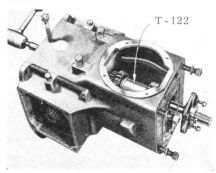

Fig. C132—Using a Case T122 spacer tool between the two sliding gears when removing the sliding gear shaft.

BEVEL GEARS, DIFFERENTIAL, FINAL DRIVE AND REAR AXLE

MAIN DRIVE BEVEL GEARS
Series C-D-L-LA-R-S

The main drive bevel pinion is integral with and is located on the transmission end of the engine clutch shaft. Refer to Fig. C133 for models with non-continuous pto, or Fig. C137A for models with continuous type pto. This shaft is also the transmission input shaft. The main drive bevel ring gear is either keyed or splined to the transmission first reduction (belt pulley) shaft, Fig. C134.

On series C, L, and R, the main drive bevel pinion can be purchased separately from the main drive bevel ring gear. On series D, LA, and S,

the main drive bevel pinion and ring gear are supplied for service only as a matched pair.

183. **PINION SHAFT BEARINGS ADJUSTMENT.** The main drive bevel pinion (clutch) shaft rotates on non-adjustable ball type bearings.

184. **BEVEL RING GEAR (FIRST REDUCTION) SHAFT BEARINGS ADJUSTMENT.** Refer to Figs. C138A through C138F for model concerned. On series C, D prior 4805354, L, LA, R and S, the bevel ring gear shaft (1st reduction or pulley shaft) rotates on taper roller type bearings. Adjust the bearings with shims (18) to re-

move all bearing play but permitting shaft to rotate without perceptible preload. These same shims, also, control the bevel ring gear-to-pinion backlash adjustment.

On series D after 4805353, the bevel ring gear shaft (1st reduction or belt pulley shaft) rotates on non-adjustable ball type bearings, as shown in Fig. C138D.

185. **PINION & RING GEAR RENEWAL.** To renew the pinion which is integral with the engine clutch shaft, proceed as in paragraphs 165, 165A and 165B. To renew the bevel ring gear which is either splined or keyed to the first reduction (belt pulley) shaft, proceed as in paragraphs 166 through 166E.

After installing the pinion shaft and bevel ring gear shaft, it will be necessary to adjust the mesh (fore and aft position of bevel pinion) and the backlash of the gears as per paragraph 186. This adjustment should be made **after**, never **before**, the bearings on the first reduction shaft have been adjusted as in paragraph 184.

186. **MESH AND BACKLASH ADJUSTMENT.** The mesh (fore and aft) position of the bevel pinion is controlled by shims and is set by using the appropriate Case cone center gauge listed in the next to last column of the

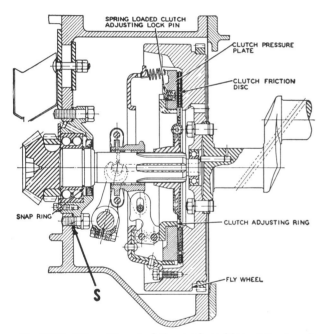

Fig. C133—Shims (S) control the mesh position of the bevel pinion which is integral with the engine clutch shaft. Series D with non-continuous type pto and over-center type clutch is illustrated.

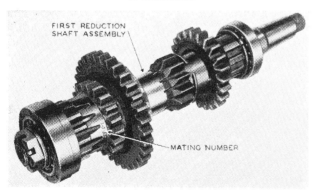

Fig. C134—First reduction gearshaft assembly as used on model D after serial 4805353. Main drive bevel ring gear not visible in this view is integral with the two spur gears at left hand end of the assembly. Series C, L, LA, R, S and early D models are similar except type of anti-friction bearings.

Fig. C135—Using a special gauge and feeler gauge to check bevel pinion mesh (fore and aft position) on a series S tractor. A similar procedure is used on series C, D, L, LA and R transmissions.

C. First reduction (belt pulley) shaft
1. Special gauge
2. Bevel pinion (integral with engine clutch shaft)
4. Feeler gauge

Transmission Main Drive Bevel Gears and Gauges

Tractor Model	Part No.	Adopted	Replaced	Gauge No.	Thickness
C-CI-CIM- CO-CV	*01632AB **01633AB	1929	1939	02890AB 02890AB1	.625
CC	*01632AB **01633AB	1929	1934	02890AB 02890AB1	.625
CC	*03509AB **03510AB	1934	1939	02890AB1	.573
CCS	*03509AB **03510AB	1938	1939	02890AB1	.573
DC-DH-DS-DEX- DOEX-DVEX DO Low Speed	†2143AA	1939 1940	1941	05694AB	.5835
D-DO-DV- DIM-DI	†2142AA	1939	1941	05694AB	.5835
D Series	†2889AA	1941	1944	05694AB	.5835
D Series	†A2889AA	1943	1944	05694AB	.5835
D Series	†3261AA †3262AA †3297AA	1943	See note***	05694AB	.5835
D Series	†3285AA	1944	1945	07628AB	.521
D Series	†3308AA	1945	1952	07628AB	.521
D Series	†4295AA	1952	Current	07628AB	.521
L	*01261AB **01260AB	1928	1932	02890AB 02890AB1	.8125
L	*01308AB2 **01260AB	1932	1940	02890AB 02890AB1	.8125
LA	†2435AA	1940	1943	05694AB 07628AB	.7715
LA	†A2435AA	1943	Current	05694AB 07628AB	.7715
R Series	*03693AB **03680AB	1935	1940	02890AB1	.500
S Series	†2578AA	1940	1952	05694AB 07628AB	.440
S Series	†4135AA	1952	Current	07628AB	.440

*Bevel pinion available separately. **Bevel ring gear available separately. ***Replacement for tractors built prior to 1944. †Composite number. Available only as a matched pair.

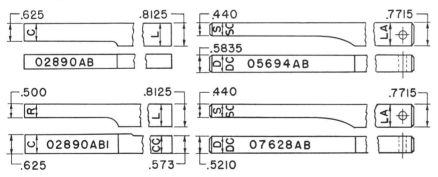

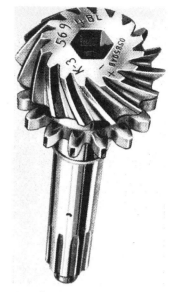

Fig. C136—Showing cone center setting and backlash specifications as marked on the transmission main drive bevel pinion.

ion as shown in Fig. C136. Amount of backlash is indicated by a number and the letters "BL" such as "4BL". The thickness of the feeler gauge to be used in conjunction with the special mesh position gauge is indicated by a plus sign and a number such as "+1". The markings will be different for each set of gears. The markings "K-3-569" are identification markings for a matched set of gears and the same number will be found on the bevel ring gear as shown in Fig. C134. The other marking "05850AB" is the part number.

186A. *Pinion Mesh Position.* First, remove transmission top cover as in paragraph 162. If the pinion mesh position requires changing, it will be necessary to detach the engine from the transmission as in paragraph 161. Then, observe the part number of the pinion and select the correct gauge to be used as indicated in the table "Transmission Main Drive Bevel Gears & Gauges." Next, insert the special gauge and a feeler gauge of the same thickness as the plus number marked on the pinion, between the end of the bevel pinion and the ground portion of the first reduction shaft as shown in Fig. C135. The pinion is correctly positioned (fore and aft position) when a light pull is required to withdraw the specified feeler gauge. If the specified feeler gauge fits loosely, the bevel pinion must be moved rearward; if gauge will not enter, the pinion must be moved forward.

To move the bevel pinion forward (increase distance between bevel pinion and first reduction shaft), add shims (S—Fig. C133 or 16—Fig. C

accompanying table. On all series except D after 4805353, first adjust the first reduction shaft bearings with shims (18) so that the shaft rotates freely, yet without perceptible preload. On series D after 4805353, the ball type bearings are non-adjustable.

With the first reduction shaft bearings adjusted, check the shaft for run-out with a dial indicator by mounting same so that the indicator button contacts the ground portion (center) of the shaft. A maximum runout of .002 is permissible. If the runout exceeds .002, it will be necessary to install another shaft.

Mesh position and backlash values are marked directly on the bevel pin-

137A) in an amount equal to the distance to be increased. To decrease the distance, remove shims. Shims are located between bevel pinion shaft bearing carrier and front face of transmission front wall. The amount of shims to be removed or added is determined by the difference between what the spacing actually is, and what it should be as marked on the pinion. Suppose the specified feeler gauge thickness is .004 as indicated by the marking "+4" on the pinion. Insert a .004 feeler between special gauge and first reduction shaft. If the .004 feeler is a loose fit but the .008 feeler fits with a slight drag, the pinion must be moved rearward .004 by removing .004 of pinion adjusting shims. Shims are available in thicknesses of .003, .005 and .012.

186B. *Backlash.* With the bevel pinion adjusted for correct mesh position, as outlined in preceding paragraph, adjust the backlash to the amount marked on the bevel pinion. On all series except series D after 4805353, to increase the backlash, remove shims (18—Figs. C138A, C138B, C138C, C 138E, or C138F) inserted under left bearing carrier (1) and insert the same shims under the right bearing carrier (9).

On series D after 4805353, the backlash is controlled by shims inserted between bearing carrier and bearing cap on left side only as shown in Fig. C138. On this model, to decrease backlash, add shims. To increase the backlash, remove shims.

1. Snap ring
2. PTO shaft gear
5. Bearing carrier
6. Engine clutch cover
7. Cone collar
8. Flywheel
9. Driven disc
10. Pilot bearing
11. Snap ring
12. Thrust collar
13. Bushing
14. Oil seal
15. Snap ring
16. Shims
17. Bushing
18. PTO drive sleeve and gear
19. Thrust washer
20. Clutch shaft
21. "O" ring
23. Retainer ring

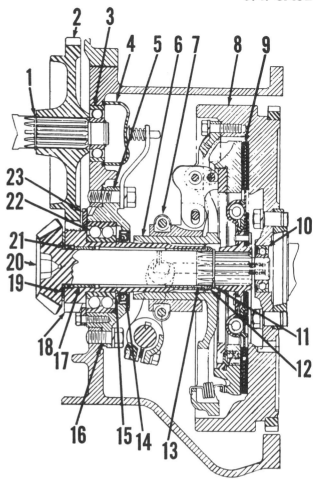

Fig. C137A—Over-Center Type engine clutch, clutch shaft, and pto drive sleeve and gear installation on series D tractors equipped with a continuous type power take-off. Series S tractors equipped with continuous pto and over-center type clutch are similar except for differences in the engine clutch. Refer to page 120 for spring loaded type clutch used on latest D and S series.

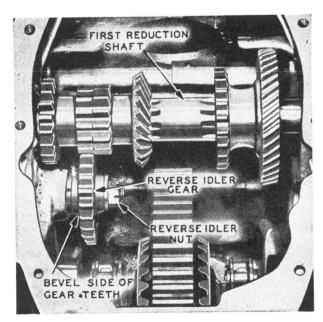

Fig. C137—Series S transmission first reduction (belt pulley) shaft installation. Series D and R are similar.

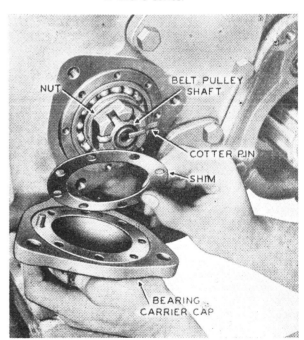

Fig. C138—Shims located between left bearing cap and bearing carrier control the bevel pinion to ring gear backlash on series D after 4805353.

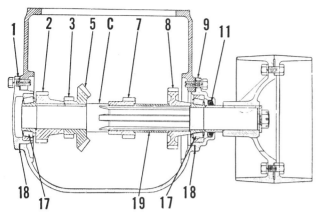

Fig. C138A—Series C transmission first reduction (belt pulley) shaft. Shims (18) control bearings adjustment and ring gear (5) to bevel pinion (clutch shaft) backlash.

C. First reduction shaft	5. Bevel ring gear	11. Oil seal (felt)
1. Bearing carrier	7. 1st gear	17. Bearing cup & cone
2. 2nd gear	8. 3rd gear	18. Shims
3. Reverse gear	9. Bearing carrier	19. Split sleeve spacer

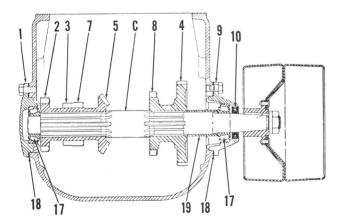

Fig. C138B—Series D prior 4805354 transmission first reduction (belt pulley) shaft. Shims (18) control bearings adjustment and ring gear (5) to bevel pinion (clutch shaft) backlash.

C. First reduction shaft	5. Bevel ring gear	11. Oil seal (felt)
1. Bearing carrier	7. 1st gear	17. Bearing cup & cone
2. 2nd gear	8. 3rd gear	18. Shims
3. Reverse gear	9. Bearing carrier	19. Split sleeve spacer
4. 4th gear	10. Oil seal (treated leather)	

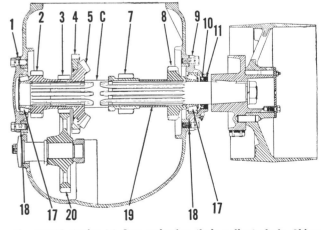

Fig. C138C—Series R and S transmission first reduction (belt pulley) shaft. Shims (18) control bearings adjustment and ring gear (5) to bevel pinion (clutch shaft) backlash.

C. First reduction shaft	4. 4th gear	10. Oil seal (treated leather)
1. Bearing carrier	5. Bevel ring gear	17. Bearing cup & cone
2. 2nd gear	7. 1st gear	18. Shims
3. Reverse gear	8. 3rd gear	19. Split sleeve spacer
	9. Bearing carrier	

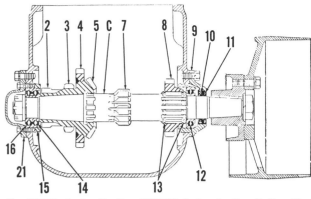

Fig. C138D—Series D after 4805353 first reduction (belt pulley) shaft. Shims (16) control ring gear (5) to bevel pinion (clutch shaft) backlash adjustment.

C. First reduction shaft	7. 1st gear	11. Bearing
2. 2nd gear	8. 3rd gear	13. Snap ring
3. Reverse gear	9. Bearing carrier	14. Bearing
4. Fourth speed gear	10. Oil seal (treated leather)	15. Bearing carrier
5. Bevel ring gear	11. Oil seal (felt)	16. Shims
		21. Carrier cap

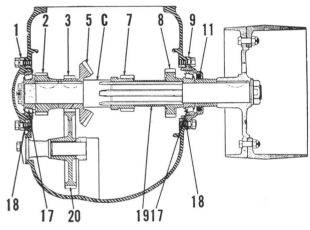

Fig. C138E—Series L first reduction (belt pulley) shaft. Shims (18) control bearings adjustment and ring gear (5) to bevel pinion (clutch shaft) backlash.

C. First reduction shaft	5. Bevel ring gear	17. Bearing cup & cone
1. Bearing carrier	7. 1st gear	18. Shims
2. 2nd gear	8. 3rd gear	19. Split sleeve spacer
3. Reverse gear	9. Bearing carrier	20. Reverse idler
	11. Oil seal (felt)	

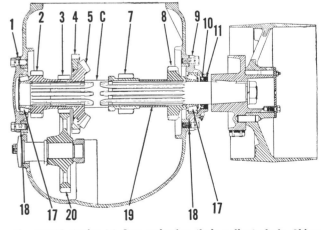

Fig. C138F—Series LA first reduction (belt pulley) shaft. Shims (18) control bearings adjustment and ring gear (5) to bevel pinion (clutch shaft) backlash.

C. First reduction shaft	5. Bevel ring gear	11. Oil seal (felt)
1. Bearing carrier	7. 1st gear	17. Bearing cup & cone
2. 2nd gear	8. 3rd gear	18. Shims
3. Reverse gear	9. Bearing carrier	19. Split sleeve spacer
4. 4th gear	10. Oil seal (treated leather)	20. Reverse idler

Series V

The main drive bevel pinion is integral with the transmission lower reduction shaft as shown in Fig. C129. The main drive bevel ring gear is riveted to the differential case as shown in Fig. C139A. Bevel pinion can be purchased separately from the bevel ring gear.

Mesh position of the main drive bevel pinion is fixed and non-adjustable. Backlash of the bevel ring gear is controlled by the same shims as are used to adjust the differential carrier bearings.

187. PINION SHAFT BEARINGS ADJUSTMENT. To adjust pinion shaft bearings, first detach transmission from torque tube, and remove transmission top cover, main (input) drive-gear, sliding gear shaft, and countershaft as in paragraphs 169A, 170, 172 and 173.

187A. The pinion shaft bearings can now be adjusted as per paragraph 175A.

2. Shims
3. Bearing carrier
4. Oil seal
5. Bearing cup & cone
6. Side gear thrust washer
7. Differential case
8. Bevel ring gear
9. Thrust washer
11. Wheel axle shaft
12. Differential side gear
13. Differential spider
14. Differential pinion

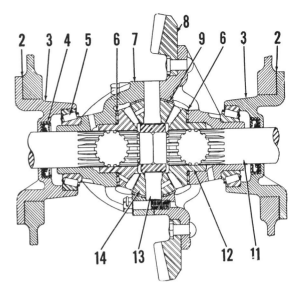

Fig. C139A—Series V tractor main drive bevel ring gear and differential unit.

188. CARRIER BEARINGS ADJUSTMENT. To adjust the differential carrier bearings, first remove transmission rear cover or hydraulic lift and/or pto housing, both wheel axle and sleeve assemblies, and both secondary brake discs. Adjust bearings by means of shims (2—Fig. C139A) inserted under bearing carriers so that the unit rotates freely, yet without perceptible end play.

188A. BEVEL RING GEAR BACKLASH. After adjusting the differential carrier bearings, adjust ring gear backlash to .006-.008 by removing

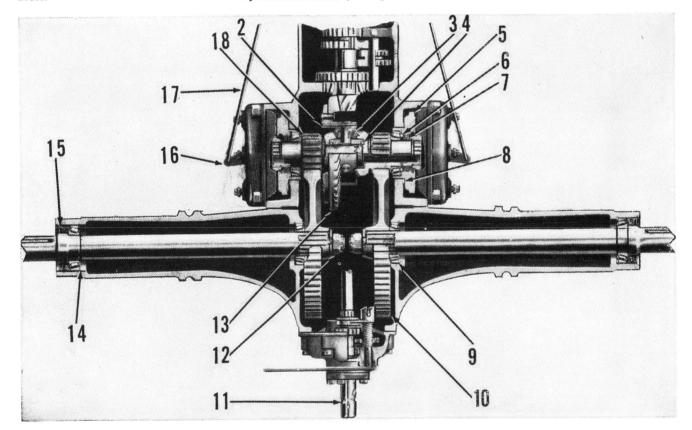

Fig. C139—Models VA, VAC and VAO showing installation of the bevel pinion, bevel ring gear, bull pinions, brakes, bull gears, wheel axle shafts, rear axle housing, and pto unit. Models after 5455000 are equipped with internal expanding brakes, Fig. C179, instead of the disc type as shown.

2. Pinion shaft retaining pin	6. Snap ring	10. Bull gear	13. Bevel ring gear	17. Brake rod
3. Differential pinion	7. Oil seal	11. PTO (external) shaft	14. Snap ring	18. Bull pinion & differential side gear
4. Thrust washer	8. Shims	12. Bull gear retaining nut	15. Oil seal	
5. Bearing carrier	9. Snap ring		16. Brake actuating lever	

Fig. C140—*Differential unit, spur ring gear, and bull pinion sprockets when viewed from transmission housing rear cover opening on series LA. Series C, D, L, R and S are similar.*

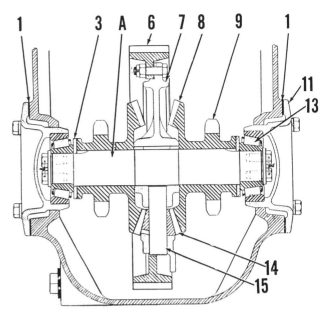

Fig. C141—*Series L differential, spur ring gear and bull pinion sprockets installation. Series C is similar except that gear (6) is mounted centrally (instead of offset) on case (7).*

A. Differential shaft	7. Differential case	13. Bearing cup & cone
1. Shims	8. Differential side gear	14. Differential pinion
3. Thrust washer	9. Bull pinion sprocket	15. Differential pinion
6. Spur ring gear	11. Bearing carrier	shaft

shims from under one carrier and inserting the same amount under the opposite carrier. To increase backlash, remove some shims from ring gear side and install the same number on the opposite side. The backlash adjustment should be made **after**, never **before**, the bearings have been adjusted.

189. PINION & RING GEAR RENEWAL. To renew the bevel pinion and to remove the bevel ring gear, follow procedures as outlined in paragraphs 175 through 175B. The procedure for bench overhaul of the differential unit, including renewal of the ring gear is outlined in paragraph 205.

190. MESH AND BACKLASH ADJUSTMENT. The bevel pinion mesh (fore and aft) position is fixed and non-adjustable. Backlash adjustment of bevel pinion-to-ring gear is outlined in paragraph 188A.

Series VA

The main drive bevel pinion is integral with the transmission countershaft as shown in Fig. C130. The main drive bevel ring gear is riveted to the differential case. Bevel pinion can be purchased separately from the ring gear. Refer to Fig. C139.

Mesh position of the main drive bevel pinion is fixed and non-adjustable. Backlash of the bevel gears is controlled by the adjustable differential carrier bearings.

191. PINION SHAFT BEARINGS ADJUSTMENT. To adjust pinion shaft bearings, first detach transmission

from torque tube, and remove transmission top cover as in paragraphs 177 and 178. The pinion shaft bearings can then be adjusted by following the procedure as outlined in paragraph 180A.

192. CARRIER BEARINGS ADJUSTMENT. On models VA, VAC and VAO, to adjust the differential carrier bearings, first detach rear axle housing from transmission housing, and remove both brake assemblies. Adjustment can be accomplished without removing the axle housing from the transmission, but when done this way, it is much more difficult to check bevel gear backlash.

On model VAH, to adjust the differential carrier bearings, first remove transmission housing rear cover or pto unit, both final drive (drop) assemblies from transmission housing and brake assemblies.

Adjust bearings by means of shims (8—Fig. C139) inserted under bearing carriers so that the unit rotates freely, yet without perceptible end play.

192A. BEVEL RING GEAR BACKLASH. After adjusting the differential carrier bearings, adjust ring gear backlash to .006-.008 by removing shims (8) from under one carrier and inserting the same amount under the opposite carrier. To increase backlash, remove some shims from ring gear side and install the same number on the

opposite side. The backlash adjustment should be made **after**, never **before**, the bearings have been adjusted.

193. PINION & RING GEAR RENEWAL. To renew the bevel pinion and to remove the bevel ring gear, follow procedures as outlined in paragraphs 180 and 180A. The procedure for bench renewal of the ring gear is outlined in paragraph 208B.

194. MESH AND BACKLASH ADJUSTMENT. The bevel pinion mesh (fore and aft) position is fixed and non-adjustable. Backlash adjustment is outlined in paragraph 192A.

DIFFERENTIAL

Series C-D-L-LA-R-S

The differential unit is of the three-pinion, open case type, and is mounted on the differential shaft which is located aft of the transmission second reduction (sliding gear) shaft. The differential carrier also carries a large spur ring gear which meshes with a smaller gear (spur pinion) on the second reduction shaft. Refer to Figs. C140 through C148A.

The differential shaft also carries both differential unit side gears which are integral with the bull pinion sprockets for the final drive chains. On models equipped with independent hand type brakes, the bull pinion sprocket and differential side gear also carry the brake shaft drive gear

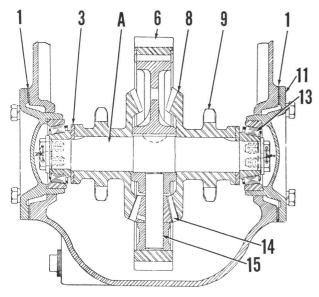

Fig. C142—Series D prior to 4511449 differential, spur ring gear and bull pinion sprocket installation. Series R is similar.

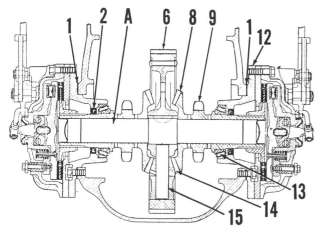

Fig. C143—Series D 4511448-5601107 differential, spur ring gear, bull pinion sprockets and single disc type brakes installation.

which is similar to that as shown in Fig. C142.

200. **BEARINGS ADJUSTMENT.** Refer to Figs. C141 through C147. The differential unit shaft bearings are shim adjusted, tapered roller type mounted in carriers which are bolted to the transmission housing. To adjust bearings, first remove transmission top cover (paragraph 162). Remove cap screws from one bearing carrier and extract shims to tighten or add shims to loosen.

201. **REMOVE & REINSTALL.** To remove the differential and spur ring gear unit, first drain transmission and clutch compartments. Detach transmission from engine as in paragraph 161, and remove transmission top cover (paragraph 162). Remove the transmission rear cover, or power take-off and rear cover, or power take-off, hydraulic lift unit and rear cover

A. Differential shaft
1. Shims
2. Oil seal
3. Thrust washer
4. Brake shaft drive gear
6. Spur ring gear
7. Differential case
8. Differential side gear
9. Bull pinion sprocket
11. Bearing carrier
12. Bearing carrier
13. Bearing cup & cone
14. Differential pinion
15. Differential pinion shaft

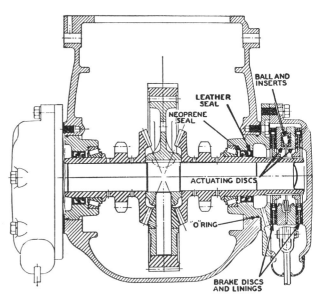

Fig. C144—Series D after 5601106 differential, spur ring gear, bull pinion sprockets and double disc type brakes installation. Series LA after 5418707 equipped with double disc brakes is similar.

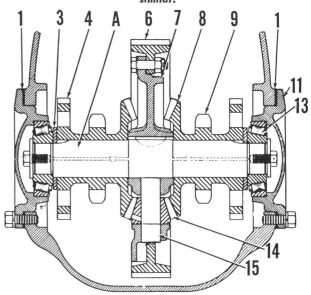

Fig. C145—Series LA prior 5418708 differential, spur ring gear, bull pinion sprocket installation on models equipped with rear wheel brakes.

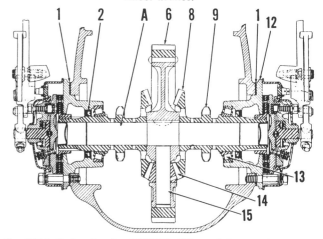

Fig. C146—Series S prior 5600001 differential, spur ring gear, bull pinion sprockets and single disc type brakes installation.

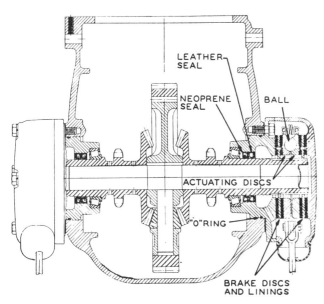

Fig. C147—Series S after 5600000 differential, spur ring gear, bull pinion sprockets and double disc type brakes installation.

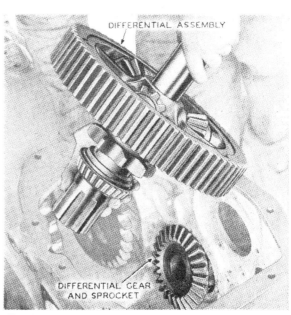

Fig. C148—Method of removing differential and spur ring gear assembly on series D after 4511448.

from models so equipped. Remove shifter rails and forks (paragraph 163), second reduction shaft (paragraph 164), clutch (bevel pinion) shaft (paragraph 165), first reduction shaft (paragraph 166 through 166E), and reverse idler and shaft (paragraph 167). On models with brake units located on the outer ends of the differential shaft, remove brake units.

Working through the transmission rear cover opening, remove the special heat treated cotter pins from final drive chain master links, and remove both final drive chains from both small sprockets. Support differential and spur ring gear unit, and remove differential shaft bearing carriers.

On series D prior 4511449, LA prior 5418708, C, L, and R, slide differential and shaft assembly toward right side of transmission housing and remove the assembly with differential side gears and bull pinion sprockets out through top cover opening by tipping left end of unit up.

On series D after 4511448 and LA after 5418707 (disc type brakes on differential shaft), slide one differential side gear unit which is integral with the bull pinion sprocket toward side of transmission housing; then slide differential and shaft assembly toward opposite side of transmission housing and remove as shown in Fig. C148.

On series S slide both differential side gear units toward sides of transmission housing; then slide differential and shaft assembly toward either side of housing. Tip the assembly, as shown in Fig. C148A, and remove same out through top cover opening.

201A. On series C, D, LA, R and S, the entire differential and shaft assembly can be reversed end for end without changing the bearings, if wear on the spur ring gear and/or bull pinion sprockets is noticeable. Reversing the assembly end for end will present new working surfaces for either the sprocket and/or spur ring gear when the transmission is in any of the forward speeds. The reversing procedure for the spur ring gear cannot be made on the series L as the spur ring gear is mounted off center (toward left side of housing) as shown in Fig. C141. However, the bull pinion sprockets can be reversed end for end and the ring gear unbolted and reversed. Reinstallation is the reverse of removal.

Oil seals, which are located in the differential bearing carriers on models equipped with differential shaft brakes, should be installed with the lip or lips facing inward. On series using a double oil seal for each side (one neoprene, and one leather type) install the neoprene seal toward the inside and the leather seal on the outside. After installing the differential unit, adjust the bearings by varying the shims.

CAUTION: When reinstalling drive chains, use new special heat treated steel cotter pins for the master link. Do not use the cotter pins a second time.

201B. To adjust tension of the final drive chains located in the transmission housing, first loosen the wheel axle shaft sleeve or final drive (drop) housing on models DCS and DS to

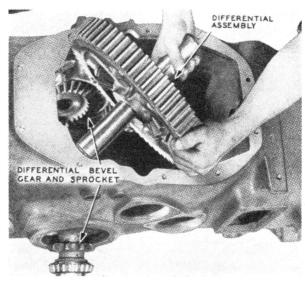

Fig. C148A—Method of removing differential and spur ring gear assembly on series S.

Fig. C149—Drive chains and bull sprockets installation when viewed from transmission housing rear cover opening on series R and S. Series C, D, L and LA are similar except for method of attaching the bull sprockets to the inner ends of the wheel axle shafts.

1. Special steel cotter pins	3. Master link
	8. Adjusting bolt

transmission housing retaining nuts or cap screws, and rotate adjusting bolts (8—Fig. C149) so as to move the final drive bull sprocket, wheel axle shaft and sleeve assembly rearward or forward. To tighten the chain, rotate the bolt in a clockwise direction. The chain adjustment is correct if the chain has a deflection of approximately one inch when checked midway between the sprockets. With adjustment completed, the head of the adjusting bolt should be set square, as shown, so that the recess in the rear cover will fit over the bolt head and prevent the bolt from turning. Retighten axle sleeve to transmission housing attaching cap screws or nuts.

202. **OVERHAUL.** On series D after 4511448, LA and S, neither the spur ring gear, case or any of the components of the differential are available separately. Renewal of either the differential pinion gears, spur ring gear, or differential case requires renewal as a complete assembly. On all other series, the spur ring gear is bolted to the differential case, and renewal of any differential components can be made after detaching the spur ring gear and removing the differential pinion shafts. Differential unit pinion gears are unbushed. Check the hardened steel thrust washers (3—Fig. C141, C142 or C145) as used on series D prior 4511449, LA prior 5418708, C, L and R, for wear. Renew the washers if excessive

end play exists between side gear and bull pinion sprocket, and bearing cone when same are installed on the shaft. The differential unit shaft can be renewed after pressing same out of differential housing. On series L only, the spur ring gear is mounted off center (toward left side) as shown in Fig. C141 and should be reinstalled as shown.

Series V

The differential unit to which is riveted the main drive bevel ring gear is of the four pinion, closed case type and is located back of the transmission housing dividing wall. Differential carrier bearings are shim adjusted, tapered roller type. The carriers are covered by their respective secondary brake discs and rear axle and sleeve assemblies, which must be removed before any work can be done on the carriers. Refer to Fig. C149A.

203. **CARRIER BEARINGS ADJUSTMENT.** To adjust differential carrier bearings and the bevel pinion-to-ring gear backlash, proceed as outlined in paragraphs 188 and 188A.

204. **REMOVE AND REINSTALL.** To remove the differential unit and bevel ring gear assembly, first drain the transmission and differential compartments. Remove both rear axle and sleeve assemblies. Remove transmission rear cover, or hydraulic lift and/or pto housing, and both secondary brake discs. Support differential unit and remove both differential bearing carriers (each held to transmission case by cap screws). Do not mix the shims. Lift differential unit and bevel ring gear assembly out through rear cover opening.

204A. Reinstallation is reverse of removal procedure. If new oil seals are installed in the differential bearing carriers, install same with lip of seal facing the differential unit.

Adjust carrier bearings and bevel pinion-to-gear backlash as outlined in paragraph 188 and 188A.

205. **OVERHAUL.** To disassemble differential unit, first place correlation marks on both halves of differential case and remove lock wire from differential case cap screws; and remove cap screws. The two halves of the differential case can now be separated. Refer to Fig. C149A.

Recommended backlash of .005 between differential side gears and pinions is controlled by the thrust washers (6 & 9). Pinion gears are unbushed. Rear axle shaft inner oil seal (located in differential bearing carrier) should be installed with lip of seal facing the differential unit.

The bevel ring gear, which is available separately from the bevel pinion, is riveted to the differential case. The preferred method of removing the rivets is by drilling. When re-riveting, temporarily bolt the ring gear to the differential case and use cold rivets, being careful not to distort the case in the process. Check trueness of ring gear back face by mounting the unit in its normal position in the differential compartment. Total run out should not exceed .002.

Series VA

The differential unit to which is riveted the main drive bevel ring gear is of the three pinion, open case type on models VA, VAC and VAO; two pinion open case type on

2. Shims	
3. Bearing carrier	
4. Oil seal	
5. Bearing cup & cone	
6. Side gear thrust washer	
7. Differential case	
8. Bevel ring gear	
9. Thrust washer	
11. Wheel axle shaft	
12. Differential side gear	
13. Differential spider	
14. Differential pinion	

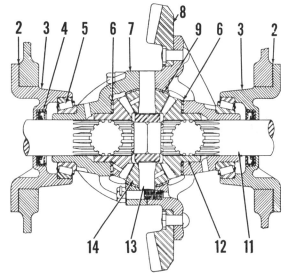

Fig C149A—Series V tractor main drive bevel ring gear and differential unit.

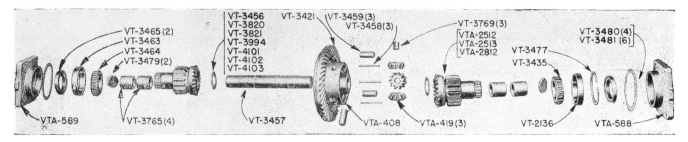

Fig. C150—Bevel ring gear, differential unit, bull pinions and components as used on models VA, VAC and VAO tractors prior 5555000.

408. Differential case	2136. Bearing cup	3456. Thrust washer	3463. Bearing cup	3479. Welch plug
419. Differential pinion	2512. Bull pinion	3457. Differential shaft	3464. Bearing cone	3480. Shims
588. Bearing carrier	3421. Bevel ring gear	3458. Retaining pin	3465. Oil seal	3765. Bushings
589. Bearing carrier	3435. Bearing cone	3459. Pinion shaft	3477. Snap ring	3769. Bushing

models VAH. Differential unit is located back of the transmission housing dividing wall. Differential carrier bearings are shim adjusted, taper roller type, mounted in carriers. Refer to Figs. C150, C150A and C150B. On models VA, VAC and VAO, carriers are covered by their respective brake units; on model VAH, the carriers are covered by their respective final drive (drop) housings.

206. BEARING ADJUSTMENT. To adjust differential carrier bearings and the ring gear backlash, proceed as outlined in paragraph 192 and 192A.

207. REMOVE AND REINSTALL. To remove the differential unit and bevel ring gear assembly, first drain the transmission and differential compartments. On models VA, VAC, and VAO, support tractor under transmission housing, and detach rear axle housing from transmission housing. On models prior 4828245, remove brake housing covers and remove nut (located on inside of brake housing) from stud attaching rear axle housing to transmission. Remove both brake assemblies from outer ends of the differential shafts.

On model VAH, support tractor under transmission housing and remove both final drive (drop) housings. Also, remove both brake assemblies, and transmission housing rear cover, or pto unit.

207A. On all models, support differential unit and remove differential bearing carriers. On models VA, VAC, and VAO, remove differential side and integral bull pinion gears. On all models, lift differential unit and bevel ring gear assembly out through rear opening in transmission housing. A groove in the transmission case will permit left end of differential shaft to be removed first.

207B. Reinstallation is reverse of removal procedure. Note on all except VAH that the left differential bearing carrier (5—Fig. C150B) which takes the ring gear thrust, contains the

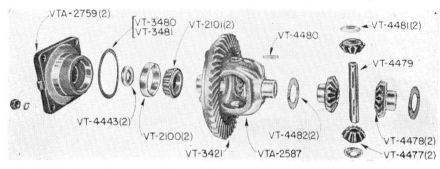

Fig. C150A—Bevel ring gear, differential unit and components as used on model VAH.

2101. Bearing cone	3421. Bevel ring gear	4477. Differential pinion	4480. Retaining pin
2587. Differential case	3480. Shims	4478. Diff. side gear	4481. Thrust washer
2759. Bearing carrier	4443. Oil seal	4479. Pinion shaft	4482. Thrust washer

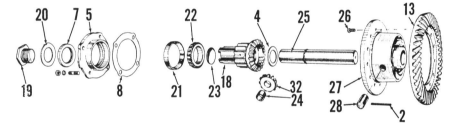

Fig. C150B—Bevel ring gear, differential unit, bull pinions and components as used on models VA, VAC, and VAO tractors after 5554999.

2. Retaining pin	13. Bevel ring gear	21. Bearing cup	25. Differential shaft
4. Thrust washer	18. Bull pinion	22. Bearing cone	26. Rivet
5. Bearing carrier	19. Brake drum retaining bolt	23. Welch plug	27. Differential case
7. Oil seal	20. Lock washer	24. Bushing	28. Pinion shaft
8. Shims			32. Differential pinion

larger (wider) bearing cup (21). The left carrier can be identified from the right carrier as the right bearing carrier has a spacer in the form of a snap ring (VT3477—Fig. C150) inserted under the bearing cup. Install both bearing carriers so that the oil drain hole is facing downward. If new oil seals (VT3465) are installed, install same with lip of seal facing the differential unit.

Adjust carrier bearings and ring gear backlash as outlined in paragraphs 192 and 192A.

208. OVERHAUL. To disassemble differential unit, bump out the pinion shaft pins (2—Fig. C150B) and remove pinion shafts which fit .001-.003 loose in their bores. The differential pinions can now be removed. Differential pinion Oilite bushings (24) used on models VA, VAC and VAO should have .003-.004 diametral clearance on the

.809-.810 diameter pinion shafts (28). On models VA, VAC and VAO units, the differential shaft (25) which is a .002-.0035 tight press fit in the differential case can be removed by pressing the shaft on the right end and out toward left end of differential case. The oil seal (7) which is located in the differential bearing carrier should be installed with the lip facing inward.

208A. The bull pinions which are integral with the differential side gears rotate on renewable bushings (VT-3765—Fig. C150) on models VA, VAC and VAO prior 5555000. Bushings require final sizing after installation to provide .0025-.004 running clearance. The bull pinions used in models VA, VAC, and VAO after 5554999 are not bushed.

208B. The bevel ring gear, which is available separately from the bevel pinion, is riveted to the differential case. The preferred method of removing the rivets is by drilling. When re-riveting, temporarily bolt the ring gear to the differential case and use cold rivets, being careful not to distort the case in the process. Check trueness of ring gear back face by mounting the unit in its normal position in the differential compartment. Total run-out should not exceed .002.

208C. Prior to reinstallation of the differential unit, adjust backlash of side gears (18—Fig. C150B) and pinions (32) by inserting a composition thrust washer (4) between side gear and differential housing to provide a .010-.015 backlash. Adjustment on one side is independent of the other side and must be made separately. Thrust washers are available in .089, .094, .099, .104, .109 and .119 thicknesses.

Reinstall bevel ring gear and differential unit as outlined in paragraph 207B.

FINAL DRIVE UNITS
(Models DCS, DS and VAH Only)

This section applies to the complete final drive (drop) units as used on models DCS, DS and VAH tractors. Refer to Fig. C151 or C156. For service information covering the wheel axle shafts and housings and drive chains located in the transmission housing as used on all other models, refer to paragraphs 225 through 241A.

Models DCS-DS

210. ADJUST WHEEL AXLE SHAFT BEARINGS. Support tractor and remove wheel and tire unit. Remove wheel axle shaft outer bearing carrier, Fig. C152. Vary the number of shims located under the outer bearing carrier to eliminate all bearing play but without binding the shaft. Removing shims reduces bearing play.

Shims located under the wheel axle shaft inner bearing carrier control the alignment of the wheel axle shaft sprocket and upper axle sprocket.

211. RENEW WHEEL AXLE BEARINGS, WHEEL AXLE, FINAL DRIVE CHAIN OR SPROCKETS. Support tractor and remove wheel and tire unit. Remove wheel axle shaft outer bearing carrier, Fig. C152. The wheel axle shaft oil seal which is located in the outer bearing carrier can be renewed at this time. Unbolt and remove the sprocket outer housing or cover, Fig. C153. Support the wheel axle shaft and sprocket assembly. Remove master link from drive chain, and re-

Fig. C151—Final drive units on models DCS and DS.

move the chain. The wheel axle shaft, wheel axle shaft outer bearing and either or both sprockets can be renewed at this time as will be seen in Fig. C153.

The wheel axle inner bearing cup can be renewed after removing the final drive housing brace and inner bearing carrier. Shims located under the inner bearing carrier, Fig. C152, control the alignment of the sprockets. Vary the shims so that both sprockets are in alignment when checked by eye or with a straightedge.

CAUTION: When reinstalling the drive chains, use new heat treated steel cotter pins for the master link. Do not use the old cotter pins a second time.

212. ADJUST UPPER AXLE SHAFT BEARINGS. Refer to Fig. C-154. Support tractor and remove wheel and tire unit. Remove wheel axle shaft outer bearing carrier. Unbolt and remove the sprockets cover or outer housing. Support the wheel axle shaft and sprocket. Remove master link from drive chain and remove chain. Remove upper axle sprocket and upper rear axle bearing locknut lock. Rotate the locknut in a clockwise direction until it is flush with the end of the collar on which it is threaded.

212A. Next, remove cotter pin from upper axle nut and tighten the nut securely. Reinstall cotter pin. Rotate the upper rear axle locknut counterclockwise until all end play is eliminated from the upper axle shaft. Con-

siderable effort must be exerted to tighten the locknut sufficiently to eliminate all of the shaft end play at which time the locknut will be tight against the inner face of sprocket. Reinstall locknut lock, sprockets, chain and housing cover.

CAUTION: Use new heat treated steel cotter pins for the master link.

213. RENEW UPPER AXLE SHAFT, BEARINGS, SPROCKET OR FINAL DRIVE HOUSING. First, follow the procedure as outlined in paragraph 212 except continue to rotate the locknut until same is removed from sleeve. Remove sprocket flange from upper axle shaft. Next, remove transmission rear cover or power take-off and lift unit from rear face of the transmission housing. Remove master link from drive chain in transmission housing and remove chain. Remove adjusting bolt, Fig. C155, and bull sprocket from shaft. Support transmission housing and final drive housing separately, and remove the inner circle of nuts retaining the final drive unit to the transmission housing. Remove final drive assembly by withdrawing it outward and away from the transmission.

Procedure for renewal of the upper axle shaft and bearings is self-evident after an examination of the unit.

213A. Reinstall final drive assembly and before tightening the assembly retaining nuts, adjust the drive chains in the transmission housing by means of the adjusting bolts, as shown in Fig. C155. To tighten the chain, rotate

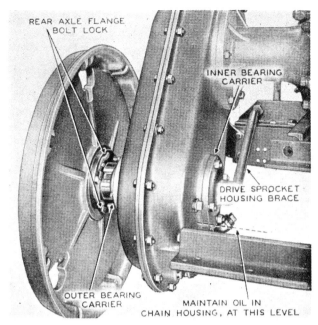

Fig. C152—Models DCS and DS final drive sprocket housing.

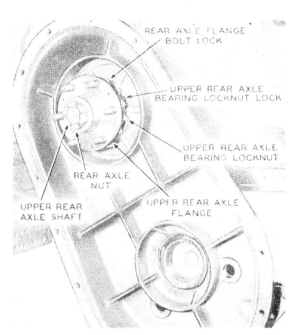

Fig. C154—Models DCS and DS final drive upper axle shaft installation.

Fig. C153—Models DCS and DS final drive sprockets and chain installation.

Fig. C155—Models DCS and DS showing drive chains installation in transmission housing when viewed from transmission rear cover opening.

the adjusting bolt in a clockwise direction. Adjustment is correct when the chain has approximately a one inch deflection. With the adjustment completed, the head of the adjusting bolt should be set square, as shown, so that the recess in the transmission rear cover will fit over the bolt head, and prevent the bolt from turning. Tighten the final drive housing to transmission housing attaching nuts, complete the reassembly of chains and wheel axle shaft; then adjust upper axle shaft bearings as outlined in paragraphs 212 and 212A.

Model VAH

215. **ADJUST WHEEL AXLE SHAFT BEARINGS.** Refer to Fig. C-156. Support tractor and remove wheel axle shaft bearing cap from inner side of final drive housing. Remove cotter pin and tighten inner bearing retaining nut (15) to eliminate all bearing play but without binding the shaft.

216. **RENEW WHEEL AXLE, BEARINGS OR BULL GEAR.** Support tractor and remove wheel and tire unit.

Remove final drive housing oil pan, and bearing retaining nut (15). Support bull gear and bump wheel axle shaft towards wheel side and withdraw axle shaft and outer bearing cone. Bull gear, take-up washer and shims will come from below. Bearing cups are positioned in housing by snap rings. Inner bearing cup can be driven out with a driver inserted from the wheel side of the housing, and the outer bearing cup with a driver inserted from the inner side.

Reassembly is reverse of disassembly. The oil seal of felt and treated

X. 1/32 inch clearance
2. Shims
6. Oil seal
7. Bull pinion shaft
8. Final drive housing
9. Brake actuating lever
11. Adjustment lock screw
12. Adjusting screw
13. Primary plate
14. Bearing carrier
16. Transmission housing
20. Differential case
20. Diff. pinion shaft
22. Differential pinion
23. Differential side gear
24. Pinion shaft
       retaining pin
26. Oil seal
27. Shims
29. Retaining pin
30. Power plate
31. Take-up washer
32. Washer
33. Shims
35. Snap ring
37. Bull gear
39. Wheel axle shaft
44. Oil seal
45. Shield
46. Snap ring

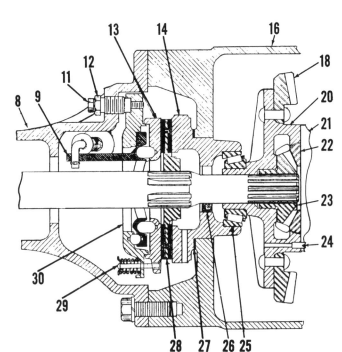

Fig. C156—Model VAH final drive (drop) housing assembly.

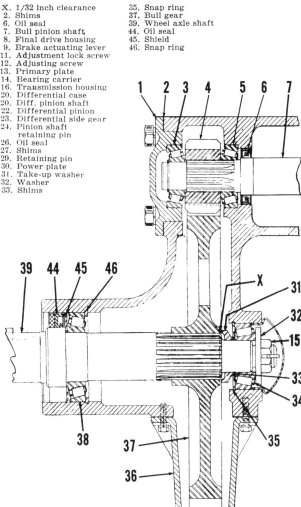

leather should be installed with the leather lip inward and the felt dirt seal out. Reinstall bull gear with long hub of same facing the wheel. Adjust the wheel axle shaft bearings with inner bearing retaining nut (15) to eliminate all bearing play but without binding the shaft. If bearing play cannot be eliminated by tightening the nut, it will be necessary to remove shim washers (33) which are located between the inner bearing cone and take-up washer (31). Next, insert a bar from lower side of final drive housing and check the bull gear for desired $\frac{1}{32}$ inch side movement (X) on the wheel axle shaft splines. If side movement of bull gear is greater than $\frac{1}{32}$, it can be reduced to this amount by adding shims (33). Recheck bearing adjustment.

217. ADJUST BULL PINION SHAFT BEARINGS. Support tractor and remove rear wheel and tire as a unit. Adjust bull pinion shaft bearings to provide zero end play and a free rolling fit with shims (2) interposed between final drive housing and

bull pinion outer bearing carrier.

218. RENEW BULL PINION, SHAFT, BEARINGS OR SEAL. To remove bull pinion shaft, first support tractor and remove wheel and tire as a unit. Lock the brake pedal in the applied position to prevent the brake middle ring (lined disc) which is splined to the inner end of the bull pinion shaft from dropping out of position. Otherwise, it will be necessary to remove the final drive (drop) housing from the transmission housing to reinstall the bull pinion shaft and brake middle ring (lined disc). Remove bull pinion shaft outer bearing carrier and shims. Remove bull pinion shaft.

Bull pinion shaft outer oil seal of leather can be renewed at this time and same should be installed with the lip facing the bull pinion. Bull pinion shaft inner oil seal which is located in the differential unit bearing carrier can be renewed after removing the final drive housing from the transmission housing. Bull pinion is splined to the bull pinion shaft and same can be pressed off of the shaft after removing

the outer bearing cone retaining snap ring.

Reassembly is reverse of disassembly. Select shims (2) between outer bearing carrier and housing to remove all bearing play without causing shaft to bind.

219. R & R ONE FINAL DRIVE (DROP) UNIT. Support tractor under transmission housing and remove wheel and tire as a unit. Remove fender and disconnect draft arm from inner side of final drive housing. Remove brake actuating rod. Support final drive housing and remove four cap screws retaining same to transmission housing. Withdraw assembly, being careful not to damage differential bearing carrier oil seal. Brake unit middle ring (lined disc), primary disc and power plate, and secondary disc (differential bearing carrier) can be renewed at this time.

DRIVE CHAINS, SPROCKETS, BULL GEARS & PINIONS

This section applies to the drive chains and sprockets located in the transmission housing on series C, D, L, LA, R and S and to the bull gears and pinions on series VA except model VAH. For service information applying to the chains and sprockets which are carried in the final drive (drop) housing on models DCS and DS or to the bull gears and pinions on model VAH, refer to paragraphs 210 through 219.

Fig. C157—Drive chains and bull sprockets installation when viewed from transmission housing rear cover opening on series R and S. Series C, D, L and LA are similar except for method of attaching the bull sprockets to the inner ends of the wheel axle shafts.

1. Special steel cotter pins   3. Master link
8. Adjusting bolt

Series C-D-L-LA-R-S

225. ADJUST DRIVE CHAINS. To adujst chains located in transmission housing on models without a power take-off unit, first remove the transmission housing rear cover; then proceed as outlined in paragraph 225C.

225A. To adjust chains located in transmission housing on models equipped with a non-continuous type power take-off unit, first remove pto output shaft and transmission housing rear cover as outlined in paragraph 275; then proceed as outlined in paragraph 225C.

225B. To adjust chains located in transmission housing on models equipped with a continuous type pto unit, first remove pto clutch and stub shaft assembly as outlined in paragraph 281. Remove clutch drum retaining snap ring shown in Fig. C157A. From models so equipped, remove the rockshaft work cylinder and hoses.

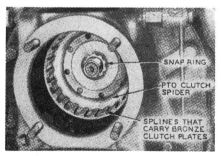

SNAP RING

PTO CLUTCH SPIDER

SPLINES THAT CARRY BRONZE CLUTCH PLATES

Fig. C157A—Continuous type pto drive shaft and drum installation when viewed from rear.

Disconnect lift links from rockshaft arms. Remove four cap screws attaching drawbar to lower side of pto and hydraulic sump housing (transmission rear cover). Lower the drawbar by loosening cap screws which attach the draft arms and pivot brackets to the drawbar frame.

On all models, remove cap screws attaching hydraulic sump housing to transmission housing. Insert two long studs into two of the retaining cap screw holes to serve as guide pins when removing the sump housing. Remove sump housing and proceed as outlined in paragraph 225C.

225C. With the transmission housing rear cover removed, jack up wheel of side to be adjusted. Loosen axle housing to transmission housing retaining nuts permitting the wheel axle housing to be shifted in its elongated holes. Turn adjusting bolt (8—Fig. C157) clockwise to shift position of axle and sleeve assembly, thereby tightening the chain. A properly adjusted chain has a deflection of approximately one inch when checked midway between the sprockets. With the chains correctly adjusted, the head of the adjusting bolt should set square

as shown, so that the recess in the rear cover will fit over head of the bolt and prevent it from turning. Retighten axle shaft housing to transmission housing retaining nuts.

226. R&R DRIVE CHAINS. To remove chains, remove transmission rear cover as outlined in paragraphs 225, 225A or 225B. Jack up rear wheel on same side chain is to be removed. Remove master link (3—Fig. C157) from drive chain. Remove the chain by threading same off the bull and pinion sprockets.

CAUTION: When reassembling drive chains, use new heat treated steel cotter pins for the master link. Do not use old cotter pins a second time. Adjust chain tension as per paragraph 225.

227. R&R BULL SPROCKET. On series C prior 4203680 and L, and models DCS and DS, the inner end of wheel axle shaft (upper axle shaft on models DCS and DS) is flanged and the bull sprocket is bolted to same as shown in Fig. C158. Sprockets on these models can be removed after removing transmission drive chains as per

2. Bull sprocket
3. Seal (felt)
13. Sleeve locknut
14. Extension spool
15. Axle flange
16. Collar

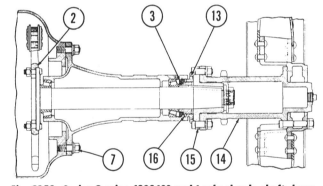

Fig. C158—Series C prior 4203680 and L wheel axle shaft, housing and bull sprocket assembly. Refer to Figs. C163 and C164 for details of axle shaft bearings adjusting sleeve.

1. Bull sprocket retaining nut
2. Bull sprocket
4. Oil seal (treated leather)
6. Oil seal retainer
8. Chain tension adjusting bolt

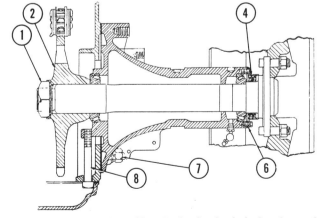

Fig. C159—Series C after 4203679 wheel axle shaft, housing and bull sprocket assembly. Wheel axle shaft bearings are adjusted by means of the bull sprocket retaining nut (1). Some series D assemblies are similar.

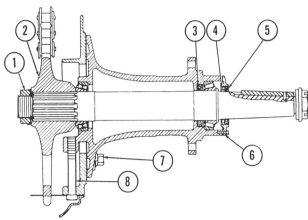

Fig. C160—Series LA wheel axle shaft, housing and bull sprocket assembly. Some series D assemblies are similar. Refer to Fig. C165 for details of the outer seal as used on LA models equipped with the 3½ inch diameter axle shafts.

1. Bull sprocket retaining nut
2. Bull sprocket
3. Seal (felt)
4. Oil seal (treated leather)
5. Seal (felt)
6. Seal retainer
8. Chain tension adjusting bolt
9. Spacer
10. Bearing cap
11. Snap ring
22. Shims

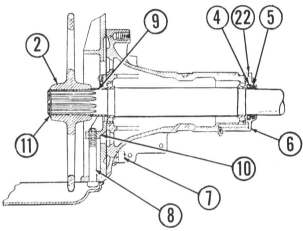

Fig. C161—Series R and S wheel axle shaft, housing and bull sproket assembly. Shims (22) control axle shaft bearings adjustment.

paragraph 226 and unbolting the sprocket from the axle shaft.

227A. On series C after 4203679-D (except models DCS and DS,) LA, R and S, inner end of wheel axle shaft is splined. Sprockets are retained in position by a large nut on all except series R and S tractors which use a snap ring. Refer to Figs. C159, C160 and C161. Removal of a bull sprocket requires removal of the chain as per paragraph 226, chain adjusting bolt and sprocket retaining nut or snap ring. Remove axle shaft and housing unit while prying the sprocket off the shaft. Remove sprocket out through transmission rear cover opening.

227B. On models which use a nut (1) to retain the bull sprocket, the nut also adjusts the wheel axle shaft bearings. Adjust wheel axle shaft bearings to eliminate all bearing play without binding the shaft.

For method of adjusting the wheel axle shaft bearings on models where the bull sprocket is retained to the axle shaft by a snap ring or where same is flanged bolted, refer to paragraphs 231 and 236.

228. R & R BULL PINION SPROCKET. Bull pinion sprockets on these models are integral with the differential unit side gears. To remove one or both bull pinion sprockets, it will be necessary to remove the differential and spur ring gear unit as outlined in paragraphs 201, 201A and 201B.

Series VA Except Model VAH

Data covered in paragraphs 229 and 230 applies to the bull pinions (integral with differential unit side gears) and bull gears on models VA, VAC and VAO. For service information on the bull pinions and bull gears located in the final drive (drop) housings on models VAH, refer to paragraphs 215 through 219.

229. R & R BULL GEAR. The rear axle housing contains the bull gears and wheel axle shafts, Fig. C162. To remove one or both bull gears, first drain transmission and rear axle housing. Remove power take-off unit and rear cover. Support transmission case and remove rear axle housing to transmission case retaining bolts. Note that upper right and lower left hand bolts are precision ground to serve as dowels.

On models prior 4828245, remove brake housing covers, and remove nut (located on inside of brake housing) from stud attaching rear axle housing to transmission case. Roll rear axle housing rearward and away from transmission housing.

229A. Remove cotter pin from final drive gear retaining nut (12). Insert a ½ inch thick steel bar between inner ends of axle shafts and back off the final drive gear retaining nut (12). This will force axle shaft out of the bull gear and axle housing.

Adjust axle shaft bearings with the bull gear retaining nut (12) to eliminate all bearing play but without binding the shaft.

230. R&R AND OVERHAUL BULL PINION. The bull pinions (18—Fig. C162) are integral with the differential unit side gears. To remove a bull pinion proceed as follows: First detach rear axle housing from transmission housing as outlined in paragraph 229. Remove brake units, and support the differential unit. Remove cap screws retaining differential bearing carrier (5) to transmission housing and remove the carrier. Withdraw the bull pinion and differential side gear.

On models VA, VAC and VAO prior 5555000 the bull pinions rotate on renewable bushings (VT-3765—Fig. C150) which require final sizing after installation to provide a .0025-.004 diametral clearance between bushings and differential shaft. Bull pinions used in models VA, VAC and VAO after 5554999 are not bushed.

Before reinstalling the bull pinion, adjust backlash between side gear and differential unit pinion (3) to .010-015 by means of the composition thrust washer (4). Thrust washers are available in .089, .094, .099, .104, .109, and .119 thicknesses. Adjustment on one side is independent of the other side and must be made separately.

230A. Reinstallation is reverse of removal procedure. Note that the left differential bearing carrier which takes the ring gear thrust contains the larger (wider) bearing cup. The left carrier can be identified from the right carrier as the right bearing carrier has a spacer in the form of a snap ring (6—Fig. C162) inserted under the bearing cup.

Install bearing carrier so that the oil drain hole is facing downward. If new oil seals (7) are installed, install same with lip of seal facing the differential unit.

Adjust differential carrier bearings and bevel ring gear backlash as outlined in paragraphs 192 and 192A.

WHEEL AXLE SHAFTS AND HOUSINGS

Series C Prior 4203680-L

231. ADJUST WHEEL AXLE SHAFT BEARINGS. On models equipped with a wheel axle extension spool (14—Fig. C158), remove rear wheel and extension spool. On series C, remove threaded sleeve lock cap screw (6—Fig. C163). On series L, remove the grease cup and slide dust shield (9—Fig. C164) toward transmission and pry up tabs of lock washer (7). Turn threaded sleeve locknut (13—Fig. C158 or 8—Figs. C163 & 164) clockwise until it is flush with the end of threaded collar (16—Fig. C158 or 10—Fig. C163) as indicated at point A in Fig. C164. Tighten wheel hub or axle flange nut to be assured wheel hub or axle flange is properly seated on axle shaft. Turn threaded sleeve locknut (8) counter-clockwise until it is tight against inner side of hub or axle flange. Considerable effort will be required to turn the

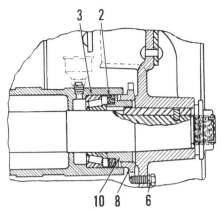

Fig. C163—Construction details of wheel axle shaft bearings adjusting sleeve as used on series C prior 4203680.

2. Seal (felt)	6. Lock bolt for
3. Outer bearing	sleeve locknut
cup & cone	8. Sleeve locknut
	10. Threaded sleeve

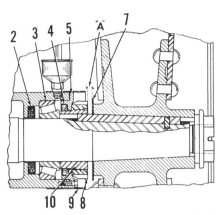

Fig. C164—Construction details of wheel axle shaft bearings adjusting sleeve as used on series L.

2. Seal (felt)	5. Seal retainer
3. Outer bearing	7. Lock washer for
cup & cone	sleeve locknut
4. Oil seal (treated	8. Sleeve locknut
leather)	9. Dust shield
	10. Threaded sleeve

threaded sleeve locknut (8) to eliminate all bearing play without binding the shaft. Lock the adjustment with cap screw (6—Fig. C163) or tab washer (7—Fig. C164) on models not equipped with a wheel axle extension spool, or with axle extension attaching bolt on models so equipped.

232. RENEW WHEEL AXLE SHAFT OR BEARINGS. Refer to Fig. C158. To renew either the axle shaft and/or bearings proceed as follows: First remove bull sprocket as outlined in paragraph 227. Remove drive chain adjusting bolt from rear face of trans-

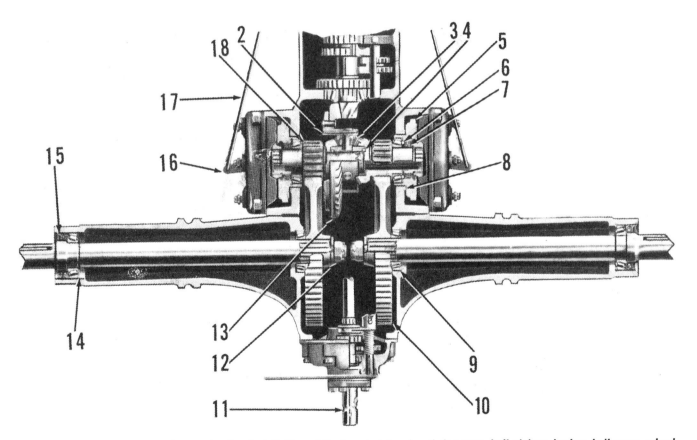

Fig. C162—Models VA, VAC and VAO showing installation of the bevel pinion, bevel ring gear, bull pinions, brakes, bull gears, wheel axle shafts, rear axle housing, and pto unit. Models after 5455000 are equipped with internal expanding shoe brakes, Fig. C179, instead of the disc type as shown.

2. Pinion shaft retaining pin	7. Oil seal	11. PTO (external shaft)	15. Oil seal
3. Differential pinion	8. Shims	12. Bull gear retaining nut	16. Brake actuating lever
4. Thrust washer	9. Snap ring	13. Bevel ring gear	17. Brake rod
5. Bearing carrier	10. Bull gear	14. Snap ring	18. Bull pinion & differential
6. Snap ring			side gear

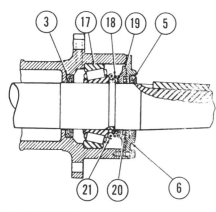

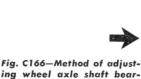

3. Seal (felt)
5. Seal (felt)
6. Seal retainer
17. Outer bearing
    cup & cone
18. Spring
19. Seal plate
20. Seal (cork)
21. Spring retainer

Fig. C165—Outer bearing and seal assembly details as used on series LA equipped with 3½ inch diameter wheel axle shafts.

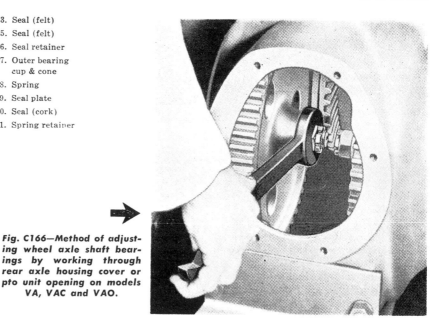

Fig. C166—Method of adjusting wheel axle shaft bearings by working through rear axle housing cover or pto unit opening on models VA, VAC and VAO.

mission housing. Remove wheel and tire as a unit, and axle housing extension spool and hub from models so equipped. Remove nuts (7) attaching axle shaft and housing to transmission housing and withdraw assembly from transmission housing.

232A. Remove wheel hub key from axle shaft. Bump wheel axle shaft on wheel hub end and remove axle shaft and inner bearing cone from housing. Inner and outer bearing cups can be removed from the housing by using a suitable puller. Felt seal (3—Fig. C158) can be renewed after removing the wheel hub or extension spool and hub from models so equipped, and threaded sleeve locknut (13) and collar (16).

Adjust wheel axle shaft bearings as outlined in paragraph 231.

Series C After 4203679-LA-D Except Models DCS & DS

233. **ADJUST WHEEL AXLE SHAFT BEARINGS.** To adjust the wheel axle shaft bearings, proceed as follows: First, remove transmission housing rear cover as outlined in paragraphs 225, 225A or 225B. Support tractor under transmission housing. Adjust axle shaft bearings by turning the bull sprocket retaining nut (1—Figs. C160 & C167B) to eliminate all bearing play without binding the shaft. Lock the adjustment by bending the tab washer.

234. **RENEW WHEEL AXLE SHAFT, BEARINGS OR SEAL.** To renew either the axle shaft or bearings (or seal on D series equipped with an axle shaft which is flanged on the outer end as shown in Fig. C167B) proceed as follows: First, remove transmission housing rear cover as outlined in paragraphs 225, 225A or 225B. Support tractor under transmission housing. Remove master link from

final drive chain, and chain tension adjusting bolt (8). Remove wheel and tire unit. Remove bull sprocket retaining nut (1) and oil seal retainer cap screws. Pry bull sprocket off of axle shaft while withdrawing the shaft. Bull sprocket will remain in transmission housing.

234A. On series D tractors equipped with an axle shaft which is flanged on the outer end, the oil seal (4—Fig. C167B can be renewed at this time. Oil seal (4—Fig. C160 or 5—Fig. C165) as used on the other models can be renewed after removing the wheel and tire as a unit, and removing the oil seal retainer.

Adjust wheel axle shaft bearings as outlined in paragraph 233.

Models DCS and DS

For information covering wheel axle shaft and upper axle shaft and bearings, refer to paragraphs 210 through 213A.

Series R-S

236. **ADJUST WHEEL AXLE SHAFT BEARINGS.** To adjust the wheel axle shaft bearings, proceed as follows: First, support tractor under transmission housing. Remove wheel and tire unit. Remove bearing retainer (6—Fig. C167A) from outer end of axle housing. Vary number of shims (22) to eliminate all bearing play, but without causing the shaft to bind.

237. **RENEW WHEEL AXLE SHAFT, BEARINGS OR SEAL.** To renew axle shaft oil seal (4—Fig.

C167A), first remove wheel and tire unit and bearing retainer (6). Oil seal is pressed into the bearing retainer and should be installed with the lip facing the bull sprocket.

237A. To renew the axle shaft, first remove rear wheel and tire unit. Remove transmission housing rear cover as outlined in paragraphs 225, 225A or 225B. Remove master link from final drive chains. Remove snap ring (11) retaining bull sprocket which is splined to the inner end of the axle shaft. Remove bearing retainer (6) from outer end of axle housing. Using a suitable puller attached to the outer end of the axle shaft or pry bar on inner end of the axle shaft, remove shaft from bull sprocket and axle housing. Inner and outer bearing cones can be removed from the axle shaft with a suitable puller. Inner bearing cup can be removed after removing the chain tension bolt (8) and inner bearing cap (10). Inner bearing cap is retained to the axle housing with cap screws.

When reinstalling the axle shaft be sure to install spacer (9) which is located between the bull sprocket hub and inner bearing cone. Adjust axle shaft bearings with shims (22) interposed between outer bearing retainer and axle housing to eliminate all bearing play but without causing the shaft to bind.

Series V

238. **ADJUST WHEEL AXLE SHAFT BEARINGS.** To adjust wheel axle shaft bearings, proceed as follows: Support tractor under trans-

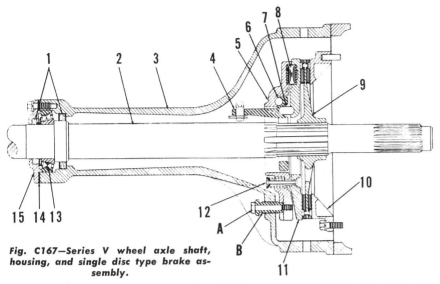

Fig. C167—Series V wheel axle shaft, housing, and single disc type brake assembly.

A. Adjustment lock screw
B. Adjusting screw
1. Seal (felt)
2. Wheel axle shaft
3. Axle shaft housing
4. Actuating lever
5. Power plate
6. Ball
7. Plunger
8. Roller and insert
9. Lined plate
10. Secondary brake plate
11. Primary plate
12. Retaining pin
13. Bearing cup & cone
14. Shims
15. Bearing retainer

axle shaft bearings as per paragraph 238.

Series VA Except Model VAH

240. ADJUST WHEEL AXLE SHAFT BEARINGS. To adjust wheel axle shaft bearings, first support tractor under transmission housing. Remove wheel and tire unit. Remove pto unit or cover from rear face of rear axle housing. Remove bull gear retaining nut cotter pin and rotate the nut as shown in Fig. C166 to eliminate all bearing play but without causing the shaft to bind.

241. RENEW WHEEL AXLE SHAFT, BEARINGS OR SEAL. Wheel axle shaft oil seal can be renewed after removing the wheel and tire unit. Install the new seal with the lip of same facing the bull gear.

241A. Wheel axle shaft can be removed without removing the axle housing but to facilitate entering shaft splines into bull gear it is advisable to R & R the axle housing. After detaching the rear axle housing from the transmission housing as outlined in paragraph 229, support the rear axle housing and remove the wheel and tire unit. Remove bull gear retaining nut cotter pin. Insert a ½ inch thick steel bar between inner ends of the axle shafts. Using a wrench, back off the bull gear retaining nut. The bull gear retaining nut will react against the steel bar and push the axle shaft out of the bull gear and housing. A similar method can be used to remove the other axle shaft.

The inner and outer bearings can be renewed at this time. Inner and outer bearing cups are positioned in the

mission housing and remove both rear wheel and tire units. Remove shims (14—Fig. C167) located between bearing retainer (15) and outer end of axle shaft housing, until both axle shafts rotate in the same direction when one shaft is rotated; then add shims (14) until the axle shafts rotate in opposite directions. This procedure will provide a slight clearance between inner ends of the axle shafts. Zero clearance may cause the inner ends of the axle shafts to weld together. Excessive clearance will result in damaged shaft oil seals.

239. RENEW WHEEL AXLE SHAFT, BEARING OR SEAL. Axle shaft outer seal of felt can be renewed

after removing the wheel and retainer.

239A. Wheel axle shaft and bearings or inner felt seal can be renewed without removing the axle housing. Lock brake pedal in the applied position so as to prevent the brake middle ring (lined disc) which is splined to the axle shaft from dropping out of position. If brake middle ring falls out of position, it will be necessary to remove the axle housing to install the axle shaft. Remove bearing retainer (15). With a suitable puller attached to the outer end of the axle shaft, remove shaft from housing. Inner felt seal can now be renewed.

Lubricate axle shaft bearing before installing shaft in housing. Adjust

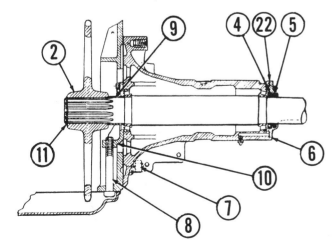

Fig. C167A—Series R and S wheel axle shaft assembly.

2. Bull sprocket
4. Oil seal (treated leather)
5. Seal (felt)
6. Bearing and seal retainer
8. Chain tension adjusting bolt
9. Spacer
10. Bearing cap
11. Snap ring
22. Shims

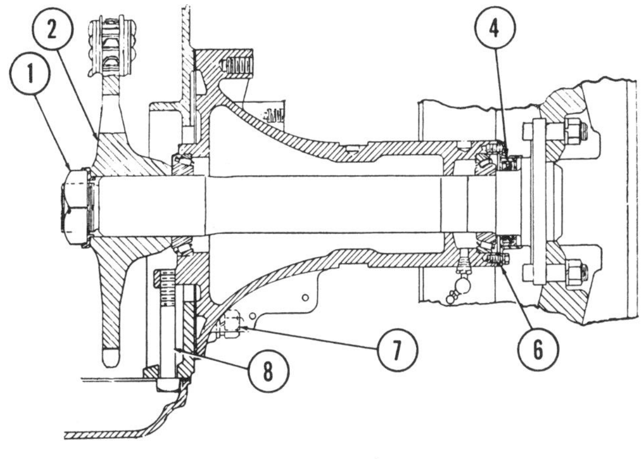

Fig. C167B—Series C after 4203679 and D wheel axle shaft, housing and bull sprocket assembly. Wheel axle shaft bearings are adjusted by means of the bull sprocket retaining nut (1).

1. Bull sprocket retaining nut
2. Bull sprocket
4. Oil seal (treated leather)
6. Oil seal retainer
8. Chain tension adjusting bolt

housing by snap rings. Removal of the cups can be accomplished with a suitable puller. Outer bearing cone can be pressed off of the shaft.

After reinstalling the axle shaft, adjust the bearings by tightening the bull gear retaining nut to eliminate all end play but without causing the shaft to bind.

Model VAH

Refer to paragraphs 215 through 219 for service information covering the final drive (drop) units.

BRAKES
CONTRACTING BAND TYPE
Series C-D Prior 4511449-L-LA-R

These service instructions apply to single hand operated installations where the drum is located on the outer end of the transmission 2nd reduction shaft as shown in Fig. C168. They also apply to the foot operated brakes where the drums are mounted externally on separate rotating brake shafts, the inner ends of which are geared to the transmission differential shaft as shown in Figs. C168A and C169.

245. **ADJUST HAND BRAKE.** To adjust the single hand brake, Fig. C168, loosen locknut; then rotate adjusting nut (6) in a clockwise direction to restore effective travel of hand lever. Lining to drum clearance should not be less than .010.

245A. **ADJUST FOOT BRAKES.** To adjust foot brakes, Fig. C168A, remove clevis pin from adjusting screw (7) then rotate adjusting screw in a clockwise direction to restore effective pedal travel. Lining to drum clearance should not be less than .015.

246. **R&R AND OVERHAUL.** The method of removing the drums is self-evident. However, in reinstalling the drum on hand brake installation, care should be exercised to prevent brinneling the roller bearings at the opposite end of 2nd reduction shaft. One method is to use a long cap screw and washer to pull the drum into place without drifting. In the case of the foot operated brakes, the reinstallation of the drum in some cases may necessitate removing the steering gear housing (which forms the transmission top front cover) to buck-up the gear end of the brake gearshaft. Another method would be to tack weld a rod to the outer end of the gearshaft to buck-up from the outside.

246A. Removal of the brake gearshaft necessitates removal of the

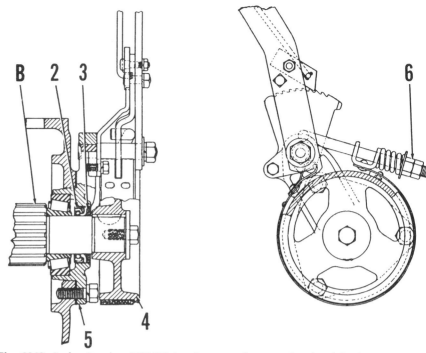

Fig. C168—Series D prior 4511449 hand operated contracting band brake which is located on the outer end of the transmission second reduction shaft. Series C, L, LA and R are similar.

B. Second reduction shaft
2. Oil seal (treated leather)
3. Seal (felt)
4. Brake drum
5. Bearing carrier
6. Adjusting rod nut

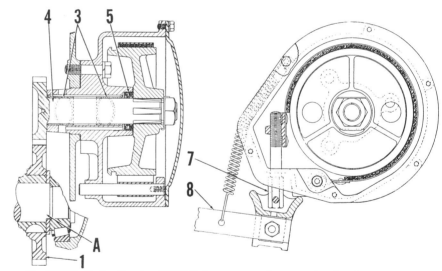

Fig. C168A—Series D prior 4511449 foot operated contracting band type brakes. Refer to Fig. C168.

A. Differential shaft
1. Drive gear
3. Bushings
4. Brake drive gear & shaft
5. Oil seal (treated leather)
7. Adjusting bolt
8. Brake pedal

transmission top front cover. Gearshaft bushings (3) should be sized after installation to provide .004-.006 running clearance. If end play of gearshaft exceeds 3/64 inch, install a shim washer (X—Fig. C169) to reduce the end play to not less than .005.

INTERNAL SHOE TYPE
Series L-LA Prior 5208165 (Mechanical Type)

The brakes as used on these models are of the mechanically actuated in- ternal expanding shoe type with the brake drums being integral with the rear wheel and hub. Refer to Fig. C170.

247. **ADJUSTMENT.** The Owners' Book method of adjusting the brakes is by shortening the pull rods. This method will eliminate excessive free travel of the brake pedals but in some cases the lever angles resulting from shortening the pull rods produces a "hard pedal". On these models, the

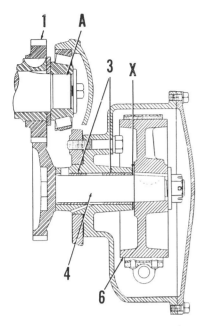

Fig. C169—Series C foot operated contracting band brakes.

A. Differential shaft
X. Shims
1. Drive gear
3. Bushing
4. Brake drive gear & shaft
6. Brake drum

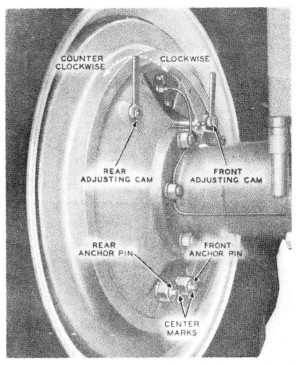

Fig. C171—Brake backing plate (left wheel) on series LA equipped with hydraulic brakes.

brake camshaft levers can be reset on the serrated camshafts to restore the original effectiveness providing the lining is not worn out. To make this adjustment proceed as follows:

Remove clevis pin from pull rod. Scribe or pencil a correlation mark on the outer end of the camshaft and the camshaft lever. Using the correlation marks as a guide, reinstall the lever to a new position on serration rearward from its former position. Pull forward on lever until lining drags on drum; then rearward until drum just rotates freely. Holding camshaft in this position, adjust length of pull rod until clevis pin enters freely without moving the pedal or the camshaft. Apply the brake heavily and note angle of camshaft lever the rod end of which should be slightly

(5-10 degrees) to the rear of perpendicular. If rod end of lever is forward of perpendicular, it should be repositioned on the camshaft one serration farther to the rear and the pull rod readjusted. Do the same to the other brake.

248. R & R AND OVERHAUL. Removal and overhaul of the brake shoes is self-evident after removing the rear tire and wheel unit.

Series LA 5208164-5418707 (Hydraulic Type)

The brakes as used on this model are of the hydraulically actuated in- *ternal expanding shoe type with the brake drums being integral with the rear wheel and hub. Refer to Fig. C172. The two hydraulic master cylinders are located on the right side of the tractor and slightly to the rear of the wheel axle shaft housing.*

249. MINOR ADJUSTMENT. To compensate for lining wear, first jack-up and support rear of tractor. Located on inner face of brake backing plate are two adjusting cams (eccentric shoe stops), as shown in Fig. C171. Rotate front cam in a clockwise direction until a slight drag is obtained when rotating the wheel; then back off the cam until the wheel revolves without

Fig. C170—Rear wheel mechanical brakes as used on series L and some LA tractors.

Fig. C172—Rear wheel hydraulic brakes as used on some series LA tractors.

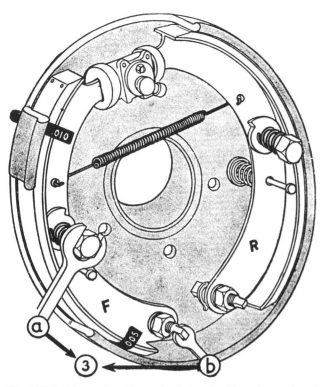

Fig. C172A—X-ray view from the brake backing plate side of the hydraulic type rear wheel brakes as used on series LA tractors. (Courtesy, United States Asbestos Div. of Raybestos-Manhattan, Inc.)

1. Boot
2. Piston
3. Cup
4. Spring
5. Wheel cylinder
6. Link

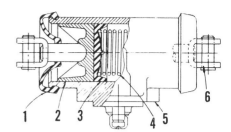

Fig. C174—Series LA hydraulic brakes wheel cylinder.

drag. Adjust the rear shoe cam in a similar manner only rotate the rear cam in a counter-clockwise direction to decrease the drum to lining clearance. Do the same to the other brake.

BLEEDER TUBE

Fig. C173—Installation of bleeder hose for purging the hydraulic brake system on series LA tractors.

250. **MAJOR ADJUSTMENT.** To adjust both the anchor pins which will be necessary only when the brakes are being relined, proceed as follows: First jack-up and support rear of tractor. Rotate both the front and rear cams (eccentric shoe stops) in a direction so as to move the shoes away from the drum. Rotate the front and rear anchor pins until marks (punch mark on end of pin) are toward the center or facing each other. Remove wheel (drum) and tire unit. Using a brake shoe centering caliper or a ring gauge, rotate the eccentric anchor pin until .004 clearance exists between the ring gauge or caliper and the heel or anchor pin end of the shoe 1½ inches from lower end of the lining. Do the same to the other shoe on the same wheel. Tighten the anchor pin lock nuts. Now, turn the adjusting cam (eccentric shoe stop) to move the shoe outward until .009 of clearance exists between the ring gauge or caliper and the toe or cam end of the shoe. The amount of clearance at heel and toe ends is not as important as obtaining the two to one ratio. Do the same to the other shoe on the same wheel. Recheck the anchor pin adjustment. Reinstall the wheel (drum) and tire unit.

250A. In an emergency when neither the ring gauge or the brake shoe centering caliper is available, the

anchor pin major adjustment can be made by referring to Fig. C172A (which is an X-ray view from the backing plate side of the brake) and proceeding as follows: With one wrench (a) on cam stop and another wrench (b) on the anchor turn each one alternately in direction of arrows until the shoe is against the drum with the same heavy drag at heel and toe. Several trials may be necessary before this condition is obtained. When position is found where both ends of lining have equal drag against the drum lock the anchor and back off the cam stop until wheel is just free. Do the same to the other 3 shoes of brake system.

251. **H Y D R A U L I C   S Y S T E M BLEEDING.** The presence of air in the system will be evidenced by a "spongy" pedal. To bleed one of the two separate systems, remove the cap screw and attach the bleeder hose as shown in Fig. C173. Loosen the bleeder screw and operate the brake pedal while the hose drains into a container which is partially filled with brake fluid. The master cylinder reservoir must be kept full of brake fluid during the bleeding operation or air may again enter the system. Continue to depress the pedal until there is a continuous flow of brake fluid without air bubbles. Tighten the bleeder screw and reinstall the cap screw. Recheck fluid supply in the master cylinder reservoir.

252. **R & R BRAKE SHOES.** Brake shoe removal is self-evident after removing the wheel and tire unit.

253. **R & R WHEEL CYLINDER.** The wheel cylinder can be removed after disconnecting the fluid supply line, removing the wheel and tire unit, and two attaching cap screws from rear face of brake backing plate.

253A. **OVERHAUL WHEEL CYLINDER.** Disassembly is self-evident after an examination of the unit and with reference to Fig. C174. After disassembling the wheel cylinder, clean all internal parts in an alcohol bath. If cylinder has been leaking, install

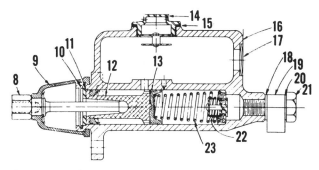

Fig. C175—Series LA hydraulic wheel brakes master cylinder.

8. Piston push rod	13. Cup	18. Gasket
9. Boot	14. Filler plug	19. Fitting
10. Stop wire	15. Gasket	20. Gasket
11. Stop washer	16. Reservoir	22. Valve assembly
12. Piston	17. Welch plug	23. Spring

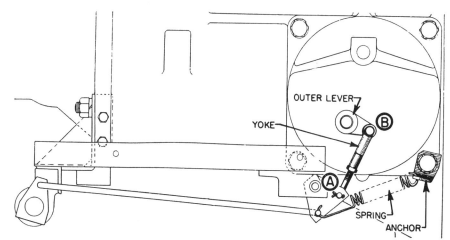

Fig. C176—On series LA equipped with hydraulic type rear wheel brakes, adjust the master cylinder piston-to-pedal push rod to provide ½ inch free travel (at pad end of pedal) before push rod contacts the piston.

new cups regardless of their good appearance. If cylinder wall is pitted, rusted or scratched, hone same but if walls do not clean up at less than .010 oversize, use a new cylinder.

254. R & R MASTER CYLINDER. The two master cylinders (one for each wheel brake unit) are mounted side by side and are located on the right side of the tractor and slightly to the rear of the wheel axle shaft housing. Removal procedure is self-evident after an examination of the unit.

Reinstall master cylinder and adjust pedal to cylinder linkage to assure the master cylinder piston returning to its stop thereby uncovering the relief ports. Adjust the linkage as follows: Referring to Fig. C176, loosen piston push rod lock nuts and rotate the piston rod to provide the pedal with ½ inch free travel (measured at foot pad end) before piston rod contacts the piston.

254A. OVERHAUL MASTER CYLINDER. Refer to Fig. C175. After disassembling the master cylinder, clean all internal parts with alcohol. If cylinder has been leaking, install new cups and valve regardless of the good appearance of these parts. If cylinder wall is pitted, rusted or scratched, hone same but if walls do not clean up at less than .010 oversize, use a new cylinder.

Some VA, VAC and VAO (Refer also to paragraph 261)

The brakes used on these models are of the mechanically actuated internal expanding shoe type. Brake drums are located on the outer ends of the bull pinion shafts. Refer to Figs. C177, C178, C179, and C180.

255. MINOR ADJUSTMENT. Refer to Figs. C177 and C178. To eliminate excessive free travel of the brake

pedal, shorten the brake rod or yoke clevis (B). Equalize the brakes by lengthening the rod on the tight brake. Recommended free travel of pedal is one inch.

Continued adjustment of the brake rods eventually results in bad lever angles and a "hard pedal". Thus, whenever the outer lever becomes perpendicular, or more than perpendicular with the brake rod or yoke, it is time to perform a major adjustment as outlined in paragraph 255A.

255A. MAJOR ADJUSTMENT. Disconnect the outer lever from its rod or yoke, and remove four cap screws and the brake housing. Housing is doweled and must be removed carefully. With brake cover and shoe unit on the bench, remove the shoe springs and the shoe. Referring to Fig. C179, remove the two nuts from the adjustable end plate and extend the shoe by moving the plate up one hole. Reassemble the shoe and reinstall the brake unit to the tractor. The shoe lin-

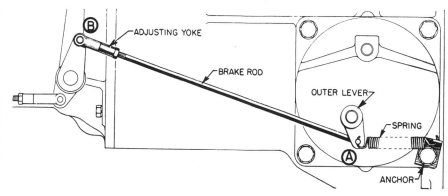

Fig. C177—Brake linkage for internal expanding shoe type brakes as installed on Model VAO tractors after 5455000.

Fig. C178—Brake linkage for internal expanding shoe type brakes as installed on some Models VA and VAC tractors.

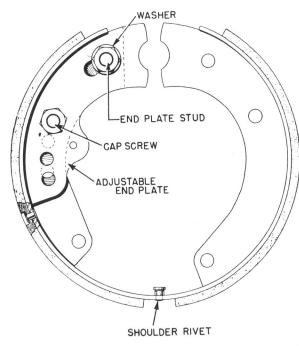

Fig. C179—Internal expanding shoe type brakes as used on some VA, VAC and VAO tractors.

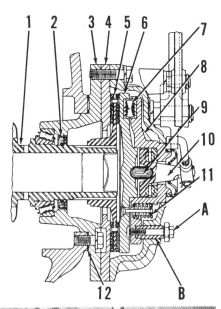

Fig. C181—Single disc type brakes as used on series D 4511449 - 5601107. Roller type inserts (7) are used on tractors prior 5201271.

A. Adjustment lock screw
B. Adjusting screw
1. Bull pinion sprocket
2. Oil seal
3. Bearing carrier
4. Brake plate (stationary)
5. Lined plate
6. Primary plate
7. Insert (roller type)
8. Power plate
9. Actuator plunger
10. Actuating lever
11. Retaining pin
12. Shims

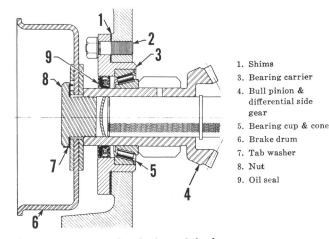

1. Shims
3. Bearing carrier
4. Bull pinion & differential side gear
5. Bearing cup & cone
6. Brake drum
7. Tab washer
8. Nut
9. Oil seal

Fig. C180—Cross-sectional view of brake drum installation on some VAC and VAO tractors.

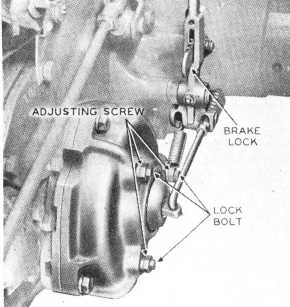

Fig. C182—Single disc type brake on series D tractors 4511449-5601107.

ing must not drag on the drum at this time.

Using a long wrench (14-20 inches) on the lever, apply the brake heavily and note angle of outer lever which should be not more than perpendicular (right angles) with yoke or brake rod. If applied position is past the perpendicular, the unit should be removed and the adjustable end plate moved into the last hole. Release the brake and with pedal in released position, adjust the rod or yoke length until clevis pin will enter freely without moving the outer lever. Equalize and synchronize the brakes by lengthening the rod on the tight brake.

256. R & R AND OVERHAUL. The brake cover and shoes unit can be removed after disconnecting the brake rod and removing the 4 cap screws retaining the cover to the transmission. The cover is doweled for positive alignment and should be removed carefully.

To remove the drum, unlock and remove the large hexagon nut, Fig. C180, from the differential shaft. Use at least 24 inches of leverage when tightening the drum nut. The oil seal can be renewed at this time without removing the differential carrier although the J. I. Case company recommends removal of the carrier.

SINGLE DISC TYPE
Series D 4511449-5601107

The foot operated brakes as used on this model are of the single disc type and are mounted on the outer

ends of the bull pinion sprockets as shown in Fig. C181.

257. ADJUSTMENT. To adjust brakes, jack-up and support rear of tractor. Loosen the three adjustment lock screws (A). Rotate the three adjusting screws (B) in a clockwise direction until same are fully bottomed. This is to assure parallelism or full contact between the primary plate (6) and lined plate (5). Then, back off each adjusting screw (B) an equal amount of approximately $\frac{1}{4}$ to $\frac{1}{2}$ of a turn or until all drag is eliminated when rotating the wheel. Lock the adjustment by tightening the lock screws (A). Equalize the brakes by backing off all of the adjusting screws on the tight brake.

258. R & R AND OVERHAUL. To remove the lined plate (5), first disconnect brake pedal linkage from the actuating lever (10). Remove four cap screws retaining brake housing cover (power and primary plates assembly) to transmission housing and remove the assembly. The lined plate which is splined to the bull pinion sprocket shaft can now be removed.

Early production D series disc brakes (tractors prior 5201271) are equipped with roller type inserts (7). Later production brakes (tractors after 5201270) are equipped with ball type inserts as shown in Fig. C183 which are also recommended for servicing the roller type. To make the change-over, disassemble the power and primary plates assembly by compressing the three springs and removing the three retaining pins (11). Remove the roller inserts and install the new ball inserts. Reassemble the power and primary plates assembly and check the actuating lever for travel as shown in Fig. C184. If there is more than 1/16 inch travel, measured at the tip of the lever, disassemble the power and primary plates assembly, and in-

sert shims between the actuating lever washer and power plate.

Series S Prior 5600001

The single disc foot operated brakes used on this model are mounted on the outer ends of the bull pinion sprockets as shown in Figs. C185 and C186.

259. ADJUSTMENT. To adjust brakes, jack-up and support rear of tractor. Loosen the three cap screws (D) which lock the power plate (11). Rotate the adjusting pinion (C) in a clockwise direction until a slight drag

Fig. C183—Ball type inserts as used between the primary and power plates on series D single disc type brakes.

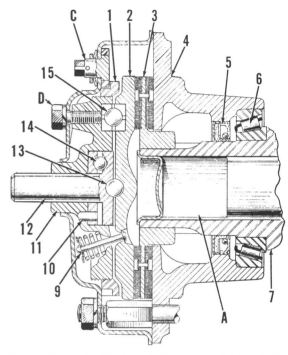

Fig. C185—Single disc type brakes as used on series S prior 5600001.

A. Differential shaft	3. Lined plate	10. Actuator lock pin
C. Adjusting pinion	4. Bearing carrier	11. Power plate
D. Adjustment lock	5. Oil seal	12. Actuator
screw	6. Bearing cup & cone	13. Ball (½ inch)
1. Adjusting ring	7. Bull pinion sprocket	14. Ball (⅜ inch)
2. Primary plate	9. Retaining pin	15. Ball and insert

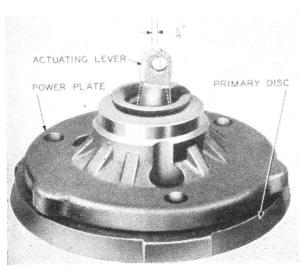

Fig. C184—Checking the free movement of the actuating lever on series D single disc type brakes.

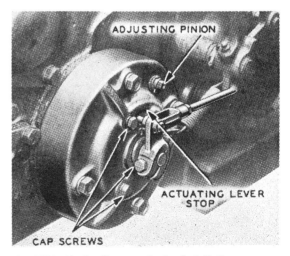

Fig. C186—Single disc type brake installation on series S prior 5600001.

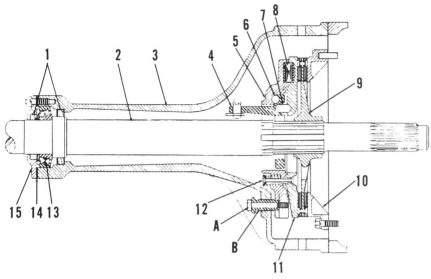

Fig. C187—Series V wheel axle shaft, housing, and single disc type brake assembly.

A. Adjustment lock screw
B. Adjusting screw
1. Seal (felt)
2. Wheel axle shaft
3. Axle shaft housing
4. Actuating lever
5. Power plate
6. Ball
7. Plunger
8. Roller and insert
9. Lined plate
10. Secondary brake plate
11. Primary plate
12. Retaining pin
13. Bearing cup & cone
14. Shims
15. Bearing retainer

is obtained when rotating the wheel; then back off the adjusting pinion one full turn. Recheck the adjustment and lock same by tightening the three cap screws (D). Do the same to the other brake. Equalize the brakes by backing off the adjusting pinion on the tight brake.

260. **R & R AND OVERHAUL.** To remove lined plate (3—Fig. C185), first disconnect the brake pedal link-

age from the actuating lever. Remove four nuts retaining the brake housing cover (primary plate, adjusting ring, and power plate assembly) to the transmission housing. The lined plate which is splined to the bull pinion sprocket shaft can now be removed.

Disassembly of the primary and power plates assembly is self-evident after an examination and with reference to Fig. C185.

**Some VA, VAC, VAH and VAO
(Refer also to paragraph 255)**

The brakes as used on these models are of the single disc type. On series V, the brake units are located on and are splined to the inner ends of the wheel axle shafts as shown in Fig. C187. On models VA, VAC, VAO prior 5455001, the disc type brake units are mounted on the outer ends of the bull pinion shafts as shown in Fig. C188. On models VAH, the disc type brake units (similar to the units shown in Fig. C188) are located on and are splined to the inner ends of the bull pinion shafts which are located in the final drive (drop) housings as shown in Fig. C156.

261. **ADJUSTMENT.** To adjust brakes, jack-up and support rear of tractor. Loosen the three adjustment lock screws (A). Rotate the three adjusting screws (B) in a clockwise direction until same are fully bottomed. This is to assure parallelism or full contact between the primary plate (11) and lined plate. Then, back off each adjusting screw (B) an equal amount of approximately one full turn or until all drag is eliminated. Lock the adjustment by tightening the lock screws (A). Equalize the brakes by backing off all of the adjusting screws on the tight brake.

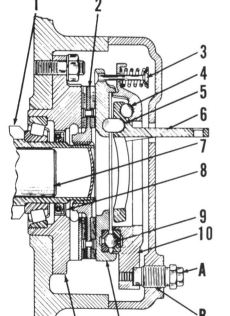

A. Adjustment lock screw
B. Adjusting screw
1. Bull pinien
2. Lined plate
3. Retaining pin
4. Ball
5. Plunger
6. Actuating lever
7. Differential shaft
8. Oil seal
9. Ball and insert
10. Power plate
11. Primary plate
12. Bearing carrier

Fig. C188—Single disc type brakes as used on some VA, VAC, and VAO tractors. Model VAH is similar except that the brake units are located on the inner ends of the final drive bull pinion shafts.

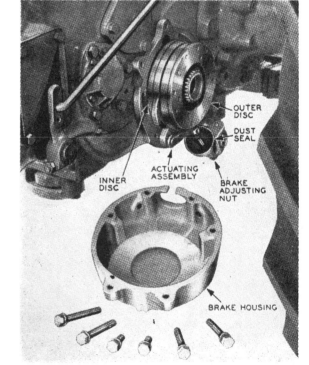

Fig. C189—Double disc type brake installation with brake housing removed on series D after 5601106. Series LA and S which are equipped with the double disc type brakes are similar.

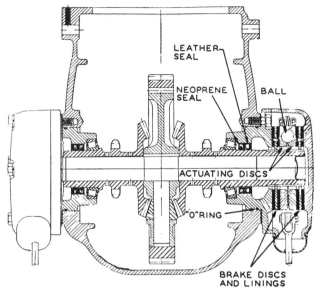

Fig. C190—Series S after 5600000 differential, spur ring gear, bull pinion sprockets and double disc type brakes installation.

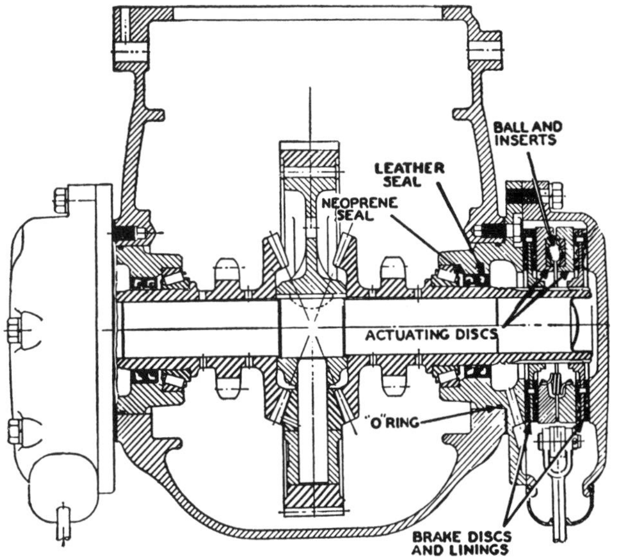

Fig. C190A—Series D after 5601106 differential, spur ring gear, bull pinion sprockets and double disc type brakes installation. Series LA after 5418707 equipped with double disc brakes is similar.

262. REMOVE & REINSTALL. To remove the lined plate on series V, first disconnect the brake pedal linkage from the brake actuator. Support rear of tractor and remove tire and wheel as a unit. Remove wheel axle shaft housing to transmission housing retaining nuts and withdraw wheel axle shaft and housing assembly. Lined plate can now be removed from the axle shaft splines.

263. On models VA, VAC and VAO, the lined plate can be removed after removing the four cap screws attaching the brake housing to the transmission housing.

264. On model VAH, the lined plate can be removed after removing the final drive (drop) housing assembly.

265. OVERHAUL. Disassembly and overhaul of the power and primary plates assembly is self-evident after an examination of the unit and with reference to Figs. C156, C187 and C188.

DOUBLE DISC TYPE
Series D After 5601106-LA After 5418707-S After 5600000

The double disc type brakes as used on these models are mounted on the outer ends of the bull pinion sprockets as shown in Figs. C189, C190, and C190A.

266. ADJUSTMENT. To adjust brakes, first jack-up and support rear of tractor. Turn adjusting nut, Fig. C189 in a clockwise direction until a slight drag is obtained when rotating the wheel; then back off the adjusting nut ¾ of a turn. Do the same to the other brake. Equalize the brakes by backing off the adjusting nut on the tight brake.

267. R & R AND OVERHAUL. To remove the lined plate, first slide the dust seal, Fig. C189, free of its groove in the brake housing. Remove six cap screws attaching the brake housing to the transmission housing, and remove the housing. Remove the brake adjusting nut. The two lined plates and actuating plates can now be removed from the splined bull sprocket shaft.

Disassembly of the actuating plates is self-evident after an examination of the unit.

Install the lined plates with the long part of the hub facing toward each other.

BELT PULLEY UNIT
Series C-D-L-LA-R-S

269. The belt pulley on these models is mounted on the right outer end of the transmission first reduction (belt pulley) shaft. Removal of the pulley is self-evident after an examination of the unit. For service information covering the transmission first reduction (belt pulley) shaft, refer to paragraphs 166 through 166E.

Series V-VA

The belt pulley unit, Fig. C191, as used on all series V tractors and on VA series with engine mounted hydraulic pump is installed on the right side of the torque tube. On VA series tractors equipped with a Kingston hydraulic pump, the belt pulley unit and the hydraulic pump are mounted on the torque tube. In all installations the pulley unit is driven from a bevel gear located on the clutch shaft.

270. GEAR ADJUSTMENT. Recommended belt pulley gear backlash of

1. Gear retaining nut
2. Pulley shaft bevel gear
3. Oil shield
4. Set screw
5. Pulley shaft sleeve
6. Snap ring
7. Seal (felt)
8. Oil seal (treated leather)
9. Bearing cup & cone
10. Pulley shaft
11. Snap ring
12. Bearing cup & cone

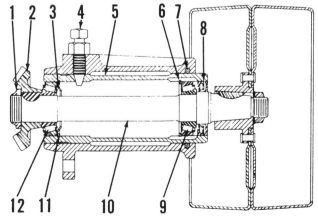

Fig. C191—Belt pulley unit as used on series V and VA tractors which are not equipped with the Kingston hydraulic piston type pump and valves unit.

.004-.008 (equal to .003-.007 when measured at rivet heads in belt pulley hub) is controlled by shims located between pulley housing and torque tube. Mesh adjustment or mesh pattern of the bevel gears is controlled by shims (10—Fig. C102) which are located between the clutch shaft rear bearing cone (26) and pulley drive gear (11).

270A. To check the bevel gears for correct mesh, first remove the steering gear housing, or steering shaft support from models so equipped. Adjust backlash to .004-.008 with shims located between pulley housing and torque tube. Looking through opening in top of torque tube, observe mesh position of the bevel gears. The heel faces of both should be flush with each other within .005 when the backlash is within the limits of .004-.008. If heel faces are not within .005 of being flush with each other after correct backlash has been obtained, it will be necessary to change the position of the clutch shaft bevel gear by removing or adding shims. If this must be done, refer

to paragraph 148A for method of removing the clutch shaft.

271. **R&R AND DISASSEMBLE.** To remove belt pulley unit, first remove the combined hydraulic pump and valves unit attached to the belt pulley housing from models so equipped. On all models remove four cap screws retaining belt pulley housing to torque tube and withdraw the unit.

271A. Disassemble belt pulley unit by removing bearing adjusting nut (1 —Fig. C191 or C192) and bumping belt pulley shaft toward outer end of belt pulley housing. Outer bearing cone will remain on belt pulley shaft. Inner and outer bearing cups will remain in pulley housing. Install pulley shaft oil seal with lip of same facing inward.

Adjust belt pulley shaft bearings with adjusting nut (1) to provide zero end play and yet permit shaft to rotate freely. Stake ends of nut into grooves of shaft when adjustment is complete.

Reinstall belt pulley unit to torque tube, and vary the number of shims,

located between pulley housing and torque tube to provide .004-.008 backlash (equal to .003-.007 when measured at rivet heads in belt pulley hub) between gear teeth.

271B. For removal of the pulley unit drive gear which is keyed to the engine clutch shaft, refer to paragraph 148A.

POWER TAKE-OFF

This data applies only to the power take-off unit external output shaft used on models with non-continuous pto, or to both the external output shaft, pto clutch and pto drive shaft used on models with continuous pto. Refer to paragraph 165 and 165B for continuous type drive gear and sleeve service data. Also, refer to paragraphs 291 through 304B for service data on the hydraulic pump and valves units.

NON-CONTINUOUS TYPE PTO

Series C-D-L-LA-R-S

There are detailed differences in the construction of the pto used on series C, D, L, and LA as compared to series R & S tractors. The accompanying illustrations, Figs. C193, C194, C195 and C196 show any necessary variations in repair procedure. The power take-off unit is placed into operation by meshing the power take-off gear with a pinion on the clutch (transmission bevel pinion) shaft.

275. **R&R AND OVERHAUL.** The power take-off external shaft can be removed as follows: Remove steering gear housing (transmission top cover) and clutch compartment hand hole covers. Working through the clutch compartment hand hole cover opening, remove power take-off shaft end cover (12—Fig. C194). Remove nut (29— Fig. C193) or snap ring (5—Fig. C194) from forward end of pto shaft. From rear face of transmission housing rear cover, remove pto shaft bearing carrier or retainer (1) and shims.

On models equipped with a mechanical type lift, unbolt same from transmission housing rear cover. On models equipped with a hydraulic type lift, remove hydraulic pump and valves unit as per paragraph 292. Working through the pump and valves unit mounting opening, disconnect pump

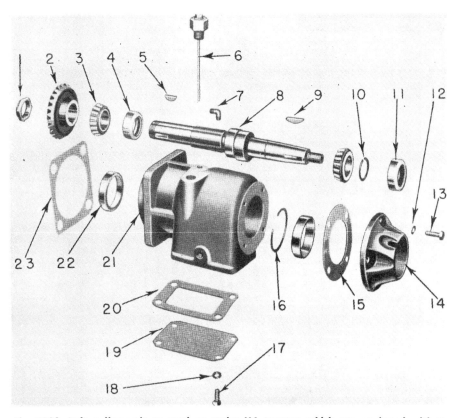

Fig. C192—Belt pulley unit as used on series VA tractors which are equipped with a Kingston hydraulic pump (piston type) and valves unit. Cams on pulley shaft (8) operate the hydraulic pump.

1. Gear retaining nut	8. Pulley shaft	15. Gasket
2. Pulley shaft bevel gear	9. Woodruff key	16. Snap ring
3. Bearing cone	10. Snap ring	20. Gasket
4. Oil seal	11. Oil seal (treated	21. Pulley shaft housing
5. Woodruff key	leather)	22. Bearing cup
6. Bayonet gauge	14. Oil seal retainer	23. Shims

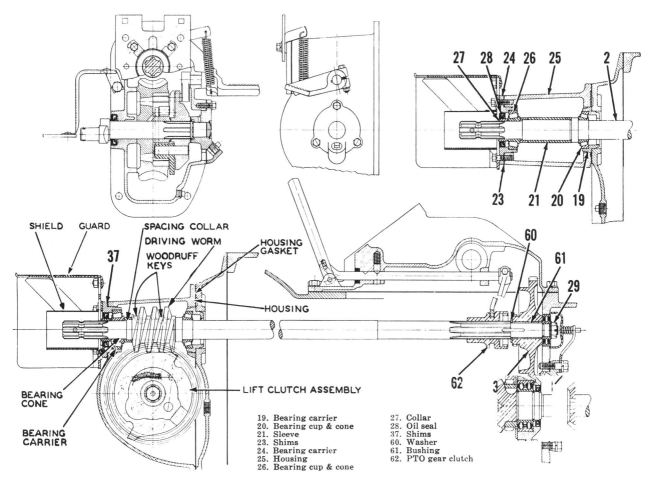

Fig. C193—Series D non-continuous type power take-off shaft installation, and mechanical lift unit. Upper right view shows a pto shaft which is designed for the installation of a mechanical lift unit. Refer to Fig. C194, for a view of a plain shaft, or Figs. C195 & C196 for a view of one with a hydraulic lift unit. Series C, L and LA are similar.

19. Bearing carrier
20. Bearing cup & cone
21. Sleeve
23. Shims
24. Bearing carrier
25. Housing
26. Bearing cup & cone
27. Collar
28. Oil seal
37. Shims
60. Washer
61. Bushing
62. PTO gear clutch

1. Bearing retainer
2. PTO shaft
3. PTO shaft gear
4. Bearing
5. Snap ring
6. Collar
7. Bearing
8. Snap ring
9. Sleeve
10. Oil seal (treated leather)
11. Gasket
12. PTO shaft cover
13. Spring
14. Shifter fork
17. Clutch (bevel pinion) shaft
18. Shims

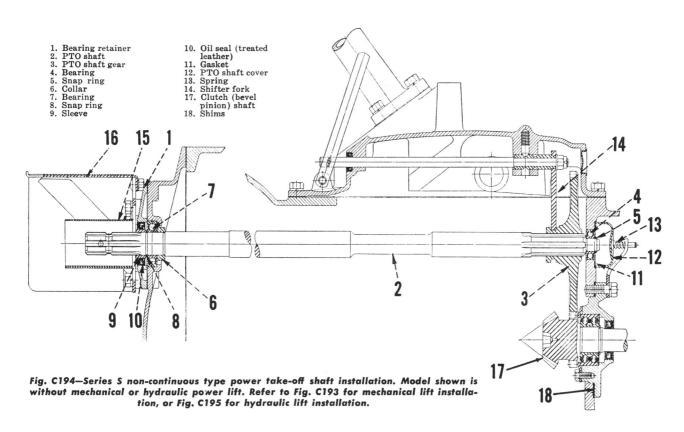

Fig. C194—Series S non-continuous type power take-off shaft installation. Model shown is without mechanical or hydraulic power lift. Refer to Fig. C193 for mechanical lift installation, or Fig. C195 for hydraulic lift installation.

drive chain by removing the locking clip from the master link, as shown in Fig. C197, and remove the drive chain.

On models without a mechanical lift, attach a suitable puller to the outer end of the pto shaft, and remove the shaft. While withdrawing the pto shaft, remove the shaft bearing, driven gear (3) and clutch out through transmission housing top cover opening. On models with a mechanical lift the use of the puller is not necessary.

275A. On series C, D, L and LA, the pto shaft gear (3) rotates on a bronze bushing which requires final sizing after installation to provide a diametral clearance of .002-.003. Install outer oil seal of treated leather so that the lip is facing the transmission gear set. On models equipped with a hydraulic type lift, install the oil seals (33—Figs. C195 & C196) so that the lips of same face away from each other.

275B. Reinstall the pto shaft by reversing the removal procedure. On units equipped with taper roller type bearings, adjust same by varying shims (23 or 37—Fig. C193) or (37—Fig. C195) to provide shaft with free rotation but zero end play.

Series V

276. **R&R AND OVERHAUL.** To remove the non-continuous type power take-off shaft, first remove cap screws retaining bearing carrier (9—Fig.

2. PTO shaft
33. Oil seals (treated leather)
34. Pump drive sprocket
38. Oil seal (treated leather)
40. Hydraulic pump & valves unit
41. Chain
42. Pump drive coupling
43. Pump driven sprocket
44. Bushing
45. Driven sprocket shaft
46. Hydraulic housing
48. Bearing carrier
49. Bearing
50. Snap ring
51. Snap ring
52. Sleeve spacer
53. Snap ring
54. Collar
55. Snap ring
64. Bearing

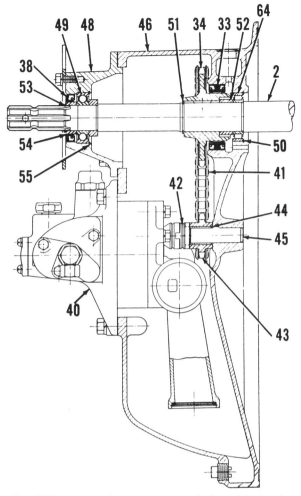

Fig. C196—Non-continuous type pto shaft and hydraulic lift pump installation on series LA.

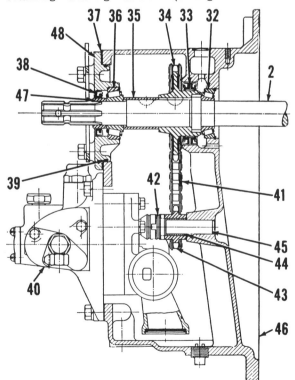

2. PTO shaft
32. Bearing cup & cone
33. Oil seals (treated leather)
34. Pump drive sprocket
35. Spacer
36. Bearing cup & cone
37. Shims
38. Oil seal (treated leather)
39. "O" ring
40. Hydraulic pump & valves unit
41. Chain
42. Pump drive coupling
43. Pump driven sprocket
44. Bushing
45. Driven sprocket shaft
46. Hydraulic housing
47. Sleeve
48. Bearing carrier

Fig. C195—Series D and S non-continuous type pto shaft and hydraulic lift pump installation.

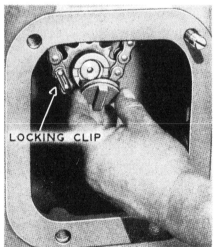

Fig. C197—Showing location of hydraulic pump drive chain master link locking clip on series D, LA and S.

C198) to transmission housing rear cover or hydraulic lift unit housing. Remove pto shifter assembly from top of transmission housing, and withdraw pto shaft rearward being careful not to drop the sleeve coupling (3) into the differential compartment of the transmission housing.

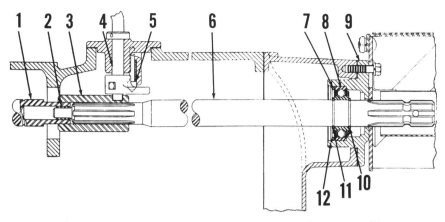

Fig. C198—Series V non-continuous type power take-off assembly.

1. Countershaft
2. Bushing
3. Coupling
4. PTO shift lever
5. Detent ball & spring
7. Spacer
8. Bearing
9. Bearing carrier
10. Snap ring
11. Snap ring
12. Snap ring

The power take-off shaft bronze pilot bushing (2) which is pressed into the end of the transmission countershaft can be renewed after removing the cover or hydraulic pump and valves unit from the rear face of the transmission housing. New bushing requires final sizing after installation to provide a .002-.003 diametral clearance. Removal procedure for the pto shaft bearing (8) is self-evident after an examination of the assembly.

Series VA

The non-continuous type pto as used on model VAH differs from the other models (VA, VAC & VAO) in that the pto shaft (3—Fig. C199) is not used. On model VAH, the pto shaft sliding gear (5) is splined to the rear end of the transmission main shaft instead of being splined and located on pto shaft (3) as shown for models VA, VAC and VAO.

277. **R&R AND OVERHAUL.** To remove the pto assembly, remove six cap screws attaching pto take-off housing to the rear face of the rear axle housing on models VA, VAC and VAO, or to the rear face of the transmission housing on model VAH. Remove the pto assembly by withdrawing the assembly rearward as shown in Fig. C200.

277A. To disassemble the pto assembly, first remove power take-off shaft oil seal retainer. Remove bearing adjusting nut (16—Fig. C199) from power take-off shaft. Buck up gear (13) and bump shaft (8) rearward. Rear bearing cone (11) will remain on shaft and both bearing cups (11) and (15) will remain in pto housing. Bump out shifter shaft pin and remove shifter fork (4) and spring. Remove drive shaft (3) and bearing (6) by withdrawing same forward and out of housing. The driven gear (13) and spacers can now be removed.

1. Transmission mainshaft
2. PTO shaft sleeve
3. PTO drive shaft
4. Shifter fork
5. PTO shaft sliding gear
6. Bearing
7. Shifter lever
8. PTO external output shaft
10. Oil seal (treated leather)
11. Bearing cup & cone (rear)
12. Snap ring
13. Driven gear
14. Snap ring
15. Bearing cup & cone (front)
16. Adjusting nut

Fig. C199—Models VA, VAC and VAO non-continuous type power take-off assembly. Model VAH is similar except that gear (5) is splined to and located on the rear end of the transmission mainshaft (1).

When reassembling the unit, insert the spring washer between the front bearing cone (15) and driven gear spacer. Adjust pto external output shaft bearings with adjusting nut (16) to provide the shaft with zero end play and yet permit same to rotate freely.

CONTINUOUS TYPE PTO
Series D-S

On these models, the continuous type power take-off unit, Fig. C201, power circuit is as follows: At all times when the engine is running, power is taken from the engine clutch cover by means of the hollow gearshaft (33) which is

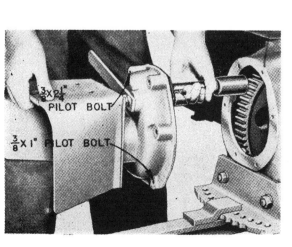

Fig. C200—Removing power take-off assembly from rear face of rear axle housing on models VA, VAC, and VAO. Model VAH is similar except that the assembly is mounted on the rear face of the transmission housing.

splined to the engine clutch cover (6) and fits over the engine clutch (bevel pinion) shaft. The spur gear portion of the hollow gearshaft meshes with gear (44) located on the forward end of the long drive shaft which through a multiple disc over-center type clutch (used to engage and disengage the pto unit) drives a short stub (external output) shaft (82). This arrangement also permits the hydraulic lift pump to operate at all times when the engine is running.

280. ADJUST PTO CLUTCH. The multiple disc clutch will require ad-justment if slippage is noticed when a power take-off implement is attached. To adjust, remove four cap screws at-taching top cover (74—Fig. C201) to the hydraulic sump housing. Disen-gage pto unit clutch by pushing the hand lever completely forward. Work-ing through the top cover opening, pull out the adjusting collar lock pin (93). While holding the lock pin out, as shown in Fig. C202, rotate the clutch adjusting collar (70—Fig. C201) clock-wise (when facing the pto shaft) one notch at a time or until a 50 pounds pull (measured at end of hand lever) is required to engage the clutch.

281. R&R CLUTCH PLATES & STUB SHAFT. The pto clutch and stub shaft, Fig. C203, can be removed as an assembly. First remove hydraulic pump relief valve and spring, Fig. C205, to provide clearance for remov-ing the pto stub shaft and assembly. Remove the pin connecting the mounted hydraulic work cylinder to the left hand rockshaft arm and allow the cylinder and arm to drop. Remove clevis pin from clutch release yoke shaft lever. Loosen cap screws retain-ing release yoke to shaft and withdraw the shaft and remove the release yoke as shown in Fig. C206. Remove four

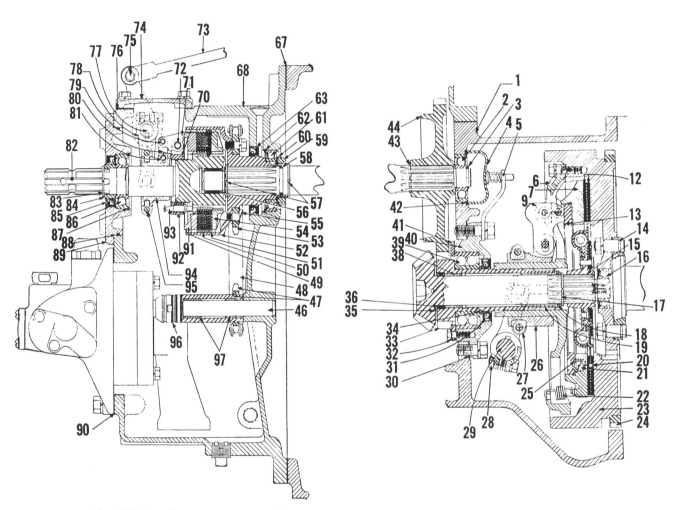

Fig. C201—Continuous type power take-off unit and hydraulic pump unit as used on series D and S.

1. Transmission housing	18. Thrust collar	35. Thrust washer (bronze)	50. Driven plate (steel)
2. Ball bearing	19. Bushing	36. "O" rings	51. Drive plate (bronze)
3. Snap ring	21. Lock pin spring	38. Bushing	52. Clutch drum
4. PTO shaft front end cover	22. Release spring	39. Ball bearing	53. Drive sprocket
5. Support	23. Flywheel	40. Bearing carrier	54. Back plate
6. Engine clutch cover	25. Adjusting lock pin	41. Retainer ring	55. Baffle
7. Clutch pressure plate	26. Sliding sleeve	42. Gasket	56. Pilot bearing
12. Lined plate	27. Clutch collar	43. Snap ring	57. Snap ring
13. Adjusting ring	29. Release yoke	44. PTO gear	58. "O" ring
15. Engine clutch (bevel pinion) shaft	29. Release yoke shaft	46. Driven sprocket shaft	59. Lock nut
16. Bearing	30. Shims	47. Driven sprocket	60. Lock nut washer
17. Snap ring	31. Oil seal	48. Chain	61. PTO drum bearing
	32. Lock washer	49. Drive plate separa-tor spring	62. Snap ring
	33. PTO drive sleeve and gear		63. Oil seal
	34. Snap ring		

67. Gasket	85. Oil seal
68. PTO and hydraulic sump housing	86. Bearing
70. Adjusting collar	87. Snap ring
71. Clutch finger	88. "O" ring
72. Finger pin	89. Bearing carrier
73. Hand control rod	90. Gasket
74. Cover	91. Clutch pressure plate
76. Gasket	92. Adjusting lock pin spring
77. Release yoke shaft	93. Adjusting lock pin
78. Release yoke	94. Sliding sleeve
80. Link	95. Clutch collar
82. PTO stub shaft	96. Coupling
83. Snap ring	97. Bushings
84. Spacer	

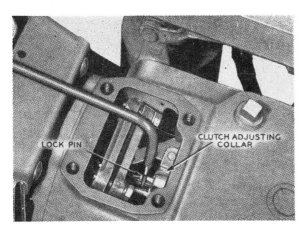

Fig. C202—Pull adjusting collar lock pin out and rotate the adjusting collar in a clockwise direction to increase pull required to engage the pto clutch on series D and S.

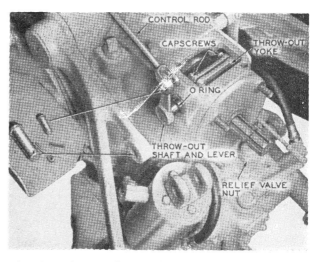

Fig. C205—First step in removing the pto clutch plates and stub shaft.

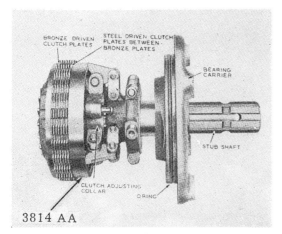

Fig. C203—PTO clutch plates and stub shaft assembly as used on series D and S tractors equipped with continuous type pto.

Fig. C206—Second step in removing the pto clutch plates and stub shaft.

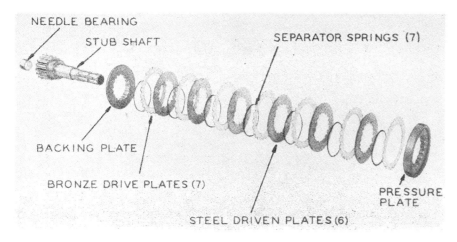

Fig. C204—Disassembled view of the continuous type pto clutch unit for D series. Series S is similar except for number of bronze and steel plates.

nuts retaining the stub shaft bearing carrier (89—Fig. C201) and withdraw stub shaft and clutch plates assembly from housing. The pto drive shaft and clutch drum (52) will remain in the housing. For removal procedure of clutch drum refer to paragraph 282B; for pto drive shaft and gear refer to paragraph 283.

281A. Reinstall stub shaft and clutch unit by reversing the removal procedure. Adjust clutch as outlined in paragraph 280.

282. OVERHAUL CLUTCH. After removing the clutch plates unit and pto stub shaft assembly, proceed as follows: Remove stub shaft snap ring (83) and press rear bearing carrier (89) off the stub shaft. Stub shaft bearing (86) and collar can be removed from the bearing carrier at this time by removing bearing snap ring (87).

Pull the clutch adjusting lock pin (93) out and rotate the collar and sleeve assembly counter-clockwise and remove same from the stub shaft. The pto clutch pressure plate, drive plates, driven plates, drive plate separator springs, and the back plate can be lifted off the stub shaft. Refer to Figs. C204 and C207A.

282A. On series D there are 7 bronze drive plates, 6 steel driven plates; on series S there are 5 bronze drive plates and 4 steel driven plates. Inspect and renew any part which shows wear.

Note: Freedom from chatter and a cleaner release will be obtained on tractors built prior to 1954 if latest plates and Automatic Transmission Fluid Type A are installed. To make the change on series D tractors discard original back plate and install new Case A7107 back plate, discard bronze plates 09037AB and install new corrugated bronze plates 09037AB1, discard dull finish steel plates 09038AB and install shiny steel plates A7102, discard engine oil from hydraulic system sump and install Automatic Transmission Fluid Type A. To make the changeover on series S tractors proceed in same manner but retain the original back plate.

The reassembly of the pto clutch will be simplified if an extra pto drive

46. Driven sprocket shaft
47. Driven sprocket
48. Chain
49. Drive plate separator spring
50. Driven plate (steel)
51. Drive plate (bronze)
52. Clutch drum
53. Drive sprocket
54. Back plate
55. Baffle
56. Pilot bearing
57. Snap ring
58. "O" Ring
59. Lock nut
60. Lock nut washer
61. PTO drum bearing
62. Snap ring
63. Oil seal
67. Gasket
68. PTO and hydraulic pump housing
70. Adjusting collar
71. Clutch finger
72. Finger pin
73. Hand control rod
74. Cover
76. Gasket
77. Release yoke shaft
78. Release yoke
80. Link
82. PTO stub shaft
83. Snap ring
84. Spacer
85. Oil seal
86. Bearing
87. Snap ring
88. "O" ring
89. Bearing carrier
90. Gasket
91. Clutch pressure plate
92. Adjusting lock pin spring
93. Adjusting lock pin
94. Sliding sleeve
95. Clutch collar
96. Coupling
97. Bushings

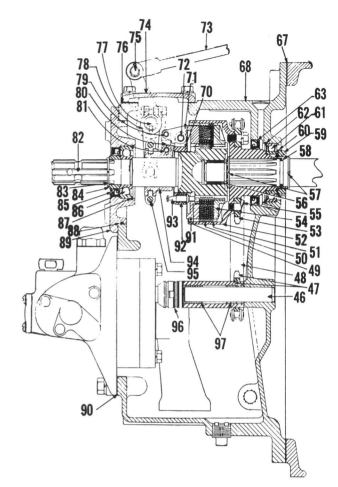

Fig. C207A—Right side sectional view of the continuous type pto clutch unit and hydraulic lift system as mounted on the rear face of the transmission housing on series D and S.

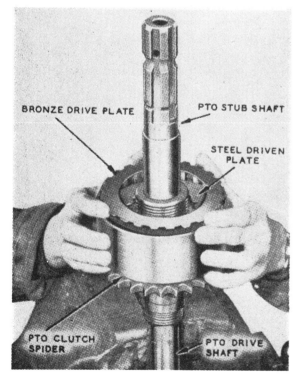

Fig. C207—Using an extra drive shaft and clutch drum when reassembling the continuous type pto clutch plates.

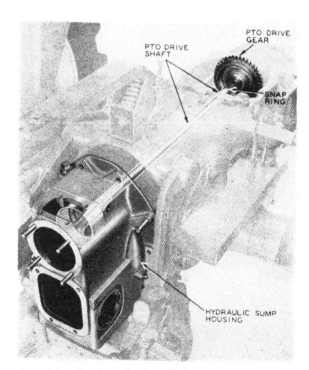

Fig. C208—Showing the installation of the pto clutch drum, drive shaft and drive gear for series D and S tractors equipped with continuous type pto.

shaft and clutch drum are used as a jig, as shown in Fig. C207. Refer to Fig. C204 for the correct order of installing the clutch plates, and separator springs. Install the sliding sleeve and adjusting collar on the stub shaft and engage the clutch. If this is not done, the bronze plates will not hold their position for easy reinstallation. If an extra clutch drum and drive shaft are not available, proceed as follows: First install the stub shaft, Fig. C207, to the pto drive shaft; then install the plates and separator springs in the order as shown in Fig. C204. Install clutch adjusting collar and throwout assembly on stub shaft.

282B. RENEW PTO CLUTCH DRUM, DRUM SEAL, OR DRUM BEARING. First remove pto clutch and stub shaft assembly as outlined in paragraph 281. Remove clutch drum retaining snap ring shown in Fig. C157A. From models so equipped, remove the rockshaft work cylinder and hoses. Disconnect lift links from rockshaft arms. Remove four cap screws attaching drawbar to lower side of pto and hydraulic sump housing (transmission rear cover). Lower the drawbar by loosening cap screws which attach the draft arms and pivot brackets to the drawbar frame.

On all models, remove cap screws attaching hydraulic sump housing. Insert two long studs into two of the retaining cap screw holes to serve as guide pins when removing the sump housing and remove sump housing.

From rear front face of hydraulic sump housing, remove lock nut (59—Fig. C207A) from clutch drum hub and lock washer (60). Using a suitable puller pull clutch drum out of bearing (61). Clutch drum, oil seal (63), bearing (61), or hydraulic pump drive sprocket (53) can be renewed at this time.

283. R&R AND OVERHAUL PTO DRIVE SHAFT & DRIVE GEAR. To remove the power take-off drive shaft and pto drive gear Fig. C208, proceed as follows: Remove hand hole plates from both sides of the transmission housing engine clutch compartment. Working through the hand hole openings, remove bearing cover (4—Fig. C201), gasket (42), and snap ring (3) from the front end of the pto shaft. Remove the transmission top cover (steering gear housing). Remove the mounted work cylinder and hoses. Disconnect the lift links from the rockshaft arms. Remove four cap screws attaching the drawbar to the lower side of the pto and hydraulic sump housing. Lower the drawbar by loosening the cap screws which retain the draft arms and pivot brackets to the drawbar frame. Remove pto and hydraulic sump housing, and drive shaft assembly by withdrawing same rearward. Remove pto drive gear and bearing from front end of shaft while withdrawing the assembly from the transmission housing.

283A. Remove pto stub shaft and clutch plates assembly as outlined in paragraph 281. Remove hydraulic pump filter, and pump and valves unit from models so equipped. Disconnect hydraulic pump drive chain. Disconnect clutch drum (spider) bearing retaining nut (59—Fig. C207A) and lock washer (60). Bump on forward end of pto drive shaft until clutch drum is free of the bearing (61) and withdraw clutch drum and pto drive shaft from housing. Clutch drum bearing (61), oil seal (63), hydraulic pump drive sprocket (53), clutch drum (52) or pto drive shaft can be renewed at this time.

Install new oil seal (63) with lip of same facing the pto clutch assembly.

283B. Reinstall the pto and hydraulic sump housing assembly by reversing the removal procedure. Adjust the pto clutch as outlined in paragraph 280.

284. R&R PTO DRIVE SLEEVE & GEAR. To remove and overhaul the pto drive sleeve and gear assembly (33—Fig. C201) follow the procedure as outlined in paragraphs 165 and 165B.

HYDRAULIC LIFT SYSTEMS

Most of the troubles encountered with the hydraulic system on modern tractors are caused by dirt or gum deposits. The dirt may enter from the outside, it may have been in the system originally, or it may show up as the result of wear or partial failure of some part of the system. The presence of gummy deposits, however, usually results from inadequate fluids or from failure to drain and renew the fluid at the recommended intervals. These principles should be kept in mind when shooting trouble on any of the existing systems and also when performing repair work on pumps, valves, and work cylinders.

Thus, when disassembling a pump or valves unit, it is good practice generally to not remove any parts which can be thoroughly inspected while they are installed. Internal parts of the pump, valve, and cylinders, when removed, should be handled with the same care as would be accorded the parts of a diesel pump or injector unit, and should be soaked or manually cleaned with

SERIES D-LA-S SINGLE & DUAL VALVE SYSTEMS

an approved solvent to remove gum deposits. Unless you practice good housekeeping (cleanliness) in your shop, do not undertake the repair of hydraulic equipment.

The combined pump and valves unit is mounted on the rear face of the hydraulic main housing (reservoir or sump) which is attached to the rear face of the transmission housing, as shown in Fig. C209. Pump is of the gear type and is chain-driven by a

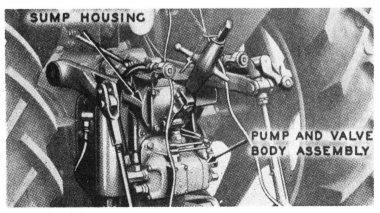

Fig. C209—Hydraulic lift system as mounted on series D and S equipped with three point (Eagle) hitch. Model shown is equipped with a dual valve control system.

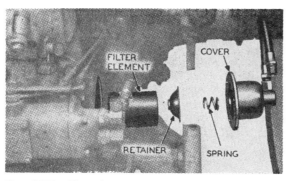

Fig. C212—The edge wound paper element type hydraulic oil filter is located on the right side of the hydraulic main housing (sump) on series D, LA and S.

Fig. C213—Removing pump and valves body unit from rear face of hydraulic main housing.

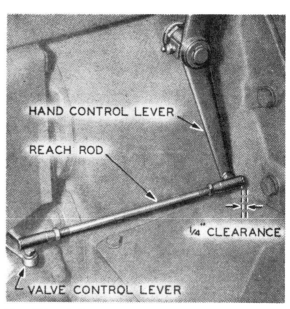

Fig. C214—With valve and hand control lever in the full lowering position, adjust connecting linkage (reach rod) to provide a ¼ inch clearance as shown.

Fig. C215—Showing pump body removal on a dual valve control type system as used on series D, LA and S. Single valve control type pump is similar except that only one valve control spool poppet is used.

sprocket located on the pto driven shaft. Refer to Figs. C195, C196 and C201. System can be equipped with either one or two control valves of the open center spool, closed port type. Refer to Figs. C217A, C223 and C227A.

The mounted work cylinders used on tractors equipped with three point (Eagle) hitch are of the single-acting type. Portable work cylinders are of the double-acting type. On models without the three point (Eagle) hitch, the mounted and portable work cylinders are of the double-acting type.

Series S tractors after serial 8040382 and D and S tractors which have the modified pto clutch use Automatic Transmission fluid Type A in the hydraulic system sump.

291. TROUBLE SHOOTING. Causes for faulty hydraulic lift system operation are outlined in the following paragraphs.

291A. UNIT WILL NOT LIFT. Insufficient oil in reservoir. Worn pump or damaged pump drive chain, paragraphs 293 & 295. Relief valve not seating, paragraphs 294 & 298. Leaking "O" ring on work cylinder piston, paragraph 305.

291B. SLUGGISH LIFTING ACTION. Worn pump, paragraph 293. Partially obstructed pump inlet screen (can be checked after removing the oil filter, paragraph 296).

291C. UNIT WILL NOT HOLD LOAD. Leaking "O" ring on work cylinder piston, paragraph 305. Damaged outlet valve, paragraph 303. Damaged return valve, paragraph 304. Damaged auxiliary relief valve, paragraph 302.

291D. HAND CONTROL LEVER WILL NOT REMAIN IN DETENT POSITION. Relief valve not seating, paragraphs 294 and 298. Control valve spool holding poppet not functioning, paragraph 293B. Work cylinder lowering speed valve adjusted for full restriction when hose line is used with a double-acting cylinder, paragraph 306.

291E. HAND CONTROL LEVER WILL NOT RETURN TO NEUTRAL. The hand control lever which operates the single acting mounted cylinder on tractors equipped with three point (Eagle) hitch will not return to neutral from the full rearward or lowering position. However, the lever should automatically return to neutral from the full forward or raise position.

Check for: Worn pump, paragraph 293. Return valve spring not installed correctly or damaged, paragraph 304. Relief valve not functioning, paragraphs 294 and 298.

292. R & R PUMP AND VALVES UNIT. Remove magnetic drain plug from lower side of hydraulic main housing (reservoir) and drain the housing. Disconnect the hose lines from the swivel elbows located on the valve heads. Remove linkage attaching valve control lever to hand control lever. Remove hydraulic oil filter as shown in Fig. C212. Remove pump and valve body unit upper right retaining cap screw and insert in its place a slotted dowel stud as shown in Fig. C213. Remove remaining attaching cap screws and remove the unit.

292A. Reinstall pump and valves unit by reversing the removal procedure. Before installing the pump and valves unit, coat the slotted surface of the pump driven sprocket with grease so as to hold the pump drive coupling in position when reinstalling the unit.

292B. After the unit is installed, adjust the control valve lever to hand lever linkage (reach rod) as shown in Fig. C214. Check and adjust the main relief valve unloading pressure as outlined in paragraph 294.

293. **OVERHAUL PUMP.** To overhaul the pump unit, first remove the pump and valves unit as per paragraph 292. Place the pump and valves unit in a vise and remove pump inlet and outlet elbows. Lift off pump cover. The pump body can now be removed as shown in Fig. C215.

To remove the pump drive shaft coupling, place the pump cover, gear and shaft assembly in a vise as shown

in Fig. C216. Using a piece of ¼ inch square stock inserted in the coupling slots, rotate the coupling in a clockwise direction. The coupling has a left hand thread.

293A. Check the valve body and the pump cover faces, and gears for scoring or wear. The total clearance between the pump gears and the body and cover faces should be .0015-.0025. Check pump gear shafts needle bearings. Needle bearings ca nbe removed, as shown in Fig. C217, using J. I. Case tool No. 3913-AA. If tool No. 3913-AA or equivalent is not available, the bearings can sometimes be removed hydraulically by filling the bearing cavity with grease and ramming downward on same using a close fitting shaft struck sharply with a ham-

mer. Bearings should be installed with the flat (lettered) end facing the pump gears and can be installed with a piloted pressing tool (J. I. Case No. 08665-AB).

293B. Inspect the spool holding poppets (detents), Fig. C218 or C219, for wear at the point or knife edge. Poppets used in dual control valve assemblies have a knife edge at their tip. If wear is evidenced, renew the poppets. Poppet springs, which have a free length of 61/64 inch, should require 18 pounds pressure to compress them to a length of $1\frac{13}{16}$ inch. Poppet springs which have a free length of $1\frac{13}{16}$ inches should require 18 pounds pressure to compress them to a length of $1\frac{3}{16}$ inches.

1. Bearing
2. Valve body
3. Dowel pin
4. Pump body
5. Pump cover
6. Pump gear & shaft
8. Outlet elbow
9. Inlet elbow
10. Pump idler gear & shaft

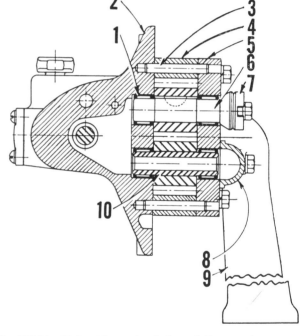

Fig. C217A—Right side sectional view of the gear type pump and valve body used on series D, LA and S.

Fig. C216—Method of removing the pump drive coupling which has a left hand thread.

Fig. C217—Removing pump shaft needle bearings with J. I. Case special puller No. 3913-AA.

Fig. C218—Valve control spool poppet (detent) as used in the single valve control systems on series D, LA and S.

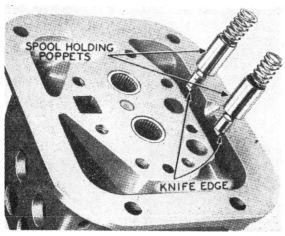

Fig. C219—Valve control spool poppets (detents) as used in the dual valve control systems on series D, LA and S.

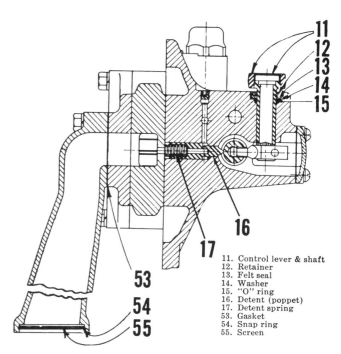

11. Control lever & shaft
12. Retainer
13. Felt seal
14. Washer
15. "O" ring
16. Detent (poppet)
17. Detent spring
53. Gasket
54. Snap ring
55. Screen

Fig. C223—Left side sectional view of a single valve control type body as used on series D, LA and S.

Fig. C221—With the pto shaft operating at 250 rpm, and the hand control lever in forward (full raise) position, the relief valve should unload at a pressure of 980-1030 psi on series D, LA and S.

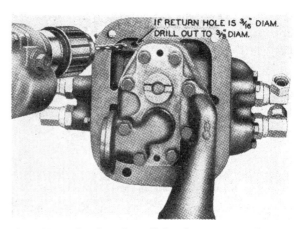

Fig. C224—Enlarging the relief valve to reservoir passage if same measures less than ⅜ inch in diameter on series D, LA, and S pumps.

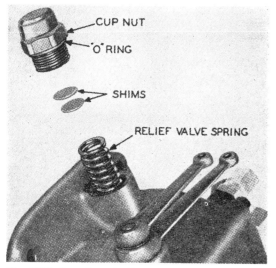

Fig. C222—Location of shims for adjusting the relief valve unloading pressure on series D, LA, and S pumps.

Fig. C225—Install pump relief valve sleeve retainer so that notch is in register with the return hole on series D, LA and S pumps.

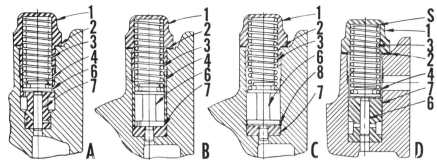

Fig. C226—Various types of adjustable relief valves as used in the series D, LA and S hydraulic systems. Latest version is shown at "D".

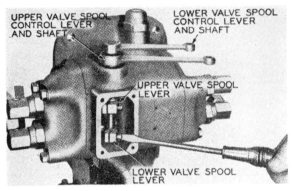

Fig. C227—Method of removing the valve control levers on a dual valve system on series D, LA and S. Removal procedure for a single valve system is similar.

S. Shims
1. Cup nut
2. "O" ring or copper gasket
3. Relief valve spring
4. Relief valve sleeve retainer
6. Relief valve
7. Relief valve sleeve (seat)
8. Snap ring

293C. When reassembling the pump, install the spool holding poppets as shown in Fig. C218 or C219. Assemble the pump by reversing the disassembly procedure. Attach pump inlet and outlet elbow and torque tighten the eight retaining cap screws to 35 ft. lbs.

294. **CHECK & ADJUST RELIEF VALVE PRESSURE.** Pump and valve unit must be installed on the tractor when checking the relief valve unloading pressure. Remove the ⅛ inch pipe plug, and install a pressure gauge of 1600 psi capacity as shown in Fig. C221. Start tractor engine and allow the pump to operate until the hydraulic fluid reaches operating temperature (95-110 deg. F); then set the engine throttle to provide a pto speed of 250 rpm. Move the hand control lever forward to the full raise position. The relief valve should unload when the pressure is within the range of 980-1130 psi. The unloading pressure can be varied by adding or removing shims, Fig. C222, which are located between the valve spring and cup nut.

295. **RENEW PUMP DRIVE CHAIN OR SPROCKET.** To renew the pump drive chain, first remove pump and valve body unit as per paragraph 292; then remove the master link locking clip from the chain, as shown in Fig.

C197, and thread chain off of drive sprocket which is located on pto shaft.

295A. The driven sprocket can now be removed by withdrawing same off sprocket shaft. Sprocket hub bushing or bushings require final sizing after installation to provide .0015-.0025 diametral clearance. To renew the sprocket shaft which is pressed into the hydraulic main housing, it will be necessary to detach the hydraulic housing (sump or reservoir) from the rear face of the transmission housing.

If it becomes necessary to service the pump drive sprocket which is either bolted to the pto clutch drum, or keyed to the pto shaft, proceed as outlined in the POWER TAKE-OFF section.

295B. Reinstall the chain by reversing the removal procedure. The drive chain is not provided with a tension adjustment.

296. **HYDRAULIC OIL FILTER.** The hydraulic system is provided with an edge wound paper element type oil filter located on the right side of the hydraulic pump housing (sump or reservoir). To remove the filter, remove 3 cap screws attaching the filter cover to the housing; then withdraw the spring, retainer and element as shown in Fig. C212. Clean the filter element by washing same in distillate or gasoline.

297. **OVERHAUL VALVE BODY & HEADS.** To overhaul valve body and heads, first remove the pump and valves unit as per paragraph 292. Procedure for bench disassembly and overhaul of the various components, which make up the valve body and heads of either the single or dual control valve type, is as follows:

298. **MAIN RELIEF VALVE.** The main relief valve used in both the single and dual control valve systems is identical in construction, function and service. Several relief valve design changes have been made and are shown in Fig. C223. The latest version (D) is supplied in kit form (No. 4443-AA) and should be used to service types (B & C). However, this kit No. 4443-AA cannot be used to service the relief valve shown as type (A).

298A. To service the relief valve, remove the cup nut and spring from the valve body. Shims located between the spring and inner side of the cup nut are used to adjust the maximum output pressure (980-1130 psi). Using a pair of long nose pliers, check the fit of the relief valve plunger in the sleeve. The plunger should have a smooth sliding fit with no noticeable side play. Check the bleed hole at the flange end of the plunger for any obstruction. Check the diameter of the relief valve return hole in the valve body. If the hole diameter measures 3/16 inch, it will be necessary to enlarge the hole to ⅜ inch after removing and discarding the relief valve sleeve.

To remove the relief valve sleeve (seat), first remove the sleeve retaining snap ring as used only on the type (C) version. Press grease in the plunger hole of the valve sleeve so as to catch any metal cuttings; then tap the valve sleeve hole with a ⅜ x 16 thread tap. Using a bolt and sleeve type puller, pull the sleeve (seat) from the valve body. If the valve sleeve is loose in the valve body and cannot be tapped, make up a clamping tool by using a piece of tubing 1⅛ inches O. D. x 1⅞ inches long and a relief valve cup nut which has the top cut off. Thoroughly wash and clean the valve body to remove the metal cuttings then enlarge the 3/16 inch hole in the valve body to ⅜ inch. Refer Fig. C224.

The new sleeve (seat) should be pressed into position, using a piloted pressing tool (Case No. 08669-AB). Install the sleeve (seat) retainer in types A, B and D versions with the same piloted pressing tool and so that the notch is registered with the valve body return hole in the valve body as shown in Fig. C225. Type (C) version is not equipped with a sleeve retainer.

298B. Reinstall spring, shims and cup nut. Make certain that the "O" ring as used on the cup nut is in good condition. The "O" ring replaces the copper washer as used on earlier units. After installing the pump and valves unit on the tractor, check and adjust the relief valve pressure as per paragraph 294.

299. VALVE CONTROL LEVER. On single valve control systems, only one control lever is used. On dual control systems, two valve control levers, Fig. C227, are used.

It is not necessary to remove the pump and valve body unit to service the valve control levers. First, remove the valve body cover plate. Working through the cover plate opening, remove the valve control shifter arm retaining machine screw. Withdraw valve control lever and shaft from valve body and valve shifter arm. On dual valve systems, the lower valve control lever and shaft, Fig. C227, must be removed before the upper valve control lever can be removed.

299A. Assemble the valve control lever felt seal and "O" ring on the lever shaft in the order as shown in Fig. C228 or C229. Install the control lever in the pump body so that the lever is parallel with the valve body mounting flange surface when the control valve spool is in the neutral position; then tighten the shifter arm machine screw. On dual control valve systems, install the second control lever in a similar manner so that it is parallel with the valve body mounting flange surface. Reinstall the valve body cover plate, and adjust hand control lever linkage as per paragraph 292B.

300. VALVE CONTROL SPOOL. To remove the valve control spool or spools, first remove the pump and valve body as per paragraph 292, and control valve lever or levers as per paragraph 299; then remove the pump cover, and pump body as outlined in paragraph 293. From pump side of the valve body, remove the spool holding poppet or poppets (detents) as shown in Fig. C218 or C219. Remove both valve heads from the valve body and withdraw the valve control spool or spools.

Inspect the spool holding poppet or poppets as outlined in paragraph 293B.

Reinstall the valve control spool by reversing the removal procedure.

301. VALVE HEADS. The left and right hand valve heads on the single valve system are similar in design as

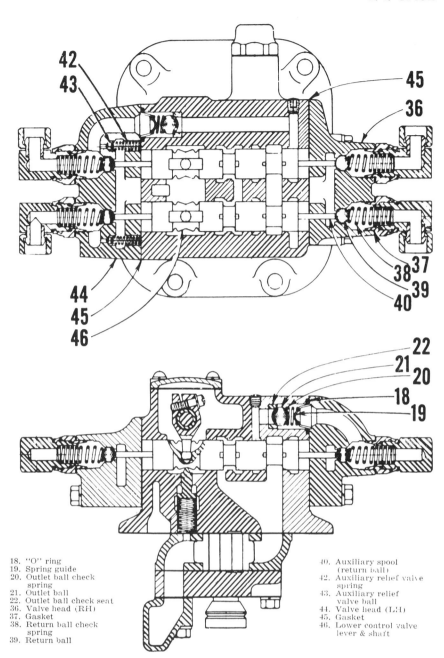

18. "O" ring
19. Spring guide
20. Outlet ball check spring
21. Outlet ball
22. Outlet ball check seat
36. Valve head (RH)
37. Gasket
38. Return ball check spring
39. Return ball
40. Auxiliary spool (return ball)
42. Auxiliary relief valve spring
43. Auxiliary relief valve ball
44. Valve head (LH)
45. Gasket
46. Lower control valve lever & shaft

Fig. C227A—Rear and lower side cross-sectional views of the gear type pump and a dual valve control body as used on series D, LA and S.

are the heads on a dual valve system. The only functional difference between the two heads is that the left head contains a non-adjustable auxiliary relief valve or valves. Both valve heads contain the outlet valve or valves and the return valve or valves. Refer to Figs. C230, C231, and C232.

Service data covering the foregoing valves are outlined in paragraphs 302 through 304B.

302. **Auxiliary Relief Valve.** The non-adjustable auxiliary relief (high pressure safety) valve or valves are located in the left head of the valve body and are preset to unload at 1500 psi. One auxiliary relief valve is used in the single valve system, Fig. C230,

and two auxiliary relief valves are used in the dual valve systems, Fig. C231.

302A. To service the auxiliary relief valve, it is not necessary to remove the pump and valves unit from the tractor. To remove the auxiliary relief valve, first disconnect work cylinder hose or hoses from left valve head. Remove cap screws attaching left head to valve body and remove the head.

302B. Check the auxiliary relief valve ball seat, which is integral with the valve head, for roughness. If the seat is not damaged badly, same can be reseated by placing the 1/4 inch auxiliary relief valve ball on the seat and tapping it lightly with a hammer and soft drift. If the auxiliary relief valve seat is damaged beyond the re-seating stage, it will be necessary to renew the valve body head, and re-

Fig. C228—Assemble the felt retainer, felt, seal and "O" ring as shown before installing the valve control lever and shaft on a single valve system.

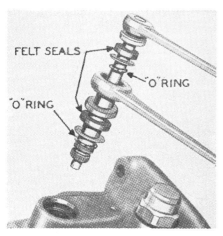

Fig. C229—Assemble the felt retainers, felts, seals and "O" rings as shown before installing the valve control levers and shafts on a dual valve system.

seat the valve by tapping it lightly with a hammer and soft drift.

The relief valve spring has a free length of 1¼ inches and should require 23 pounds pressure to compress it to a length of $1\frac{1}{16}$ inch.

303. **Outlet Valve.** To service this valve, it will be necessary to remove the pump and valve body unit from the tractor as per paragraph 292; then remove the valve head from the pump body. Refer to Fig. C230 or C231.

303A. Leakage at the outlet valve (ball type) can be checked by placing the valve body in a vise so that one of the head ends is facing up and in a level position. Place the outlet valve ball on the seat and fill the outlet valve port with gasoline. If the seat leaks, the valve can be reseated by tapping the ball with a hammer and soft drift. Valve seats which are damaged beyond the reseating stage can be removed by tapping the seat with a $\frac{9}{16}$ x 18 tap and using a screw type puller.

Install the new seat with the chamfer end up and reseat same by tapping the outlet valve ball with a hammer and soft drift.

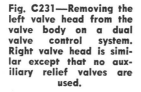

Fig. C231—Removing the left valve head from the valve body on a dual valve control system. Right valve head is similar except that no auxiliary relief valves are used.

303B. Reinstall the outlet valve ball, spring and guide. Install valve head and torque tighten the head retaining cap screws to 35 foot pounds.

304. **Return Valve.** To service this valve, it will not be necessary to remove the pump and valve body unit from the tractor. Disconnect work cylinder to valve head hoses and remove cap screws retaining the valve head to the valve body. Refer to Fig. C232. Return valve ball and spring can be removed after removing the swivel elbow from the valve head. The valve seat is integral with the valve head.

304A. Leakage at the return valve (ball type) can be checked by placing the head in a vise and in a level position. Place the return valve ball on the seat and fill the port with gasoline. If the seat leaks, the valve can be reseated by tapping the ball with a hammer and soft drift.

If the valve seat is damaged beyond the reseating stage, it will be necessary to renew the valve body head.

304B. Reinstall the return valve ball and spring (small diameter end contacting the valve ball) by reversing the removal procedure; then reinstall

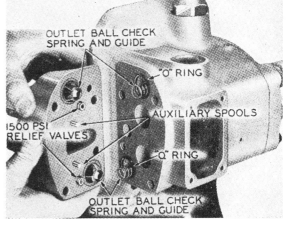

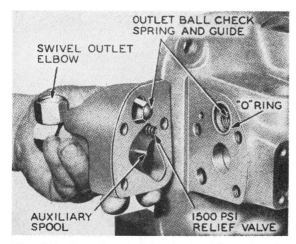

Fig. C230—Removing the left valve head from the valve body on a single valve control type. Right valve head is similar except that no auxiliary relief valve is used.

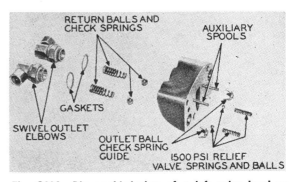

Fig. C232—Disassembled view of a left valve head as used on a dual valve control system. Single valve control system valve head is similar.

the valve head and torque tighten the head retaining cap screws to 35 foot pounds.

305. WORK CYLINDERS. Refer to Figs. C233 and C234. Method of removal is self-evident. To disassemble the work cylinder, first remove 4 cap screws (8) from the cylinder head clamp (9). Push the cylinder head (14) inward slightly, and remove snap ring (10); then withdraw piston rod assembly from cylinder. Wash all parts and thoroughly inspect same for wear, scoring and scratch marks.

Reassemble the work cylinder by reversing the disassembly procedure. Renew all "O" rings and install the lip type seal (11) with the lip facing toward the rod end of the cylinder.

306. SPEED REGULATOR. On tractors equipped with three point (Eagle) hitch, a speed regulating valve, Fig. C236, is inserted in the hose line between the mounted work cylinder and pump to control the lowering speed of the single-acting cylinder.

To retard the lowering speed, loosen the adjusting screw stop screw, then rotate the adjusting screw in a clockwise direction.

Method of removal and disassembly of the speed regulating valve is self-evident.

307. OVERHAUL THREE POINT (EAGLE) HITCH. The three point hitch, as used on series D and S tractors only, consists of two draft arms (attached to lower sides of transmission housing), rockshaft (mounted in top rear portion of transmission housing), and two rockshaft arms which are connected to their respective draft arms by a lift link and combination lift link and leveling screw.

Overhaul data covering the various components which make up the three point hitch system are outlined in the following paragraphs.

307A. DRAFT ARMS. Procedure for the removal and disassembly of the draft arms is self-evident after an examination of the unit.

After draft arms are installed, check the distance between the draft arms. This distance should measure 28 inches when checked from the center of one arm to the center of the other arm and at the latch end. The distance can be varied by adding or removing shims which are located between the transmission housing and draft arm brackets. Refer to Fig. C237.

307B. ROCKSHAFT AND BUSHINGS. To remove the rockshaft, first remove one rear wheel and tire unit, and fender. Disconnect lift links and work cylinder from rockshaft arms. Remove machine screws and washer retaining rockshaft arms to the rockshaft, and remove the arms. Bump the rockshaft toward the side where the wheel and tire unit were removed and remove the rockshaft. Refer to Fig. C238.

Rockshaft bushings supplied for service are pre-sized and should be installed with a sizing arbor having an outside diameter of 2.377 inches.

Reinstall the rockshaft and install the right rockshaft arm end plate and machine screw. Install left rockshaft arm so that it is in the same plane as

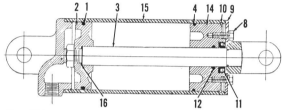

Fig. C233—Cross-sectional view of a single-acting type mounted work cylinder as used on series D, LA and S tractors equipped with three point (Eagle) hitch.

1. Piston "O" ring
2. Piston
3. Piston rod
4. Cylinder head "O" ring
9. Cylinder head clamp
10. Snap ring
11. Lip seal
12. "O" ring
14. Cylinder head
16. Gasket

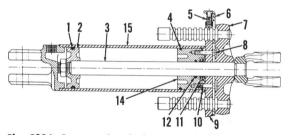

Fig. C234—Cross-sectional view of a double-acting type portable work cylinder as used on series D, LA and S tractors. Mounted work cylinder as used on series D, LA and S without three point hitch is similar.

1. Piston "O" ring
2. Piston
3. Piston rod
4. Cylinder head "O" ring
5. Detent
6. Stop bar latch
7. Stop bar
9. Cylinder head clamp
10. Snap ring
11. Lip seal
12. "O" ring
14. Cylinder head

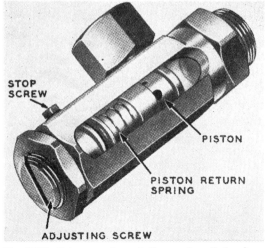

Fig. C236—Cutaway view of the work cylinder lowering speed regulator valve.

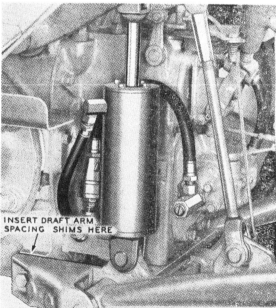

Fig. C237—Showing the location of shims for adjusting the draft arms on series D and S tractors equipped with three point hitch.

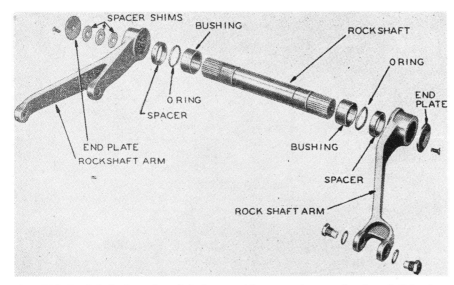

Fig. C238—Exploded view of rockshaft assembly as used on series D and S tractors equipped with three point hitch.

the right arm. Vary the number of shim washers located between the outer left end of the rockshaft and end plate to remove all rockshaft end play and yet permit same to rotate freely.

307C. LEVEL SCREW. Procedure for the disassembly and reassembly of the level screw, is self-evident after an examination of the unit.

When reassembling the level screw, turn the gear shaft adjusting nut until all gear shaft end play is removed; then back off 1/6 turn and tighten lock (jam) nut.

SERIES V COMBINED PUMP, VALVES & WORK CYLINDER UNIT

The combined pump, valves and single-acting work cylinder unit is attached to the rear face of the transmission housing. The piston type pump receives its drive from a cam which is keyed to the power take-off shaft. Refer to Fig. C240.

310. **TROUBLE SHOOTING.** Causes for faulty hydraulic lift system operation are outlined in the following paragraphs.

310A. UNIT WILL NOT HOLD LOAD. Release valve (43) not seating. The valve can be reseated by tapping the ball in its seat with a hammer and soft drift.

310B. SLOW LIFTING ACTION. Worn pump. Leaking work cylinder piston packing (26). Insufficient oil in reservoir.

310C. NOISY OPERATION. Broken piston return spring (12). Damaged rocker arm (11).

311.**REMOVE AND REINSTALL.** To remove the combined pump, valves

and work cylinder unit from the tractor, first drain the transmission housing. Disconnect hand control lever and work cylinder linkage. Remove cap screws attaching the unit to the rear face of the transmission housing, and lift off the unit.

312. **DISASSEMBLE AND OVERHAUL.** The need and procedure for the disassembly will be determined by an examination of the unit and with reference to Fig. C240.

Wash all parts and thoroughly inspect same for wear. The relief valve (16), check valve (22), release valve (43), discharge valve (54), and inlet valve (57) can be reseated by tapping the ball with a hammer and soft drift.

312A. SPEED REGULATOR. The release valve (43) which controls the lowering speed of the work cylinder is adjustable. To decrease the lowering speed, rotate the set screw (46) in (clockwise direction). To increase the lowering speed, turn the set screw out (counter-clockwise).

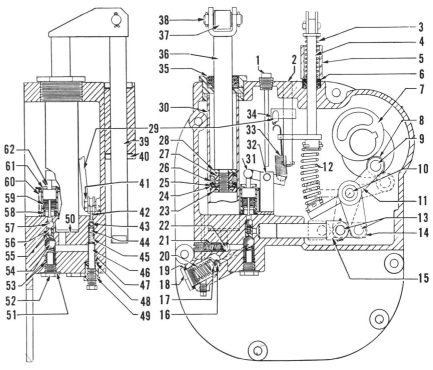

Fig. C240—Series V combined hydraulic pump, valves and work cylinder unit which is mounted on the rear face of the transmission housing.

1. Bayonet gauge	22. Check valve	43. Release valve
2. Guide pin	23. Work cylinder piston	44. Spring seat
3. Control rod spring	24. Spring washer	45. Release valve spring
4. Control rod	25. Plain washer	46. Adjusting screw
5. Control rod stop	26. Piston packing	47. Packing
6. Oil seal	27. Cylinder packing nut	48. Packing washer
7. Operating cam	28. Retainer	49. Packing nut
8. Bearing	29. Release rod	50. Packing gland gasket
9. Pin	30. Work cylinder	51. Valve plug washer
10. Rocker arm shaft	31. Inlet valve release arm	52. Discharge valve plug
11. Rocker arm	32. Release valve fulcrum	53. Valve spring
12. Plunger return spring	33. Inlet valve release spring	54. Discharge valve
13. Plunger pin	34. Release pick-up block	55. Spring seat
14. Connecting block	35. Packing gland	56. Inlet valve spring
15. Pump plunger	36. Work cylinder rod	57. Inlet valve
16. Relief valve	37. Work cylinder yoke	58. Screen
17. Spring seat	38. Yoke pin	59. Release plunger spring
18. Relief valve plug	39. Work cylinder guide	60. Screen and plunger
19. Gasket	40. Felt seal	spring housing
20. Relief spring	41. Release lever bolt	61. Inlet valve cylinder
21. Check valve spring	42. Valve release lever	62. Inlet valve release plunger

VA SERIES WITH PUMP & VALVES UNIT ON BP

The hydraulic system as installed on early series VA tractors consists of a pump which is of the dual piston type, valves unit which is located in the pump housing, single-acting work cylinder, yoke assembly (raising bail) and a retracting spring for the yoke assembly. The high pressure pump (rated working pressure of 4500 psi) is driven by two cams which are integral with the belt pulley shaft. The system reservoir is formed by the belt pulley unit housing. Refer to Figs. C241 and C242.

315. **TROUBLE SHOOTING.** Causes for faulty hydraulic lift system operation are outlined in the following paragraphs.

315A. **UNIT WILL NOT LIFT.** Insufficient oil in reservoir. Worn pump. Leaking packing on work cylinder piston. Relief valve not seating; check as per paragraph 322A. Check pump lockout screw as per paragraph 323.

315B. **SLUGGISH LIFTING ACTION.** Worn pump. Air in system.

315C. **UNIT WILL NOT HOLD LOAD.** Leaking packing on work cyl-

inder piston. Faulty sliding valve; check as per paragraph 322A.

316. **R&R PUMP AND VALVES UNIT.** To remove the pump and valves unit from the belt pulley housing, first place the work cylinder in the retracted position. Remove cover plate (20—Fig. C243) from pump housing and drain the hydraulic oil. Disconnect the pump to work cylinder hose (8—Fig. C241), control rod (2) and intake line (10) at the pump. Remove belt pulley to pump housing oil inlet tube. Remove four cap screws attaching pump and valves unit to belt pulley housing and remove the unit.

Reinstall the pump and valves unit by reversing the removal procedure, and bleed the system as outlined in paragraph 317. Check and adjust pump lock out screw (6) as per paragraph 323.

317. **HYDRAULIC SYSTEM BLEEDING.** The hydraulic system should be purged whenever the system has been disconnected and/or drained and refilled with operating fluid. To bleed the system, first disconnect work cylinder to pump hose (8) at the pump; then open the bleeder valve (7) or remove the 1/8 inch slotted pipe plug

which covers the bleed port. Work hand control lever (1) up and down a few times until hydraulic oil flows from the bleeder valve or pipe plug opening. Reinstall the pipe plug or close the bleeder valve.

Recheck the system for air by pushing the hand control lever forward. Oil should flow from the pump housing to work cylinder hose connection when the lever is pushed forward. If this does not occur, repeat the purging procedure by removing the pipe plug or opening the bleeder valve and operating the hand control lever.

319. **CHECK AND ADJUST RELIEF VALVE PRESSURE.** Pump and valves unit must be installed on the tractor when checking the relief valve unloading pressure. Disconnect the pump to work cylinder hose (8) at the pump, and install a pressure gauge of 5000 psi capacity. Start tractor engine and allow pump to operate until the hydraulic fluid reaches operating temperature. Move hand control lever to the full raise position. The relief valve should unload when the pressure is within the range of 4350-4500 psi. The unloading pressure can be varied by adding or removing shims which are located between outer end of relief valve spring

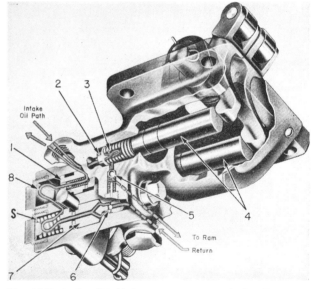

Fig. C242—Series VA high pressure system hydraulic power lift pump and valves unit assembly.

1. Oil return control piston	3. Piston spring	6. Check Valve ball
2. Check ball cage	4. Piston	7. Relief valve
	5. Sliding valve ball	8. Spring seat plug

Fig. C241—Series VA high pressure system pump and valves unit are mounted on the lower side of the belt pulley housing.

1. Control lever	4. Lifting yoke	8. High pressure hose
2. Control rod	5. Belt pulley housing	9. Hydraulic pump
3. Filler and breather plug	6. Lock-out screw	10. Intake line
	7. Bleeder valve	11. Pulley shaft cap

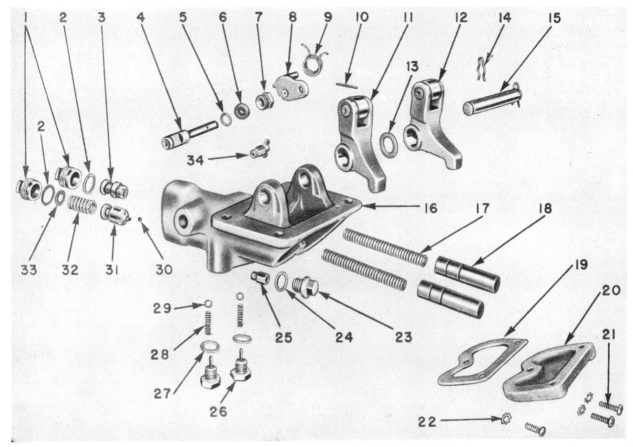

Fig. C243—Series VA high pressure system hydraulic power lift pump with late style straight type pistons (18) and valves assembly—exploded view. Shims which control the relief valve unloading pressure and are located between items (32 & 33) are not shown.

1. Spring seat plug	7. Packing nut	13. Spacer	20. Pump housing cover	29. Sliding valve ball (9/16)
2. Spring seat	8. Control lever	14. Shaft retaining pin	23. Relief valve plug	30. Check valve ball
3. Oil return control piston	9. Control lever spring	15. Rocker arm shaft	25. Relief valve seat	31. Relief valve
4. Control shaft	10. Groov pin	16. Pump body	26. Spring seat plug	32. Relief valve spring
5. Packing nut washer	11. Rocker arm (RH)	17. Piston spring	27. Gasket	33. Spring seat plug pilot
6. Control rod packing	12. Rocker arm (LH)	18. Piston	28. Ball valve spring	34. Bleeder valve

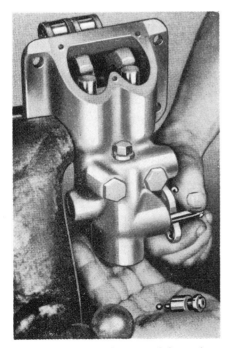

Fig. C244—Rotate the control lever in a downward direction to remove the relief valve and check valve ball on a series VA high pressure pump.

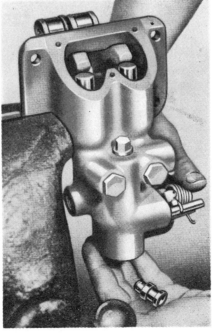

Fig. C245—Rotating the control lever in an upward direction to remove the oil control piston on a series VA high pressure pump.

(32—Fig. C243) and spring seat plug pilot (33).

320. DISASSEMBLE PUMP AND VALVES UNIT. Procedure for bench disassembly of the pump is as follows: First remove relief valve spring seat plug, relief valve spring (32) and shims. Place the pump and valves body in a vise; then rotate the control lever in a downward direction and remove relief valve and check valve ball as shown in Fig. C244. It may be necessary to tap the body lightly to jar the relief valve and check ball from the pump and valves body.

Remove the oil return control piston plug; then rotate the control lever in an upward direction and remove the oil return control piston (3—Fig. C243) as shown in Fig. C245.

Remove the two spring seat plugs (26—Fig. C243), gaskets (27), springs (28) and sliding valve balls (29). Remove shaft retaining pin (14), rocker arm shaft (15) and rocker arms (11 & 12). Place correlation marks on both pump pistons to assure correct reinstallation, and remove pump pistons (18) and springs (17).

321. OVERHAUL. The ball check cage (2—Fig. C242) located in the bottom of each plunger bore is not renewable on pumps built prior to 1948. Pumps used in the 1948 production and after are equipped with renewable ball check cages. The later type pumps which are equipped with the renewable ball check cages can be identified by the stamped marks "48-X" on the pump and valves body. Pump and valves h o u s i n g, and pistons are matched at manufacture and are not available for service. Any derangement of the housing or pistons is corrected by renewal of the entire pump and valves unit.

To check the condition of the ball check cages, place the pump and valves body in a vise so that the open ends of the piston bores are up; then fill the bores with gasoline. If the ball check cages leak, renew same on pumps which are marked "48-X" or renew the complete pump and valves unit on models which are not marked "48-X".

Remove relief valve plug (23—Fig. C243). Thoroughly wash and clean the pump and valves housing and components with gasoline. Do not disassemble the oil return control piston. If defects are observed, it will be necessary to renew the complete piston assembly. Check the small end of the relief valve for wear or damage. Check the rocker arms and shaft for wear and roughness and reject any arm which has more than .010 clearance on the shaft.

Check relief valve seat (25) for grooves, pitting or other leak produc-

ing conditions. The relief valve seat can be removed by bumping same out of pump and valves housing with a 5/16 inch diameter brass rod as shown in Fig. C246.

322. REASSEMBLE & TEST PUMP AND VALVES UNIT. Reassemble the pump by reversing the disassembly procedure and with reference to Figs. C243, C248, and C249. Reseat the two sliding valve balls (29—Fig. C243) by placing the ball on the seat and tapping same with a hammer and soft drift.

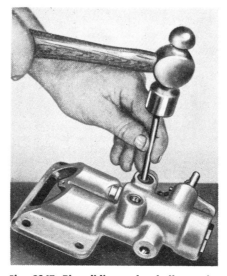

Fig. C247—The sliding valve balls can be reseated by tapping the balls with a hammer and soft drift on a series VA high pressure type pump.

Fig. C249—Rotate the control lever in a downward direction to install the oil control piston.

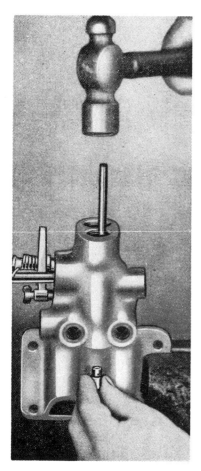

Fig. C246—Using a 5/16 inch diameter rod to remove the relief valve seat on a series VA high pressure type pump.

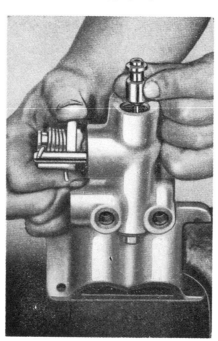

Fig. C248—Rotate the control lever in an upward direction to install the relief valve and check valve ball.

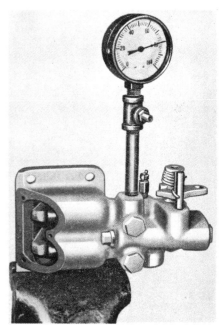

Fig. C250—Install pressure gauge as shown to make a pump and valves unit bench test for faulty sliding valve balls or relief valve.

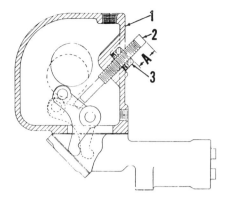

Fig. C251—The pump will be inoperative when the pump lockout screw (2) is turned in so that dimension "A" measures 3/8-1/2 inch. To place pump into operation, back out lockscrew (2) until dimension "A" measures 1 1/8-1 1/4 inches.

322A. A pump and valves unit bench test to check for faulty sliding valve balls (29—Fig. C243) and/or relief valve (31) can be made as follows: First place the pump and valves unit in a vise as shown in Fig. C250. Fill the pump to work cylinder passage in pump body with oil and install a pressure gauge (100 psi capacity, Case Tool No. T117 or equivalent) as shown. (It will be noted in the illustration that provisions are made for applying air pressure to the gauge.) Apply air pressure to the gauge until the gauge registers 60 to 90 psi. If the gauge reading is maintained for several minutes, the sliding valve balls and relief

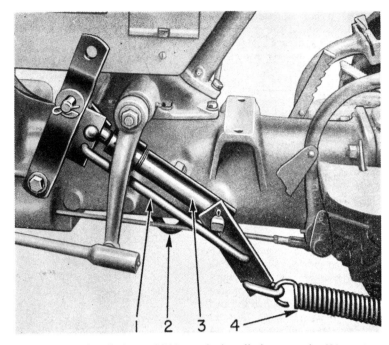

Fig. C254—Work cylinder and lifting yoke installation on series VA tractors.
1. Retracting link   2. High pressure line   3. Work cylinder   4. Retracting spring

valve are properly seated. If pressure drops to zero, either the sliding valve balls or relief valve are at fault. Check and reseat the sliding valve balls as per paragraph 322, or renew the relief valve and seat.

323. **PUMP LOCKOUT SCREW.** Provision is made for placing the pump in a non-operating position by means of a lockout screw (6—Fig. C241) or (2—Fig. C251) located on the belt pulley housing. The lockout position is obtained by turning the lockout screw (2) in until distance (A) measures 3/8-1/2 inch. Turning the screw in forces the pump rocker arms away from the cams.

To place pump in the operating position, back-out the lockout screw until distance (A) measures 1 1/8 to 1 1/4 inches; then tighten the lock nut.

324. **WORK CYLINDER.** Refer to Figs. C252, C253, and C254. Method of removing and disassembling the single-acting work cylinder is self-evident.

It is advisable to renew the leather piston cup or chevron packings whenever a unit is disassembled. On models having a leather piston cup, install the cup so that the open end of same faces the oil supply. On models having chevron type packings, the packing should be installed with the open end of the "V" towards the oil supply.

Reassemble the work cylinder by reversing the disassembly procedure.

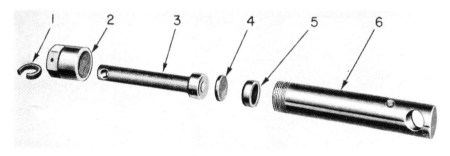

Fig. C252—Single-acting type work cylinder equipped with a leather piston cup as used on some early series VA tractors.

1. Felt washer	3. Cylinder shaft and piston	5. Leather piston cup
2. Cylinder cap	4. Packing washer	6. Cylinder

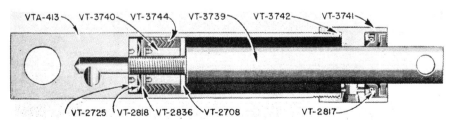

Fig. C253—Single-acting type work cylinder equipped with a chevron type packing as used on some early series VA tractors.

VT 2708. Washer	VT 2818. Spring washer	VT 3741. Cylinder cap
VT 2725. Packing nut	VT 2836. Washer	VT 3742. Gasket
VT 2817. Oil seal	VT 3740. Packing sleeve	VT 3744. Packing

VA SERIES SYSTEM WITH GEAR TYPE PUMP ON ENGINE

Data in paragraphs 325 through 341 pertains to the low pressure system used on later production series VA tractors. Earlier tractors were equipped with a high pressure system utilizing a piston type pump mounted on the belt pulley housing. In the maximum version of the low pressure system, the various assemblies are as shown in Fig. C255.

The system is available in these 4 additional versions:

2. Pump, rockshaft and depth control valve only.

3. Pump, one way control valve and remote cylinder only.

4. Pump, two way control valve and remote cylinder only.

5. Pump, rockshaft, depth control valve, one way control valve and remote cylinder only.

The depth control unit mounted on the underside of the rockshaft housing is of the two-plunger type. The quadrant for the control lever is provided with two permanent stops; one to limit the total lift and one to limit the total drop. An adjustable stop is also provided for selective operation. A full pressure relief valve located in the depth control valves housing by passes the hydraulic fluid to the reservoir when system pressure exceeds 800 psi. After tractor serial 5458558 the system includes an oil filter mounted in the tractor torque tube.

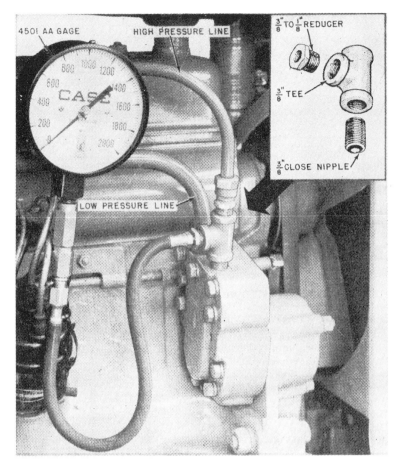

Fig. C256—Series VA tractor. Using Case #4501AA gauge to check relief valve setting and pump pressure. A hydraulic snubber is inserted in the gauge line.

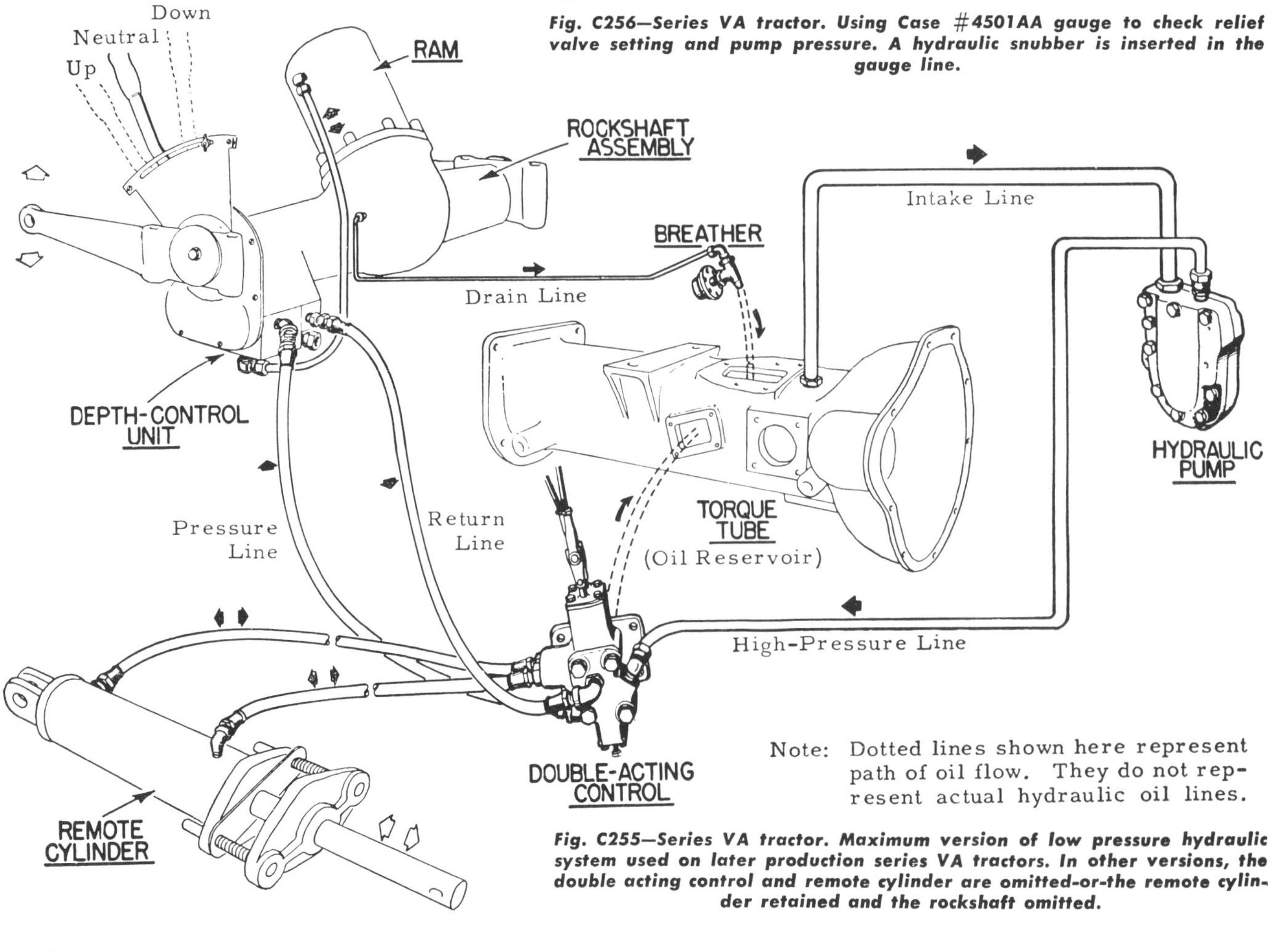

Note: Dotted lines shown here represent path of oil flow. They do not represent actual hydraulic oil lines.

Fig. C255—Series VA tractor. Maximum version of low pressure hydraulic system used on later production series VA tractors. In other versions, the double acting control and remote cylinder are omitted—or the remote cylinder retained and the rockshaft omitted.

Hydraulic fluid original fill capacity is 8 quarts. Recommended oil is SAE 20 for temperatures above 60 degrees F; SAE 5W for temperatures below 32 degrees F and SAE 10 for temperatures between 32 and 60 degrees Fahrenheit.

325. **TESTING.** To test system pressure, connect a pressure gauge to pump as shown in Fig. C256. Gauge shown is Case No. 4501AA which reads 0 to 2000 psi and has a hydraulic snub-

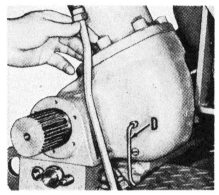

Fig. C257—Series VA tractor. Remove drain line (D) to check for leaking piston seal in rockshaft cylinder.

ber in series between hose and gauge. With engine operating at 1425-1525 rpm, actuate the depth control several times and note gauge reading in full lift position. Reading should be 800-850 psi. If reading is lower than 800 either the pump is at fault or there is a derangement of the relief valve in the depth control valve unit. If reading is higher than 850, the trouble is most likely in the relief valve.

One method of determining which unit is at fault is to insert a shut-off valve between the ⅜ inch tee and the pressure line and to momentarily take the relief valve out of the circuit.

CAUTION: Do not operate more than a few seconds with pump outlet obstructed. To do so will at least blow the gasket and possibly cause serious pump damage.

326. **WON'T HOLD IN RAISED POSITION.** If the pump is known to be O.K. but system will not hold an implement in the raised position proceed

as follows: Disconnect drain (breather) line (D—Fig. C257) from bottom of rockshaft cylinder. With engine running and an implement hooked to system place control lever in "raise" position. If fluid gushes out drain (breather) line connection, the trouble is located in the cylinder. If implement does not stay in raised position and only a slight amount of fluid is emitted at the line connection look for a faulty pressure control valve in the depth control valve unit. Refer also to paragraph 336A.

327. **PUMP.** Hydraulic pump used on tractors with magneto ignition is internally similar to the pump used on tractors with battery ignition although the covers are different. In each case the pump is driven by the engine camshaft gear. The two gears inside the pump are straddle mounted on replaceable needle bearings. These gears and the pump body are matched at manufacture and are not furnished

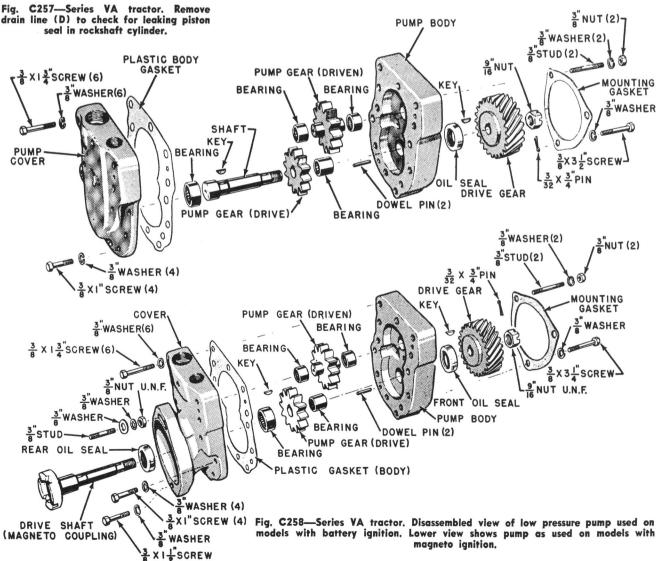

Fig. C258—Series VA tractor. Disassembled view of low pressure pump used on models with battery ignition. Lower view shows pump as used on models with magneto ignition.

as repair items. Any derangement of gears or body is corrected by renewal of the entire pump unit.

328. R&R AND OVERHAUL. Procedure for removal of the pump from the engine is self-evident after examining the installation. After unit is removed clean exterior to remove all dirt, grease, etc.

When disassembling the pump, mount same carefully in a vise and bump cover to body dowels into body far enough to ease the tightness, then remove the cover cap screws. Use a soft hammer to bump cover off body. Do not pry cover off body. Clean all parts in solvent.

Inspect needle bearings and shafts for signs of failure. If rollers fall out of cages, install new bearings even though no signs of wear are apparent.

Bearings in body and cover can be hydraulically removed by partially filling the bearing cavity with cup grease, inserting a close fitting shaft into the cavity and striking the end of shaft sharply with a hammer. J. I. Case tool T-133 is available for this job. The Case company also supplies an oil seal pilot sleeve, a seal and bearing driver and an oil seal driver washer. All of these items are available as a pump overhauling tool kit under their parts number VTA1728. Use Fig. C258 as a guide when reassembling.

Install all bearings from the inside pressing them into place with properly fitting piloted arbors. Always enter the unstamped end of the bearings as indicated in Fig. C259. Oil seal in cover of pumps used on magneto ignition should be installed with lip facing toward the magneto end of gear shaft. The similar seal in the pump body of both models should be installed with the lip facing inward.

When pump gear is pressed on to shaft there should be .002 clearance for end play as shown in Fig. C260. When ready to install the pump cover, use a new plastic gasket and tighten the cover screws to 35 foot pounds torque. When installing the helical drive gear to outside end of shaft as shown in Fig. C261, swing the pump through an arc to make certain there is

no dragging of internal gear against inner face of pump body. Pump should rotate freely when drive gear is seated. If pump binds when drive gear is seated and tightened but is free when nut is loose install a thin shim washer between inner face of gear and outer shoulder on shaft.

329. **ADJUST DEPTH CONTROL.** To adjust the depth control it is necessary to first remove the cover and cam housing from the control unit. To remove the cover, place the control in position that will lower the implement so as to remove all oil from the hydraulic cylinder. Remove right hand fender, disconnect right hand lift link from lift arm and lift arm from rockshaft as shown in Fig. C262. Remove and retain the shims. Remove screws from cover and withdraw cover and operator's control lever as shown in Fig. C263.

329A. Remove the cam housing as shown in Fig. C264. Remove the pressure control valve plunger (PV) and also the valve and spring from the valve housing. Reassemble these parts as in Fig. C265 and check the gap (over-travel) existing between the bottom of the plunger and top of spool on valve. If gap is less than $\frac{1}{8}$ inch insert thin shim washers Case No. VT20206 at top of spring until $\frac{1}{8}$ gap

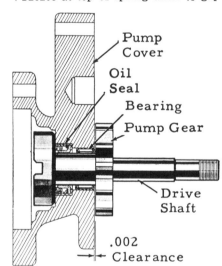

Fig. C260—Series VA tractor. Pump internal driving gear should be positioned on shaft to obtain .002 end clearance in assembled body.

is obtained. Reinstall the plunger unit and the cam housing as in Fig. C266.

330. Refer to Fig. C266. Check to see that actuating lever (A) can be moved slightly about its pivot without con-

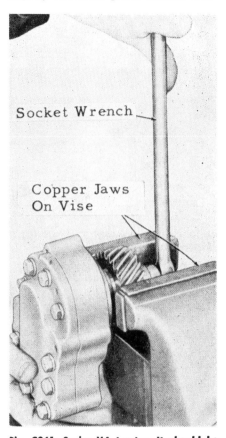

Fig. C261—Series VA tractor. It should be possible to securely tighten the external helical drive gear without binding the internal pump gears.

Fig. C262—Series VA tractor. To adjust depth control valves it is necessary to remove right lift arm, control cam and cover. The shims shown are used to adjust friction of control lever.

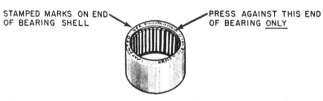

STAMPED MARKS ON END OF BEARING SHELL   PRESS AGAINST THIS END OF BEARING ONLY

Fig. C259—Series VA tractor. Press against the stamp marked end when installing needle bearings.

tacting either one of the valve plungers. If actuating lever has no free movement adjust the tappet at the relief valve end as shown until a gap of .005 to .010 is obtained at that end before the tappet at opposite end first contacts its plunger.

331. Refer to Fig. C267. Temporarily reinstall operator's control lever

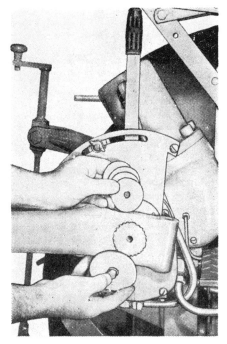

Fig. C262A—Series VA tractor. Outer face of right hand lift arm will extend beyond outer face of rockshaft. Install enough of the smaller shim washers so that when retaining cap screw is fully tightened, shaft will rotate freely with minimum end play.

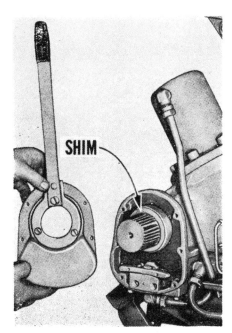

Fig. C263—Series VA tractor. Cam cover removed showing shims which control neutral free travel of control lever.

and cam over end of rockshaft as shown. Grasping the cam with one hand and the lever with the other hand urge the lever and cam in toward the valve plungers as far as possible. With cam in this neutral position some clearance should exist between it and the actuating lever at points "A" and "B" but cam should travel not more than 1/16 inch in either direction (forward or backward) before contacting the actuating lever. If free travel is more than 1/16 remove a shim from the location shown in Fig. C263 until this amount of free travel is obtained.

332. Refer to Fig. C268. Temporarily install the right hand lift arm and move the control unit to the full "raise" position as shown. Place a screwdriver as shown and pry outward on actuating lever at that end. This should cause the control valve

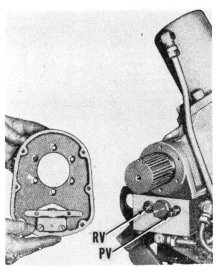

Fig. C264—Series VA tractor. Depth control cam housing removed to check over-travel of pressure control valve plunger.

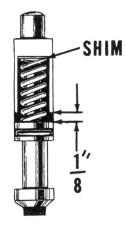

Fig. C265—Series VA tractor. First step in adjusting depth control valve system is to obtain ⅛ inch over-travel of pressure valve plunger before it contacts spool on valve.

plunger at opposite end of lever to move inward (over-travel) an additional 1/16 inch approximately. If over-travel (inward movement) is less than 1/16 adjust the tappet screw at "PV" end of lever until this condition is obtained.

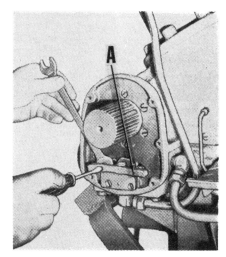

Fig. C266—Series VA tractor. Actuating lever (A) should have some free travel (tappet gap) about its pivot.

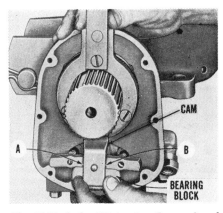

Fig. C267—Series VA tractor. Forward and backward free travel of cam should not exceed 1/16 inch in each direction.

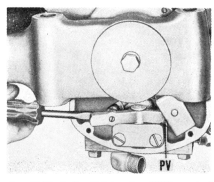

Fig. C268—Series VA tractor. With control lever in full raise position, the plunger of seated pressure control valve operated by "PV" end of lever should have 1/16 over-travel when opposite end of actuating lever is pried outward as shown.

Fig. C269—Series VA tractor. Make final adjustment of tappet clearance at the "RV" end of actuating lever.

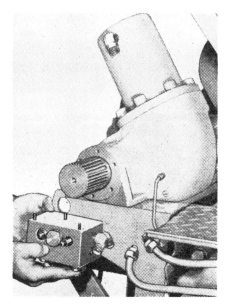

Fig. C270—Series VA tractor. Depth valve control unit being withdrawn after removing control lever cam cover and housing and right hand lift arm.

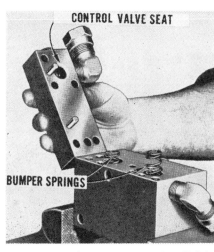

Fig. C271—Series VA tractor. The slightly heavier bumper spring of depth control valve unit is for the pressure control valve.

333. Return the control lever to neutral position as in Fig. C269 and recheck free movement of actuating lever as shown. If there is no free movement adjust the tappet at "RV" end until approximately .005 gap is obtained at one tappet before opposite tappet contacts its plunger.

333A. Remove right hand lift arm. Reinstall control lever, cam and control cover, but not the lever quadrant. Move control lever rearward until same is in horizontal position. Install a shim Fig. C262 (if a shim was previously removed) at outer face of control lever hub. Temporarily install right hand lift arm and with same pushed on to splines all the way, check freedom of movement of control lever by moving same up and down slightly. If control lever does not move freely or has more than .010 end play, remove or add shim washers behind the lift arm. The smaller diameter shim washers Fig. C262A fit inside the spline bore of lift arm on outer end of rockshaft and should be varied in number to hold arm securely to rockshaft but without any end thrust.

334. **R&R AND OVERHAUL DEPTH VALVE UNIT.** To remove the depth control valve unit first remove the control cover as described in paragraph 329, then on VAH remove the cam housing (held by 6 screws). On all models disconnect the oil lines from

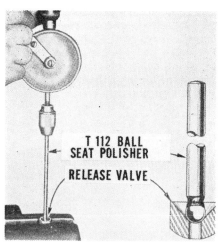

Fig. C273—Series VA tractor. A rod with ball end and a hand drill are used to reseat the ball portion of the two-stage release valve in depth control valve unit.

valve block. Remove cap screws from under side of unit and lift same from tractor as shown in Fig. C270.

With unit clamped in vise remove the cap from the body as shown in Fig. C271 being careful not to damage the lead gasket. The slightly heavier of the two bumper springs is for the pressure control valve. The procedure for the further disassembly of the unit is self-evident after examining the unit and referring to Fig. C272.

335. RELEASE VALVE. Early production units are equipped with a simple ball type release valve comprising

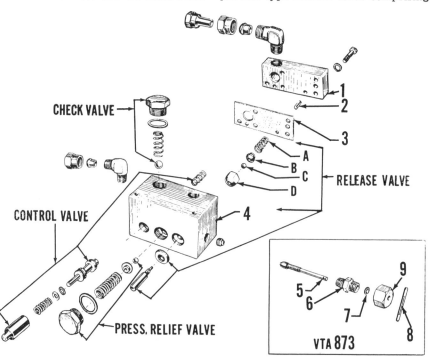

Fig. C272—Series VA tractor. Disassembled view of depth control valve unit used on later production tractors. Items shown boxed comprise the flow restrictor attachment Case No. VTA873. Latest relief valve is composed of items (A) through (D); earlier valves included items (A) and (C) only.

only items (A) and (C). On these a leaking valve can be reseated by dropping the ball into the seat in the body and tapping the new or old ball lightly with a soft metal rod and a hammer. The two stage release valve used in later production includes the seat (B), poppet valve element (D) and items (A) and (C). Reseating of the ball to its seat in the poppet valve is accomplished by fine lapping with the Case No. T112 ball seat polisher (a similar tool can be made by welding a new ball to the end of a piece of $\frac{3}{16}$ drill rod) and a breast drill as shown

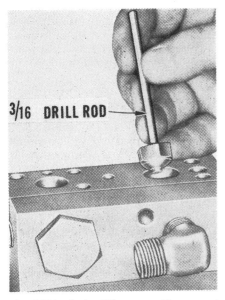

Fig. C274—Series VA tractor. The poppet portion of two-stage release valve can be reseated or polished (burnished) after fitting it to a piece of drill rod as shown.

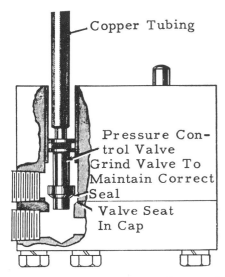

Fig. C275—Series VA tractor. Because of production changes in valve and seat angles, a new pressure control valve or body cap should always be reseated before final assembly, by direct lapping as shown. Copper tubing is used as an arbor for oscillation of valve on seat.

in Fig. C273. To reseat the poppet portion of this valve, form a piece of $\frac{3}{16}$ drill rod to fit the poppet as shown in Fig. C274; then use a breast drill to oscillate the assembly on seat in valve block. Use a fine grade lapping compound to obtain a perfect line seat. To obtain a polish on seating surfaces use tallow instead of abrasive.

336. PRESSURE CONTROL VALVE. When a new pressure control valve or a new valve housing or cap is to be installed, same should be reseated by lapping with abrasive. This procedure is recommended for the reason that changes have been made in the seat angle and it is possible to encounter a condition where the angle of the valve is not the same as the angle of the seat in the cap. Lap to a sharp line contact using valve grinding compound to obtain the line contact. To obtain a polish use tallow. Lapping should be done with the cap assembled to the block as shown in Fig. C275. A clean rag should be inserted in the recess below the seat in the cap before starting the lapping operation and the entire assembly should be thoroughly washed and dried after lapping is completed.

To facilitate the lapping operation fit a piece of ⅜ copper tubing over the valve as shown and use a breast drill to oscillate the valve on its seat.

336A. LEAK TEST. To determine if pressure and release valves in depth valve unit are leaking attach the iron pipe, tee, tire valve and pressure gauge unit (shown in Fig. C250) to the inlet port of the assembled depth control valve. The inlet port is indicated at "X" in Fig. C275A. Fill iron pipe with oil then put 40 psi air pressure into circuit by connecting an air supply to tire valve. If valves are leaking the gauge pressure reading will drop off rapidly and oil will appear at outlet ports of valve unit. If valves are sealing satisfactorily the gauge will hold at 40 psi for 2 minutes or more.

337. CONTROL VALVE FOR REMOTE CYLINDER. When VA model tractors are equipped to operate a remote cylinder the system will include a double-acting control valve shown in Fig. C276. The cap screws "A" and "B" can be adjusted to regulate the speed at which the control releases. Turning the screws into the body slows the speed of release. Screws should be adjusted to same length.

To adjust the control rods, first remove clevis pins from same and with

Fig. C275A—Series VA tractor. To seat pressure and release valves of depth control valve unit, exert fluid or air pressure on same at inlet port (X).

control lever in neutral adjust rod lengths so that clevis pins can be reinserted without moving either the plungers or the lever from their neutral positions. Exploded view of valve shown in Fig. C277 indicates procedure for disassembly. Spool valves (11) should be bright and clean. To correct a leaking two-stage check valve renew the ball or ball and seat if reseating does not prove effective.

337A. Check all tractors carefully to make sure that both of the plungers move freely when the control lever is moved through its full travel. Some installations may be encountered where the inner plunger does not travel into the full release position when the control lever is moved forward to the full raised position. The corrective procedure for most cases of misalignment is to renew the mounting plate on the torque tube.

On individual tractors prior to serial 5370000 where the lifting capacity at rated engine speed is not satisfac-

Fig. C276—Series VA tractor. Double-acting valve used for control of remote work cylinder. Release speed is adjusted by means of the screws "A" and "B".

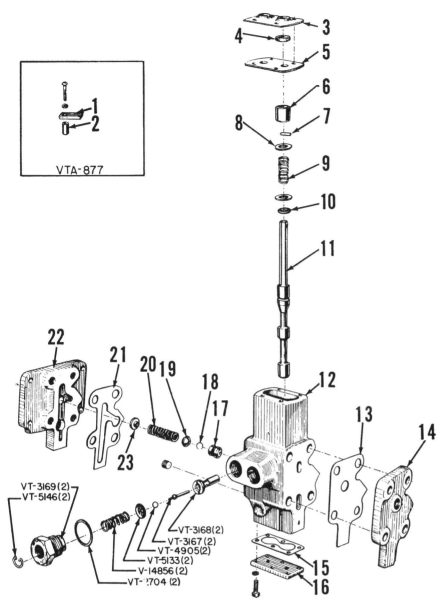

ling depth control cam cover and control lever and before installing the lever quadrant rotate control lever rearwards until same is horizontal. Temporarily install right hand lift arm after inserting a shim washer, Fig. C262, if a washer was previously removed.

Push lift arm all the way in then check to see if control arm is free but has not more than .010 end play when moved up and down half an inch each way. Obtain this condition by adding or removing shims from back face of lift arm. The smaller diameter shim washers, Fig. C262A, fit inside the spline bore of the right hand lift arm on outer end of rockshaft and should be varied in number to hold arm securely to rockshaft but without any end thrust.

Caution—Be sure rockshaft housing attachment bolts are correct length front and rear. Bolts that are too long for the front holes will interfere with the final drive gears. Front bolts 1¾ inches long are for early housings; bolts of 2 inches length are used at front on reinforced housings.

340. REMOTE CYLINDER. Procedure for removal of remote cylinder is self-evident. To disassemble cylinder first remove the cap screws (6—Fig. C280) which holds the limit stop (8) to the cylinder head (12); then extract the snap ring (7) from open end of cylinder. Wash all parts, renew seals and "O" rings. Renew piston rod and/-or cylinder sleeve if rubbing surfaces of same are scratched, flaked or pitted.

Fig. C277—Series VA tractor. Disassembled view of double-acting valve used for control of remote cylinder.

1. Valve stop	9. Spring	17. Relief valve seat
2. Valve stop bushing	10. "O" ring	18. Relief valve ball
3. Dust seal shield	11. Lift valve	19. Washer
4. Felt dust seal	12. Control body	20. Relief valve spring
5. Top cover	13. Gasket	22. Adapter
6. Spacer	14. End cover	23. Spring pressure
7. Pin	15. Gasket	spacer
8. Washer	16. Bottom cover	

tory, it usually can be made so by eliminating a possible restriction in the two stage check valves. Kits containing the component parts, Fig. C277, for this installation are available from Case under part #VTA874.

338. ROCKSHAFT AND CYLINDER. The rockshaft hydraulic cylinder can be overhauled without removing rockshaft from tractor. Procedure for removal of cylinder is evident from an examination of Fig. C278. If cylinder piston cup is to be renewed, the cup should be centered in cylinder before being anchored to the piston. This

is done by placing the cup and plate on the piston with the screws started but not tightened. Insert piston and cup assembly into cylinder with open end down as shown in Fig. C279; then tighten the screws securely. Remove piston from cylinder and install same in operating position. On later production, an "O" ring is used instead of the leather cup.

339. Procedure for removal, disassembly and overhaul of the rockshaft and lift arms is self-evident. All splines on rockshaft are cut so that no mistake can be made in assembling the lift arms and the center arm. After instal-

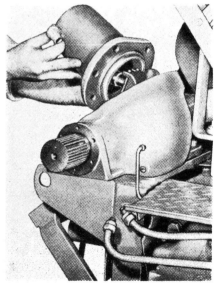

Fig. C278—Series VA tractor. Removing rockshaft cylinder and piston from housing.

341. OIL FILTER. On VA tractors after serial 5458558, the hydraulic system includes an edge wound paper cylinder type filter mounted at the discharge end of the system in the torque tube. The filter element should be removed and cleaned by washing in distillate or gasoline each time the oil is changed.

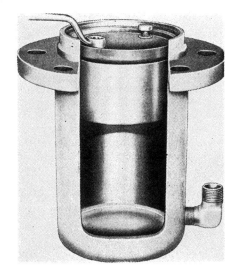

Fig. C279—Series VA tractor. Method of centering rockshaft piston cup before installation. Later pistons are sealed with an "O" ring.

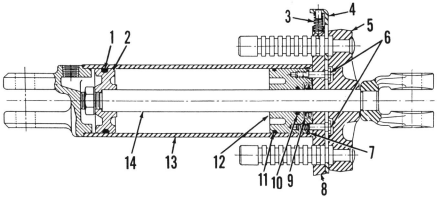

Fig. C280—Series VA tractor. To disassemble remote cylinder remove screws (6) and snap ring (7).

1. "O" ring	5. Limit stop	10. "O" ring
2. Piston	7. Snap ring	11. "O" ring
3. Detent	8. Limit stop guide	12. Cylinder head
4. Limit stop latch	9. Seal	14. Rod and yoke

MECHANICAL LIFT SYSTEM

Series C-D-R-S

345. R&R AND OVERHAUL. The mechanical type lift unit can be removed from the tractor without removing the pto shaft. To remove the lift unit, first remove pto shaft guard shield, and bearing carrier (15-Fig. C281). Using a suitable puller, remove pto shaft outer bearing cone (2), being careful not to dislodge the ball bearing from the forward end of the pto shaft. Remove cap screws attaching the lift housing to the rear face of the transmission housing rear cover. Support pto shaft and thread lift unit off of worm (4) by rotating the pto shaft while removing the lift unit. Worm (4) can now be removed by using a suitable puller.

The need and procedure for disassembly will be determined by an inspection of the unit and with reference to Fig. C281.

Reinstall the lift unit by reversing the removal procedure. Adjust pto shaft taper roller bearings by varying the number of shims (1) inserted between lift unit housing and bearing carrier to provide zero end play and a free rolling fit.

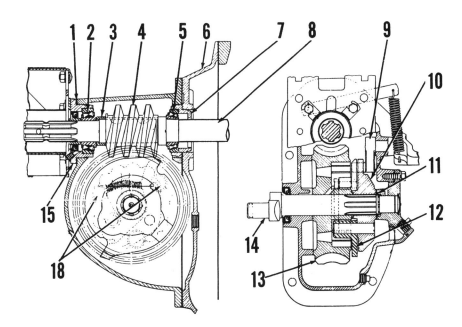

Fig. C281—Series S mechanical type power lift unit. Series C, D and R are similar.

1. Shims	7. Bearing carrier	12. Clutch driving dog
3. Spacer	8. PTO shaft	13. Worm gear
4. Worm	9. Release lever dog	14. Lift crank
6. Transmission housing rear cover	10. Driving plate	15. Bearing carrier
	11. Snap ring	18. Clutch auxiliary dog

APPENDIX

Pertaining to the spring loaded type engine clutch as used in the late series, non orchard type D and S tractors.

CLUTCH ADJUSTMENT
Series D

141A. Refer to Fig. C300. To check the adjustment of the linkage, first disconnect the clutch lever from the throwout rod; then check the free movement of the clutch lever. If the free movement is less than ⅜ inch when measured at the throwout rod clevis pin hole, it will be necessary to recondition the clutch unit to restore the free travel.

If clutch lever travel is more than ⅜ inch (an indication that the clutch lined plate is in good condition), push the clutch lever forward until free travel is eliminated. Place clutch pedal against platform stop and adjust throwout rod clevis until the rod is approximately ⅛ inch shorter than is necessary to connect it to the clutch lever. Reconnect the throwout rod to the clutch lever. Pedal should be against platform stop and have not less than 5/16 inch free movement.

Check the position of the pedal assisting spring. The spring should be located approximately ⅜ inch rearward of the center of clutch pedal pivot pin as shown in Fig. C300. To obtain this position, loosen nuts attaching spring hanger to transmission housing and move spring hanger either forward or rearward in its elongated mounting holes until the ⅜ inch measurement is obtained.

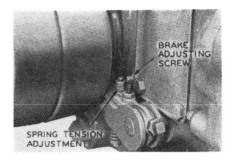

Fig. C301—Belt pulley brake as installed on series D and S tractors.

141B. After adjusting the engine clutch linkage, check the belt pulley brake to be sure it places a slight drag on the pulley when the clutch pedal free travel is taken up. Pulley should be free to rotate when clutch pedal is against its stop. To adjust the pulley brake, rotate brake adjusting screw, Fig. C301, counter-clockwise to apply the brake earlier. Adjust belt pulley brake spring to a point where the pulley brake prevents the pulley from slipping when the clutch is fully disengaged.

Fig. C302—Clutch pedal linkage as installed on a series S tractor equipped with a spring loaded type clutch.

Series S

141C. Refer to Fig. C302. To check adjustment of the linkage, disconnect clutch lever from throwout rod; then check free movement of clutch lever. If free movement is less than ⅜ inch when measured at throwout rod clevis pin hole, it will be necessary to recondition the clutch unit to restore free travel.

If clutch lever free travel is more than ⅜ inch (an indication that the clutch lined plate is in good condition), vary the length of the throwout rod to obtain a ⅜-½ inch free pedal travel as shown.

141D. Check pulley brake adjustment and adjust same as outlined in appendix paragraph 141B.

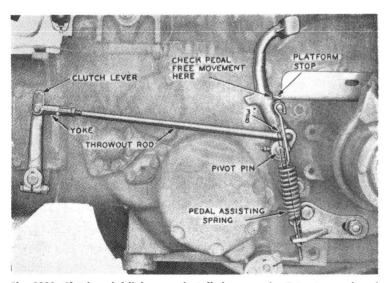

Fig. C300—Clutch pedal linkage as installed on a series D tractor equipped with a spring loaded type clutch.

R & R CLUTCH UNIT
Series D-S

144C. To remove clutch assembly, first detach engine from transmission housing as outlined in paragraph 40A; then proceed as follows: Mark clutch cover and flywheel to assure correct alignment when reinstalling. Install three ⅜ x 2 inches long cap screws on series S or ⅜ x 2¼ inches long cap screws on series D as shown in Fig. C303 so as to partially compress the clutch pressure plate springs. This procedure will facilitate both removal and reinstallation. Remove cap screws attaching clutch cover to flywheel and lift assembly from flywheel.

144D. When reinstalling clutch assembly use a dummy shaft or clutch lined plate aligning tool to obtain and maintain alignment of the lined plate splines. Use three cap screws (⅜ x 2 inches for series S or ⅜ x 2¼ inches

for series D) in the tapped holes of the pressure plate to keep the springs compressed until the assembly is attached to the flywheel.

144E. Remove the three special cap screws and adjust the release levers as shown in Fig. C304. Adjust release levers so that the release bearing contacting surface of each lever is flush with the hub of the clutch cover as shown.

144F. Reconnect engine to transmission housing and adjust the clutch pedal linkage as outlined in this appendix paragraph 141A or 141C; and belt pulley brake, paragraph 141B.

OVERHAUL CLUTCH UNIT
Series D-S

145C. Disassembly of the unit is self-evident after an examination and reference to Figs. C305, C306 and C307.

145D. Refer to Fig. C306. Clutch

units used in early production D series tractors are equipped with 15 pressure plate springs (Case part No. 09540AB) which require 195-205 pounds to compress each spring to a length of 1 13/16 inches. Later production D series clutches are equipped with 15 pressure plate springs (Case Part No. A7533) which require 245-255 pounds to compress each spring to a length of 1 13/16 inches. The heavier springs which are painted red for identification should be used when servicing early production units equipped with Case part No. 09540AB springs.

145E. Clutch units used in early production S series tractors are equipped with 12 pressure plate springs which require 140-150 pounds pressure to compress each spring to a length of 1 7/16 inches. Later production S series clutches are equipped with 15 pressure plate springs which

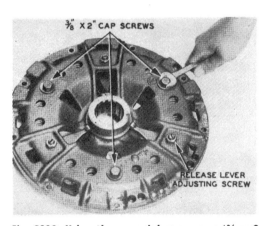

Fig. C303—Using three special cap screws (⅜ x 2 inches on series S or ⅜ x 2¼ inches on series D) to partially compress the clutch pressure plate springs.

Fig. C304—Adjust release levers of spring loaded type clutch so that the release bearing contacting surface of each lever is flush with the hub of the clutch cover.

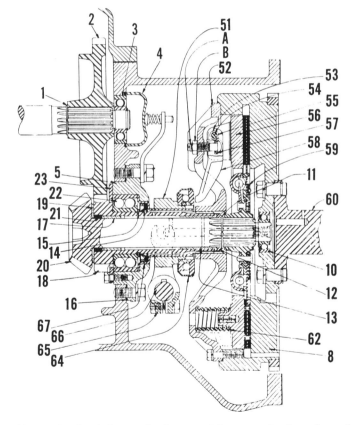

Fig. C305—Spring loaded type clutch as used in non orchard versions of the series D tractors. Non orchard type series S tractors are equipped with a similar clutch.

A. Adjusting screw	15. Snap ring	52. Clutch cover
B. Lock nut	16. Shim	53. Retaining ring
1. Snap ring	17. Bushing	54. Release lever pin
2. PTO drive shaft gear	18. PTO drive sleeve and gear	55. Lock spring
5. Bearing carrier	19. Thrust washer	56. Pressure plate
8. Flywheel	20. Clutch shaft	57. Release lever block
10. Pilot bearing	21. "O" ring (2 used)	58. Release lever
11. Snap ring	23. Retainer ring	59. Lined plate
12. Thrust collar	51. Release bearing sleeve	60. Crankshaft
13. Bushing		64. Release bearing
14. Oil seal		

have a similar rating. When servicing clutches with 12 springs, install three additional springs as shown in Fig. C307.

145F. Check condition of clutch release bearing, clutch shaft pilot bearing, pressure plate, clutch cover, friction face of flywheel, and release levers. Thickness of a new lined plate assembly is .340-.357.

Release levers are adusted after attaching the clutch assembly to the flywheel. Adjust release levers so that the bearing contacting surface of each lever is flush with the hub of the clutch cover as shown in Fig. C304.

CLUTCH SHAFT & SEALS

147B. The clutch shaft, as used in these series, is integral with the transmission main drive bevel pinion. For clutch shaft removal and overhaul procedures, refer to paragraphs 165, 165A and 165B which start on page 54.

Fig. C306—Spring loaded type clutch as used on series D tractors.

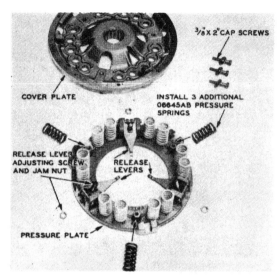

Fig. C307—Spring loaded type clutch as used on Series S tractors.

NOTES

CASE

Models ■ 200B ■ 300 ■ 300B ■ 350 ■ 400B ■ 500B ■ 600B

Previously contained in I & T Shop Service Manual No. C-11

SHOP MANUAL

J. I. CASE

SERIES 200B-300-300B-350-400B-500B-600B

Tractor serial number is stamped on dash name plate.

Engine serial number is stamped on side of cylinder block.

BUILT IN THESE VERSIONS

GASOLINE	LOW COST	LP-GAS	DIESEL	
210B, 310, 310B, 350, 410B, 510B, 610B	310, 310B	310, 310B, 410B, 510B, 610B	300, 300B	Utility, non-adjustable front axle
310B, 410B	310B	310B, 410B	300B	Utility, adjustable front axle
211B, 311, 311B, 312, 312B, 351, 411B, 511B, 611B	311, 311B, 312, 312B	311, 311B, 312, 312B, 411B, 511B, 611B	301, 301B, 302, 302B	General purpose dual wheel tricycle
211B, 311, 311B, 312B, 351, 411B, 511B, 611B	311, 311B, 312B	311, 311B, 312B, 411B, 511B, 611B	301, 301B, 302B	General purpose adjustable front axle
312, 312B, 351, 511B, 611B	312, 312B	312, 312B, 511B, 611B	302, 302B	General purpose single wheel tricycle

INDEX (By Starting Paragraph)

3

CONDENSED SERVICE DATA

Tractor Series	200B	300 & 300B Non-Diesel	300 & 300B Diesel	350	400B	500B	600B
GENERAL							
Engine Make	Own	Own	Continental	Own	Own	Own	Own
Engine Model	G126	G148	GD157	G164	G148	G164	G164
Cylinders, Number of	4	4	4	4	4	4	4
Bore — Inches	3⅛	3⅜	3⅜	3 9/16	3⅜	3 9/16	3 9/16
Stroke — Inches	— 4⅛ —			4⅜	— 4⅛ —		
Displacement—Cubic Inches	126.5	148	157	164	148	164	164
Compression Ratio—Gas & Diesel	7.5:1	7.25:1	15.5:1	7.25:1	7.25:1	7.25:1	7.25:1
Compression Ratio—LP-Gas or High Altitude	....	8.5:1	....	8.5:1	8.5:1	8.5:1	8.5:1
Pistons Removed From	Above	Above	Above	Above	Above	Above	Above
Main Bearings, Number of	3	3	3	3	3	3	3
Main & Rod Bearings Adjustable	No	No	No	No	No	No	No
Cylinder Sleeves—Type	Wet	Wet	Wet	Dry	Wet	Dry	Dry
Forward Speeds	4, 8 or 12	4, 8 or 12	4, 8 or 12	4, 8 or 12	4 or 8	4, 8 or 12	4 or 8
Generator & Starter Make	— Auto-Lite —				— Auto-Lite —		
Ignition Magneto	Case 41	Case 41	....	Case 41	Case 41	Case 41	Case 41
Battery Terminal Grounded	Positive	Positive	Positive	Positive	Positive	Positive	Positive
TUNE-UP							
Firing Order	1-3-4-2	1-3-4-2	1-3-4-2	1-3-4-2	1-3-4-2	1-3-4-2	1-3-4-2
Valve Tappet Gap	0.014C	0.014C	0.014H	0.014C	0.014C	0.014C	0.014C
Inlet Valve Face Angle	30°	30°	44°	30°	30°	30°	30°
Inlet Valve Seat Angle	30°	30°	45°	30°	30°	30°	30°
Exhaust Valve Face Angle	45°	45°	44°	45°	45°	45°	45°
Exhaust Valve Seat Angle	45°	45°	45°	45°	45°	45°	45°
Distributor Breaker Gap	0.020	0.020	....	0.020	0.020	0.020	0.020
Magneto Breaker Gap	0.010	0.010	....	0.010	0.010	0.010	0.010
Ignition Timing	— See Paragraph 82 or 83 —						
Injection Timing	....	....	26° BTC	....	....	....	....
Plug Electrode Gap	0.025	0.025	....	0.025	0.025	0.025	0.025
Carburetor Make, Gas or Low Cost	M-S	M-S	....	M-S	M-S	Zenith	M-S
Carburetor Make, LP-Gas	....	**	....	....	Ensign	Ensign	Ensign
Engine Loaded RPM	1900	***	1750	1900	2000	2000	2250
Engine High Idle, No Load RPM	2050	#	1950	2050	2125	2125	2400
PTO RPM, No Load	624	##	593	572	584	588	658
PTO RPM @ Engine Loaded RPM	533*	533*	533	530	541†	533††	541†
SIZES—CAPACITIES—CLEARANCES							
(Clearances in Thousandths)							
Crankshaft Journal Diameter	2.2495	2.2495	2.3745	2.6235	2.2495	2.6235	2.6235
Crankpin Diameter	1.937	1.937	2.062	2.061	1.937	2.061	2.061
Camshaft Journal Diameter—Front	1.7495	1.7495	1.809	1.7495	1.7495	1.7495	1.7495
Camshaft Journal Diameter—Center	1.7495	1.7495	1.746	1.7495	1.7495	1.7495	1.7495
Camshaft Journal Diameter—Rear	1.7495	1.7495	1.68375	1.7495	1.7495	1.7495	1.7495
Piston Pin Diameter	0.8592	0.8592	1.1251	0.99905	0.8592	0.99905	0.99905
Valve Stem Diameter—Intake	0.341	0.341	0.341	0.341	0.341	0.341	0.341
Valve Stem Diameter—Exhaust	0.3386	0.3386	0.3386	0.3386	0.3386	0.3386	0.3386
Main Bearings Running Clearance	1.5-2.0	1.5-2.0	0.9-3.6	1.5-2.0	1.5-2.0	1.5-2.0	1.5-2.0
Crankshaft End Play	3-5	3-5	4-6	3-7	3-5	3-7	3-7
Camshaft End Play	3-7	3-7	3-7	3-7	3-7	3-7	3-7
Rod Bearings Running Clearance	1.5-2.0	1.5-2.0	0.6-3.1	1.5-2.0	1.5-2.0	1.5-2.0	1.5-2.0
Piston Skirt Clearance	— See Paragraphs 27 and 28 —						
Camshaft Bearings Diameter Clearance	2-4	2-4	3-4	2-4	2-4	2-4	2-4
Cooling System—Gallons	3.0	3.0	3.0	3.0	3.0	3.0	3.0
Crankcase Oil—Quarts, with Filter	5.0	5.0	6.0	5.0	5.0	5.0	5.0
Transmission—Quarts							
Utility	17	17	17	44	17	44	44
General Purpose	28	28	28	44	28	44	44
Torque Tube—Quarts							
4-Speed	14	14	14	14	16	14	16
8 or 12-Speed	12	12	12	12	16	12	16
TIGHTENING TORQUES—Ft.-Lbs.							
Cylinder Head	60-65	60-65	70-75	95-100	60-65	95-100	95-100
Connecting Rods	45-50	45-50	35-40	45-50	45-50	45-50	45-50
Main Bearings	90-100	90-100	85-95	90-100	90-100	90-100	90-100

*At 1750 engine rpm.

**Marvel-Schebler on 300; Ensign on 300B.

***1750 on 300; 1900 on 300B.

#1875 on 300; 2050 on 300B.

##572 on 300; 624 on 300B.

†At 1970 engine rpm.

††At 1925 engine rpm.

FRONT SYSTEM

Tricycle Models

1. Dual front wheels are mounted on a horizontal axle which is retained to the vertical spindle by a nut (N—Fig. C750) for models 211B, 301, 301B, 311, 311B & 411B. On models 302, 302B, 312, 312B, 351, 511B and 611B, the horizontal axle is retained by a snap ring. To adjust the wheel bearings, tighten the spindle nut until a slight drag is obtained when wheel is man-

ually turned. Then back-off the nut until drag is just removed and install the cotter pin.

1A. A sectional view showing the installation of models 302, 302B, 312, 312B, 351, 511B and 611B fork mounted single front wheel is shown in Fig. C750A. Axle is carried in taper roller bearings which can be adjusted to provide a noticeable drag by shims located between the bearing cones and the fork bearing blocks. Shims are 0.003 and 0.005 thick.

STEERING KNUCKLES

Models 210B-300-300B-310-310B-410B, Non-Adjustable Axle

2. An exploded view of a typical non-adjustable type front axle is shown in Fig. C751. To remove the steering knuckles, support front of tractor and remove wheel and hub assemblies. Disconnect tie-rod and drag

link and remove plugs (17). Remove screws (22) and bump king pins (16) from axle and knuckles. Ream the king pin bushings (15), after installation, to provide a clearance of 0.001-0.003 for the king pin. When reassembling, renew thrust bearings (21) if their condition is questionable.

Models 211B-301-301B-302B-311-311B-312B-351-411B-511B-611B

3. Exploded views of the general purpose adjustable front axles are shown in Fig. C752 and C752A. To remove the steering knuckles, support front of tractor, remove steering arm (13) and withdraw knuckle. Bushings (15) are pre-sized to provide a clearance of 0.003-0.007 for the knuckles. Be sure to renew thrust bearings (21) if their condition is questionable. When reassembling, vary the number of 0.010 thick shims (1) to remove all spindle end play without causing any binding.

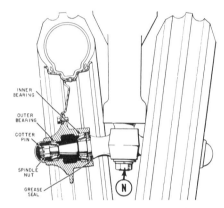

Fig. C750—Models 211B, 301, 301B, 311, 311B and 411B dual wheel tricycle front wheel installation. Wheel bearings should be adjusted to zero end play without binding. Models 302, 302B, 312, 312B, 351, 511B and 611B are similar except the horizontal axle is retained by a snap ring.

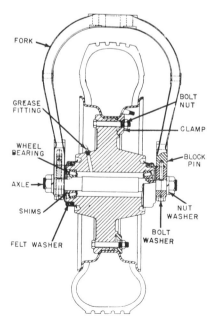

Fig. C750A — Sectional view of the fork mounted single front wheel used on models 302, 302B, 312, 312B, 351, 511B and 611B.

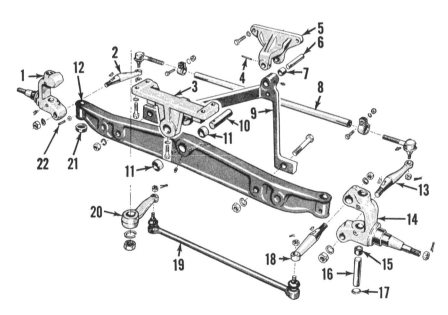

Fig. C751—Exploded view of a typical non-adjustable front axle used on models 210B, 300, 300B, 310, 310B and 410B utility tractors. Detailed differences in construction on some models will be evident after an examination of the installation.

1. Right knuckle	8. Tie rod tube	16. King pin
2. Right tie rod arm	9. Radius rod	17. Plug
3. Front pivot bracket	10. Axle pivot pin	18. Drag link arm
4. Roll pin	11. Bushing	19. Drag link
5. Rear pivot bracket	13. Left tie rod arm	20. Center steering arm
6. Radius rod pivot pin	14. Left knuckle	21. Thrust bearing
7. Bushings (2 used)	15. King pin bushings	22. Screw

Models 300B-310B-410B,
Adjustable Axle

3A. An exploded view of the adjustable front axle is shown in Fig. C752B. To remove the steering knuckles, support front of tractor, remove steering arm (13) and withdraw knuckle. Bushings (15) are pre-sized to provide a clearance of 0.003-0.007 for the knuckles. Be sure to renew thrust bearings (21) if their condition is questionable. When reassembling, vary the number of 0.010 thick shims (1) to remove all spindle end play without causing any binding.

Models 350-510B-610B

3B. An exploded view of the heavy-duty, non-adjustable front axle is shown in Fig. C752C. Knuckles (14) are carried in Torrington needle bearings (22). Before removing old bearings, note their exact position and install new ones in the same location. Install upper bearing, using Case arbor A-8354 or equivalent. Install lower bearing, using Case arbor A-8351 or equivalent. When installing the steering arms, tighten the retaining nut to remove all spindle end play without causing binding.

TIE-RODS AND DRAG LINK
All Models So Equipped

4. Tie-rod ends are of the renewable automotive type; whereas, the drag link and ends on some models are an integral unit. On models with two tie-rods, adjust the length of each rod an equal amount to provide a toe-in of $\frac{1}{8}$-$\frac{3}{16}$ inch. On utility models with one tie-rod, turn the tie-rod either way as required to obtain the recommended toe-in of $\frac{1}{8}$-$\frac{3}{16}$ inch.

AXLE PIVOT PINS AND BUSHINGS
General Purpose Models

5. Radius rod is integral with the axle center member. To renew the axle and radius rod pivot pins and bushings, support front of tractor, and remove cap screws retaining rear pivot bracket (5—Fig. C752 or C752A) and front pivot bracket to tractor. Raise front of tractor enough to clear the pivot brackets and roll axle and wheels assembly forward. Extract roll pin (4) and remove pivot pin (6). Remove bolt (17) and extract pivot pin (10). The need and procedure for further disassembly is evident. Bushings (7) are pre-sized to provide a clearance of 0.002-0.004 for the pivot pin (6). Bushings (11) are pre-sized to provide a clearance of 0.002-0.004 for the pivot pin (10).

Before installing the pivot bracket (3—Fig. C752), back-off the lever stud mesh adjusting screw (41—Fig. C756) to insure full seating of pivot bracket; then, when assembly is complete, adjust the lever stud mesh position as outlined in paragraph 7E.

Utility Models

6. Radius rod is bolted to axle main member. To remove the radius rod pivot pin and bushings, remove roll pin (4—Fig. C751, C752B or C752C) and bump pin (6) from pivot bracket. Unbolt and remove radius rod from axle main member. Bushings (7) are pre-sized to provide a clearance of 0.002-0.004 for the pivot pin (6).

To remove the axle pivot pin and bushings, unbolt pivot bracket (5) from tractor and remove the bolt retaining pivot pin (10) to front pivot bracket (3). Remove pivot pin (10) and allow axle to drop down enough to renew bushings (11). After installation, ream bushings (11) to provide a clearance of 0.001-0.003 for the pivot pin (10).

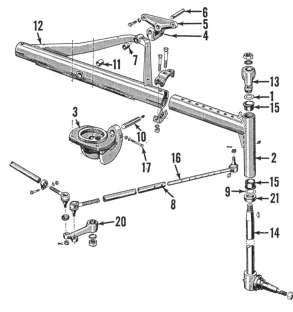

Fig. C752—Typical exploded view of the general purpose adjustable front axle used on models 211B, 301, 301B, 311 and 311B. The 411B is similar except for the details of the rear pivot bracket (5).

1. Shims (0.010)
2. Axle extension
3. Front pivot bracket
4. Roll pin
5. Rear pivot bracket
6. Radius rod pivot pin
7. Bushings
8. Tie rod tube
9. Washer
10. Axle pivot pin
11. Bushings
12. Axle center member
13. Knuckle steering arm
14. Knuckle
15. Bushings
16. Tie-rod extension
20. Center steering arm
21. Thrust bearing

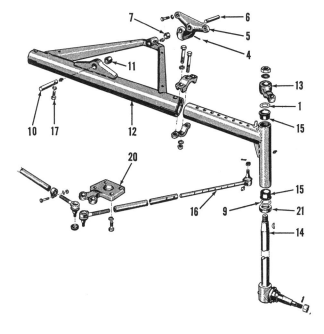

Fig. C752A — Exploded view of adjustable axle used on models 302B, 312B, 351, 511B and 611B. Refer to legend for Fig. C752.

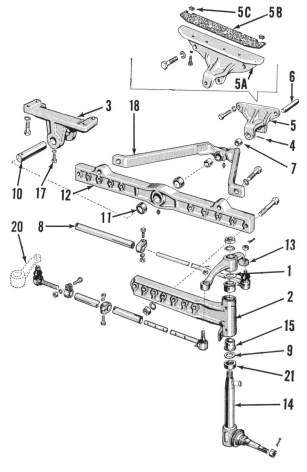

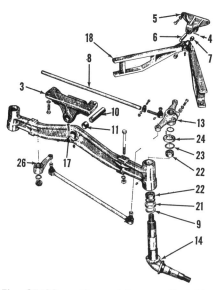

Fig. C752B — Exploded view of adjustable axle used on models 300B, 310B and 410B.

1. Shims (0.010)
2. Axle extension
3. Front pivot bracket
4. Roll pin
5. Rear pivot bracket (300B & 310B)
5A. Rear pivot bracket (410B)
5B. Gasket (410B)
5C. Gasket (410B)
6. Radius rod pivot pin
7. Bushings
8. Tie rod tube
9. Washer
10. Axle pivot pin
11. Bushings
12. Axle center member
13. Knuckle steering arm
14. Knuckle
15. Bushings
17. Bolt
18. Radius rod
20. Center steering arm
21. Thrust bearing

Fig. C752C — Heavy-duty non-adjustable front axle used on model 350. Steering knuckles (14) are carried in needle bearings (22).

3. Front pivot bracket
4. Roll pin
5. Rear pivot bracket
6. Radius rod pivot pin
7. Bushings
8. Tie-rod tube
9. Washer
10. Axle pivot pin
11. Bushings
13. Knuckle steering arm
14. Knuckle
17. Bolt
18. Radius rod
21. Thrust bearing
22. Needle bearing
23. Oil seal
24. Dust shield
26. Center steering arm

STEERING SYSTEM

The steering gear on early series 300 is a two-stud type; whereas, on other models the unit is a three-stud cam and lever type. On all models the wormshaft is carried in a separate housing on top of the front axle support.

For the purposes of this section, the steering gear unit will include, in addition to the worm (cam) shaft and lever, the lever shaft and bearings. On tricycle models, the lever shaft is catalogued a steering spindle.

7. **ADJUSTMENT.** On axle type tractors, the steering gear unit is provided with adjustments which control the worm (cam) shaft end play and the lever stud mesh position. Tricycle type tractors as well as adjustable axle models 302B, 312B, 351, 511B and 611B have these same adjustments plus an end play adjustment for the vertical steering spindle.

7A. WORM (CAM) SHAFT END PLAY. Remove the steering wormshaft and housing assembly as outlined in paragraph 7F. Remove the wormshaft rear bearing cover and vary the number of shims between cover and housing until the shaft rotates freely without binding. Shaft end

play not exceeding 0.002 is permissible. Shims are available in thicknesses of 0.002, 0.003 and 0.010. Refer to Fig. C753, C754 and C754A.

7B. STEERING SPINDLE END PLAY. (TRICYCLE MODELS 211B, 301, 301B, 311, 311B and 411B). Remove steering wormshaft and housing as-

Fig. C753 — Sectional view of wormshaft housing showing the shaft thrust bearings and adjusting shims. Upper end of lever shaft is carried in bushing (B) on some models.

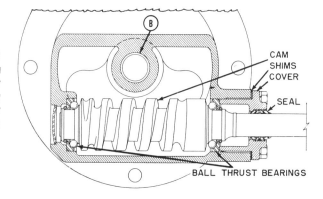

sembly as outlined in paragraph 7F. Mark the relative position of steering lever (50 or 50A—Fig. C755) with respect to the steering spindle (47) and remove the steering lever. Unlock the bearing adjusting nut (51) and with front end of tractor jacked up, tighten the nut to remove all spindle end play without causing any binding tendency. After adjustment is complete, lock the nut in position with tab washer (52). Before installing the wormshaft housing, back-off the mesh adjusting screw (41) to insure full seating of housing; then, when assembly is complete, adjust the stud mesh as outlined in paragraph 7E.

7C. STEERING SPINDLE END PLAY. (MODELS 302, 302B, 312, 312B, 351, 511B and 611B). Refer to Fig. C755A. Remove cover from top of gear housing, extract cotter pin and with front end of tractor jacked up, tighten nut (51) to remove all end play without causing any binding tendency. After adjustment is complete, install the cotter pin and cover.

7D. LEVER STUD MESH. The lever stud mesh (or backlash) on all axle models except 302B, 312B, 351, 511B and 611B can be adjusted as outlined in paragraph 7E. The same applies to tricycle models and axle models 302B, 312B, 351, 511B and 611B providing the steering spindle does not have excessive up and down play indicating loose or damaged roller bearings. If the spindle end play is O. K., proceed as in the following paragraph. If end play is excessive, first adjust the spindle bearings as in paragraph 7B or 7C.

7E. Raise front of tractor and locate mid-point or center position of steering gear by rotating steering wheel from full right to full left-turn position, then back half-way. Tighten the mesh adjusting screw (41—Fig. C755, C755A or C756) until a slight drag is felt only when turning the steering gear through this mid or high point position. The gear unit will be and should be free when off the mid-point.

7F. OVERHAUL. To overhaul the steering gear, drain cooling system and remove hood, grille, air cleaner and radiator. Remove radiator mounting plate and baffle plate from front support. Remove the frame left side rail as shown in Fig. C757 and remove the cap screw retaining the steering shaft bracket to the torque tube. Remove the roll pin (P) retaining the front universal joint to the wormshaft and bump the joint yoke off the

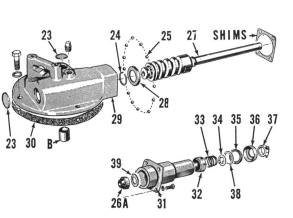

23. Plugs
24. Retainer ring
25. Bearing balls
26A. Oil seal
27. Worm (cam) shaft
28. Bearing cup
30. Gasket
31. Housing bearing cover
32. Bearing assembly
33. Spring
34. Spring seat
35. Felt seal
36. Seal cup
37. Clamp
38. Washer
39. Gasket

Fig. C754 — Exploded view of the steering worm (cam) shaft, housing and components typical of that used on all models except 302, 302B, 312, 312B, 351, 511B and 611B. Some early models were not fitted with items (38) and (39).

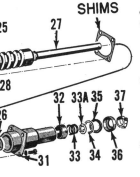

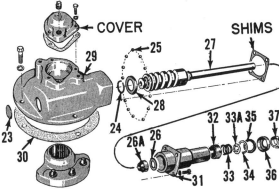

Fig. C754A — Exploded view of the steering worm (cam) shaft, housing and components used on models 302, 302B, 312, 312B, 351, 511B and 611B.

23. Expansion plug
24. Retainer ring
25. Bearing balls
26. Seal gasket
26A. Oil seal
27. Worm (cam) shaft
28. Bearing cup
29. Housing
30. Gasket
31. Housing bearing cover
32. Bearing assembly
33. Spring
33A. Spring seat
34. Washer
35. Felt seal
36. Seal cup
37. Clamp

Fig. C755—Models 211B, 301, 301B, 311, 311B and 411B dual wheel tricycle pedestal, vertical steering spindle, steering lever and associated parts. The two-stud lever (50A) is used only on some models of the 300 series.

38. "O" ring	48. Woodruff key
39. Adjusting plug	50. Steering lever
40. Pedestal	(three stud)
41. Set screw	50A. Steering lever
42. Jam nut	(two stud)
43. Bearing cup	51. Adjusting nut
44. Bearing cone	52. Lock plate
45. Thrust washer	53. Bearing cone
46. Oil seal	54. Bearing cup
47. Steering spindle	55. Oil seal

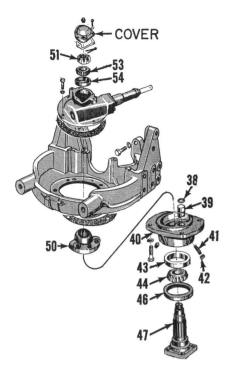

Fig. C755A—Models 302, 302B, 312, 312B, 351, 511B and 611B pedestal, vertical spindle, steering lever and associated parts. Refer to legend under Fig. C755.

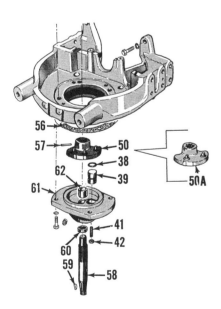

Fig. C756—Steering lever, shaft and support typical of that used on axle type tractors except 302B, 312B, 351, 511B and 611B. The two-stud lever (50A) is used only on some models of the 300 series.

38. "O" ring
39. Adjusting plug
41. Set screw
42. Jam nut
50. Steering lever (three stud)
50A. Steering lever (two stud)
56. Gasket
57. Roll pin
58. Lever shaft
59. Woodruff key
60. Oil seal
61. Lever shaft support
62. Bushing

shaft. On diesel models only, unbolt the radius rod pivot bracket from torque tube and loosen the screws retaining the front support to the engine. On models 302, 302B, 312, 312B, 351, 511B and 611B, remove cover from top of housing, extract cotter pin and remove nut (51—Fig. C755A). On all models, remove the cap screws retaining wormshaft housing to the front support and remove the housing assembly as shown in Fig. C758.

7G. Procedure for bench overhaul of the removed unit is evident after an examination of Fig. C754 or C754A. If ball bearings are lumpy after they have been washed in solvent, remove the snap rings (24) and inspect. The inner race for each bearing is integral with the wormshaft which should be renewed if pitted or otherwise damaged. Outer races and individual balls (28 of them) are catalogued separately. Bushing (B) used on all models except 302, 302B, 312, 312B, 351, 511B and 611B, is located by a chamfer which is obvious upon examination and bushing should be burnished after installation to provide a clearance of 0.0005-0.0025 for upper end of lever shaft. On late models, seal (26A) is positioned by a flange on O.D. of seal and a washer (39) is used between seal and housing. On early models, seal is not fitted with a flange and seal should be pressed into housing until rear of seal is flush with rear face of seal boss in housing. Lip of seal faces toward front of tractor on all models. When reassembling, vary the number of shims between bearing cover (31) and housing until the wormshaft rotates freely without binding. Shaft end play not exceeding 0.002 is permissible. Shims are available in thicknesses of 0.002, 0.003 and 0.010. When unit is assembled, lay it aside for subsequent installation.

NOTE: Both right hand and left hand cams have been used with the 3-stud levers. Be sure the proper one is used. Cams can be identified by the last two digits of the number stamped on them. For right hand cams, the last two digits are "23." For left hand cams, the last two digits are "22."

7H. On axle type tractors except 302B, 312B, 351, 511B and 611B, remove the center steering arm, extract roll pin (P—Fig. C758) and withdraw the lever shaft. If bushing (62—Fig. C756) is damaged, install a new one and ream it to provide a clearance of 0.001-0.003 for the lever shaft. Install oil seal (60) with lip of same facing upward and be sure that lower surface is slightly above lower surface of housing to avoid seal damage from interference with center steering arm. Renew the steering lever (50 or 50A) if any of the studs are worn or damaged.

The roll pin hole in replacement steering levers is drilled through one side only. When installing the lever, align hole in lever with hole in shaft; then, using the holes as a pilot, drill straight through other side of lever with a ¼-inch drill. Be sure to remove chips from housing.

7J. On tricycle type models except 302, 302B, 312, 312B, 351, 511B and 611B, mark the relative position of

Fig. C757—Side view of a typical non-diesel engine with side rail partially removed. Steering shaft front universal joint is retained to wormshaft by roll pin (P).

Fig. C758 — Steering wormshaft and housing assembly removed from front support. Upper end of lever shaft is supported in bushing (B) on all models except 302, 302B, 312, 312B, 351, 511B and 611B.

lever (50 or 50A—Fig. C755) with respect to the lever shaft (spindle) and remove the lever. Remove front wheels and bearing adjusting nut (51). Withdraw the steering spindle. The need and procedure for further disassembly is evident. Install oil seals (46 and 55) with lips facing upward. Tighten the bearing adjusting nut (51) to remove all spindle end play without causing any binding tendency. After adjustment is complete, lock the nut in position with tab washer (52). Renew the steering lever (50 or 50A) if any of the studs are worn or damaged.

7K. On models 302, 302B, 312, 312B, 351, 511B and 611B, mark the relative position of lever (50—Fig. C755A)

with respect to the lever shaft (upper spindle) and remove the lever. Raise front of tractor and withdraw spindle (47). When reassembling, install seal (46) with lip facing upward. Renew lever (50), if any of the studs are worn or damaged.

7L. Before installing the wormshaft housing on all models, back-off the mesh adjusting set screw (41—Fig. C755, C755A or C756) to insure full seating of housing. On models 302, 302B, 312, 312B, 351, 511B and 611B, adjust the spindle end play as in paragraph 7C. On all models, adjust the stud mesh as outlined in paragraph 7E.

POWER STEERING SYSTEM

The power steering attachment for gasoline utility models is of the linkage booster type and is powered by either an Eaton belt driven pump with a separately mounted reservoir or a Barnes gear driven pump with an integral reservoir. The power steering attachment for gasoline and LPG general purpose tractors uses the same Barnes pump as

the utility tractors and a Char-Lynn booster motor is incorporated in the steering linkage.

Note: The maintenance of absolute cleanliness of all parts is of utmost importance in the operation and servicing of the hydraulic power steering system. Of equal importance is the avoidance of nicks and burrs on any of the working parts.

LUBRICATION AND BLEEDING

8. The J. I Case Company specifies that only Automatic Transmission Fluid Type A should be used in the power steering system. On systems using a belt driven pump, fluid level should be maintained at full mark on reservoir dip stick. On systems with

POWER STEERING SYSTEM TROUBLE SHOOTING CHART

	Loss of Power Assistance	Power Assistance In One Direction Only	Unequal Turning Radius	Erratic Steering Control	Fluid Foaming Out of Reservoir
Binding, worn or bent mechanical linkage	★		★	★	
Insufficient fluid in reservoir	★				
Faulty pump drive belt	★			★	
Low pump pressure	★				
Internal leakage in booster motor	★				
Sticking or binding valve spool	★	★		★	
Improperly positioned steering shaft	★	★			
Damaged or restricted hose or tubing	★	★			★
Pump to valve hose lines reversed	★				
Wrong fluid in system	★			★	★
Improperly adjusted tie rods			★	★	
Faulty steering lever timing			★		
Air in system				★	★
Faulty flow control valve in pump				★	
Plugged filter element					★
Internal leak in valve					★
Faulty cylinder	★			★	

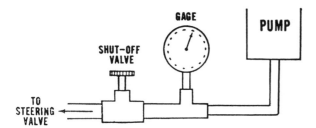

Note: To avoid heating the steering fluid excessively, don't hold wheels in this test position for more than 30 seconds.

If the pressure is higher than specified, the pump relief valve is probably stuck in the closed position. If the pressure is less than specified, turn the wheels to straight ahead position; at which time, the gage reading should drop considerably. Now slowly close the shut-off valve and retain in closed position only long enough to observe the gage reading. Pump may be seriously damaged if valve is left in closed position for more than 10 seconds. If the gage reading increases to 850-900 psi for models with a belt driven pump or 1050-1150 psi for models with a gear driven pump, with valve closed, the pump and relief valve are O. K. and the trouble is located in the control valve, the power cylinder or connections.

While a low pressure reading with the shut-off valve closed indicates the need of pump or relief valve repair, it does not necessarily mean the remainder of the system is in good condition. After overhauling the pump and/or relief valve, recheck the pressure reading to make sure the control valve, cylinder and connections are in satisfactory operating condition.

Refer to paragraph 8C for data concerning the flow control and relief valve used on belt driven pumps or paragraph 8D for data concerning the relief valve on gear driven pumps.

FLOW CONTROL AND RELIEF VALVE

Models With Belt Driven Pump

8C. This combination valve can be removed from the pump by removing the outlet hose and valve cap fitting.

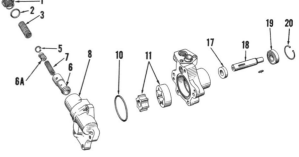

Fig. C758A — Pressure gage and shut-off valve installation for trouble-shooting the power steering system.

a gear driven pump, fluid level should be maintained 1-inch below filler opening.

On models with a belt driven pump, filter element in fluid reservoir should be renewed every 1000 hours; or, more often in severe dust conditions. Spring in reservoir cover must exert a downward sealing pressure on filter cartridge to assure filtration of fluid. Note: A plugged filter element can be the cause of fluid bubbling out of the filler cap air vent.

On models with a gear driven pump, the reservoir breather should be removed and thoroughly cleaned in a suitable solvent every 150 hours of operation; or, more often in severe dust conditions. Loss of oil from the reservoir is an indication that reservoir is too full or that breather is not functioning properly.

To bleed both systems, fill the reservoir, start engine and run same until normal operating temperature is reached. With engine running at high idle speed, cycle the power steering system several times to bleed out any trapped air and refill the reservoir.

TROUBLE SHOOTING

8A. The accompanying table lists troubles which may be encountered in the operation of the power steering system. The procedure for correcting most of the troubles is evident; for those not readily remedied, refer to the appropriate subsequent paragraphs.

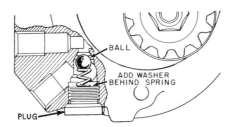

Fig. C758C—Cut-away view showing relief valve installation on gear driven pump.

SYSTEM OPERATING PRESSURE

8B. A pressure test of the hydraulic circuit will disclose whether the pump or some other unit in the system is malfunctioning. To make such a test proceed as follows:

Connect a pressure test gage and a shut-off valve in series with the pump pressure line as shown in Fig. C758A. Note that shut-off valve is connected in the circuit between the gage and the control valve. Open the shut-off valve and run the engine at idle speed until oil is warmed. With engine running at high idle speed, rotate front wheels to extreme right or extreme left turn position and note gage reading which should be in the range of 850-900 psi for models with a belt driven pump or 1050-1150 psi for models with a gear driven pump.

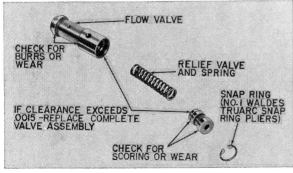

Fig. C758B—Exploded view of the combination flow control and relief valve used on belt driven pumps.

Fig. C758D—Exploded view of the belt driven power steering pump. Rotors (11) are available as a matched set only.

1. Valve cap fitting	6. Flow control valve
2. Seal ring	6A. Relief valve
3. Flow control valve spring	7. Relief valve spring
5. Snap ring	8. Pump cover
	10. "O" ring

11. Rotors
17. Seal
18. Pump shaft
19. Bearing
20. Snap ring

The spring and valve can then be lifted out without removing the pump from the engine. Valve can be disassembled by removing the snap ring (Fig. C758B) from flow valve. Remove any slight roughness or scratching from outside surface of flow valve and from its bore in pump cover, using crocus cloth. Check clearance of relief valve in flow control valve bore and if greater than 0.0015, install a new valve assembly as only the spring is available separately.

If valves are free in their bores and clearances are not excessive, but the system pressure test indicates a malfunctioning relief valve, renew the relief valve spring.

When reassembling, reverse the disassembly procedure and bleed the system as in paragraph 8.

RELIEF VALVE

Models With Gear Driven Pump

8D. If the system pressure test indicates a malfunctioning relief valve, remove the side panel, disconnect fluid lines and remove pump. Remove the relief valve plug (C758C) and extract the spring and ball. Thoroughly clean the ball and seat as required. Shim washers can be added behind the spring to increase the relief valve cracking pressure.

PUMP

Belt Driven Type

8E. R&R AND OVERHAUL. Procedure for removal and installation of the pump is self-evident. Order of disassembly is as follows: Thoroughly clean the pump and remove the adapter cover from top of pump body.

Remove pulley, separate the pump cover from the pump housing and lift the gasket out of the groove in the housing. Mark outer face of each rotor with chalk so that rotors will be reinstalled same way. Remove the outer (driven) rotor (Fig. C758D) from the housing.

To remove rotor shaft, extract the bearing snap ring (retainer) from pulley end of housing then press or bump shaft and bearing out of housing. Remove flow control and relief valve assembly from pump cover. Do not wash the sealed ball bearing.

Check pump body and cover faces for wear; also check the bushings in the body and cover. If pump body or bushing in body is worn, install a new body and bushing assembly. If the cover or bushing in cover is worn install a new cover and bushing assembly. None of the bushings are available separately.

Partially reassemble the pump and check clearance between rotors at teeth, with feeler gages as shown in Fig. C758E and between outer rotor and body insert as shown in Fig. 758F. If clearance between rotors exceeds 0.008; or if clearance between outer rotor and insert in pump body exceeds 0.006, install a new pump body and rotors assembly. Also check end clearance between end of rotors and body face as shown in Fig. C758G. If end

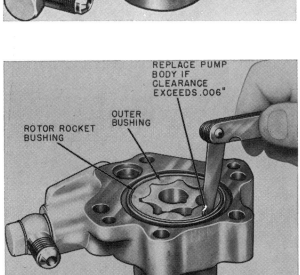

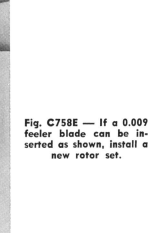

Fig. C758E — If a 0.009 feeler blade can be inserted as shown, install a new rotor set.

Fig. C758F — If a 0.007 feeler blade can be inserted between driven rotor and pump body insert, install a new pump body and rotors assembly.

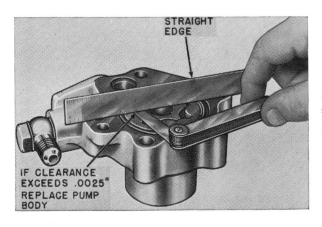

Fig. C758G—If a 0.0035 feeler blade can be inserted between faces of rotors and underside of straight-edge, install a new pump body and rotors assembly.

clearance exceeds 0.0025, install a new body and rotors assembly.

Always install new "O" rings and seals. Rotor shaft oil seal should be assembled with lip of same nearest the rotors.

Gear Driven Pump

8F. Component parts of the gear driven pump are not catalogued by the J. I. Case Company. Failure of the pump is corrected by renewing the complete pump and reservoir assembly.

CYLINDER AND VALVE
Gasoline Utility Models

8G. A partially exploded view of the cylinder and valve unit is shown in Fig. C758H. The outer sleeve and valve spool assemblies can be easily removed for servicing, but the piston and cylinder unit cannot be disassembled.

To disassemble the unit as shown, disconnect the linkage and oil lines and remove the three assembly bolts. Remove ball stud shield and be sure to remove grease fitting (3) from outer sleeve. Withdraw the sleeve and spool assembly from valve housing. Slide the ball stud forward in the sleeve as far as possible and drive the

lock pin from the stop screw. Turn the spool and stop screw from the inner sleeve and remove the spring, spring stop, ball seats and ball stud. The need for further disassembly is evident and the overhaul procedure is conventional.

When reassembling the unit, be sure that all parts are absolutely clean and lubricate same with Automatic Transmission Fluid, Type A. Slide one of the ball stud seats into the inner sleeve and position inner sleeve in outer sleeve. Slide ball stud into sleeve; then, install the other ball stud seat, spring and spring stop. Using a small punch, inserted through tie-rod end of outer sleeve, push inner sleeve, with seats and stud, to flanged end of outer sleeve.

Push the valve spool bolt through the torque adjusting stop screw (6) and turn the stop screw into inner sleeve as far as it will go. Back-off the stop screw until the nearest hole in same lines up with slots in inner sleeve and install the lock pin (5) as shown in Fig. C758J. Move ball stud and inner sleeve assembly toward tie-rod end of outer sleeve and make certain that lock pin is not preventing the assembly from traveling its full stroke.

Install spool spacer on bolt, position spool into housing with "O" rings and large and small bushings in position and make certain that large bushing and "O" ring is toward outer sleeve. Place the flange plate on end of spool, slide spool bolt through spool and tighten the lock nut. Using the three assembly bolts and two small "O" rings, secure valve assembly to ram cylinder. Assemble the remaining parts and install the tie-rod end three full turns past end of slot in outer sleeve.

When installing the unit on tractor, tighten the jam nut so that one thread on rod is extending beyond the nut as shown in Fig. C758K.

BOOSTER MOTORS
Gasoline & LPG General Purpose Tractors

8H. An exploded view of the components making up the installation of the Char-Lynn power steering booster motor are shown in Fig. C758L. None of the internal parts of the motor are catalogued by the J. I. Case Company. If the motor becomes inoperative, the complete unit must be renewed.

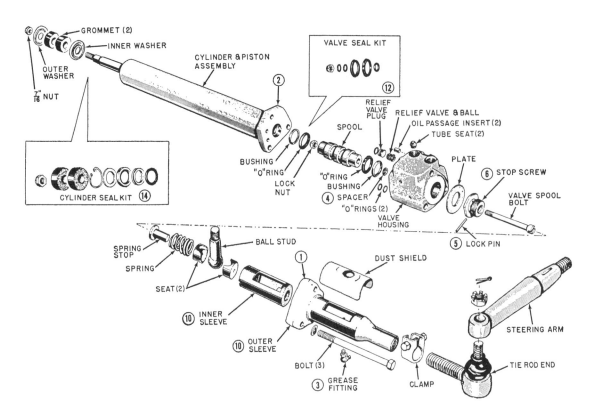

Fig. C758H—Partially exploded view of the gasoline utility power steering cylinder and valve unit. Component parts of the cylinder and piston unit are not available.

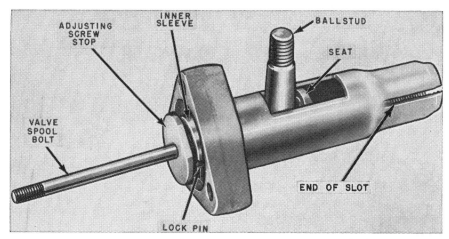

Fig. C758J—View showing proper assembly of outer sleeve, screw stop, lock pin and associated parts.

The procedure for removing the motor from the tractor is obvious, but when installing the unit it is extremely important that extra care be exercised to maintain exact alignment of the input and output shafts. Also, check the following:

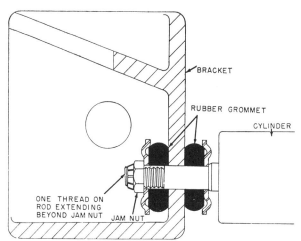

Fig. C758K—Power steering cylinder is properly installed when one thread extends beyond the nut as shown.

   a. Steering gear adjustment, paragraphs 7 through 7E.

   b. Steering gear lubricant level.

   c. Front wheel toe-in.

   d. Lubrication of tie-rod and drag link ends.

   e. Adjustment of front wheel bearings.

   f. Front tire inflation pressure.

Install the frame side rail, using the proper spacer between rear end of rail and torque tube. On the 211B, 311B and 511B, use one ½-inch thick spacer. On the 411B and 611B, use one 3/16-inch thick spacer.

Install the ⅝x1¼-inch bolt through side rail and booster motor mounting bracket, but notice that the lower bracket hole is used on the 411B and 611B and that the upper hole is used on all other models. With all mounting bolts finger tight, check the alignment of all parts and tighten the bolts securely. Make certain that upper steering shaft turns freely in its sleeve.

Fill and bleed the power steering system as in paragraph 8; then, with engine running at 1000 rpm, turn the steering wheel to the left and right several times to check the operation of the unit. The steering unit should "center" (return slightly from extreme end positions). Never hold steering wheel in extreme positions for more than 30 seconds.

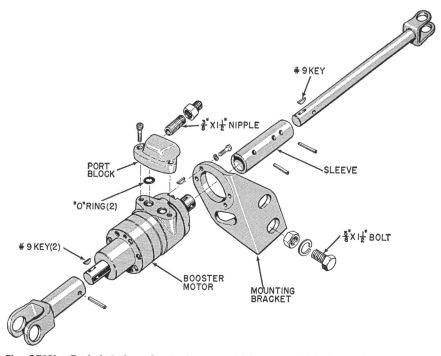

Fig. C758L—Exploded view of only the parts which are available for servicing the Char-Lynn power steering booster motor. None of the internal parts of the motor are catalogued by J. I. Case. On series 211B and 311B the port block is in the up position as shown. On series 411B, 511B and 611B, the booster motor is rotated 180 degrees with the port block down.

ENGINE AND COMPONENTS

R&R ENGINE WITH CLUTCH
All Models

9. To remove the engine and clutch as a unit, first drain cooling system and if engine is to be disassembled, drain oil pan.

Remove the hood, grille, air cleaner and radiator. Remove both of the frame side rails, unbolt the steering shaft intermediate support bracket and disconnect steering shaft at U-joint. Support tractor under torque tube and on axle type tractors, unbolt the radius rod pivot bracket. Unbolt the front support from engine and roll the front support, wheels and axle assembly forward and away from tractor.

Disconnect fuel line and remove throttle rod and fuel tank. On models with underneath muffler, disconnect exhaust pipe from manifold. Disconnect wiring harness from generator and starting motor and remove starting motor. Disconnect the oil pressure gage line from cylinder block and the heat indicator sending unit from cylinder head. On non-diesels, disconnect wire from coil and choke wire from carburetor. On diesels, disconnect the fuel shut-off wire at injection pump. On all models, swing engine in a hoist, then unbolt and remove engine from torque tube.

The extent to which the engine is disassembled subsequent to its removal will, to some extent, govern the installation procedure. In general, however, the engine can be installed by reversing the removal procedure. Frame side rail cap screws and engine to torque tube bolts should be tightened to a torque of 115 ft.-lbs.

CYLINDER HEAD
Non-Diesel Models

10. To remove the cylinder head, first drain cooling system and on series 300 and 350, raise the hood panel. On other models, remove hood. Disconnect wires from coil and remove the valve cover, rocker arms assembly and push rods. Disconnect the radiator support brackets and water outlet casting from front of head. Disconnect the spark plug wires and re-move the spark plugs. Disconnect the heat`indicator sending unit from head and remove manifold. Remove the cylinder head retaining stud nuts and lift cylinder head from engine.

When reassembling, install head gasket with side marked "TOP" facing up. Tighten the head retaining stud nuts in the sequence shown in Fig. C759 and to a torque of 60-65 ft.-lbs. for series 200B, 300, 300B and 400B; 95-100 ft.-lbs. for series 350, 500B and 600B. Other pertinent torque specifications are as follows:

Manifold stud nuts.......30 ft.-lbs.
Manifold to carburetor
   bolts ....15 ft.-lbs.
Spark plugs ....34 ft.-lbs.
Rocker arm bracket
   nuts ....20 ft.-lbs.

Retighten the cylinder head nuts to the specified torque value after engine is warmed to operating temperature. Valve tappet gap should be adjusted to 0.014 cold. Firing order is 1-3-4-2.

Diesel Models

11. To remove the cylinder head, first drain cooling system, raise the hood panel and proceed as follows: Disconnect the head light wires at snap couplers on each side of battery. Open grille and disconnect intake pipe from air cleaner. Remove both of the engine side panels. Remove cap screws retaining hood and sheet metal assembly to tractor, spread side sheets sufficiently to clear dash, slide the complete assembly forward and remove same from tractor.

Remove the valve cover, rocker arms assembly and push rods. Note: When removing push rods, snap them side-ways out of the tappet sockets. This method serves to break the hydraulic connection and permits lifting the push rods out and leaving the cyl-indrical type tappets in place. Disconnect the radiator support brackets and unbolt water pump from front of head. Remove the air inlet and exhaust manifold. Disconnect the high pressure and return lines from injection nozzles and immediately cap the fuel line connections to eliminate the entrance of dirt. Disconnect oil lines and remove the oil filter. Remove the number four cylinder injection nozzle and disconnect the heat indicator sending unit from head. Disconnect the necessary fuel lines and unbolt the fuel filter mounting bracket from head. Remove the cylinder head retaining stud nuts and lift cylinder head from engine.

When reassembling, tighten the stud nuts progressively from center outward and to a torque of 70-75 ft.-lbs. Tighten the manifold stud nuts to a torque of 25-30 ft.-lbs. and the nozzle retaining screws to a torque of 14-16 ft.-lbs.

Retighten the cylinder head nuts to the specified torque value after engine is warmed to operating temperature. Valve tappet gap should be adjusted to 0.014 hot. Firing order is 1-3-4-2.

VALVES AND SEATS
Non-Diesel Models

12. Intake and exhaust valves are not interchangeable and the intake valves seat directly in cylinder head with a face and seat angle of 30 degrees and a desired seat width of 0.066-0.078. Intake valve stem diameter is 0.3406-0.3414.

Exhaust valves should have a face and seat angle of 45 degrees and a stem diameter of 0.3382-0.3390. Exhaust valves on all models except some early production standard altitude gasoline and low cost fuel burning engines seat on renewable seat inserts. Desired seat width is 0.064-0.078.

Fig. C759 — Non-diesel engine cylinder head nut tightening sequence.

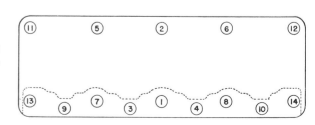

Valve seat inserts can be removed by using a suitable puller. Chill new inserts in dry ice before pressing them into position and make certain that seat bottoms in counterbore. Maximum allowable seat run-out is 0.002.

Valve tappet gap is 0.014 cold.

Diesel Models

13. Intake and exhaust valves are not interchangeable and the intake valves seat directly in cylinder head with a face angle of 44 degrees and a seat angle of 45 degrees. Exhaust valves seat on renewable seat inserts with a face angle of 44 degrees and a seat angle of 45 degrees. Desired width of the intake and exhaust valve seats is $\frac{1}{16}$-$\frac{3}{32}$-inch. Seats can be narrowed, using 15 and 70 degree stones. Valve stem diameter is 0.3406-0.3414 for the intake, 0.3382-0.3390 for the exhaust.

Exhaust valve seat inserts can be removed by using a suitable puller. Chill new inserts in dry ice before pressing them into position and make certain that seat bottoms in counterbore. Maximum allowable seat run-out is 0.002.

Valve tappet gap is 0.014 hot.

VALVE GUIDES AND SPRINGS

Non-Diesel Models

14. Intake and exhaust valve guides should be pressed out from bottom of head so they emerge at top of head.

On the 300, press new guides into top of cylinder head until top of guide protrudes $\frac{31}{32}$-inch above spring counterbore as shown in Fig. C760. On the 200B, 300B, 350, 400B, 500B and 600B, refer to Fig. C760A and press new guides into top of head until guide protrusion above the spring seat counterbore is $\frac{31}{32}$-inch for the exhaust, $1\frac{1}{32}$ inches for the intake. Ream the guides after installation to an inside diameter of 0.3422-0.3432. This provides a normal stem to guide clearance of 0.0008-0.0026 for the intake, 0.0032-0.0050 for the exhaust.

14A. Intake and exhaust valve springs are not interchangeable. Approximate valve spring free length is 2⅜ inches for the intake, $2\frac{3}{16}$ inches for the exhaust. Renew any spring which is rusted, discolored or does not meet the following pressure test specifications:

Intake Valve Spring

Pounds test @ 1½ inches......110
Pounds test @ 1⅞ inches....... 53

Exhaust Valve Spring

Pounds test @ $1\frac{11}{32}$ inches......110
Pounds test @ $1\frac{11}{16}$ inches....... 53

Install valve springs with closed or dampening coils against cylinder head.

Diesel Models

15. Intake and exhaust valve guides are interchangeable. Press old guides out from bottom of head so they emerge at top of head. Press new guides into combustion chamber side of head. Distance from port end of intake valve guide to face of valve seat is 2⅜ inches. Distance from port end of exhaust valve guide to face of valve seat is $1\frac{15}{32}$ inches. Refer to Fig. C761. Ream the guides after installation to an inside diameter of 0.3422-0.3432. This provides a normal stem to guide clearance of 0.0008-0.0026 for the intake, 0.0032-0.0050 for the exhaust. Desired stem to guide clearance is 0.0015 for the intake, 0.004 for the exhaust.

Suggested maximum wear limits pertaining to valve stems and guides are as follows:

Valve guide I. D.............0.3447
Stem diameter, Intake........0.3386
              Exhaust 0.3362
Stem to guide max.
    clearance, Intake.........0.0046
              Exhaust 0.007

Note: Diameter wear limits are effective only when total clearance does not exceed maximum limits.

16. Intake and exhaust valve springs are interchangeable. New valve springs should test 58-64 pounds when compressed to a length of 1⅞ inches and 115-123 pounds when compressed to a length of 1.521 inches. Used valve springs, however, will function satisfactorily providing they test at least 52 pounds when compressed to 1⅞ inches and 105 pounds when compressed to 1.521 inches. Renew any spring which is rusted, discolored or does not meet the test specifications.

Install valve springs with closed or dampening coils against cylinder head.

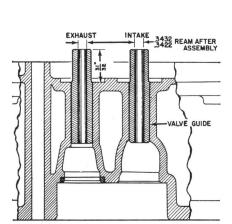

Fig. C760—Sectional view of the 300 non-diesel cylinder head, showing proper installation of valve guides.

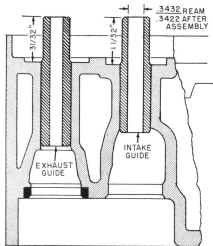

Fig. C760A—Sectional view of 200B, 300B, 350, 400B, 500B and 600B cylinder head showing proper valve guide installation.

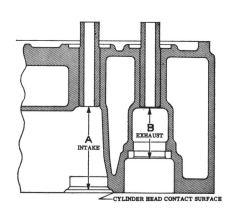

Fig. C761 — Sectional view of diesel cylinder head, showing proper installation of valve guides.

VALVE ROTATORS
Models So Equipped

17. Positive type exhaust valve rotators are used on some models. Normal servicing of the rotators consists of renewing the units. It is important, however, to observe the valve action after engine is started. The valve rotator action can be considered satisfactory if the valve rotates a slight amount each time the valve opens. Sectional views of a typical rotator are shown in Fig. C762. The J. I. Case Co., recommends that rotators be kept with the original valve. It is also recommended that rotator be renewed if valve is renewed.

ROCKER ARMS
Non-Diesel Models

18. An exploded view of the camshaft, rocker arms and valve train is shown in Fig. C763. To remove the rocker arms and shaft assembly, raise hood panel on series 300 and 350; remove hood on other models. Disconnect wires from coil and remove the

rocker arm cover. Remove the supporting bracket retaining bolts and lift assembly from engine. The procedure for subsequent disassembly will be evident from an examination of the unit. Stem contacting end of rocker arms can be reground if they are cupped. Rocker arm bushings are available separately for field installation in rocker arms which are in otherwise good condition. Replacement rocker arms are factory fitted with bushings.

When installing new bushings, make certain that oil hole is at top as shown in Fig. C764 and ream the bushing to an inside diameter of 0.624-0.625. This will provide a clearance of 0.001-0.003 for the 0.622-0.623 diameter rocker arm shaft.

There are four right hand and four left hand rocker arms and they must be properly assembled to shaft. Also, make certain that oil holes in shaft are toward valve stem contacting end of rocker arms as shown in Fig. C765. Tighten the shaft bracket retaining bolts and nuts to a torque of 20 ft.-lbs.

Diesel Models

19. An exploded view of the camshaft and valve train is shown in Fig. C766. To remove the rocker arms and shaft assembly, raise the hood panel and remove the rocker arm cover. Remove the retaining bolts and lift as-

sembly from cylinder head. Note: If push rods are removed, snap them side-ways out of the tappet sockets. This method serves to break the hydraulic connection and permits lifting the push rods out and leaving the cylindrical type tappets in place. The procedure for subsequent disassembly will be evident from an examination of the unit.

Rocker arm bushings are not available as separate replacement parts. If bushing to shaft clearance exceeds 0.005, the rocker arm and/or shaft should be renewed.

There are four right hand and four left hand rocker arms and they must be properly installed to shaft.

VALVE TAPPETS (CAM FOLLOWERS)
All Models

20. Non-diesel engines are fitted with mushroom type tappets, whereas diesel engines are fitted with cylindrical type. Tappets operate directly in the machined bores in the cylinder block and are available in standard size only. Any tappet can be removed after first removing the camshaft as outlined in paragraph 24 for non-diesels, paragraph 25 for diesels. The 0.5615-0.5620 diameter tappets have a normal clearance of 0.0005-0.002 in the block bores.

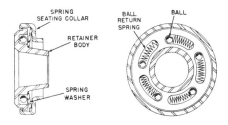

Fig. C762 — Sectional views of the positive type exhaust valve rotators used on some models. Rotator units should be renewed if exhaust valves fail to rotate.

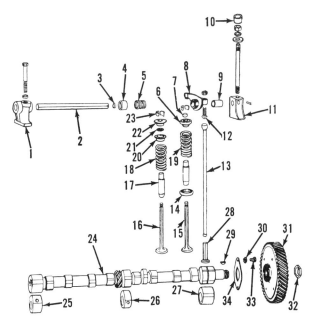

Fig. C763 — Non-diesel camshaft, rocker arms and valve train. Nut (32) is locked by a tab plate.

1. Center bracket
2. Rocker arm shaft (2 used)
3. Shaft plug
4. Spacer
5. Spring
6. Rotator (some models)
7. Exhaust valve cap (some models)
8. Rocker arm
9. Bushing
10. Spacer
11. Shaft bracket
12. Adjusting screw
13. Push rod
14. Exhaust valve seat (some models)
15. Exhaust valve
16. Intake valve
17. Valve guide
18. Valve spring, intake
19. Valve spring, exhaust
20. Gasket retainer
21. Gasket
22. Spring retainer
23. Keepers
24. Camshaft
25. Rear bushing
26. Center bushing
27. Front bushing
28. Cam follower
29. Key
30. Washer
31. Cam gear
32. Nut
33. Screw
34. Thrust plate

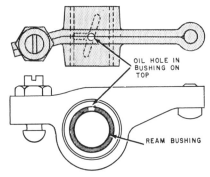

Fig. C764 — Non-diesel rocker arm showing proper installation of the renewable type bushings.

Fig. C765 — Non-diesel rocker arms showing proper location of the shaft oil holes.

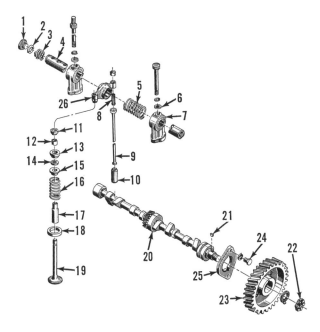

Fig. C766 — Diesel engine camshaft, rocker arms and valve train.

1. Shaft plug
2. Gasket
3. Short spring
4. Rocker arm shaft
5. Long spring
6. Copper washer
7. Support bracket
8. Adjusting screw
9. Push rod
10. Cam follower
11. Valve stem cap
12. Keepers
13. Spring retainer
14. Gasket (intake)
15. Gasket retainer (intake)
16. Valve spring
17. Valve guide
18. Exhaust valve seat
19. Valve
20. Camshaft
21. Key
22. Nut
23. Cam gear
24. Screw
25. Thrust plate
26. Rocker arm

TIMING GEAR COVER AND CRANKSHAFT FRONT OIL SEAL

All Models

21. To remove the timing gear cover, first drain cooling system.

Remove the hood, grille, air cleaner and radiator. Remove both of the frame side rails and remove the cap screw retaining the steering shaft bracket to torque tube. Remove the roll pin retaining the steering shaft front universal joint to the steering worm (cam) shaft and bump the U-joint yoke rearward and off the shaft. Support tractor under torque tube and on axle type tractors, unbolt the radius rod pivot bracket. Unbolt the front support from engine and roll the front support, wheels and axle assembly forward and away from tractor.

Remove fan belt and starting crank jaw and carefully pry the fan belt pulley from end of crankshaft. Disconnect linkage from governor lever and remove the fan blades. Remove the cap screws retaining oil pan to timing gear cover and loosen the remaining oil pan cap screws. On diesel models, remove 'the nuts retaining the injection pump drive gear housing to timing gear cover and loosen the others several turns. On all models, remove the cap screws retaining timing gear cover to cylinder block, carefully separate oil pan gasket from timing gear cover and remove cover from engine. Crankshaft front oil seal can be renewed at this time. Install seal with lip of same facing timing gears. Install a new pulley dust seal.

Before installing the timing gear cover, make certain that the locating dowels are properly positioned. Tighten the cover retaining screws to a torque of 25 ft.-lbs.

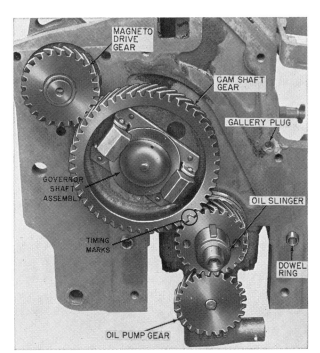

Fig. C767 — Non-diesel engine timing gear train. On battery ignition models, the magneto drive gear is not used.

TIMING GEARS

Non-Diesel Models

22. Timing gear train consists of the camshaft gear, crankshaft gear, oil pump drive gear and on magneto equipped engines, the magneto drive gear. Refer to Fig. C767. Backlash between camshaft gear and crankshaft gear should be 0.003. To remove the camshaft and crankshaft timing gears, first remove the timing gear cover as outlined in paragraph 21.

22A. CAMSHAFT GEAR. Remove the governor shaft assembly and the nut retaining the gear to camshaft. Using a suitable puller as shown in Fig. C768, remove the gear. Refer to paragraph 75 for information concerning the governor weight unit. Before installing

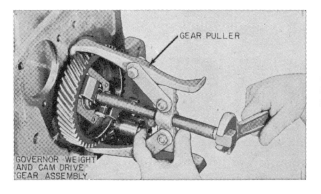

Fig. C768 — Using a suitable puller to remove the non-diesel engine camshaft gear.

the camshaft gear, check the camshaft end play which should be 0.003-0.007. If end play is not as specified, renew the camshaft thrust plate. Remove the oil pan and buck-up camshaft while gear is being drifted on.

Make sure that oil hole in camshaft and in hub of gear are open and clean and install gear so that oil holes are in register. Mesh double punch marked tooth space in cam gear with the single punch marked tooth on crank gear as shown in Fig. C767. Install the gear retaining nut and locking tab washer and tighten the nut to a torque of 80-90 ft.-lbs. Bend tab of lock washer against flat of nut.

On magneto equipped engines, mesh the magneto drive gear with the camshaft gear as outlined in paragraph 84.

22B. CRANKSHAFT GEAR. Crankshaft gear is a press fit on the shaft and can be removed after timing gear cover is off by removing the oil slinger and using a suitable puller with screws of same threaded into the two tapped holes in the gear.

On series 200B, 300, 300B and 400B, crankshaft end play of 0.003-0.005 is controlled by the thrust washer, shims and thrust plate shown in Fig. C769. Before installing the crankshaft gear, install an improvised spacer as shown in Fig. C770, tighten the starting crank jaw until spacer is tight against the thrust plate and check the crankshaft end play. If end play is not as specified, add or deduct shims as required. Shims are available in thicknesses of 0.002 and 0.010.

Heat crankshaft gear in oil and drift same into position so that single punch marked tooth on crankshaft gear is meshed with double punch marked tooth space on camshaft gear as shown in Fig. C767.

22C. MAGNETO AND OIL PUMP DRIVE GEAR. To renew the magneto drive gear, refer to paragraph 84. To renew the oil pump drive gear, refer to paragraph 43.

Diesel Models

23. Timing gear train consists of the camshaft gear, crankshaft gear, oil pump drive gear and the injection pump drive gear.

Recommended timing gear backlash is 0.002.

To remove the camshaft and crankshaft timing gears, first remove the timing gear cover as outlined in paragraph 21.

23A. CAMSHAFT GEAR. Remove the governor shaft assembly and the nut retaining gear to camshaft. Remove the injection pump and injection pump drive gear housing as outlined in paragraph 69. Using a suitable puller with screws of same threaded into the two tapped holes in the gear, remove gear from camshaft.

Before installing the camshaft gear, check the camshaft end play which should be 0.003-0.007. If end play is not as specified, renew the camshaft thrust plate. Remove the oil pan and buck up camshaft while gear is being drifted on and mesh double punch marked tooth space on camshaft gear with single punch marked tooth on crankshaft gear as shown in Fig. C771. Install the gear retaining nut and washer and tighten the nut to a torque of 70 ft.-lbs.

23B. CRANKSHAFT GEAR. Crankshaft gear is a press fit on crankshaft and can be removed after timing gear cover is off by removing the oil slinger and using a suitable puller as shown in Fig. C772.

Crankshaft end play of 0.004-0.006 is controlled by the thrust washer, shims and plate shown in Fig. C773. Before installing the crankshaft gear, install an improvised spacer in a manner similar to that shown in Fig. C770, tighten the starting crank jaw until spacer is tight against shim pack and check the crankshaft end play. If end play is not as specified, add or deduct shims as required. Shims are available in thicknesses of 0.002 and 0.008.

Heat crankshaft gear in oil and drift same into position so that single punch marked tooth on crankshaft gear is meshed with double punch marked tooth space on camshaft gear as shown in Fig. C771.

23C. INJECTION PUMP DRIVE GEAR. The procedure for removing the injection pump drive gear is evident after removing the injection pump as in paragraphs 69 & 70.

23D. OIL PUMP DRIVE GEAR. To renew the oil pump drive gear, refer to paragraph 43.

Fig. C769 — Recommended crankshaft end play of 0.003-0.005 on 200B, 300, 300B and 400B non-diesel engines is controlled by the thrust washer, plate and shims shown.

Fig. C770—Using starting crank jaw and an improvised spacer to check the crankshaft end play on series 200B, 300, 300B and 400B.

Fig. C771 — Diesel engine timing gear train showing the punched timing marks. The injection pump drive gear (not shown) is driven by the camshaft gear.

Fig. C772 — Removing the diesel engine crankshaft gear. Installation of gear will be easier if gear is heated.

Check the camshaft and bushings against the values which follow:

Camshaft journal
diameter 1.749-1.750
Bushing inside
diameter 1.752-1.753
Journal running
clearance 0.002-0.004
Thrust plate
thickness ....:0.147-0.149

If the camshaft bushings require renewal, it will be necessary to perform the additional work of detaching (splitting) engine from clutch housing and removing the flywheel and the "Hubbard" plug located behind the camshaft rear bearing. Install the bushings as shown in Fig. C774 and align ream the bushings after installation to an inside diameter of 1.752-1.753. Be sure to coat rim of "Hubbard" plug with sealing compound before installation. Make sure the camshaft thrust plate is in good condition, install the thrust plate and check the camshaft end play which should be within 0.003-0.007. Install the camshaft gear as outlined in paragraph 22A.

Retime the ignition as outlined in paragraph 82 or 83.

CAMSHAFT AND BEARINGS
Non-Diesel Models

24. The camshaft is carried in three steel backed, tin base, babbitt lined bushings. All camshaft journals are of the same diameter. Camshaft end play is controlled by a thrust plate at front of shaft.

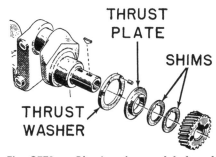

Fig. C773 — Diesel engine crankshaft end play is controlled by thrust washers and a series of shims.

To remove the camshaft, first remove the timing gear cover as outlined in paragraph 21. Remove the governor shaft assembly and the nut retaining gear to camshaft. Using a suitable puller as shown in Fig. C768, remove the gear. Disconnect wire from coil and remove the rocker arm cover. Remove the rocker arms assembly, push rods and on models so equipped, remove the ignition distributor. Remove the oil pan. Push the tappets (cam followers) up into their bores as far as they will go. If normal gum deposits will not hold the tappets up, apply cup grease to the flanged portion. Remove the two cap screws retaining the camshaft thrust plate to the cylinder block and remove camshaft from engine, being careful not to damage the bushings.

Diesel Models

25. The three camshaft bearing journals ride directly in the machined bores in the cylinder block. Camshaft end play is controlled by a thrust plate at front of shaft.

To remove the camshaft, first remove the timing gear cover as outlined in paragraph 21. Remove the governor shaft assembly and the nut retaining gear to camshaft. Remove the injection pump and injection pump drive gear housing as outlined in paragraph 69. Using a suitable puller with screws of same threaded into the two

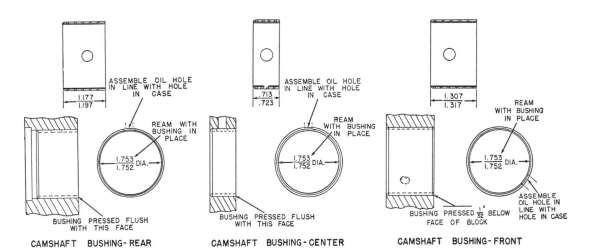

Fig. C774 — Illustration showing proper installation of camshaft bushings on non-diesel engines. Bushings must be align reamed after installation.

tapped holes in the gear, remove gear from camshaft. Remove the rocker arm cover, rocker arms assembly and push rods. Note: When removing push rods, snap them sideways out of the tappet sockets. This method serves to break the hydraulic connection and permits lifting the push rods out and leaving the cylindrical type tappets in place. Remove the oil pan and push the tappets up to clear the camshaft lobes. Remove the two cap screws retaining the camshaft thrust plate to the cylinder block and remove camshaft from engine.

Check the camshaft and bearing bores against the values which follow:

Camshaft journal diameter:
No. 1 (front).........1.8085-1.8090
No. 2 ....1.7455-1.7465
No. 3 ....1.6835-1.6840

Renew the camshaft if journal diameters are less than:
No. 1 (front) 1.8075
No. 2 ....1.7445
No. 3 ....1.6820

Camshaft bearing bore diameters in cylinder block are as follows:
No. 1 (front) 1.8120-1.8125
No. 2 ....1.7495-1.7500
No. 3 ....1.6870-1.6875
Journal running
  clearance ....0.003 -0.004

Before installing the camshaft gear, check the camshaft end play which should be 0.003-0.007. If end play is not as specified, renew the camshaft thrust plate. Install camshaft gear as outlined in paragraph 23A.

Retime the injection pump as outlined in paragraph 70.

CONNECTING ROD AND PISTON UNITS

All Models

26. Connecting rod and piston units are removed from above after removing cylinder head, oil pan and the connecting rod bearing caps. Be sure to remove ridge from unworn portion of sleeves before withdrawing the rod and piston units.

When reinstalling, the cylinder numbers on the rod and the cap should be in register and face toward camshaft side of engine. Tighten the connecting rod bolts to a torque of 45-50 ft.-lbs. on non-diesels, 35-40 ft.-lbs. on diesels.

PISTONS, RINGS AND SLEEVES
Non-Diesel Models

27. Pistons are fitted with three compression rings and one oil control ring. Check the pistons, sleeves and rings against the values which follow:

Compression ring end gap..0.010-0.020

Top compression ring side
  clearance
    200B, 300, 300B, 400B......0.003
    350, 500B, 600B............0.004

Other compression ring side
  clearance
    200B, 300, 300B, 400B......0.002
    350, 500B, 600B............0.003

Oil control ring end gap...0.010-0.018

Oil control ring side clearance..0.0015

Cylinder sleeve diameter
    200B 3.1223-3.1228
    300, 300B, 400B 3.3750-3.3765
    350, 500B, 600B..........3.5625

Piston skirt clearance can be considered satisfactory if a spring scale pull of 5-10 lbs. is required to withdraw a ½-inch wide, 0.003 thick feeler ribbon from between piston skirt (90° to piston pin) and cylinder sleeve. When installing replacement piston rings, be sure to follow closely the instructions in the package. Note: Both normal (square ends) and high pressure (tongue and groove ends) oil ring expander springs may be contained in re-ring packages. The decision as to which springs to install is up to you, but be sure to install either the complete set of normal springs OR the complete set of high pressure springs. NEVER attempt to install two springs behind one ring. Assemble compression rings on piston so that the word "TOP" or the dot on ring is in the up position.

The procedure for renewing the wet type cylinder sleeves is outlined in paragraph 29; for dry type sleeves used in the 350, 500B and 600B engines, refer to paragraph 29A.

Diesel Models

28. Pistons are fitted with three ⅛-inch wide compression rings and one ¼-inch wide oil control ring. Check the pistons, sleeves and rings against the values which follow:

Top compression ring
  end gap 0.008-0.016
Top compression ring side
  clearance 0.003-0.0045

Other compression ring
  end gap 0.010-0.015

Other compression ring
  side clearance 0.002-0.0035

Oil control ring end gap..0.008-0.016

Oil control ring side
  clearance 0.002-0.0035

Cylinder sleeve bore
  diameter 3.375-3.377

Renew sleeve if bore
  diameter exceeds 3.385

Piston skirt clearance can be considered satisfactory if a spring scale pull of 5-10 lbs. is required to withdraw a ½-inch wide, 0.004 thick feeler ribbon from between piston skirt (90° to piston pin) and cylinder sleeve. Tapered piston rings must be installed with word "TOP" facing up.

The procedure for renewing the wet type cylinder sleeves is outlined in paragraph 29.

R&R CYLINDER SLEEVES
All Models With Wet Sleeves

29. The wet type cylinder sleeves can be removed from cylinder block by using a suitable puller after connecting rod and piston units are out. Coolant leakage at bottom of sleeves is prevented by two packing rings and at the top by the cylinder head gasket. Before installing new sleeves, thoroughly clean the mating and sealing surfaces of cylinder block and sleeves. Then insert sleeves (without sealing rings) and check the sleeve stand-out above the cylinder block. Recommended stand-out is 0.001-0.004.

Excessive sleeve stand-out could be caused by foreign material under the sleeve flange; or on diesels, there may be too many shims under the sleeve flange. Insufficient stand-out on diesels can be corrected by making up and installing shims under the sleeve flange.

After making the trial installation, remove the sleeves and install packing rings dry in sleeve grooves. Make certain that packing rings are not twisted, then lubricate the packings with petroleum jelly, carefully lower sleeves into cylinder block and press in place with hand pressure.

Fill the cylinder block water jacket with cold water and check for leaks adjacent to the packing rings.

All Models With Dry Sleeves

29A. The dry type cylinder sleeves used in series 350, 500B and 600B can be removed by using a suitable puller after connecting rod and piston units are out. Before installing new sleeves, thoroughly clean and polish the cylinder block bores. A cylinder hone can be used for this purpose, but be careful not to enlarge the block bores. Thoroughly clean counter-bores at top of block.

Wash O. D. of sleeves in a petroleum solvent. Coat O. D. of sleeves and I. D. of block bores with brake fluid and slide sleeve into block bores as far as they will go with hand pressure only. If mating surfaces are clean and if sleeve is started straight, it should be possible to push the sleeves approximately two inches into block bores. Then, using Case sleeve pusher G15033, or equivalent, and a rawhide hammer, tap sleeves in place.

CAUTION: Under no circumstances should excessive pressure, hammering or force be used to install the sleeves. If difficulty is encountered, chill the sleeves in cracked ice and slip them in place. When properly installed, sleeves will stand-out 0.001-0.004 above cylinder block head gasket surface.

PISTON PINS

Non-Diesel

30. The full floating type piston pins are retained in the piston bosses by snap rings and are available in standard size only, as follows:
200B, 300, 300B, 400B....0.8591-0.8593
350, 500B, 600B..........0.9939-0.9992

When installing bushing in connecting rod, be sure to align the oil holes and ream the bushing to an inside diameter of 0.8596-0.8598 for series 200B, 300, 300B and 400B; 0.9993-0.9996 for series 350, 500B and 600B. Piston pin should have a 0.0003 press fit in piston and a clearance in connecting rod bushing of 0.0003-0.0007 for series 200B, 300, 300B and 400B; 0.0001-0.0007 for series 350, 500B and 600B. Heating pistons in hot water will facilitate installation of the pins.

Diesels

31. The 1.1250-1.1252 diameter full floating type piston pins are retained in the piston bosses by snap rings and are available in standard size only. When installing bushing in connecting rod, be sure to align the oil holes and

ream the bushing to an inside diameter of 1.1253-1.1255 to provide a standard piston pin diametral clearance of 0.0001-0.0005. Piston pin should have a 0.0001-0.0003 press fit in piston. Heating pistons in hot water will facilitate installation of the pins.

CONNECTING RODS AND BEARINGS

All Models

32. Connecting rod bearings are of the steel-backed, non-adjustable, slip-in precision type, renewable from below after removing oil pan and bearing caps.

When installing new bearing shells, make certain that the bearing shell projections engage the milled slot in connecting rod and bearing cap and that cylinder numbers are in register and face toward camshaft side of engine. Tighten the connecting rod bolts to a torque of 45-50 ft.-lbs. on non-diesels, 35-40 ft.-lbs. on diesels.

On all models, connecting rod bearing inserts are available in standard size as well as various undersizes. On some non-diesel models, semi-finished bearing inserts are also available.

Check the connecting rods, bearing inserts and crankshaft crankpins against the values which follow:

Crankpin diameter:
    Non-diesel—
    200B, 300, 300B, 400B..1.9365-1.9375
    350, 500B, 600B........2.0605-2.0615
    Diesel 2.0615-2.0625

Rod side play:
    Non-diesel 0.004-0.006
    Diesel 0.006-0.010

Rod bearing running clearance:
    Non-diesel 0.0015-0.002
    Diesel 0.0006-0.0031

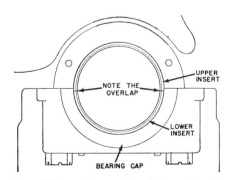

Fig. C775 — Proper installation of main bearing inserts on 200B, 300, 300B and 400B non-diesel engines. Notice that upper half of bearing insert extends below parting line of crankcase and bearing cap.

CRANKSHAFT AND MAIN BEARINGS

Non-Diesel Models

33. Crankshaft is supported in three non-adjustable, slip-in, precision type main bearing inserts which can be renewed after removing oil pan, oil pump and main bearing caps. Note: Bearing caps should be marked before removal so they can be reinstalled in the same position.

Normal main bearing running clearance is 0.0015-0.002. If running clearance is excessive, renew the bearing inserts and/or regrind crankshaft.

On all models, main bearing inserts are available in standard size as well as various undersizes. On some models, semi-finished bearing inserts are also available.

On series 200B, 300, 300B and 400B, upper and lower bearing inserts are not interchangeable and must be checked carefully before installation. When properly installed, the bearing upper half will extend $\frac{1}{16}$-inch beyond parting line of bearing cap and cylinder block at both sides as shown in Fig. C775 and an ear on liner will engage groove in cylinder block. Ends of lower liner will be $\frac{1}{16}$-inch below face of bearing cap and ear on liner will engage groove in cap. Install bearing cap carefully to allow the upper bearing half to enter bearing cap properly without interference or damaging the liner ear. Make certain also that oil passage in cylinder block is not covered when installing the rear main bearing inserts.

Tighten the main bearing bolts to a torque of 90-100 ft.-lbs.

On series 200B, 300, 300B and 400B, normal crankshaft end play of 0.003-0.005 is controlled by the thrust washer, shims and thrust plate shown in Fig. C769. To adjust the end play, refer to paragraph 22B.

On series 350, 500B and 600B normal crankshaft end play 0.003-0.007 is controlled by the flanged center main bearing inserts.

34. To remove the crankshaft, first remove the engine as outlined in paragraph 9 and proceed as follows: Remove the flywheel and crankshaft rear oil seal retainer. Remove the fan belt and starting crank jaw and carefully pry pulley from front end of crankshaft. Remove the timing gear cover, oil pan and oil pump. Remove the main and rod bearing caps and lift crankshaft from engine.

Check the crankshaft and main bearing inserts against the values which follow:

Series 200B-300-300B-400B

Crankpin diameter
(new) 1.9365-1.9375
Crankpin (0.020 under)..1.9165-1.9175
Crankpin (0.022 under)..1.9145-1.9155
Crankpin (0.040 under)..1.8965-1.8975
Main journal diameter
(new) 2.2490-2.2500
Mains (0.020 under).....2.2290-2.2300
Mains (0.022 under).....2.2270-2.2280
Mains (0.040 under).....2.2090-2.2100
Main bearing running
clearance 0.0015-0.002
Main bearing
bolt torque ....90-100 ft.-lbs.
Rod bearing
bolt torque ....45-50 ft.-lbs.

Series 350-500B-600B

Crankpin diameter
(new) 2.0605-2.0615
Crankpin (0.002 under)..2.0585-2.0595
Crankpin (0.020 under)..2.0405-2.0415
Crankpin (0.022 under)..2.0385-2.0395
Crankpin (0.040 under)..2.0205-2.0215
Main journal diameter
(new) 2.6230-2.6240
Mains (0.002 under).....2.6210-2.6220
Mains (0.020 under).....2.6030-2.6040
Mains (0.022 under).....2.6010-2.6020
Mains (0.040 under).....2.5830-2.5840
Main bearing running
clearance 0.0015-0.002
Main bearing
bolt torque ....90-100 ft.-lbs.
Rod bearing
bolt torque ....45-50 ft.-lbs.

On all models, bearing journals should be reground if tapered more than 0.001 or if out-of-round more than 0.005.

Refer to paragraph 38 when installing the crankshaft rear oil seal and to paragraph 43 when installing the oil pump.

Diesel Models

35. Crankshaft is supported in three non-adjustable, slip-in, precision type main bearing inserts which can be renewed after removing oil pan, oil pump and main bearing caps.

Normal main bearing running clearance is 0.0009-0.0036. If running clearance exceeds 0.0046, renew the bearing inserts and/or regrind the crankshaft. Bearing inserts are available in standard size as well as undersizes of 0.002, 0.020 and 0.040. When installing new bearing shells, make certain that the bearing shell projections engage

the milled slot in cylinder block and bearing cap. Tighten the main bearing bolts to a torque of 85-95 ft.-lbs.

Normal crankshaft end play of 0.004-0.006 is controlled by the thrust washer, shims and plate shown in Fig. C773. To adjust the end play, refer to paragraph 23B.

36. To remove the crankshaft, first remove the engine as outlined in paragraph 9 and proceed as follows: Remove the clutch, flywheel and crankshaft rear oil seal retainer. Remove the timing gear cover, oil pan and oil pump. Remove the main and rod bearing caps and lift crankshaft from engine.

Check the crankshaft and main bearing inserts against the values which follow:

Crankpin diameter
(new) 2.0615-2.0625
Crankpin (0.020 under)..2.0415-2.0425
Crankpin (0.040 under)..2.0215-2.0225
Main journal diameter
(new) ....2.374 -2.375
Mains (0.020 under).....2.354 -2.355
Mains (0.040 under).....2.334 -2.335
Main bearing running
clearance 0.0009-0.0036
Main bearing
bolt torque ....85-95 ft.-lbs.
Rod bearing
bolt torque ....35-40 ft.-lbs.

Bearing journals should be reground if new bearing inserts do not have the desired running clearance.

Refer to paragraph 40 when installing the crankshaft rear oil seal and to paragraph 43 when installing the oil pump. Adjust the crankshaft end play to 0.004-0.006 as outlined in paragraph 23B.

CRANKSHAFT OIL SEALS
Non-Diesel Models

37. **FRONT SEAL.** The crankshaft front oil seal is located in the timing gear cover and can be renewed after removing the complete front support and axle assembly and the crankshaft pulley. Be sure to measure depth of old seal and install new one to the same depth.

38. **REAR SEAL.** Crankshaft rear oil seal is a one-piece type located in a retainer bolted to rear face of cylinder block as shown in Fig. C776. To renew the seal, first detach (split) engine from torque tube as outlined in

paragraph 90 or 94Q and remove the flywheel. Remove the two cap screws retaining the oil pan to seal retainer and loosen the remaining oil pan cap screws. Unbolt and remove seal and retainer assembly from cylinder block. Install new seal in retainer and cement gasket to retainer. When installing seal over crankshaft flange, use J. I. Case seal sleeve No. T-106 for series 200B, 300, 300B and 400B, or seal sleeve No. G15028 for series 350, 500B and 600B, to avoid damaging the seal lip. Refer to Fig. C777. If seal sleeves are not available, shim stock can be used providing care is exercised. Install top and bottom two retainer screws finger tight, then using J. I. Case aligning tool No. G13500 for series 200B, 300, 300B and 400B, or tool No. G13506 for series 350, 500B and 600B, align the seal retainer with respect to the crankshaft and tighten the three screws to a torque of 6-8 ft.-lbs. Refer to Fig.

Fig. C776 — Rear view of the 200B, 300, 300B and 400B non-diesel engine cylinder block showing the crankshaft rear oil seal installation. Series 350, 500B and 600B are similar.

Fig. C777 — Using Case seal sleeve No. T-106 to install the non-diesel engine crankshaft rear oil seal on series 200B, 300, 300B and 400B. A similar procedure is used on series 350, 500B and 600B, but the seal sleeve part No. is G15028.

C778. Remove the aligning tool, install the remaining screws and tighten same to a torque of 6-8 ft.-lbs. Notice that the top three screws are 1-inch long whereas, the lower four are ¾-inch long.

In the absence of an aligning tool, install all of the retainer screws finger tight, turn crankshaft several revolutions to center the seal, then tighten the screws to a torque of 6-8 ft.-lbs.

Diesel Models

39. **FRONT SEAL.** The crankshaft front oil seal is located in the timing gear cover and can be renewed after removing the complete front support and axle assembly and the crankshaft pulley. Be sure to note the exact location of the old seal and install new one in the same position.

40. **REAR SEAL.** Crankshaft rear oil seal is a one-piece type located in a retainer bolted to rear face of cylinder block and rear main bearing cap. To renew the seal, first detach (split) engine from torque tube as outlined in paragraph 90 and remove the clutch and flywheel. Remove the oil pan and the two cap screws securing the seal retainer to the rear main bearing cap. Unbolt seal retainer from cylinder block and renew the seal.

OIL PAN
All Models

41. On tricycle type tractors, the oil pan can be removed in a conventional manner without removing any other parts. On utility type tractors, remove the radius rod and on adjustable axle types, remove the axle and wheels assembly before attempting to remove the oil pan.

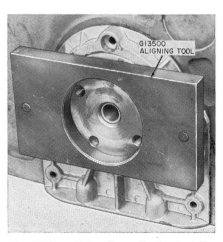

Fig. C778 — Using Case aligning tool No. G13500 to align the non-diesel engine crankshaft rear oil seal retainer on series 200B, 300, 300B and 400B. A similar procedure is used on series 350, 500B and 600B, but the aligning tool part No. is G13506.

FLYWHEEL
All Models

42. To remove the flywheel, first detach (split) engine from torque tube as outlined in paragraph 90 or 94Q. The remainder of the removal procedure is evident.

To install a new flywheel ring gear, heat same to approximately 550 degrees F.

OIL PUMP AND PRESSURE RELIEF VALVE
All Models

43. **PUMP.** Oil pump is gear driven by the engine crankshaft gear and is bolted to lower face of the front main bearing cap. To remove the oil pump, remove the oil pan, remove the two retaining nuts and remove pump assembly from engine, but do not lose the shims located between oil pump and main bearing cap.

To disassemble the pump, remove the pump cover (32—Fig. C779) and extract idler gear (36). Extract cotter pin (35) and remove retainer (39), spring (40) and relief valve (38). Further disassembly of the pump would be pointless, since the body, driving gear, driven gear and shafts are available as an assembled unit only. Renew the assembly if any defects are noted.

Non-diesel engine oil pump specifications are as follows:

Pumping gears
backlash 0.001-0.002
Pumping gears end play
in body 0.0035-0.0065
Pumping gears radial clearance in pump body......0.002-0.005
Relief valve spring free
length ....1$\frac{15}{16}$ inches
Relief valve spring pressure test ....4½-5 lbs. @ 1⅜ inches

Diesel engine oil pump specifications are as follows:

Pumping gears
backlash 0.001-0.005
Pumping gears end play
in body 0.002-0.004
Pumping gears radial clearance in pump body......0.001-0.003
Relief valve spring
free length ....1$\frac{13}{16}$-2$\frac{1}{16}$ inches
Relief valve spring pressure test ....30-40 lbs. @ 1⅜ inches

When reassembling, be sure that cupped side of retainer cap (39) is toward the relief valve spring. Install pump by reversing the removal procedure and vary the number of shims between pump and main bearing cap to provide a backlash between crankshaft gear and oil pump drive gear of 0.005-0.010 for non-diesels, 0.004-0.008 for diesels. **Tighten the nuts to a torque of 90-100 ft.-lbs. on non-diesels, 85-95 ft.-lbs. on diesels.**

44. **RELIEF VALVE.** Normal oil pressure with engine warm and running at high idle speed is 14-20 psi on non-diesels, 30-40 psi on diesels. Pressure is maintained by relief valve and spring (38 & 40—Fig. C779). If oil pressure is not as specified, remove the oil pan, cotter pin (35), spring (40) and valve (38). Make certain that valve (38) is free in the body bore and renew the spring if it does not test 4½-5 lbs. when compressed to a height of 1⅜ inches on non-diesels, 30-40 lbs. when compressed to a height of 1⅜ inches on diesels. Retainer (39) must be installed with cupped side toward spring. The J. I. Case Co. specifies that insufficient oil pressure should never be corrected by installing shim washers under the spring.

Fig. C779 — Exploded view of non-diesel engine oil pump. Pump body, drive gear, driven gear and shaft are available as an assembled unit only. Diesel engine oil pumps are similarly constructed.

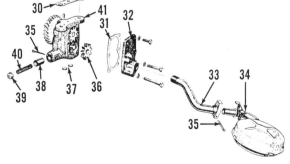

30. Shims
31. Gasket
32. Cover
33. Suction pipe
34. Intake screen
35. Cotter pin
36. Idler gear
37. Hubbard plugs
38. Relief valve
39. Spring retainer
40. Relief valve spring
41. Body and gears assembly

CARBURETOR

Non-Diesels (Except LP-Gas)

45. Gasoline and low cost fuel burning engines are fitted with Marvel-Schebler and Zenith carburetors, as follows:

200B	TSX 635
300 and 300B gasoline	TSX 635
300 and 300B low cost	TSX 680
350	TSX 696
400B	TSX 749
500B	62AJ 10
600B	TSX 749

Carburetor adjustments are conventional and calibration data are as follows:

TSX 635
Float setting	¼-inch
Repair kit	286-1101
Gasket set	16-592
Inlet needle and seat	233-536
Idle jet	49-101L
Nozzle	47-589
Power jet	49-279

TSX 680
Float setting	¼-inch
Repair kit	286-1129
Gasket set	16-592
Inlet needle and seat	233-536
Idle jet	49-285
Nozzle	47-589
Power jet	49-369

TSX 696
Float setting	¼-inch
Repair kit	286-1152
Gasket set	16-592
Inlet needle and seat	233-536
Idle jet	49-101L
Power jet	49-301
Nozzle	47-712

TSX 749
Float setting	¼-inch
Repair kit	286-1236
Gasket set	16-594
Inlet needle and seat	233-536
Idle jet	49-279
Nozzle	47-365
Power jet	49-165

62AJ10-12255
Float setting	1 39/64 inches
Repair kit	K12255
Gasket set	C181-65
Inlet needle and seat	C81-1-45
Economizer jet	C52-1-0
Well vent jet	C77-18-18
Main discharge jet	C66-68-80
Idle jet	C55-7-12
Main jet	C52-6-30

LP-GAS SYSTEM

(Applies to Series 300 Using Marvel-Schebler Components)

Note: Refer to page 30 for series 300B, 400B, 500B and 600B using Ensign Equipment.

The 300 series tractors are available with a factory installed LP-Gas system, using Marvel-Schebler equipment. Like other LP-Gas systems, this system is designed to operate with the fuel supply tank not more than 80% filled (18.4 gallons). It is important when starting the engine to open the vapor or liquid withdrawal valves on the supply tank SLOWLY; if opened too fast, the excess flow valves will automatically shut off the fuel flow. If this happens, merely close the withdrawal valves until pressure is equalized, then open the valves slowly.

Marvel-Schebler components of the system are as follows:

Magnetic shut-off valve	171-551
Liquid filter	4-521
Regulator	VT-725
Carburetor	TSG-7

A schematic drawing of the system is shown in Fig. C780.

CAUTION: LP-Gas expands readily with any decided increase in temperature. If tractor must be taken into a warm shop to be worked on during extremely cold weather, make certain that fuel tank is as near empty as possible. LP-Gas tractors should never be stored or worked on in an unventilated space.

ADJUSTMENTS
Series 300

46. Before starting the engine the first time, or after installing a new carburetor and/or regulator, refer to Fig. C781 and C782 and make the following preliminary adjustments:

a. Loosen the lock nut and turn the idle adjusting screw (Fig. C781) on the fuel regulator clockwise until it bottoms lightly on its seat, then check the number stamped on the lower left hand corner of the regulator name plate. On regulators having Part No. 6-595 stamped on plate, turn the idle adjusting screw counter-clockwise 22 turns and tighten the lock nut only finger tight. On regulators having Part No. 6-733 stamped on the plate, turn the idle adjusting screw counter-clockwise 8½ turns and tighten the lock nut only finger tight. Do not use a wrench on the lock nut or damage to the rubber gasket will result.

b. Loosen the lock nut and turn the load adjustment needle (Fig. C782) on the carburetor clockwise until it bottoms lightly on its seat, then turn it counter-clockwise approximately 1¼ turns.

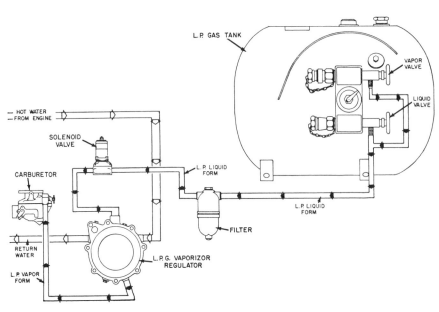

Fig. C780 — Schematic drawing showing the relative location of the 300 series LP-Gas system components.

Fig. C781 — Side view of the 300 series LP-Gas engine showing the solenoid, carburetor and regulator installation.

Fig. C782 — Series 300 LP-Gas carburetor showing the load adjusting screw and throttle stop screw.

Note: The preceding adjustments are only approximate settings. Final accurate settings should be made after engine is warm as follows:

46A. **IDLE ADJUSTING SCREW.** Remove the ¼-inch pipe plug from the intake manifold and connect a vacuum gage to the pipe plug opening as shown in Fig. C783. Start engine and bring to operating temperature. Place the hand throttle in the slow idle position and adjust the throttle stop screw (Fig. C782) to obtain an engine slow idle speed of 500 rpm. Loosen the lock nut and turn the idle adjusting screw on regulator to obtain the highest possible vacuum reading (approximately 19-20½ lbs.). Retighten the jam nut finger tight. Do not use a wrench on the lock nut or damage to the rubber gasket will result.

In the absence of a vacuum gage, the following procedure may be used: Place the hand throttle in the slow idle position and adjust the throttle stop screw to obtain an engine slow idle speed of 500 rpm. Turn the idle adjusting screw on regulator either way as required to obtain the highest rpm, then readjust the throttle stop screw to obtain a slow idle speed of 500 rpm.

46B. **LOAD ADJUSTMENT.** With engine at operating temperature and choke button fully in, set the throttle control in wide open position so that engine is running at approximately 1875 rpm. Loosen the lock nut on the load adjusting needle (Fig. C782) and turn the needle clockwise until the engine speed drops due to leanness. Then, turn the load needle counterclockwise until engine speed drops due to richness. Now, turn the needle clockwise to obtain the highest rpm and tighten the lock nut.

FUEL FILTER
Series 300

47. An exploded view of the liquid fuel filter is shown in Fig. C784. The filter is designed to stop the passage of scale, rust or other foreign material into the remainder of the system.

Note: If frost collects on outside of filter body, it is clogged and must be cleaned.

To clean the filter, close both of the fuel tank withdrawal valves, remove the assembly bolt and withdraw the filter base, gasket and element. If element appears to be in good condition, wash same in clean gasoline and dry with compressed air. If element contains considerable imbedded dirt, install a new element.

SOLENOID VALVE
Series 300

48. When ignition switch is turned on, the solenoid operated valve (Fig. C781) opens the fuel line to the regulator. When the switch is turned off, the valve is closed and the fuel supply is shut-off. Should the valve leak or become inoperative, it will be necessary to install a complete new unit as component parts are not sold separately. Before removing the unit, make certain that both fuel tank withdrawal valves are closed.

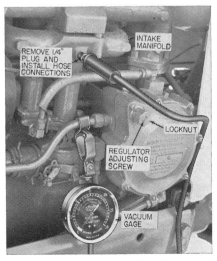

Fig. C783—Vacuum gage installation used when setting the 300 series idle adjusting screw.

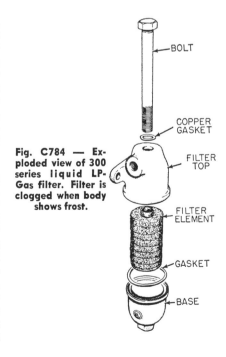

Fig. C784 — Exploded view of 300 series liquid LP-Gas filter. Filter is clogged when body shows frost.

VAPORIZER-REGULATOR
Series 300

49. **HOW IT OPERATES.** Refer to Fig. C785. When the ignition switch is turned on, the solenoid winding is energized, opening the solenoid valve and permitting fuel to enter the primary cavity through the inlet valve which is held open by the primary lever spring. When the unit is at rest, the spring forces the diaphragm inward, applying pressure to the primary lever and holding the valve open. As the liquid fuel enters the cavity, it must pass through a series of passages which accelerate vaporization. As the pressure builds up in the cavity, the diaphragm is forced outward, compressing the spring. The valve is returned to its seat by the pressure of spring, forcing the primary valve to follow the outward movement of the diaphragm. The diaphragm maintains the pressure in the cavity at approximately 2½-5 psi. Before delivery to the engine, a further reduction of pressure occurs in the second stage pressure regulator.

Air flow through the venturi of the adapter creates a vacuum over the slot through which gas is admitted to the engine. This reduces the pressure in the secondary cavity causing the diaphragm to move and apply pressure to the secondary valve lever, opening the secondary valve. Since the passage covered by the valve is connected to the cavity, gas from this passage will flow into the cavity and maintain a constant supply of fuel to meet engine demands.

An idle adjusting screw is provided on the regulator for the purpose of increasing or decreasing the pressure on the spring. As the screw is rotated, the spring retainer moves in the housing to alter the spring pressure and provide the initial fuel required for best idle performance.

The water cavity is connected to the cooling system of the engine. Engine cooling water circulates through this cavity at all times to prevent the regulator valves from becoming inoperative due to the refrigerating effects of the liquid gas freezing any moisture which might be in the fuel. Since vaporization is accomplished in the primary cavity and primary cavity passages, the regulator is designed so that the heat is applied at these locations. Fins in the water cavity provide additional heating surfaces and direct the movement of water over the area more uniformly.

Since the gas is already vaporized when it enters the secondary cavity, and further heating would only expand the gas unnecessarily, a plate forms a wall for the water cavity and produces an air cavity. Holes in the regulator casting connect this air cavity with the outside air, forming an air pocket which prevents excess heat in the water cavity from penetrating through the plate into the wall of the secondary cavity.

50. **R&R AND OVERHAUL.** Before removing the regulator, close the fuel tank withdrawal valves, start engine and let it run until it dies due to lack of fuel. Disconnect wire from solenoid, drain cooling system and disconnect water hoses from regulator. Disconnect fuel lines, then unbolt and remove regulator from its mounting. Refer to Fig. C786 and proceed as follows:

Remove the cap, primary diaphragm spring and all ten of the screws and lock washers from the primary cover. Tap the edge of the primary cover lightly with a soft-faced hammer until the cover begins to separate from the primary body. Lift the cover carefully away from the body. NOTE: Exercise care when separating the primary cover from the primary body to prevent damage to the primary diaphragm.

Remove the seven screws and lock washers that hold the primary body to the secondary body, tap the pri-

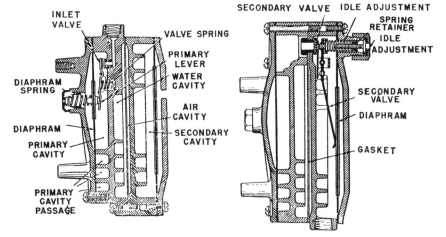

Fig. C785—Sectional views of Marvel-Schebler LP-Gas vaporizer-regulator used on Case 300 series tractors. Refer to paragraph 49 for operating description.

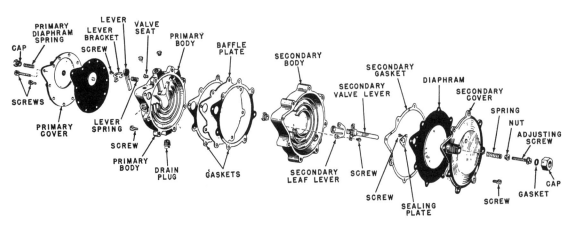

Fig. C786—Exploded view of 300 series LP-Gas vaporizer-regulator.

mary body with a soft-faced hammer until it becomes loosened from the gasket and carefully separate the primary body from the secondary body. Remove the baffle plate and both gaskets. Gaskets will probably adhere to the parts and it will be necessary to scrape off the broken pieces. Remove the cap and gasket from the secondary cover. Remove the adjusting screw and nut assembly and lift out the spring. Remove the eight screws and lock washers that hold the secondary cover to the secondary body, tap around the edge of the cover with a soft-faced hammer to loosen it, and carefully separate the secondary cover from the secondary body to avoid damaging the diaphragm assembly. Remove the three screws, sealing plate and diaphragm assembly from the secondary cover. Remove the two screws and lock washers to free the secondary valve lever, and secondary valve leaf lever from the secondary body. Remove the two screws and lock washers to free the primary lever bracket and primary lever assembly from the primary body, then remove the primary lever spring and the primary valve seat.

Wash all parts except diaphragms and valves in gasoline or solvent. If necessary, the parts can be immersed and stubborn materials removed by brushing with a small paint brush. Rinse in clean solution and dry with compressed air. The diaphragms will normally be clean, since it is difficult for foreign material to enter the regulator; however, if dirty, wipe the affected areas with a cloth moistened with the cleaning solvents. The valves can be cleaned by wiping the surfaces with a clean, dry cloth.

Visually check all parts for damage such as cracks in housings and covers. Any obviously damaged parts should be renewed. Inspect the diaphragms for signs of chafing, tears and worn spots. Also, check for excessive wrinkling which will result if the diaphragms have been stretched through extended use. If the condition of diaphragms is questionable, renew them. Check the valves and operating levers for obvious wear and misalignment. It must be remembered that the valves contact and leave their seats with extreme sensitivity and, if the valve faces are grooved, scratched, or scored, they must be renewed. Check the secondary

leaf lever assembly carefully for signs of distortion or cracks. If there is any sign of damage, renew this part. Check for wear on the knife edge of the primary lever bracket. If this edge is worn, it will interfere with operation of the primary lever and the bracket must be renewed. Check all springs for signs of fatigue and distortion. If springs are bent or distorted in any way, renew them. The primary lever and primary diaphragm springs should always be renewed unless the regulator has had very little service.

Install the secondary valve seat in the secondary body and tighten firmly in place. Place a gasket in position on each side of the baffle plate, and lay the baffle plate, with gaskets in place, on the secondary body. Place the primary body in position against the baffle plate and install the seven screws and lock washers. Tighten the screws progressively in diametrically opposite pairs. Install the primary valve seat in the primary body. Tighten the seat carefully, using a thin-wall socket to prevent damage to the seat. (John Crane plastic seal or similar thread seal should be used.) Do not use shellac or similar compounds. Place the primary lever spring in the recess of the primary body. Engage the primary lever assembly in the primary lever bracket and attach the bracket to the primary body with the two screw and lock washer assemblies. Make certain the spring is engaged with the projection on the lever. Operate the lever by hand while observing the action of the valve cap. The valve should operate smoothly without binding and the valve cap should always come to rest squarely against the valve seat. Place the diaphragm assembly carefully on the primary body, insert the ten screw and lock washer assemblies through the holes in the primary cover and carefully position the cover on the primary body. Apply enough pressure on the primary cover to flatten out the diaphragm, and while holding the diaphragm in a free position, start the screws into the threads in the primary body. Rotate all screws approximately two turns. Press the primary cover down firmly on the primary body and tighten all screws in diametrically opposite pairs. IMPORTANT: Make sure the diaphragm does not wrinkle. Install and tighten the screw and lock washer assembly. Insert the primary diaphragm spring into the recess of

the retainer cap and install the spring and cap in the primary cover. Tighten the cap firmly. Place the secondary valve leaf lever assembly in position in the secondary body. Assemble the secondary valve lever on top of the leaf lever and install the screw and lock washer assemblies. Check to make sure that the lever assembly is not binding, and that the leaf lever has not been warped or distorted by installation of screws. The valve on the leaf lever should fit the seat squarely. Adjust the secondary valve lever to extend ⅛-inch (plus or minus 1/64″) above the face of the secondary body. (See Fig. C787.) If an adjustment is necessary, bend at the location indicated.

Place the secondary diaphragm assembly in position against the secondary cover and align all holes carefully. Place the diaphragm sealing plate in position against the diaphragm and install and tighten the three screws. Insert the eight screw and lock washer assemblies through the secondary cover, making certain the screws pass freely through the holes in the diaphragm. Place the secondary gasket on the screws. Position the secondary cover, with the diaphragm in place, carefully against the secondary body, making certain that the diaphragm is laying flat and smooth. Start the eight screws into the threads of the secondary body a few turns, leaving the secondary cover loose enough against the body for the diaphragm to assume its normal position. Insert the valve spring through the opening in the cover, making certain that the end of the spring engages the seat of the secondary valve lever assembly. Start the retainer on the idle adjusting screw and rotate it until the retainer is approximately half way on the adjusting screw head. Grasp

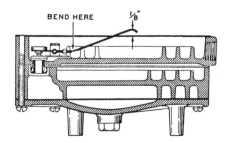

Fig. C787 — Secondary lever can be bent at location shown so that lever extends ⅛-inch above face of secondary body.

the idle adjusting screw and insert the retainer into the hexagonal guide in the cover. Make certain the retainer slides freely in the guide. Lubricate the rubber gasket and press it on the head of the screw. Then press the cap over the screw and tighten the cap into place, holding idle adjusting screw with a screwdriver to keep it from turning while the cap is being tightened. NOTE: The cap should be tight enough to apply sufficient pressure to the rubber gasket, but free enough to permit the screw to be rotated with a screwdriver. This will prevent the adjusting screw from turning through vibration. Tighten all eight screw and lock washer assemblies progressively in diametrically opposite pairs to secure the cover to the body. Install all drain plugs and reducers removed during disassembly.

Before installing the regulator, it may be checked and adjusted by using the simple test hookup as follows:

Attach an air hose to the inlet opening of the regulator. This is the opening in which the solenoid valve is attached. The air source should have a pressure between 60 and 100 psi. Refer to Fig. C787A. Remove the ⅛" pipe plug from the regulator and connect a low reading pressure gage to the opening from which the plug was removed. Connect a rubber hose to the vapor outlet of the regulator and in-sert the end of the hose into a bottle of water. The end of the hose should be approximately ¼" below the water level. Loosen the idle adjusting screw lock nut on regulator, and turn the idle adjusting screw clockwise—to the right—until it bottoms lightly on the seat. NOTE: Check the number stamped on the lower left hand corner of the regulator name plate. On the regulators having Part No. 6-595 stamped on plate, turn idle adjusting screw counter-clockwise 22 turns. On regulators having Part No. 6-733 stamped on the plate, turn idle adjusting screw counter-clockwise 8½ turns. Now, turn on the air and note the reading on the pressure gage which should be between 2½ and 5 psi. Rotate the idle adjusting screw until a slow bubbling (20 to 30 bubbles per minute) effect is produced. This setting will assure approximate correct starting mixture. Disconnect the regulator from the test rig, assemble regulator to tractor and make final idle speed and mixture adjustment as in paragraphs 46A and 46B.

CARBURETOR
Series 300

50A. With the fuel tank withdrawal valves closed, the procedure for removing and overhauling the carburetor is conventional.

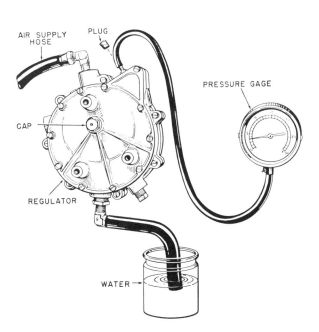

Fig. C787A — Simplified test stand hook-up which can be used to adjust the 300 series regulator when unit is off tractor.

LP-GAS SYSTEM

(Applies to Series 300B-400B-500B-600B Using Ensign Equipment)

Note: Refer to page 26 for Series 300 using Marvel-Schebler equipment.

The 300B, 400B, 500B and 600B series tractors are available with a factory installed LP-Gas system, using Ensign equipment. Like other LP-Gas systems, this system is designed to operate with the fuel supply tank not more than 80% filled (18.4 gallons). It is important when starting the engine to open the vapor or liquid withdrawal valves on the supply tank SLOWLY; if opened too fast, the excess flow valves will automatically shut off the flow of fuel. If this happens, merely close the withdrawal valves until pressure is equalized, then open the valves slowly.

Ensign components of the system are as follows:

Series 300B

Liquid filter	6257
Carburetor	MG1-1"-4216
Regulator	W-4207

Series 400B

Liquid filter	6257
Carburetor	XG-1¼"-353
Regulator	W-4361

Series 500B-600B

Liquid filter	6257
Carburetor	XG-1¼"-4364
Regulator	W-4372

A schematic drawing of a typical system is shown in Fig. C788.

CAUTION: LP-Gas expands readily with any decided increase in temperature. If tractor must be taken into a warm shop to be worked on during extremely cold weather, make certain that fuel tank is as near empty as possible. LP-Gas tractors should never be stored or worked on in an unventilated space.

ADJUSTMENTS
Series 300B-400B-500B-600B

51. Before starting the engine the first time, or after installing a new carburetor and/or regulator refer to Fig. C788A, C788B or C788C and make the following preliminary adjustments:

a. Turn the idle adjusting screw on regulator clockwise until it bottoms lightly on its seat; then, back it off ½-turn for series 300B, 1½ turns for series 400B, 1⅛ turns for series 500B and 600B.

b. Loosen lock nut and turn the load adjusting screw on the carburetor clockwise until it bottoms lightly on its seat; then, back it off 1⅜ turns for series 300B, 2⅛ turns for series 400B, 2½ turns for series 500B and 600B.

c. Loosen lock nut and turn the starting adjustment on carburetor clockwise until it bottoms lightly on its seat; then, back it off ⅞-turn for series 300B, ⅝-turn for series 400B, and ¾-turn for series 500B and 600B.

Note: The preceding adjustments are only approximate settings. Final accurate settings should be made after engine is warm as follows:

51A. IDLE ADJUSTING SCREW. Start engine and bring to operating temperature. Place hand throttle in the slow idle position and adjust the throttle stop screw (idle speed adjustment) on carburetor to obtain an engine slow idle speed of approximately 500 rpm. Turn idle adjusting screw on regulator either way as required to obtain the smoothest idle; then, readjust the throttle stop screw if necessary.

51B. LOAD ADJUSTMENT. With engine at operating temperature and choke button fully in, set the throttle control in wide open position so that engine high idle, no-load speed is:

300B . . . . .2050 rpm
400B and 500B. . . . . .2125 rpm
600B . . . . .2400 rpm

Loosen the lock nut on the load adjusting screw and turn the screw clockwise until the engine speed drops due to leanness. Then, turn the load screw counter-clockwise until engine speed drops due to richness. Now, turn the needle clockwise until engine runs smoothly. Tighten the lock nut.

FUEL FILTER

Series 300B-400B-500B-600B

51E. An exploded view of the liquid fuel filter is shown in Fig. C788E. The filter is designed to stop the passage of scale, rust or other foreign material into the remainder of the system.

Note: If frost collects on outside of filter bowl, it is probably clogged and must be cleaned.

To clean the unit, close both of the fuel tank withdrawal valves, remove the stud nut and withdraw the bowl and element. Thoroughly clean all parts and when reassembling, be sure to use new gaskets. If element contains considerable imbedded dirt, install a new element.

REGULATOR

Series 300B-400B-500B-600B

51F. **HOW IT OPERATES.** In the Ensign model W regulator, fuel from the supply tank enters the regulating unit (A—Fig. C788F) at tank pressure and is reduced from tank pressure to

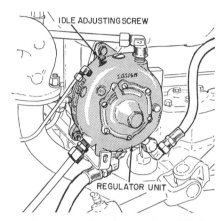

Fig. C788A—Typical installation of Ensign model W LP-Gas regulator, showing the installation of the idle adjusting screw.

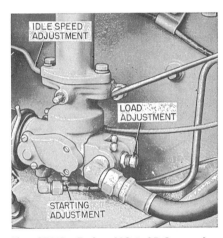

Fig. C788B—Ensign MG-1 LP-Gas carburetor installation on series 300B.

Fig. C788C—Ensign XG LP-Gas carburetor installation on series 400B, 500B and 600B.

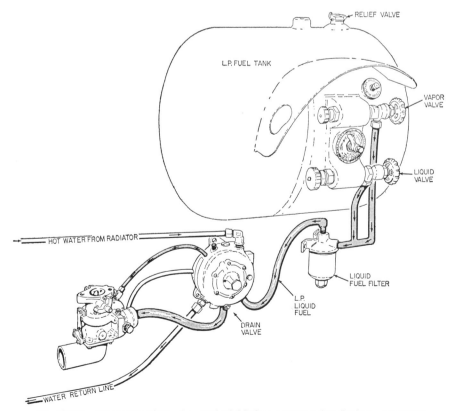

Fig. C788—Schematic view of a typical LP-Gas system using Ensign equipment.

31

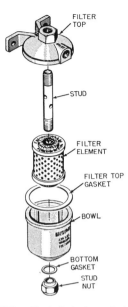

Fig. C788E — Exploded view of the Ensign liquid LP-Gas fuel filter No. 6257, used on series 300B, 400B, 500B and 600B.

approximately 4 psi at the inlet valve (C) after passing through the strainer (B). Flow through the inlet valve is controlled by the adjacent spring and diaphragm. When the liquid fuel enters the vaporizing chamber (D) via the valve (C) it expands rapidly and is converted from a liquid to a gas by heat from the water jacket (E) which is connected to the coolant system of the engine. The vaporized gas then passes at a pressure slightly below atmospheric pressure via the outlet valve (F) into the low pressure chamber (G) where it is drawn off to the carburetor via outlet (H). The outlet valve is controlled by the larger diaphragm and small spring.

Fuel for the idling range of the engine is supplied from a separate outlet (J) which is connected by tubing to a separate idle fuel connection on the carburetor. Adjustment of the carburetor idle mixture is controlled by the idle fuel screw (K) and the calibrated orifice (L) in the regulator. The balance line (M) is connected to the air inlet horn of the carburetor so as to reduce the flow of fuel and thus prevent over-richening of the mixture which would otherwise result when the air cleaner or air inlet system becomes restricted.

51G. TROUBLE-SHOOTING. The following data should be helpful in shooting trouble on the LP-Gas tractor.

SYMPTOM. Engine will not idle with idle mixture adjustment screw in any position.

CAUSE AND CORRECTION. A leaking valve or gasket is the cause of the trouble. Look for a leaking outlet valve caused by deposits on valve or seat. To correct the trouble, wash the valve and seat in gasoline or other petroleum solvent.

If the foregoing remedy does not correct the trouble, check for a leak at the inlet valve by connecting a low reading (0-20 psi) pressure gage at point (R—Fig. C788F). If the pressure increases after a warm engine is

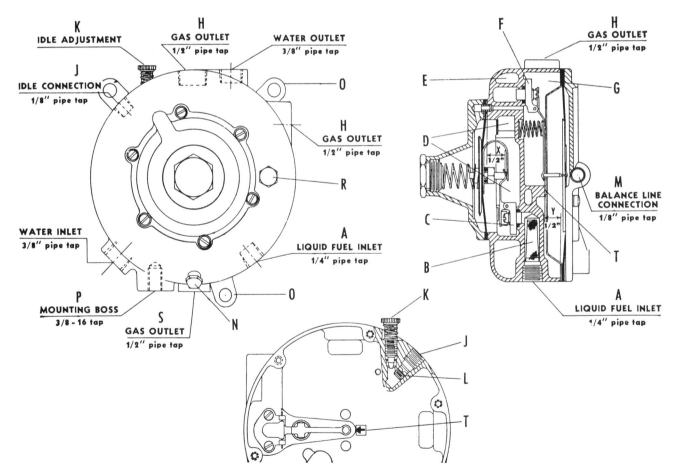

Fig. C788F—Model W Ensign LP-Gas regulator of similar construction to that used on series 300B, 400B, 500B and 600B. For exploded view refer to Fig. C788G.

A. Fuel inlet	E. Water jacket	J. Idle connection
B. Strainer	F. Outlet valve	L. Orifice (idling)
C. Inlet valve	G. Low pressure chamber	M. Balance line connection
D. Vaporizing chamber	H. Gas outlet	T. Boss or post

stopped, it proves a leak in the inlet valve. Normal pressure is 3½-5 psi.

SYMPTOM. Cold regulator shows moisture and frost after standing.

CAUSE AND CORRECTION. Trouble is due either to leaking valves, or the valve levers are not properly set. For information on setting of valve lever, refer to paragraph 51H.

51H. **REGULATOR OVERHAUL.** Remove the unit from the engine and completely disassemble, using Fig. C788G as a reference. Thoroughly wash all parts and blow out all passages with compressed air. Inspect each part carefully and discard any which are worn.

Before reassembling the unit, note dimension (X—Fig. C788F) which is measured from the face on the high pressure side of the casting to the inside of the groove in the valve lever when valve is held firmly shut as shown in Fig. C778H. If dimension (X) which can be measured with Ensign gage No. 8276 as shown in Fig. C788H or with a depth rule is more or less than ½-inch, bend the lever until this setting is obtained.

The casting (C — Fig. C788J) is marked with an arrow to assist in setting the low pressure valve lever. Be sure to center the lever on the arrow before tightening the screws which retain the valve block. The top of the lever should be flush with the top of the two posts (P).

CARBURETOR
Series 300B-400B-500B-600B

51J. With the fuel tank withdrawal valves closed, the procedure for removing and overhauling the carburetor is conventional.

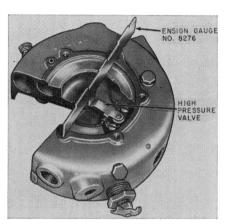

Fig. C788H — Using Ensign gage 8276 to set the fuel inlet valve lever to the dimension as indicated at "X" in Fig. C788F.

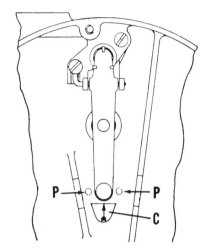

Fig. C788J — Location of posts (P) and arrow (C) for the purpose of setting the fuel inlet valve lever.

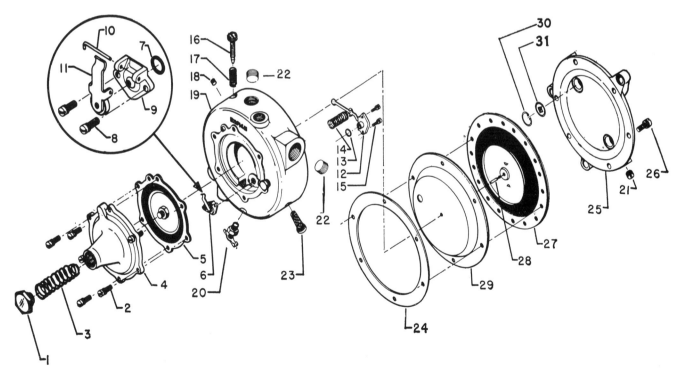

Fig. C788G—Exploded view of the Ensign Model W LP-Gas regulator.

1. Spring retainer	11. Inlet diaphragm lever	23. Strainer
3. Inlet diaphragm spring	12. Outlet valve assembly	24. Partition plate
4. Regulator cover	13. "O" ring	25. Back cover plate
5. Inlet pressure diaphragm	14. Outlet diaphragm spring	27. Outlet pressure diaphragm
6. Inlet valve assembly	16. Idle adjusting screw	28. Push pin
7. "O" ring	17. Idle screw spring	29. Partition plate
9. Valve seat	18. Bleed screw	30. Retainer (snap) ring
10. Pivot pin	19. Regulator body	31. Compensator
	20. Drain cock	

DIESEL FUEL SYSTEM

The Diesel fuel system consists of three basic units; the fuel filters, injection pump and injection nozzles. Refer to Fig. C789. When servicing any unit associated with the fuel system, the maintenance of absolute cleanliness is of utmost importance. Of equal importance is the avoidance of nicks or burrs on any of the working parts.

Probably the most important precaution that service personnel can impart to owners of Diesel powered tractors is to urge them to use an approved fuel that is absolutely clean and free from foreign materials. Extra precaution should be taken to make certain that no water enters the fuel storage tanks. This last precaution is based on the fact that all Diesel fuels contain some sulphur. When water is mixed with sulphur, sulphuric acid is formed and the acid will quickly erode the closely fitting parts of the injection pump and nozzles.

FILTERS AND BLEEDING

52. **BLEEDING.** Refer to Fig. C790. To bleed series 300 system which incorporates a hand priming pump, first make certain that the fuel tank is full and open the fuel shut-off valve located at lower right front corner of tank. Carefully clean dirt from bleed

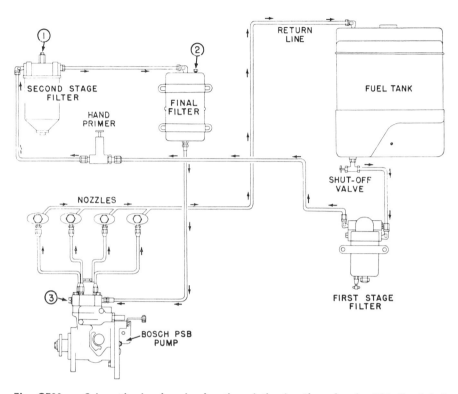

Fig. C790 — Schematic drawing showing the relative location of series 300 diesel fuel system components. Numbered items show bleeder valves. Series 300B diesels are similar except a fuel transfer pump is mounted on the injection pump and hand primer is not used.

Fig. C789 — Side view of series 300 diesel engine showing the fuel system components. Series 300B diesels are similar except a fuel transfer pump is mounted on the injection pump and the hand primer is not used.

plugs (1, 2 and 3). Unscrew bail nut from hand priming pump and with plug (1) removed from second stage filter, slowly operate the hand primer until a solid bubble-free flow of fuel comes from the bleeder hole; then re-install and tighten bleed plug (1). Remove plug (2) from final filter and operate the hand primer until a solid bubble-free flow of fuel comes from the bleeder hole; then reinstall and tighten bleed plug (2). Remove plug (3) from injection pump and operate the hand primer until a solid bubble-free flow of fuel comes from the bleeder hole; then while still operating the hand primer, install and tighten plug (3).

On series 300B, the hand priming pump is not used, but a gear driven fuel transfer pump is mounted on the injection pump. The procedure for bleeding this system is similar to the earlier models, except the engine should be cranked with the starting motor during the bleeding process.

53. FUEL TANK. Fuel tank of any Diesel engine using "cracked" type fuels should be cleaned periodically to remove gum and varnish deposits. These deposits can be removed using a 50-50 mixture of denatured, uncolored alcohol and benzol.

54. FIRST STAGE FILTER. Every fifty hours or more often in extreme conditions, clean dirt from exterior of filter, close the fuel shut-off valve and open drain valve on filter. Remove the filter body, unscrew filter nut and remove the brass edge-type filter. Clean element in clean diesel fuel and blow dry. Do not use linty rags for cleaning. Gum or varnish deposits can be removed by soaking in a suitable solvent. After assembly, bleed the system as in paragraph 52.

55. SECOND STAGE FILTER. The cartridge or inner element of this filter must be discarded every 500 hours or more often in extreme conditions. The procedure for renewing the element is evident. After assembly, bleed the system as in paragraph 52.

56. FINAL STAGE FILTER. This sealed type filter should be discarded and a new one installed every 2000 hours or more often in extreme conditions. The procedure for renewing the filter is evident. After assembly, bleed the system as in paragraph 52.

DIESEL (INJECTORS) NOZZLES

Users of the I&T SHOP SERVICE are warned to play safe when servicing the fuel injectors.

When testing or adjusting fuel injectors, do not place any part of your body in front of the injector nozzle. Fuel spray from an injector has sufficient penetrating power to puncture the flesh and destroy tissue. Should the fuel enter the blood stream it may cause blood poisoning.

In the event the skin is punctured from the discharge, immediately wash the injured part with boric acid solution, support the injured member with a splint and sling and get the patient to a physician as soon as possible.

57. LOCATING FAULTY INJECTOR. If one engine cylinder is misfiring, it is reasonable to suspect a malfunctioning injector in that cylinder. As in shorting out spark plugs in a spark ignition engine, the faulty injector is the one which, when its feed line is loosened (to deprive it of fuel), least affects the running of the engine. Remove the suspected injector from the engine as outlined in paragraph 58.

To check the diagnosis, install a new or rebuilt injector or one from a cylinder which is firing regularly. If the cylinder fires regularly with the other injector, the condemned unit should be serviced as per paragraphs 59 through 62.

It is recommended that whenever an injector is suspected of being faulty that the mating energy cell (pre-combustion chamber) be also removed and serviced as outlined in paragraph 73.

58. R&R INJECTOR. Before loosening any lines, wash the connections and adjacent parts with fuel oil or

kerosene. After disconnecting the high pressure and leak off lines, cap the exposed lines and fittings with plastic plugs to prevent the entrance of dirt. Remove injector retaining screws and withdraw the injector using a pry bar with a right angle claw if necessary. Remove the solid copper gasket.

Thoroughly clean the injector recess in the cylinder head before reinserting injector. Make certain that the old copper gasket (1—Fig. C791) has been removed (it may have stuck in the head recess seat). Install a new gasket and carefully insert the injector using care not to strike the nozzle tip against any hard surface. Torque the clamping nuts to 14-16 foot-pounds.

59. CLEANING & INSPECTING INJECTORS. Hard or sharp tools, emery cloth, crocus cloth, grinding compounds or abrasives of any kind should NEVER be used in the cleaning of injectors.

Injectors which do not check O.K. on the test stand or by visual testing, should be disassembled, inspected and cleaned as follows: Carefully clamp injector in a soft jawed vise and remove nozzle cap nut. Remove spray nozzle unit consisting of the valve (V —Fig. C793) and the tip (T). If valve cannot be readily withdrawn from tip with the fingers, soak the assembly in acetone, carbon tetrachloride or equivalent until parts are free.

The nozzle valve and body (tip) are a mated (non-interchangeable) unit and should be handled accordingly. Neither part is available separately.

All surfaces of the nozzle valve should be mirror bright except the contact line of the beveled seating surface. Polish the valve with mutton tallow used on a soft cloth or felt pad. The valve may be held by its stem in a revolving chuck during this operation. A piece of soft wood soaked in oil, or a brass wire brush will be helpful in removing carbon from the valve.

If valve shows any dull spots on the sliding surfaces, or if a magnifying glass inspection shows any nicks, scratches or etchings, discard the valve and the nozzle body; or, send them to an authorized station for possible overhaul.

The inside of the nozzle tip (body) can be cleaned by forming a piece of soft oil soaked wood to a taper of the same angle as the valve seat in the

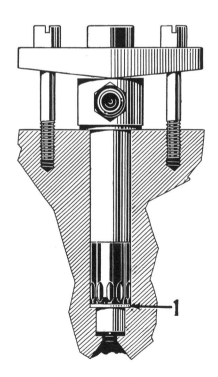

Fig. C791 — Sectional view showing typical injection nozzle installation. Whenever the nozzle has been removed, always renew the copper gasket (1).

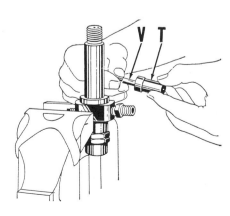

Fig. C793 — Removing spray nozzle tip (T) and valve (V).

body. Some mechanics use an ignition distributor felt oiling wick instead of the soft wood for cleaning the pintle seat in the tip. In this procedure, the mechanic shapes the end of the wick to the seat contour and coats the formed end with tallow for polishing.

The orifice at the end of the tip can be cleaned with a wood splinter. Outer surfaces of the nozzle should be cleaned with a brass wire brush and a soft cloth soaked in carbon solvent. Any erosion or loss of metal at or adjacent to the tip orifice warrants discarding the nozzle body and valve.

The lapped surface of the nozzle body (tip), where it contacts the mating surface of the holder, must be flat and shiny. Any discoloration of these sealing surfaces should be removed by cleaning with tallow on a lapping plate using a figure 8 motion. If a magnifying glass inspection shows any nicks, scratches, etching or discoloration after cleaning, discard the injector assembly or send the holder and nozzle to an authorized station for re-lapping.

60. NOZZLE HOLDER. Clean the exterior of the holder (except the lapped sealing surface at the end of same) using a brass wire brush and carbon solvent. The lapped sealing surface must be flat and shiny. Any discoloration should be removed by cleaning with tallow on a lapping plate using a figure 8 motion. If a magnifying glass inspection shows any etching, scratching or discoloration after cleaning, send the holder and nozzle assembly to an authorized station for re-lapping.

The cap nut for the holder should be cleaned of deposits using a brass wire brush and carbon solvent. Shallow irregularities of the gasket contacting lower surface, which would provide a possible leakage path across the sealing face, should be removed by reduction lapping using emery cloth laid on a flat surface. If the irregularities cannot be corrected in a reasonably short time, use a new nut.

61. SPINDLE & SPRING. If the shop is not equipped with a nozzle tester, further disassembly of the injector should not be attempted. If a tester is available, remove the spindle and spring (Fig. C794) and wash parts in clean fuel. If spindle is bent, if end which contacts the valve stem is cracked or broken or if attached spring seat is worn, renew the spindle and seat. Reject the spring if same is rusted or shows any surface cracks. Reject upper seat if worn.

62. REASSEMBLY OF INJECTOR. After all of the component parts have been cleaned and checked, lay all of them in a clean container filled with clean diesel fuel. Withdraw the parts

from the container as needed and assemble them while they are still wet. Observe the following points before and during assembly.

Nozzle valve should have a minimum clearance (free fit) in the tip or body. If valve is raised approximately ⅜ of an inch off its seat, it should be free enough to slide down to its seat without aid when the assembly is held at a 45 degree angle.

If parts have been properly cleaned, all sliding and sealing surfaces of same will have a mirror finish. Any scratches, etching, etc., on these surfaces warrants renewal; or, sending the mating pieces to an authorized service station for possible repairs.

After injector is assembled, set the opening pressure and check the spray pattern and leakage as outlined in paragraphs 64 through 66.

63. **TEST & ADJUST INJECTORS.** A complete job of testing and adjusting the injector requires the use of an approved nozzle tester. The nozzle should be tested for leakage, spray pattern and opening pressure.

Note: Only clean, approved testing oil such as Bosch TSE 76141 should be used in the tester tank. Other approved oils are available from the large oil companies and these are usually to be diluted 3 parts oil to one of water white kerosene before using.

64. OPENING PRESSURE. While operating the tester handle, observe the gage pressure at which the spray occurs. The gage pressure should be 1750-1850 psi. If the pressure is not as specified, remove the nozzle protecting cap, exposing the pressure adjusting screw and lock nut. Loosen the lock nut and turn the adjusting screw as shown in Fig. C795 either way as required, to obtain an opening pressure of 1750-1850 psi. Note: If a new pressure spring has been installed in a nozzle holder, adjust the opening pressure to 1900-1950 psi so as to compensate for subsequent fall off in pressure of the new spring.

Note: When testing and adjusting a set of injectors from the same engine, it is desirable to have the opening pressure as near alike as possible.

65. LEAKAGE. The nozzle valve should not leak at any pressure below 1450 psi. To check for leakage, actuate the tester handle slowly and as the gage needle approaches 1450 psi, ob-

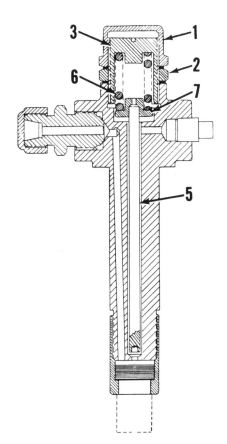

Fig. C794—Sectional view of pintle type nozzle typical of that used on 300 and 300B diesel engines.

1. Cap nut	5. Spindle
2. Jam nut	6. Spring
3. Adjusting screw	7. Lower spring seat

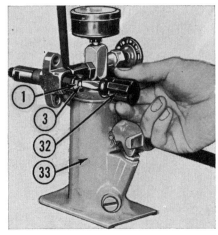

Fig. C795 — Adjusting the nozzle opening pressure, using a suitable nozzle tester.

1. Nut	32. Screw driver
3. Adjusting screw	33. Nozzle tester

serve the nozzle tip for drops of fuel. If drops of fuel collect at nozzle valve orifice at pressures less than 1450 psi, the nozzle valve is not seating properly. Some slight over-flow at leak off port is normal but if greater than on a new nozzle, it indicates wear between the cylindrical surfaces of the nozzle valve and the body.

66. SPRAY PATTERN. Operate the tester handle slowly until nozzle valve opens; then, move the lever with light quick strokes at the rate of approximately 100 strokes per minute. If the nozzle has a ragged spray pattern as shown in top view of Fig. C796, the nozzle valve should be serviced. It is not necessary that the nozzle produce an audible chatter.

PSB INJECTION PUMP

The subsequent paragraphs will outline ONLY the injection pump service work which can be accomplished without the use of special, costly pump testing equipment. If additional service work is required, the pump should be turned over to an official Diesel service station for overhaul. Inexperienced service personnel should never attempt to overhaul a Diesel injection pump.

67. LUBRICATION. With the exception of the upper portion of the injection pump plunger, the pump is dependent on the engine crankcase oil for lubrication. The J. I. Case Company recommends that engine crankcase be drained and refilled at least every 60 hours.

68. CHECK AND RESET TIMING. Recommended injection timing is 26 degrees before top center. To check the timing, proceed as follows: This check presumes that the engine will start and run but you wish to make sure that the timing is to specifications. Remove pump window cover as shown in Fig. C797. Crank engine until the number one (front) piston is coming up on compression stroke and continue cranking engine until pointer in window on right side of torque tube is exactly in register with the 26 degree BTDC mark on flywheel. At this time, the beveled (red) tooth on the pump distributor gear should be visible in the timing window.

Remove plug from the timing port in upper surface of drive shaft housing. Sighting through the timing plug port, observe the position of the

pointer on pump body with respect to the mark on pump shaft hub. If marks are not in register as shown, the timing should be corrected as outlined in the next paragraph.

If the pointer is not in register with hub mark, proceed as follows: Drain cooling system, raise the hood panel and disconnect the head light wires at snap couplers on each side of battery. Open grille and disconnect intake pipe from air cleaner. Remove cap screws retaining hood and sheet metal assembly to tractor, spread side sheets sufficiently to clear dash, slide the complete assembly forward and remove same from tractor. Remove air cleaner and radiator. Remove both of the frame side rails and remove the cap screws retaining the steering shaft bracket to torque tube. Remove the roll pin retaining the steering shaft front universal joint to the steering worm (cam) shaft and bump the U-joint yoke rearward and off the shaft. Support tractor under torque tube and on axle type tractors, unbolt the radius rod pivot bracket from torque tube. Unbolt the front support from engine and roll the front support, wheels and axle assembly forward and away from tractor. Remove the small cover from front of the injection pump drive housing, loosen the three clamp screws retaining gear to hub and using a ¾-inch socket, turn hub until line on hub registers with pointer and tighten the three clamp screws.

69. REMOVE PUMP. To remove the injection pump, close the fuel shut-off valve, disconnect the governor linkage and remove fuel lines from pump. Unbolt and remove pump from drive housing. Drive gear housing can be renewed at this time. To install the pump, refer to the following paragraph.

70. INSTALL AND TIME PSB PUMP. To install a new or repaired PSB injection pump, proceed as follows: Remove the timing window cover from side of pump. Crank engine until number one piston is coming up on compression stroke and continue cranking until the pointer in window on right side of torque tube is exactly in register with the 26 degree BTDC mark on flywheel.

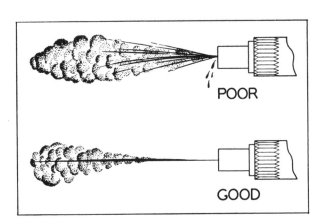

POOR

GOOD

Fig. C796 — Injection nozzle spray patterns. Lower view shows the spray pattern of a good nozzle.

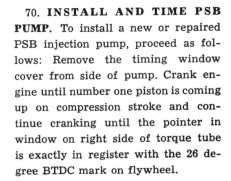

Fig. C797—PSB injection pump installation. Injection timing is 26 degrees B.T.C.

RED TOOTH

WINDOW COVER REMOVED

Turn the pump cam shaft in the normal direction until the beveled (red) tooth is visible in the pump window and the pointer on pump housing is in register with mark on shaft hub as shown in Fig. C797. Install pump to engine. Timing pointer may go off register with line on coupling, as gear teeth go into mesh. After the pump mounting cap screws are tightened, recheck the pump timing as outlined in paragraph 68.

Reinstall the fuel lines and linkage.

DIESEL GOVERNOR

An exploded view of the diesel engine governor weight unit and actuating linkage is shown in Fig. C798.

71. ADJUST ENGINE SPEED. Recommended engine speeds are as follows:

Slow idle . . . . 600 rpm
Fast idle (no-load) . . . . . . .1950 rpm
Full load . . . . .1750 rpm

Before attempting to adjust the governed speeds, first make certain that governor fly-weights are completely "in" when the injection pump is in the wide open or full fuel delivery position. To make this adjustment, first shut off the engine and move the throttle control hand lever to the wide open position. Remove pin (1—Fig. C798) and adjust the length of rod (3) with clevis (2) so that pin (1) will slide freely into position; then, shorten the rod 1½ turns and install the clevis pin.

To adjust the governed speed, back off the bumper spring adjusting screw (4). Loosen nuts (5) enough to permit movement of bracket (6) and move the speed control hand lever to the position where the recommended high idle (no load) engine speed is obtained. Move bracket (6) until it just touches stop (X) on spade (7) and tighten nuts (5). Turn the bumper spring adjusting screw (4) **in** only enough to eliminate surge. If surge cannot be eliminated without increasing the engine speed, it will be necessary to readjust the engine high idle, no load speed.

With the throttle closed, turn screw (8) either way as required to obtain an engine slow idle speed of 600 rpm.

72. OVERHAUL. The governor is accessible for removal and/or overhaul after removing the timing gear cover as outlined in paragraph 21. Refer to Fig. C798. Check condition of governor shaft, bearing (16) and bumper spring (19). If excessive wear is noted, renew worn parts. Bumper spring should be renewed if it shows signs of being set or is distorted. Pull shaft and race assembly (22) out of camshaft and check plunger for wear. Remove the retaining nut and remove weight unit (21). Check plate and weight units for wear and renew complete unit if necessary. Reinstall parts in reverse order of removal and renew oil seals where necessary. Note: Governor shaft pivots at lower end on a steel ball (18). Be sure ball is in position when reassembling. Adjust the governed speeds as outlined in paragraph 71.

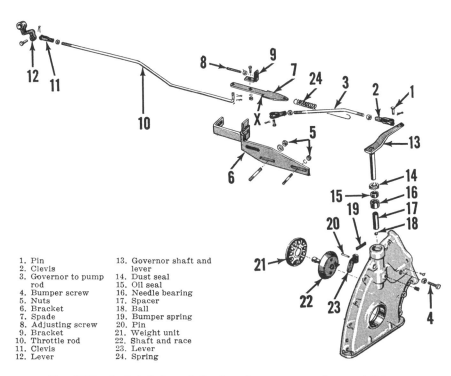

1. Pin
2. Clevis
3. Governor to pump rod
4. Bumper screw
5. Nuts
6. Bracket
7. Spade
8. Adjusting screw
9. Bracket
10. Throttle rod
11. Clevis
12. Lever
13. Governor shaft and lever
14. Dust seal
15. Oil seal
16. Needle bearing
17. Spacer
18. Ball
19. Bumper spring
20. Pin
21. Weight unit
22. Shaft and race
23. Lever
24. Spring

Fig. C798—Exploded view of diesel engine governor and control linkage.

ENERGY CELLS

73. R&R AND CLEAN. The J. I. Case Company recommends, that whenever an injector is removed for servicing, the mating energy cell be removed and checked or serviced. It is stated that in almost every instance where a carbon-fouled or burned energy cell is encountered, the cause is traceable to either a malfunctioning injector, incorrect fuel or incorrect installation of the energy cell. Manifestations of a fouled or burned unit are misfiring, exhaust smoke, loss of power and/or pronounced detonation (knock) at low engine speeds. It is necessary to remove manifold before removing No. 2 or 3 energy cell.

Fig. C799 — Side view of diesel engine showing governor to injection pump linkage.

To remove the complete energy cell, first remove the threaded plug (97—Fig. C801) and cap (98). Using a pair of thin nosed pliers, grip the tip of the energy cell cap (99) and pull cap out of chamber. Cell (100) can be removed by screwing a puller bolt into cell and pulling cell from chamber.

Fig. C800—Side view of diesel engine showing location of energy cells. Manifold must be removed to gain access to Nos. 2 and 3 cells.

The energy cell can also be bumped out by first removing the injection nozzle. Then, bumping cell out with a brass drift inserted through the injection nozzle opening.

The removed parts can be cleaned in an approved carbon solvent. After parts are cleaned, visually inspect them for cracks and other damage. Renew any damaged parts. Inspect the mating surfaces of the cell body and the cell cap for being rough and pitted. The surfaces can be reconditioned by lapping with valve grinding compound. Make certain that the energy cell seating surface in cylinder head is clean and free from carbon deposits.

When installing the energy cell, tighten the threaded plug enough to insure an air tight seal.

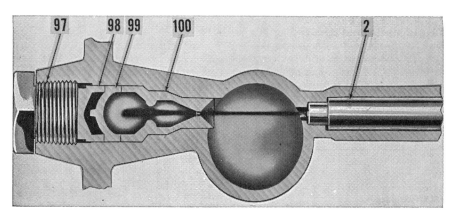

Fig. C801 — Energy cell (pre-combustion chamber) is located in each cylinder opposite to the nozzle.

NON-DIESEL GOVERNOR

A schematic view of the governor and actuating linkage is shown in Fig. C802.

74. ADJUST ENGINE SPEED. Recommended engine speeds are as follows:

Slow idle .... 500 rpm

Fast idle (no load)
  200B, 300B, 350...........2050 rpm
  300 ....1875 rpm
  400B, 500B ....2125 rpm
  600B ....2400 rpm

Full load speed
  200B, 300B, 350...........1900 rpm
  300 ....1750 rpm
  400B, 500B ....2000 rpm
  600B ....2250 rpm

Before attempting to adjust the governed speeds, free-up and align all linkage to remove binding or lost motion. Place the speed control hand lever and carburetor throttle valve in wide open position and adjust the length of governor lever to carburetor throttle link rod so as to slightly preload the governor lever. This is done by adjusting the carburetor link rod $\frac{1}{16}$-inch longer than is necessary to connect the carburetor throttle valve arm to the governor arm.

With engine at operating temperature, move throttle control to wide open position, loosen jam nut (Fig. C802) and turn adjusting nut either way as required to obtain the recom-

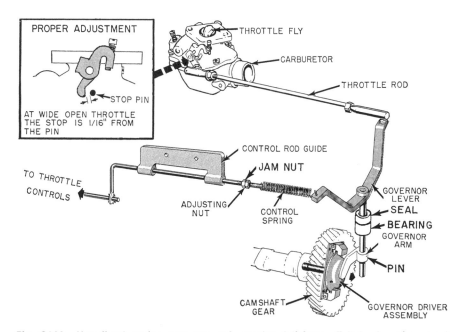

Fig. C802—Non-diesel engine governor and associated linkage. Full load engine speed should never exceed the values listed in the text.

mended high idle no-load speed. If engine tends to surge, vary the length of the governor arm to carburetor link rod slightly and recheck.

With the throttle in the closed position, turn the throttle stop screw to obtain a slow idle speed of 500 rpm.

75. **OVERHAUL.** The governor is accessible for removal and/or overhaul after removing the timing gear cover as outlined in paragraph 21. Refer to Fig. C802. Check condition of governor shaft, bearing and seal. Renew any worn or damaged parts. Pull shaft and race out of camshaft and check plunger for wear. Remove the retaining nut and withdraw the weight (driver) unit and camshaft gear. If any wear is noted, rivet a new assembly to the gear. When reassembling, remove the oil pan and buck-up camshaft while gear is being drifted on. Make sure that oil hole in camshaft and in hub of gear are open and clean and install gear so that oil holes are in register. Mesh gears so that timing marks are in register and tighten the nut to a torque of 80-90 ft.-lbs. Bend tab of lock washer against flat of nut. When assembly is complete, adjust the governed speeds as in paragraph 74.

COOLING SYSTEM

RADIATOR
All Models

76. To remove the radiator, drain cooling system and remove the hood, grille and air cleaner. On series 400B and 600B, disconnect the "Case-O-Matic" oil cooler lines. The remainder of the removal procedure is evident.

THERMOSTAT
All Models

78. On non-diesel engines, the thermostat is located in the upper radiator hose. On diesel engines, the thermostat is located in the engine water outlet casting.

WATER PUMP
Non-Diesel Models

79. The procedure for removing the water pump is evident after removing the radiator as outlined in paragraph 76. To disassemble the pump, refer to Fig. C803, use a suitable puller and remove impeller from pump shaft. Extract retainer wire and press shaft and bearing assembly from pump housing. Press shaft out of hub.

When reassembling, press seal in housing and fan hub on pump shaft. Press shaft and bearing assembly into housing and install the retaining ring. Press impeller on shaft until end of impeller is flush with end of shaft.

Diesel Models

80. The procedure for removing the water pump is evident after removing the radiator as outlined in paragraph 76. To disassemble the pump, refer to Fig. C804 and proceed as follows: Remove screw (9) from the impeller hub and using a suitable puller, remove impeller from shaft. Remove snap ring from impeller and extract the seal assembly (13). Pull fan hub from shaft. Remove snap ring (23) and press bearing and shaft assembly (22) forward and out of pump housing (16).

Seal contacting surface of bushing (15) must be smooth and flat. If surface cannot be reconditioned, renew the bushing. Lubricate face of seal and use thick soap suds on seal and shaft when reassembling.

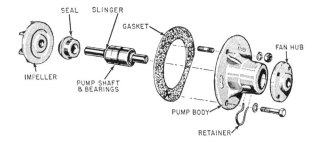

Fig. C803 — Exploded view of water pump used on non-diesel engines. Coolant leakage from body usually indicates a worn seal.

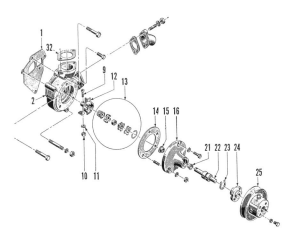

Fig. C804 — Exploded view of diesel engine water pump.

1. Gasket
2. Body
9. Screw
10. Nut
11. Lockwasher
12. Impeller
13. Seal assembly
14. Gasket
15. Bushing
16. Housing
21. Slinger
22. Shaft and bearing assembly
23. Snap ring
24. Hub
25. Pulley assembly

IGNITION AND ELECTRICAL SYSTEM

DISTRIBUTOR

Non-Diesel Models

81. Auto-Lite battery ignition distributor is standard equipment on all non-diesel engines, with magneto ignition being available as optional equipment. Spark plug electrode gap is 0.025. Firing order is 1-3-4-2.

Distributor applications are as follows:

200B, 350, 400B, 500B
   600B IAD-6003-2F
300, 300B IAD-6003-2J

Distributor test specifications are:
IAD-6003-2F
  Breaker contact gap..........0.020
  Contact pressure ....17-20 oz.
  Condenser capacity ..0.23-0.26 mfd.
  Cam angle, degrees.............42

Advance data in distributor degrees and distributor rpm:

    0° @ 275 rpm
    1° @ 300 rpm
    8° @ 475 rpm
    12° @ 800 rpm
    13° @ 875 rpm

IAD-6003-2J

Breaker contact gap...........0.020
Contact pressure ....17-20 oz.
Condenser capacity...0.23-0.26 mfd.
Cam angle, degrees.............42

Advance data in distributor degrees and distributor rpm:

    0° @   275 rpm
    1° @   330 rpm
    7° @   680 rpm
    12° @   965 rpm
    13° @ 1025 rpm

82. INSTALLATION AND TIMING. To install the distributor, crank engine until number one piston is coming up on compression stroke and continue cranking until the static ignition timing mark on flywheel rim is in register with pointer in timing port on side of torque tube.

Specified static ignition timing mark is as follows:

200B ....DC
300 LP-Gas ....IGN, $\frac{3}{32}$" below pointer
Other 300 ......DC, $\frac{7}{32}$" before pointer
300B LP-Gas ..IGN, $\frac{3}{32}$" below pointer
Other 300B ....IGN
350 ....IGN
400B LP-Gas ..IGN, $\frac{3}{32}$" below pointer
400B Gasoline ....IGN
500B LP-Gas ..IGN, $\frac{3}{32}$" below pointer
500B Gasoline ....IGN
600B LP-Gas ..IGN, $\frac{3}{32}$" below pointer
600B Gasoline ....IGN

With the rotor arm in the number one firing position, install the distributor, rotate distributor until breaker contacts are just starting to open and tighten the clamp.

Fig. C805 — When engine is in position to fire number one cylinder, slot in magneto drive shaft will be in a horizontal position.

MAGNETO
Non-Diesel Models

82A. J. I. Case Model 41 magneto is optionally available on all non-diesel models. Breaker contact gap is 0.008-0.012. Condenser capacity is 20 mfd., plus or minus 10%. Lag angle is 25 degrees. Spark plug electrode gap is 0.025. Firing order is 1-3-4-2.

83. INSTALLATION AND TIMING. To install and time the magneto, crank engine until number one piston is coming up on compression stroke and continue cranking until the static ignition timing mark on flywheel rim is in register with pointer in timing port on side of torque tube.

Specified static ignition timing mark is as follows:

200B ....DC
300 ....DC, $\frac{7}{32}$" before pointer
300B ....DC
350 ....IGN
400B ....DC
500B ....DC
600B ....DC

With the timing marks properly positioned, the slot in magneto drive shaft should be in a horizontal position as shown in Fig. C805. If the slot is not in the proper position, it will be necessary to remesh the drive gear with the camshaft gear as in paragraph 84. With magneto in an upright position, insert a spark plug wire in number one terminal (upper right) on distributor cap. Holding the other end of the wire ⅛-inch from the magneto frame, turn the impulse coupling in normal direction of rotation until a spark occurs.

Without disturbing the position of either the magneto or flywheel, install magneto, push top of same toward engine and tighten the retainer screw finger tight. Crank engine one complete revolution and again align the static timing marks. Slowly rotate top of magneto away from engine until the impulse coupling just snaps and tighten the screw and bolt securely.

84. MAGNETO ADAPTER. Remove magneto, then unbolt and remove the adapter assembly from engine. Remove pin (43—Fig. C806) and bump shaft rearward and out of gear and housing. If bushings are excessively worn, install new ones and ream them to provide a shaft diametral clearance of 0.002-0.004. Install seal (48) with lip facing the drive gear.

Install the unit so that drive slot is horizontal as shown in Fig. C805 when the static ignition timing marks are positioned as in paragraph 83. Install and retime the magneto.

ELECTRICAL UNITS
All Models

85. Positive battery terminal is grounded on all models. Auto-Lite electrical units are used and their applications are as follows:

Generator

200B, 300 non-diesel....GHD-6002A
300 diesel, 300B diesel..GHH-6002A
Other 300B, 350, 400B, 500B, 600B, no power steering....GHH-6002B
Other 300B, 350, 400B, 500B, 600B, with power steering..GHH-6002S

Regulator

200B, 300 non-diesel....VRR-4105A
All other models........VRT-4105

Starting Motor

200B, 300 non-diesel.......MZ-4176
300 diesel, 300B diesel.....ML-4352
300B non-diesel MBG-4105
400B MDL-6019
350, 500B, 600B..........MDL-6006

Electrical unit test specifications are as follows:

GHD-6002A Generator

Brush spring tension......26-46 oz.
Field draw
  Volts 5.0
  Amperes 2.0-2.2
Motoring draw
  Volts 5.0
  Amperes 4.8-5.4
Output
  Volts 8.0
  Maximum amperes 19.0

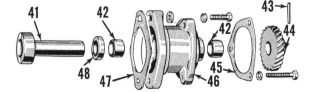

Fig. C806 — Exploded view of magneto adapter.

41. Drive shaft
42. Bushing
43. Pin
44. Drive gear
45. Gasket
46. Housing
47. Gasket
48. Seal

GHH-6002A, B, S Generators
Brush spring tension......26-46 oz.
Field draw
  Volts 10.0
  Amperes 1.6-1.7
Motoring draw
  Volts 10.0
  Amperes 2.9-3.3
Output
  Volts 15.0
  Maximum amperes 14.7

VRR-4105A Regulator
Cut-out relay
  Air gap, inches.......0.031-0.034
  Point gap, inches......0.015 min.
  Closing voltage 6.3-6.8
  Opening amperes 4.0-6.0
Voltage regulator
  Air gap, inches......0.048-0.052
  Voltage setting, range......6.9-7.3
  Voltage setting, adjust to.....7.1

VRT-4105 Regulator
Cut-out relay
  Air gap, inches............0.032
  Point gap, inches.........0.015
  Closing voltage 13.0-13.8
  Opening voltage 3.0-5.0

Voltage regulator
  Air gap, inches.......0.048-0.052
  Voltage setting, range....14.1-14.7
  Voltage setting, adjust to.....14.4

MBG-4105 Starter
Volts 12
Brush spring tension......42-53 oz.
No load test
  Volts 10.0
  Maximum amperes 55
  Minimum rpm 5200
Lock test
  Volts 4.0
  Maximum amperes 235
  Minimum torque ......5.2 ft.-lbs.

MDL-6006 and MDL-6019 Starters
Volts 12
Brush spring tension......31-47 oz.
No load test
  Volts 10.0
  Maximum amperes 60
  Minimum rpm 3200
Lock test
  Volts 4.0
  Maximum amperes 225
  Minimum torque ......6.0 ft.-lbs.

ML-4352 Starter
Volts 12
Brush spring tension......42-53 oz.
No load test
  Volts 10.0
  Maximum amperes 80
  Minimum rpm 3900
Lock test
  Volts 4.0
  Maximum amperes 750
  Minimum torque 26.0

MZ-4176 Starter
Volts 6
Brush spring tension......42-53 oz.
No load test
  Volts 5.0
  Maximum amperes 68
  Minimum rpm 4000
Lock test
  Volts 2.0
  Maximum amperes 280
  Minimum torque 4.4

ENGINE CLUTCH

Tractors are equipped with a single plate clutch as standard equipment. On series 300, a heavy-duty, double plate clutch is optionally available.

ADJUSTMENT

All Models Except Series 400B-600B

86. Adjustment to compensate for lining wear is accomplished by adjusting the clutch pedal linkage, NOT by adjusting the position of the clutch release levers. The clutch is properly adjusted when the clutch rod free travel is $\frac{1}{16}$-inch when measured at point where rod enters torque tube. Adjustment is accomplished by varying the length of the clutch rod. Refer to Fig. C807.

R&R AND OVERHAUL

All Models Except Series 400B-600B

87. To remove the clutch, first detach (split) engine from torque tube as outlined in paragraph 90, unscrew the cover to flywheel screws evenly

Fig. C807—Clutch is properly adjusted when clutch rod has 1/16-inch free movement as shown.

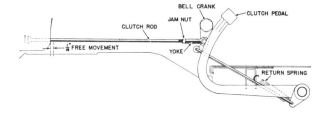

ENGINE CLUTCH
(Pressure Spring Test Specifications)

Tractor Series	Clutch Size	Pressure Spring Part No.	Pressure Spring Free Length	Lbs. Test		Height Inches
200B	9¼	G46058	$2\frac{25}{32}$	185	@	$1\frac{13}{16}$
300 Single Plate...........	9¼	G46058	$2\frac{25}{32}$	185	@	$1\frac{13}{16}$
300 Double Plate	8½	2825	$2\frac{25}{32}$	185	@	$1\frac{13}{16}$
300B	9¼	G46058	$2\frac{25}{32}$	185	@	$1\frac{13}{16}$
350	10	G47027	$2\frac{15}{16}$	152	@	$1\frac{9}{16}$
350	11	G47013	$2\frac{11}{16}$	142	@	$1\frac{11}{16}$
500B	10	G47027	$2\frac{15}{16}$	152	@	$1\frac{9}{16}$
500B	11	G47013	$2\frac{11}{16}$	142	@	$1\frac{11}{16}$

to avoid damaging the cover bracket and remove the cover assembly and lined plate. On series 300 dual plate clutches, lift out the inner ring and inner lined plate.

To disassemble the clutches, mark position of cover bracket and pressure plate so they may be reassembled in the same relative position. Place cov-er assembly on the bed of an arbor press and depress inner ends of re-lease levers as far as they will go without forcing. After removing ad-justing screws (7—Figs. C808, C809, C810 or C811), lock nuts and washers, slowly release pressure on release lev-ers and remove clutch cover bracket. To remove release levers (8), grind off one end of the pivot pin (9), be-ing careful not to damage the cover bracket.

Pressure spring test specifications are listed in the accompanying table. Friction faces of pressure plate, inner ring on dual clutches and flywheel must be smooth and flat. Driven plate linings can be renewed if plate is in otherwise good condition.

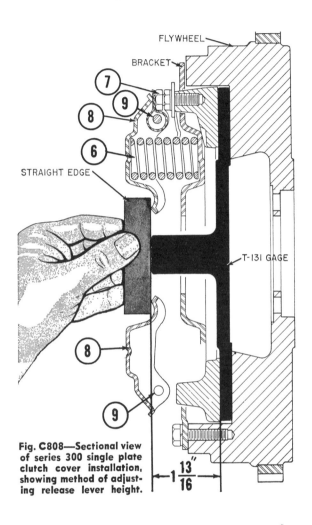

Fig. C808—Sectional view of series 300 single plate clutch cover installation, showing method of adjust-ing release lever height.

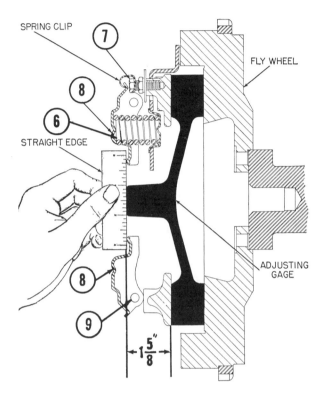

Fig. C809—Sectional view of series 300 double plate clutch cover installation, showing method of adjusting release lever height.

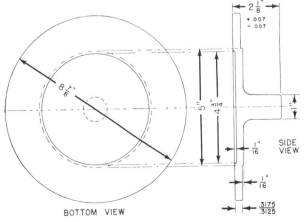

Fig. C808A—Dimensions which can be used to make special gage for adjusting the series 300 single plate clutch release lever height.

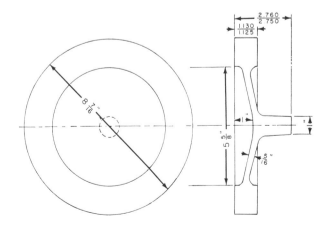

Fig. C809A—Dimensions which can be used to make special gage for adjusting series 300 double plate clutch release lever height.

When installing release levers, drive the pivot pins all the way in and spread the end with a chisel to hold them in place. Place pressure plate and pressure springs in position, depress inner ends of release levers as far as they will go without forcing and install washers, nuts and adjusting screws (7). Adjust the release lever height as in the following paragraphs.

87A. SERIES 300 SINGLE PLATE CLUTCHES. Bolt cover assembly and a new lined plate (long hub toward flywheel) to flywheel and turn the release lever adjusting screws until distance from release bearing contacting surface of each release lever is

$1\frac{13}{16}$ inches from friction face of pressure plate as shown in Fig. C808. NOTE: Variations in lever height must not exceed 0.004.

Lever height can also be checked by installing Case gage No. T-131 in place of the lined plate as shown. Gage can be made, using the dimensions shown in Fig. 808A.

NOTE: When reinstalling the single plate clutch, make certain that long hub of driven plate is toward flywheel.

87B. SERIES 300 DOUBLE PLATE CLUTCHES. Install a new front driven plate (long hub toward flywheel), inner ring, new rear driven plate (long hub toward rear) and cover assembly. Turn the release lever adjusting screws until distance from release bearing contacting surface of each release lever is $1\frac{5}{8}$ inches from friction face of

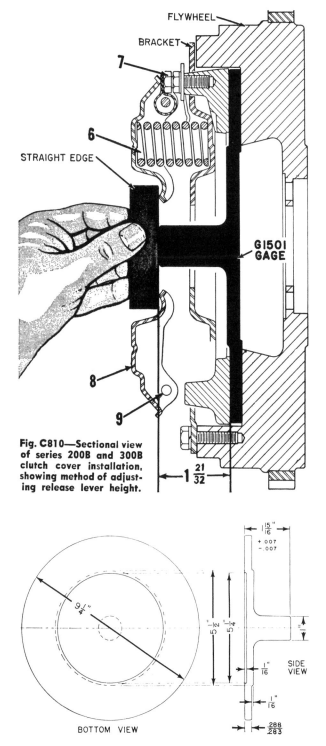

Fig. C810—Sectional view of series 200B and 300B clutch cover installation, showing method of adjusting release lever height.

Fig. C810A—Dimensions which can be used to make special gage for adjusting the series 200B and 300B clutch release lever height.

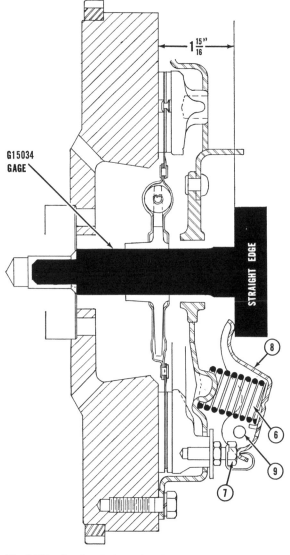

Fig C811—Sectional view of series 350 and 500B clutch cover installation, showing method of adjusting release lever height.

pressure plate as shown in Fig. C809. NOTE: Variations in lever height must not exceed 0.004.

Lever height can also be checked by installing Case gage in place of the lined plate as shown. Gage can be made, using the dimensions shown in Fig. C809A.

NOTE: When reinstalling the double plate clutch, make certain that front plate is installed with long hub toward flywheel and rear plate with long hub toward rear.

87C. SERIES 200B AND 300B. Bolt cover assembly and a new lined plate to flywheel and turn the release lever adjusting screws until distance from release bearing contacting surface of each release lever is $1\frac{21}{32}$ inches from friction face of pressure plate as shown in Fig. C810. NOTE: Variations in lever height must not exceed 0.004.

Lever height can also be checked by installing Case gage No. G1501 in place of the lined plate as shown. Gage can be made, using the dimensions shown in Fig. C810A.

87D. SERIES 350 AND 500B. Bolt cover assembly and a new lined plate to flywheel and turn the release lever adjusting screws until distance from

release bearing contacting surface of each release is $1\frac{15}{16}$ inches from friction face of flywheel as shown in Fig. C811. NOTE: Variations in lever height must not exceed 0.004.

Lever height can also be checked by using a straight edge and installing Case gage No. G15034 as shown.

ENGINE CLUTCH RELEASE BEARING

All Models Except Series 400B-600B

88. The engine clutch release bearing used on models with a single plate clutch differs from that used on models with a double plate clutch. Be sure that proper bearing is used. The procedure for renewing the bearing is evident after detaching (splitting) engine from torque tube as outlined in paragraph 90. Refer to Fig. C812. Be sure that small diameter of return spring is toward rear.

ENGINE CLUTCH SHAFT

All Models Except Series 400B-600B

89. To remove the engine clutch shaft, first detach (split) engine from torque tube as outlined in paragraph 90. Withdraw the clutch release bearing and disconnect the clutch release rod from the pedal. Disconnect the flexible tachometer drive shaft, then unscrew and remove the tachometer drive gear and sleeve assembly from the drive housing. Remove the three cap screws retaining the drive housing to torque tube and remove the drive housing and attached parts as a unit as shown in Fig. C813. On models so equipped, unbolt and remove belt pulley assembly from torque tube and be careful not to damage or lose the shims located between pulley

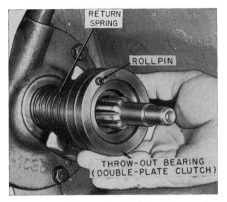

Fig. C812—Series 300 double plate engine clutch release bearing installation. Bearing installation on models with a single plate engine clutch is similar.

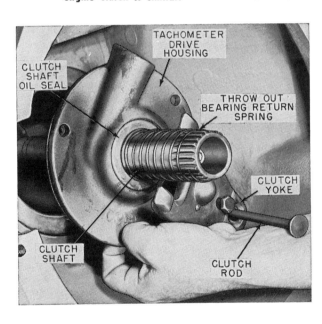

Fig. C813 — Removing tachometer drive housing and attached parts from front wall in torque tube on all series except 400B and 600B.

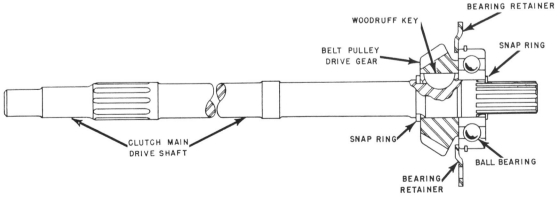

Fig. C814—Engine clutch shaft, belt pulley drive gear and associated parts removed from torque tube on all series except 400B and 600B. The belt pulley drive gear is not used on models with a shuttle transmission.

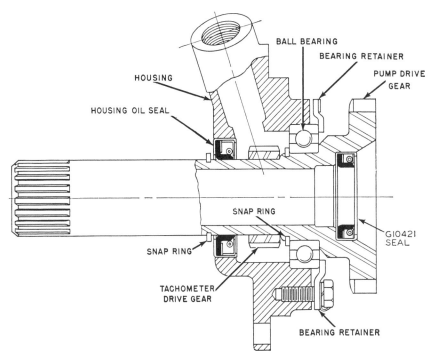

HOUSING

HOUSING OIL SEAL

BALL BEARING

BEARING RETAINER

PUMP DRIVE GEAR

SNAP RING

SNAP RING

TACHOMETER DRIVE GEAR

G10421 SEAL

BEARING RETAINER

Fig. C815 — Sectional view of tachometer drive housing, showing the installation of the hydraulic pump drive gear on all series except 400B and 600B. Always inspect seal in rear of pump drive gear when the engine clutch shaft is removed.

housing and torque tube. On models with live power take-off, engage the hand operated transmission clutch.

On all models, unlock and remove the three cap screws retaining the clutch shaft rear bearing retainer to center wall in torque tube and withdraw the clutch shaft assembly shown in Fig. C814. Remove snap ring from end of shaft and press off the bearing and belt pulley drive gear. Check all parts and renew any which are damaged or worn. NOTE: On models with a shuttle transmission, a spacer is used instead of the belt pulley driving bevel gear.

When reinstalling the assembled clutch shaft, tighten the rear bearing retainer cap screws to a torque of 15 ft.-lbs. Inspect the oil seal located in rear end of the hydraulic pump drive gear and renew same if damaged or worn. Refer to Fig. C815. Install seal with a suitable driver until same bottoms in the gear.

When reinstalling the tachometer drive housing assembly over the clutch shaft, use shim stock as a sleeve to avoid damaging the clutch shaft seal. Be sure to use sealing compound on the housing gasket. Turn housing slightly to center re-

lease yoke and tighten the housing retaining screws to a torque of 25-30 ft.-lbs. Use a new "O" ring and install the tachometer drive gear and sleeve. Install the clutch release bearing.

TRACTOR SPLIT
All Models Except Series 400B-600B

90. To detach (split) engine from clutch housing, drain cooling system and remove hood.

Remove both of the frame side rails and remove the cap screw retaining the steering shaft bracket to torque tube. Remove the roll pin retaining the steering shaft front universal joint to the steering worm (cam) shaft and bump the U-joint yoke rearward and off the shaft. Support both halves of tractor separately and on axle type tractors, unbolt the radius rod pivot bracket.

Disconnect fuel line and remove throttle rod and fuel tank. On models with underneath muffler, disconnect exhaust pipe from manifold. Disconnect wiring harness from generator and starting motor and remove starting motor. Disconnect the oil pressure gage line from cylinder block and heat indicator sending unit from cylinder head. On non-diesels, disconnect wire from coil and choke wire from carburetor. On diesels, disconnect the fuel shut-off wire at injection pump. Unbolt engine from torque tube and separate the tractor halves.

When reassembling, tighten the engine to torque tube bolts and the frame side rail cap screws to a torque 115 ft.-lbs.

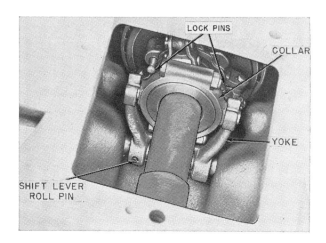

LOCK PINS

COLLAR

YOKE

SHIFT LEVER ROLL PIN

Fig. C816 — Transmission clutch installation as viewed through upper rear opening in torque tube.

TRANSMISSION CLUTCH

The hand operated, over-center, dry-type transmission clutch is used on series 200B, 300, 300B, 350 and 500B with live power take-off only.

ADJUSTMENT

All Models With Live PTO Except Series 400B and 600B

91. To adjust the clutch, proceed as follows: On models with a four-speed transmission, remove top rear cover from torque tube. On models with a twelve-speed transmission, place the "Tripl-Range" control lever in neutral and remove the "Tripl-Range" control cover assembly. Move the clutch hand lever rearward to the disengaged position and turn mainshaft (Fig. C816) until the spring loaded lock pins are visible as shown. Note that only one of the pins is engaged in the "in" position. To tighten the clutch, pull the engaged lock pin outward and turn the main shaft in a counter-clockwise direction. Continue this procedure until clutch lever goes over-center with a distinct snap. This should require 45-60 lbs. pull at end of hand lever.

R&R AND OVERHAUL

All Models With Live PTO Except Series 400B and 600B

92. To remove the transmission clutch, first detach (split) torque tube from transmission as outlined in paragraph 93. Bump roll pin (Fig. C817) out of fork and control lever, pull lever toward left until Woodruff key is exposed, extract key and remove control lever and throwout yoke. Pull clutch assembly out of drum and withdraw drum from the engine clutch shaft.

To disassemble the unit, pull the adjusting lock pins outward and completely unscrew the adjusting collar from the clutch hub. The need and procedure for further disassembly is evident. Thoroughly clean all parts and renew any which are visibly damaged or worn. Be sure that control lever oil seal in torque tube is in good condition.

Use Fig. C817A as a guide during reassembly and when installing the clutch drum, tap same with a soft faced hammer to be sure that drum is fully forward on the clutch shaft splines. Install clutch and throwout yoke; then, using care not to damage the lever oil seals, push control lever part way into yoke and install Woodruff key. Continue pushing lever inward until the yoke roll pin can be installed. Engage clutch and wire the

control lever in the engaged position until torque tube is connected to transmission.

Adjust the clutch as outlined in paragraph 91.

TRACTOR SPLIT

All Models Except Series 400B-600B

93. To detach (split) torque tube from transmission, drain oil from hydraulic reservoir in torque tube and disconnect the engine clutch release rod and speedometer drive cable. On models with a four-speed transmission, remove top rear cover from torque tube. On models with a twelve-speed transmission, place the "Tripl-Range" control lever in neutral and remove the "Tripl-Range" control cover assembly. On models with shuttle gears, remove the shuttle control cover. Disconnect the tail light wire and the hydraulic system control rods and lines. On models with live pto, wire the transmission clutch control lever in the engaged position Remove underneath muffler. Support both halves of tractor separately, then unbolt and separate the tractor halves.

When reassembling, tighten the torque tube to transmission bolts to a torque of 80 ft.-lbs.

Fig. C817 — Rear view of torque tube showing the installation of the transmission clutch. The clutch is a multiple disc over-center type.

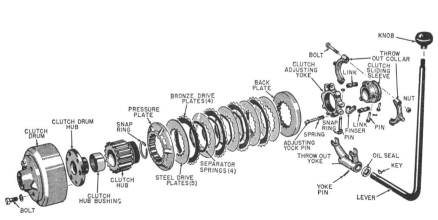

Fig. C817A—Exploded view of the hand operated, over-center type transmission clutch.

"CASE-O-MATIC"

(Torque Convertor System)

The "Case-O-Matic" drive on series 400B and 600B tractors consists of the oil reservoir (located in torque tube), oil supply pump, oil filter, oil cooler unit (integral with radiator), control valve and clutch pressure regulator unit, convertor pressure regulating unit and torque convertor unit which includes two hydraulically operated clutches.

When disassembling any part of the "Case-O-Matic" system, it is good practice generally to not remove any parts which can be thoroughly inspected while installed. All parts of the system, when removed, should be handled with the same care as would be accorded the parts of a diesel injection pump or nozzle unit and should be soaked or cleaned with an approved solvent to remove gum deposits. Unless you practice good housekeeping (cleanliness) in your shop, do not undertake the repair of the "Case-O-Matic" system.

An important precaution that service personnel should impart to owners of "Case-O-Matic" tractors is to urge them to service the breather, filter and reservoir as in paragraph 94B.

HOW IT OPERATES
Series 400B-600B

94. When the engine is started, the oil supply pump draws oil out of the torque tube reservoir and forces it through the filter, the cooling unit in the radiator and on to the control valve. At this point, the clutch pressure regulator valve obstructs the flow of oil, building up pressure on the main drive (multiple plate) clutch.

When the engine is running at low idle speed, a small volume of oil is pumped through the system and due to oil bleeding off for lubrication, low pressure may be built up in both the clutch and the converter.

When engine speed is increased, pressures are increased by means of the clutch pressure regulator and the converter pressure regulator. At governed engine speed, proper pressures are 110-125 psi on the clutches, and 30-40 psi in the torque converter unit.

When the "Case-O-Matic" spool valve is closed (clutch pedal depressed), the oil flow and pressure to the main drive clutch is shut-off permitting the operator to shift the transmission into gear in the usual manner.

With the transmission gears engaged, and the engine turning at sufficient rpm, opening the "Case-O-Matic" spool valve (releasing pedal), will permit the normal engine torque, multiplied by the converter to be applied to the transmission input shaft.

When the direct drive spool valve is opened (lever raised), oil under pressure is directed to the direct drive clutch which locks the drive and driven portions of the torque converter together, thus making a direct drive connection from the engine to the transmission.

TROUBLE SHOOTING
Series 400B-600B

94A. The following items should facilitate locating troubles in the "Case-O-Matic" system.

CONVERTOR OIL TEMPERATURE TOO HIGH. Could be caused by:

a. Faulty temperature gage. If gage is faulty it may indicate excessive temperature even though the actual oil temperature is O.K.

b. Tractor being operated in "Case-O-Matic" drive too long or at near "stall" speed. Stall speed is the point at which the torque convertor is unable to drive the tractor even though the engine is running at full speed. In which case, the engine is delivering full power to the convertor and all of this energy is released in the form of heat. To correct the trouble, shift tractor to neutral and allow oil to cool. Then, use a lower transmission gear.

c. Hydraulic lift system control valve may be stuck in partly open position causing pump to by-pass and heat the oil.

d. Restricted oil lines or hoses.

e. Dirty or clogged "Case-O-Matic" oil filter.

f. Dirt or trash in grille screen or radiator core.

g. Clogged radiator.

h. Faulty thermostat.

i. Oil level in reservoir too high.

j. Faulty engine timing, causing engine to overheat.

TRACTOR WILL NOT START LOAD IN "CASE-O-MATIC" OR ACTION IS SLOW AND WILL NOT START LOAD SMOOTHLY. Could be caused by:

a. Insufficient oil in reservoir.

b. Insufficient clutch pressure.

c. Insufficient convertor pressure.

d. Air leak in pump suction tube or gasket.

e. Improper oil.

f. Faulty pump or pump relief valve.

g. Restricted "Case-O-Matic" oil filter.

ENGINE PULLS DOWN OR LUGS WHEN OPERATING IN "CASE-O-MATIC" DRIVE. Could be caused by:

a. If engine is properly tuned and tractor is not being operated in too high gear range, the trouble is due to oil level in reservoir being too high and flooding the flywheel housing.

TRACTOR WILL NOT PULL LOAD IN DIRECT DRIVE. Could be caused by:

a. Faulty control valve linkage adjustment, preventing direct drive spool from being moved far enough to allow full flow of oil to direct drive clutch.

b. Convertor pressure too high due to a stuck spool or incorrectly installed gasket under convertor pressure regulator.

NOISE OR SQUEAL IN CONVERTOR LINES WHEN ENGINE IS STARTED. Could be caused by:

a. Cold or improper oil.

b. Faulty control valve linkage adjustments, preventing full opening of spools.

c. Air in control valve or oil lines. This can usually be corrected by depressing and releasing the foot pedal a few times.

WORKING FLUID, BREATHER, FILTER
Series 400B-600B

94B. It is recommended that torque convertor fluid level (in torque tube) be checked once each day when oil is warm and engine is running at low idle speed. Add oil to full mark on dip stick.

The torque tube (reservoir) breather should be removed, disassembled and cleaned in tractor fuel every 200 hours or more often in extremely dusty conditions. Dry same with compressed air before installation.

Fig. C818—Changing the "Case-O-Matic" oil filter element.

Under normal operating conditions, the "Case-O-Matic" oil filter (Fig. C818) should be serviced every 500 hours of operation and the filter element should be renewed every 1000 hours, as follows: Remove grille screen, then unbolt and remove the filter cap. Withdraw filter element and thoroughly clean filter body and cap. Wash element in tractor fuel and dry same with compressed air. If filter element shows any visible damage, renew same. When reassembling, be sure "O" ring on cap is properly positioned and tighten the retaining screws evenly and securely. After engine is started and oil is warm, check for fluid leaks at filter.

Recommended working fluid for the "Case-O-Matic" system (same as hydraulic lift system) is S.A.E. No. 10-W motor oil (MS-DG Grades). In extremely cold weather, use S.A.E. No. 5-W (MS-DG) motor oil. Approxi-

mate torque tube capacity is 16 U. S. quarts. Oil should be changed every 1000 hours as follows:

With oil warm, remove the two ½-inch P.T. plugs (Fig. C818A) from torque tube and allow to drain. Have an assistant turn engine over slowly until the ¼-inch plug (A), located in convertor unit, aligns with the ½-inch plug holes; then, remove the ¼-inch plugs (A and B). CAUTION: Do not lose these plugs from end of Allen wrench as they are removed. Renew the "Case-O-Matic" filter element.

The working fluid for the hydraulic power lift system is the same as the "Case-O-Matic" system. It is therefore important that the hydraulic system oil filter element be serviced when draining and refilling the torque tube oil reservoir, as follows:

Remove cover cap from filter compartment of the hydraulic control housing but be careful not to damage the "O" ring. Remove the filter cartridge from housing and examine same for damage, especially at each end.

NOTE: Fluid passes through the filter from inside to outside, therefore dirt and foreign material will be found on inside surface of cartridge. Thoroughly wash the filter with clean tractor fuel and blow dry with LOW PRESSURE air. When installing the element, be sure it seats properly on the spring loaded cap. Be sure "O" ring is in good condition, install cover cap and tighten the screws securely.

After all oil is drained, reinstall the ¼ and ½-inch plugs and fill reservoir with 16 quarts of the specified lubricant. Start engine and allow to run until oil is warm. Then recheck oil level.

LINKAGE ADJUSTMENTS
Series 400B-600B

94C. Operating linkage for the "Case-O-Matic" control valves was changed slightly during the 1958 production; this, however, does not change the adjustment procedure.

Before making any adjustments, refer to Fig. C818B and make certain that the "Case-O-Matic" spool interlock plate overlaps the direct drive spool interlock plate as shown. Disconnect linkage from spools and proceeds as follows:

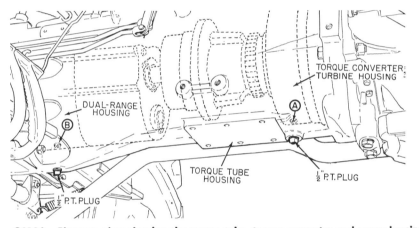

Fig. C818A—Phantom view showing the torque tube, torque converter and range housing drain plugs.

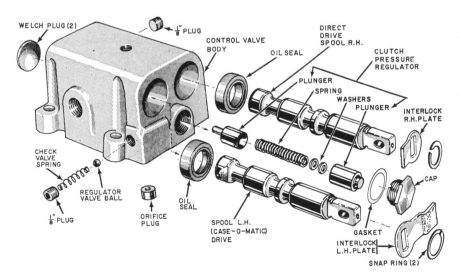

Fig. C818B—Exploded view of "Case-O-Matic" control valve and clutch pressure regulator valve.

Position the direct drive spool (R.H. side of valve housing) so the detent ball is resting in the groove with the spool in the extended (rear) position. Attach control arm to inner spool and hand control linkage and make certain that spool travels from one detent to the other as hand lever is moved. If action is not satisfactory, check for improper assembly or excessively worn linkage. Back-off the spool stop bolt several turns. Refer to Fig. C818C or C818D. Position the "Case-O-Matic" valve spool (L.H. side of valve housing) so the detent ball is resting in the groove with the spool in the extended (rear) position. Adjust the $\frac{5}{16}$x1½-inch spool stop bolt to definitely stop rearward travel of spool at this point and tighten the jam nut. Attach linkage to spool and, with spool still against the $\frac{5}{16}$-inch stop bolt, adjust the linkage with the threaded yoke until the flat arm on right end of clutch lever is in a vertical position and attach linkage to clutch lever. Depress foot pedal and adjust the ⅜-inch foot pedal stop bolt to make definite contact with clutch lever when detent ball seats in groove with spool in the inward (forward) position.

PRESSURE TESTS AND ADJUSTMENTS

Series 400B-600B

Note: All oil used to operate the "Case-O-Matic" system must pass through the system oil filter; it is therefore essential that filter element be clean before making any pressure checks. Refer to paragraph 94B. It is equally important that the control spool linkage be properly adjusted as in paragraph 94C before checking the clutch and convertor pressures.

94D. GENERAL DATA. Most cases of improper "Case-O-Matic" operation can be traced to malfunctions within either the convertor pressure regulator or the clutch pressure regulator. More cases of improper operation are the result of equalized pressures than any other single factor. High pressure, 110-125 psi, must be maintained on the clutches to keep them from slipping under load. Low pressure, 30-40 psi, is required in the torque convertor for proper operation as well as for cooling and lubrication. Therefore, if the high pressure is lowered or the low pressure is raised to a point where they are nearly equal, the unit cannot operate in a normal manner.

The convertor pressure regulator unit controls the pressure in the convertor; therefore, low convertor pressure is usually due to a weak or broken spring, or the valve is stuck in the open position. Should this valve stick in the closed position, the result will be high pressure in the convertor.

The "Case-O-Matic" control valve unit is the most likely place to find the cause of any equalization of pressures.

The clutch pressure regulator valve located immediately below the two spool valves in the control body is the one item upon which the most depends insofar as pressure controls are concerned. Should the innermost plunger of this valve become stuck in the open position, low clutch pressure will result and there will be very little difference between clutch pressure and convertor pressure. Should it become stuck in the closed position, little or no oil will enter the torque convertor resulting in low convertor pressure and high clutch pressure. Should the spring between the two plungers be broken or weak, the result will be low pressure on the clutches and nearly equal pressure in the convertor. Should the outer plunger of this valve be stuck in the outer position, low clutch pressure will result. Should the orifice that meters oil to the clutch pressure regulator be plugged or covered by the gasket, low clutch pressure will result.

Should the clutch pressure regulator valve (ball) be held off its seat or leaking badly, the clutch will "grab" when the pedal is released.

94E. PUMP PRESSURE. The spur gear type oil pump incorporates a non-adjustable pressure relief valve within the pump housing. This valve is set to open and by-pass oil when the pump output pressure reaches approximately 250 psi. If, however, the remainder of the system is operating properly, full pump capacity will not be used and the working pressure will be well under 200 psi. For all practical purposes, the pump relief valve is merely a safety valve insofar as the "Case-O-Matic" drive is concerned.

To check the pressure available to operate the "Case-O-Matic" system, first start engine and run same until "Case-O-Matic" oil is at normal operating temperature. Disconnect oil line from filter body and connect a suitable pressure gage as shown in Fig. C818E. J. I. Case gage No. 4501AA which incorporates a snubber to handle flash pressure is suitable. With engine running at rated rpm, there should be at least 130 psi pressure at this point. Insufficient pressure could be caused by:
a. Plugged filter.
b. Restricted oil lines.
c. Faulty pump relief valve spring or ball.
d. Faulty pump.
e. Air leak in pump suction tube.
f. Faulty suction tube gasket.

Excessive pressure could be caused by a sticking pump pressure relief valve or a faulty relief valve spring.

Note: If an ordinary gage (no snubber) is used, reduce the engine speed to about 1100 rpm.

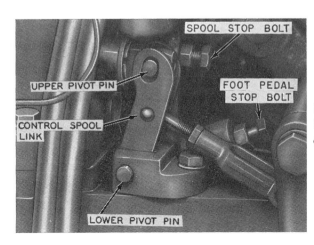

Fig. C818C — "Case-O-Matic" control valve linkage used on early production tractors.

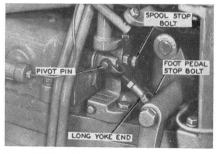

Fig. C818D—"Case-O-Matic" control valve linkage used on late production tractors.

94F. CLUTCH PRESSURE. To check the "Case-O-Matic" clutch pressure, first run engine until "Case-O-Matic" oil is at normal operating temperature. Disconnect the tractor oil pressure gage line from "T" fitting on side of control valve and connect a suitable pressure gage as shown in Fig. C818F. With engine running at rated rpm and the main clutch engaged (pedal released), the operating pressure should be 100-125 psi. If the pressure is not as specified, refer to paragraph 94D for possible causes. If the clutch pressure regulator valve requires cleaning and/or overhauling, refer to paragraph 94L.

94G. CONVERTOR PRESSURE. To check the "Case-O-Matic" torque convertor pressure, first run engine until "Case-O-Matic" oil is at normal operating temperature. Connect a suitable pressure gage to the plug opening in top of control valve housing as shown

in Fig. C818G. With engine running at rated rpm, the convertor pressure should be 30-40 psi. If the pressure is not as specified, refer to paragraph 94D for possible causes. If the convertor pressure regulator valve requires cleaning and/or overhauling, refer to paragraph 94K.

PUMP

Series 400B-600B

94H. REMOVE AND REINSTALL. To remove the "Case-O-Matic" oil pump, first remove the hood and rear quarter panels unit and disconnect fuel line and gage wire from fuel tank. Then unbolt and remove fuel tank from tractor. Disconnect battery cables and remove battery and battery mounting tray. Uncouple steering shaft at one of the universal joints and unbolt the steering shaft intermediate sup-

Fig. C818E — Pressure gage installation for checking pressure available to operate the "Case-O-Matic" system.

Fig. C818F — Pressure gage installation for checking the "Case-O-Matic" system clutch pressure.

port bracket. Disconnect the speedometer cable, tachometer cable at adapter housing, the "Case-O-Matic" heat indicator sending unit and oil line from the torque convertor pressure regulator valve housing and pressure gage oil line at "T" fitting. Disconnect the throttle rod and direct drive control linkage. Remove cap screws retaining the steering support mounting bracket to torque tube and lay steering support, instrument panel and attached units forward on engine.

Disconnect the "Case-O-Matic" oil lines, and the hydraulic lift system oil lines and control rods. Remove cap screws around perimeter and one Allen screw near the center of the adapter housing and, using a hoist and lifting eye, lift the complete assembly straight up and out of torque tube as shown in Fig. C818H. Use care during this operation not to distort the pump suction tubes.

Remove cap screws retaining the "Case-O-Matic" pump to lower side of adapter housing and carefully work the pump from its locating dowels.

When reassembling, reverse the disassembly procedure and tighten the adapter housing retaining screws evenly and securely.

94J. OVERHAUL. With pump removed as outlined in paragraph 94H, remove the 1/4x3/8" self-locking screw (Fig. C818J) and, using a suitable puller, remove the combination tachometer and pump drive gear. Extract Woodruff key from pump shaft. Remove cap screws retaining pump cover to pump housing and disassemble the pump as shown in Fig. C818K. Bushings, shafts, gears and machined surfaces of housing and cover should be carefully inspected and any damaged, worn or questionable parts should be renewed. If gears are not a tight press fit on the shafts, renew the worn part. Thoroughly wash the components of the relief valve in a suitable solvent and dry with compressed air. If relief valve ball or ball seat in valve body show any signs of scoring or pitting, renew both parts.

Reassemble the pump by reversing the disassembly procedure and apply a light coat of sealing compound to mating surfaces of housing and cover. Tighten the cover retaining screws to a torque of 6 ft.-lbs.

If snap ring and washer (Fig. C818J) have been removed from outer end of drive gear hub, they must be installed before positioning the drive gear on the pump shaft. Tighten the self-locking screw securely.

CONVERTOR PRESSURE REGULATOR

Series 400B-600B

An exploded view of the convertor pressure regulator valve is shown in Fig. C818L.

The valve is calibrated to maintain 30-40 psi pressure in the torque convertor. Low convertor pressure will result if spring is weak or broken or if spool is stuck in the open position. High pressure will result if spring is too strong or if spool is stuck in the closed position.

94K. **R&R AND OVERHAUL.** On models with no hydraulic power lift system and on models with a single control valve hydraulic lift system, the convertor pressure regulator valve can be removed without removing the fuel tank. On models with a dual valve hydraulic lift system, it is necessary to first remove the fuel tank.

In either case, disconnect the oil line and oil temperature gage sending unit from valve housing, then unbolt valve housing from adapter housing.

Remove the end plug retaining snap ring and using a screw driver through hole in bottom of housing, push spool, spring and end plug from housing bore. Thoroughly clean all parts in a suitable solvent and renew any which are damaged beyond repair. Spring should require 10-12 lbs. to compress it to a height of ¾-inch. If spool and spool bore show any roughnes or if spool has any binding tendency in the housing bore, use a fine grade of lapping compound to provide a free sliding fit. When reassembling the valve, make certain that all parts are clean and be sure to use a new "O" ring.

When installing the valve, use a new gasket, without sealer.

Fig. C818G — Pressure gage installation for checking the torque convertor pressure.

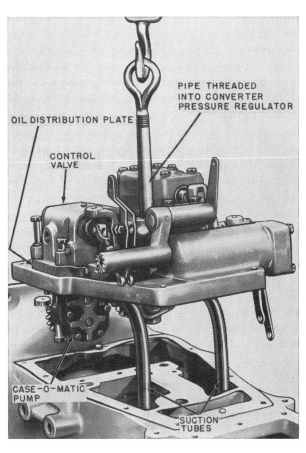

Fig. C818H — Removing "Case-O-Matic" and hydraulic lift system adapter housing from torque tube. Care must be exercised not to distort the pump suction lines.

CONTROL VALVE AND CLUTCH PRESSURE REGULATOR

Series 400B-600B

94L. **R&R AND OVERHAUL.** To remove the control valve assembly, first remove the hood and rear quarter panels unit and disconnect the fuel line and gage wire from fuel tank. Then unbolt and remove fuel tank from tractor. Remove pins retaining linkage to control valve spools and disconnect the oil lines. Remove cap screws retaining control valve body to adapter housing and remove control valve, using care not to lose the detent balls and springs. Remove the oil distribution plate so that all oil passages can be thoroughly cleaned.

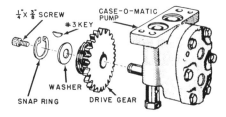

Fig. C818J — Drive gear and associated parts exploded from "Case-O-Matic" oil pump.

Refer to Fig. C818B, note position of interlock plates and remove spools from valve body. Remove pressure regulator cap and gasket and withdraw plungers and spring. Remove check valve plug, spring and ball.

Wash all parts in petroleum solvent and dry with compressed air; being sure to blow out all bores and passages in the valve body including the small pressure regulator orifice. The plug through which the orifice is drilled will seldom need to be removed; however, a tag wire should be used to be sure orifice is open and clean. Check the orifice plug to be sure it is "bottomed" in the hole or until there is at least 1/8-inch clearance between it and the gasket when the valve assembly is bolted in place.

Carefully inspect pressure regulator plungers, spools and bores for wear or rough areas. Plungers and spools must move freely in the bores.

The pressure regulator spring should require 29 pounds to compress it to a height of $1\frac{21}{32}$ inches. If less than 27 pounds will compress the spring to the $1\frac{21}{32}$-inch length, the spring should be renewed. Note: If the clutch pressure test indicated low pressure and the spring is O.K., two washers may be added between the spring and plunger. Never use more than two washers for shims. Do not use a flat slug instead of a washer as it will not allow oil to pass through the small bleed hole in the plunger. Examine the clutch pressure plate regulator valve to be sure the ball and seat are smooth and form a tight seal. The purpose of this ball check type valve is to allow an immediate release of pressure from the clutch pressure regulator when the "Case-O-Matic" spool is closed (pedal depressed).

When reassembling, be sure mating surfaces of valve body and oil distribution plate, also mating surfaces of oil distribution plate and adaptor housing are dry and free of oil. Place new gaskets between adaptor housing and oil distribution plate and between oil distribution plate and control valve, but do not use any type of sealer on gaskets. A flat sheet metal plate can be used to retain detent springs and balls while positioning the valve assembly. Install all four bolts and tighten to a torque of 35 ft.-lbs.

Always recheck clutch pressure when tractor is assembled.

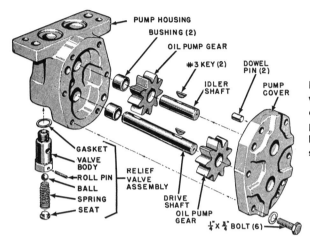

Fig. C818K — Exploded view of "Case-O-Matic" oil pump and the pump pressure relief valve. Relief valve cracking pressure is approximately 250 psi.

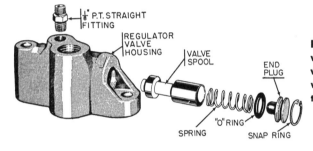

Fig. C818L — Exploded view of the torque convertor pressure regulator valve. The valve maintains a convertor pressure of 30-40 psi.

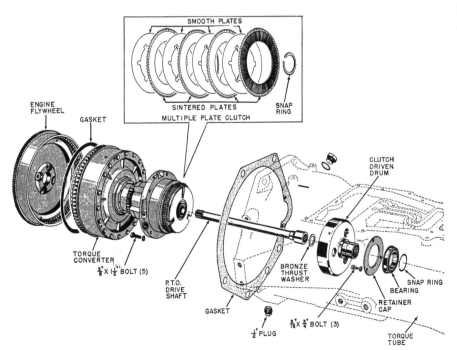

Fig. C818M—Partially exploded view of torque convertor, main clutch, clutch drum and associated parts.

Fig. C818N — Installation of jack screws used to compress main clutch plates during removal and reinstallation of the torque convertor.

MAIN CLUTCH

Series 400B-600B

94M. R&R PLATES. To remove the main clutch plates, Fig. C818M, first remove the torque convertor unit as outlined in paragraph 94P. Remove the two jack screws shown in Fig. C818N, extract the clutch plate retaining snap ring (Fig. C818M) and lift clutch plates from the convertor unit. Inspect the clutch plates in the conventional manner and renew any damaged or worn parts.

When reassembling, first make certain that the pto drive shaft is installed in the convertor unit so that clutch plates will be properly centered. Then, install clutch plates and snap ring. Using a spare clutch drum, align the clutch plates and install the two jack screws shown in Fig. C818N. Tighten the jack screws enough to prevent clutch plates from becoming misaligned when convertor is installed in torque tube; then, lift-off the spare clutch drum.

94N. R&R CLUTCH DRUM. To remove the clutch drum (Fig. C818M), first remove the torque convertor unit as outlined in paragraph 94P. Turn the drum until the three holes in same align with the three cap screws which retain the bearing retainer to wall in torque tube. Remove the three cap screws and withdraw the drum and bearing unit straight forward and out of torque tube. The need and procedure for further disassembly is evident.

TORQUE CONVERTOR UNIT

Series 400B-600B

The torque convertor unit includes the single stage torque convertor, the single disc direct drive clutch, the "Case-O-Matic" and hydraulic system pump drive gear and the multiple disc main clutch. The main clutch is the only section of the convertor unit which has the individual parts catalogued by the J. I. Case Co. If any part of the torque convertor unit, other than the main clutch, is malfunctioning, the complete convertor unit must be renewed.

94P. REMOVE AND REINSTALL. To remove the torque convertor unit, first detach (split) engine from torque tube (convertor housing) as outlined in paragraph 94Q. Remove battery and battery mounting tray. Disconnect the speedometer cable, tachometer cable at adapter housing, the "Case-O-Matic" heat indicator sending unit and oil line from the torque convertor pressure regulator valve housing and pressure gage oil line at "T" fitting. Disconnect any interferring linkage, then unbolt and remove the steering support, instrument panel and attached units from torque tube.

Disconnect the hydraulic lift system oil lines and control rods. Remove cap screws around perimeter and one Allen screw near the center of the adapter housing and using a hoist and lifting eye, lift the complete adapter housing assembly straight up and out of torque tube as shown in Fig. C818H. Use care during this operation not to distort the pump suction tubes. Work-

ing through opening in top of torque tube, install two $\frac{5}{16}$x1½-inch jack screws as shown in Fig. C818N, and tighten the screws enough to prevent clutch plates from becoming misaligned when convertor is removed. Attach Case special loading tool to convertor and working through opening in top of torque tube, remove cap screws retaining convertor hub to torque tube. Pull torque convertor unit straight forward and remove same from torque tube as shown in Fig. C818P.

Before installing the convertor unit, refer to Fig. C818M and make certain that pto drive shaft is properly installed in the convertor unit and that the bronze thrust washer is positioned in hub of clutch drum. A small amount of gun grease will facilitate holding the thrust washer in place.

Slide convertor unit straight into torque tube, align oil passages in convertor hub with those in torque tube and slide unit into place. Tighten the convertor hub bolts to a torque of 20 ft.-lbs. and remove the two clutch plate jack screws. Be sure to renew the rubber gasket between convertor and flywheel before joining the tractor halves.

TRACTOR SPLIT (ENGINE FROM TORQUE TUBE)

Series 400B-600B

94Q. To detach (split) engine from torque tube (convertor housing), first drain cooling system and torque tube and remove hood.

Remove both of the frame side rails, uncouple steering shaft at one of the universal joints and unbolt the steering shaft intermediate support bracket. Support both halves of tractor separately and on axle type tractors, unbolt the radius rod pivot bracket from torque tube.

Disconnect fuel line and fuel gage wire and remove throttle rod and fuel tank. On models with underneath muffler, disconnect exhaust pipe from manifold. Disconnect wiring harness from generator and starting motor and remove starting motor. Disconnect the oil pressure gage line from cylinder block and heat indicator sending unit from cylinder head. Disconnect wire

Fig. C818P — Using special Case loading tool to remove torque convertor unit from torque tube.

from coil and choke wire from carburetor. Disconnect the "Case-O-Matic" system oil lines. Unbolt engine from torque tube and carefully separate the tractor halves.

When reassembling, be sure to renew the rubber gasket between convertor and flywheel before joining the tractor halves. Tighten the engine to torque tube bolts and the frame side rail screws to a torque of 115 ft.-lbs.

TRACTOR SPLIT (TORQUE TUBE FROM TRANSMISSION)

Series 400B-600B

94R. To detach (split) torque tube (convertor housing) from transmission, drain oil from torque tube and disconnect the speedometer drive cable and regulator oil line from torque tube rear cover. On models with a four-speed transmission, remove top rear cover from torque tube. On models with an eight-speed transmission, place the dual range control lever in neutral and remove the control cover assembly. On models with shuttle gears, remove the shuttle control cover. Disconnect the tail light wire and the hydraulic system control rods and lines. Remove underneath muffler. Support both halves of tractor separately, then unbolt and separate the tractor halves.

When reassembling, tighten the torque tube to transmission bolts to a torque of 80 ft.-lbs.

"TRIPL-RANGE" GEAR BOX

The "Tripl-Range" gear box (Fig. C820), which is optionally available on series 200B, 300, 300B, 350 and 500B, is bolted to front of the standard transmission and provides twelve forward and three reverse speeds. Tractors equipped with the transmission clutch shown have live power take-off. The transmission clutch is not used on models with non-continuous power take-off.

Inspect all parts and renew any which are damaged or worn. Length of interlock plug must be 1.698-1.702. Mesh lock springs should have a minimum free length of one-inch and require approximately 38 lbs. to compress them to a height of $\frac{13}{16}$-inch. NOTE: Springs used on some very early models of the 300 series have a free length of 1⅜ inches. When these springs are encountered, discard them and install the shorter springs.

CONTROL COVER

95. **R&R AND OVERHAUL.** Remove dip stick from cover, place the control lever in neutral, remove the four retaining cap screws and lift cover from torque tube. Refer to Fig. C819. Unwire and remove set screws retaining shifter forks to the rails. Using a punch and hammer, bump the rails forward and remove the rails, steel balls, springs and forks. Inspect corners of the oil baffle to make certain that they are tight and will not permit oil to circulate under the breather. The fork lever is keyed to the control lever and retained by a roll pin. If roll pin is removed, it will have to be bent back along surface of cover.

Fig. C819 — Exploded view of the "Tripl-Range" control cover. The metal baffle attached to cover prevents oil from circulating under the breather.

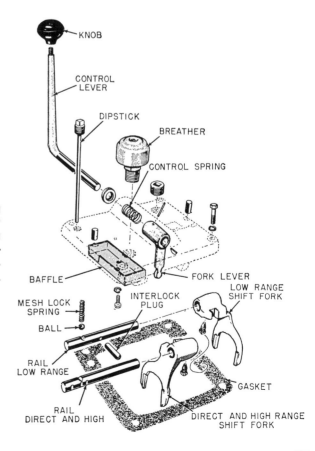

KNOB
CONTROL LEVER
DIPSTICK
BREATHER
CONTROL SPRING
BAFFLE
MESH LOCK SPRING
BALL
INTERLOCK PLUG
FORK LEVER
LOW RANGE SHIFT FORK
RAIL LOW RANGE
RAIL DIRECT AND HIGH
GASKET
DIRECT AND HIGH RANGE SHIFT FORK

When reassembling, tighten the fork retaining set screws to a torque of 35 ft.-lbs. and secure with safety wire. After cover is installed, move the hand control lever to the high range position and make certain that the mesh lock ball has fallen into notch in rail. Failure to lock in high range may indicate that rail is striking torque tube. In which case, rework the rail as shown in Fig. C821. Tighten the cover bolts to a torque of 35 ft.-lbs.

GEAR UNIT

95A. REMOVE AND REINSTALL. To remove the "Tripl-Range" gear box, first detach (split) torque tube from transmission as outlined in paragraph 93. Refer to Fig. C822. Remove nuts retaining the gear housing to front of transmission and remove the complete "Tripl-Range" gear assembly. Do not lose the shims located in front of the mainshaft front bearing cup. These shims control the transmission main shaft bearing adjustment. Refer to paragraph 102A.

95B. When reassembling, shellac both sides of the gasket and install gasket on transmission as shown in Fig. C823. Notice that gasket is installed so that reverse idler shaft hole is covered. If gasket is reversed, oil will transfer from torque tube to transmission. Renew oil seals (Fig. C823A) in carrier if their condition is questionable. Install the shims against the main shaft front bearing cup and using shim stock or Case sleeve No. G13501 over the splines of the main shaft, install the oil seal car-

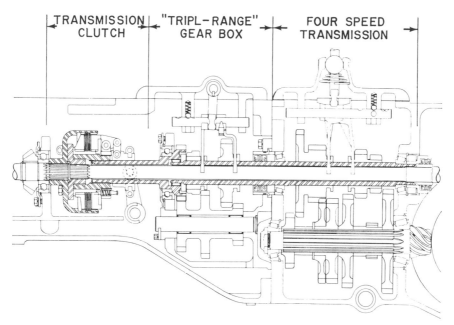

Fig. C820—Sectional view of torque tube and transmission assembly on models with continuous power take-off and "Tripl-Range" gear box. Models so equipped have twelve forward and three reverse speeds.

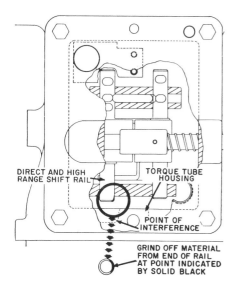

Fig. C821 — Failure of "Tripl-Range" control lever to lock in high range usually indicates the need of reworking the direct and high range shift rail.

Fig. C822—"Tripl-Range" gear box installation on front of transmission. Housing retaining nuts should be tightened to a torque of 35-40 ft. lbs.

rier. Install the gear box and both sliding gears. Notice that shifter fork slots in both sliding gears face rearward as shown in Fig. C824. Tighten the housing retaining nuts to a torque of 35-40 ft.-lbs.

96. **OVERHAUL.** To overhaul the "Tripl-Range" gear box assembly, first remove the unit as outlined in paragraph 95 and proceed as follows: Disassemble the unit as shown in Fig. C825, thoroughly clean all parts and renew any which are damaged or worn. New countershaft and input

gear needle bearings can be installed by using a piloted driver against marked end of bearing. Install cluster gear in housing and assemble one bronze thrust washer at each end of gear. Install countershaft, front bushing and lock, then tighten the bushing retaining screws to a torque of 15 ft.-lbs. Be careful not to loosen expansion plug located at rear of countershaft. An oil leak at this point will permit transfer of lubricant from torque tube to transmission.

Move cluster gear rearward and front thrust washer against the front

bushing. Then, using a feeler gage, check the clearance between gear and front thrust washer. If clearance is not between the limits of 0.010-0.0125, vary the number of shims under the front bushing. Shims are available in thicknesses of 0.005 and 0.0075.

Install input gear bearing with snap ring, then install the input gear and its retaining snap ring. Tighten the input gear bearing retainer screws to a torque of 15 ft.-lbs. and install the safety wire.

When installing the "Tripl-Range" unit, refer to paragraph 95A.

Fig. C823 — Front view of transmission with "Tripl-Range" gear box removed. Gasket must be installed properly as shown.

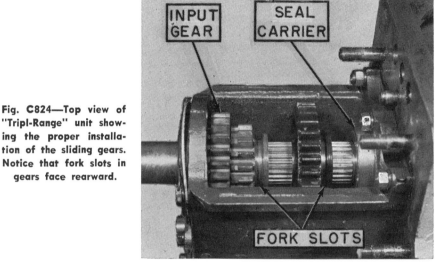

Fig. C824—Top view of "Tripl-Range" unit showing the proper installation of the sliding gears. Notice that fork slots in gears face rearward.

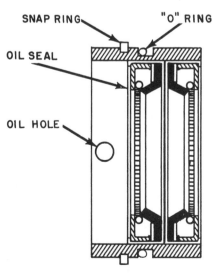

Fig. C823A — Sectional view of seal carrier, showing installation of seals and "O" ring. Seals must not cover the two oil holes in carrier.

Fig. C825 — Exploded view of "Tripl-Range" gear box. Cluster gear end play of 0.010-0.0125 is controlled by shims under the countershaft front bushing.

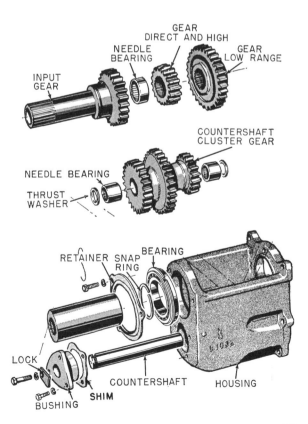

DUAL RANGE SHUTTLE GEAR BOXES

The dual range shuttle gear boxes (Fig. C826) are bolted to front of the standard transmission and provides eight forward and eight reverse speeds.

NOTE: On series 400B and 600B with dual range gears and no shuttle gears, refer to paragraphs beginning with paragraph 99A.

CONTROL COVER

97. R&R AND OVERHAUL. Remove dip stick from cover, place the control lever in neutral and on series 400B and 600B, disconnect the regulator valve oil line. Remove the four retaining cap screws and lift cover from torque tube. Refer to Fig. C827. Unwire and remove set screws retaining shifter forks to the rails. Using a punch and hammer, bump the rails forward and remove the rails, steel balls, springs and forks. The fork levers may be keyed or splined to the control levers; but in either case they are positioned by roll pins. If roll pins are removed, they will have to be bent back along surface of cover.

Inspect all parts and renew any which are damaged or worn. Mesh lock springs should have a minimum

free length of one-inch and require approximately 38 lbs. to compress them to a height of $1\frac{3}{16}$-inch. NOTE: Springs used on some very early models of the 300 series have a free length of $1\frac{3}{8}$ inches. When these springs are encountered, discard them and install the shorter springs.

When reassembling, tighten the fork retaining set screws to a torque of 35 ft.-lbs. and secure with safety wire. Install the cover, making certain that both forks are engaged and tighten the retaining screws to a torque of 35 ft.-lbs.

GEAR UNIT

98. REMOVE AND REINSTALL. To remove the shuttle gear box, first detach (split) torque tube from transmission as outlined in paragraph 93 or 94R. Refer to Fig. C826. Unbolt and remove the shuttle gear housing from the low range gear housing. Remove but do not lose shims (Fig. C828) which are located in front of the low range countershaft collar. Remove the snap ring located in front of the low range shift collar carrier then unbolt

and remove the low range housing from transmission. Oil seal carrier can be removed from the transmission main shaft after removing its retaining snap ring; but do not lose the shims located in front of the mainshaft front bearing cup. These

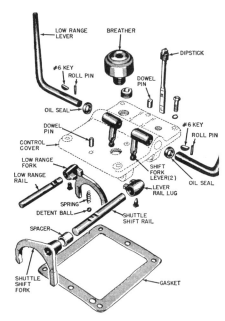

Fig. C827 — Exploded view of shuttle control cover. Fork retaining set screws should be tightened to a torque of 35 ft.-lbs.

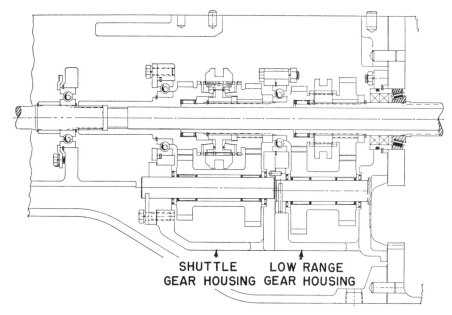

Fig. C826 — Sectional view of torque tube showing the dual range shuttle gear box installation. The unit is bolted to front of the regular transmission and provides eight forward and eight reverse speeds.

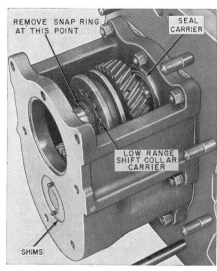

Fig. C828 — Low range gear housing installation on front of transmission.

shims control the transmission main shaft bearing adjustment. Refer to paragraph 102A. When installing the oil seal carrier over the main shaft splines, use a sleeve or shim stock to avoid damaging the seal lips. Be sure to install the seal carrier retaining snap ring and make certain that the oil passages are properly positioned.

98A. When reassembling, shellac both sides of the gasket and install gasket on transmission as shown in Fig. C829. Notice that gasket is installed so that reverse idler shaft hole is covered. If gasket is reversed, oil will transfer from torque tube to transmission. Install the low range gear housing and components and be sure that oil grooves (Fig. C830) in low range output gear and shift collar carrier are toward rear. Tighten the retaining nuts to a torque of 35 ft.-lbs. Select the proper thickness of snap ring (Fig. C828) to give the parts a

minimum of end play and install the snap ring, using care not to damage the main shaft splines. Available snap ring thicknesses are 0.076, 0.080, 0.083, 0.087 and 0.091.

Install shims (Fig. C828) in front of the low range countershaft collar, then install the shuttle gear housing and tighten the retaining screws and nuts to a torque of 35 ft.-lbs.

99. **OVERHAUL.** To overhaul the shuttle and low range gear box assemblies, first remove the units as outlined in paragraph 98. Refer to Fig. C831. Unstake and remove nut (26) from reverse idler stud, then using a brass drift, bump the stud (22) rearward and out of gear (23). Gear will then drop down enough to permit removal of input gear. Remove retainer (1) and withdraw the input gear. Notice that short screw is used in lower hole of retainer. Remove snap ring (10) in front of synchronizer unit, but use care not to damage splines on the low range input gear (31). Remove synchronizer unit, output gear (12) and low range input gear (31). Lift out the reverse idler gear (23). Remove the countershaft

Fig. C829 — Front view of transmission with gasket properly installed.

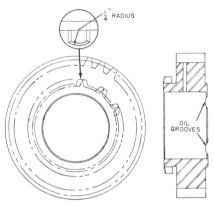

Fig. C830 — Views of low range output gear. Oil grooves must be toward rear.

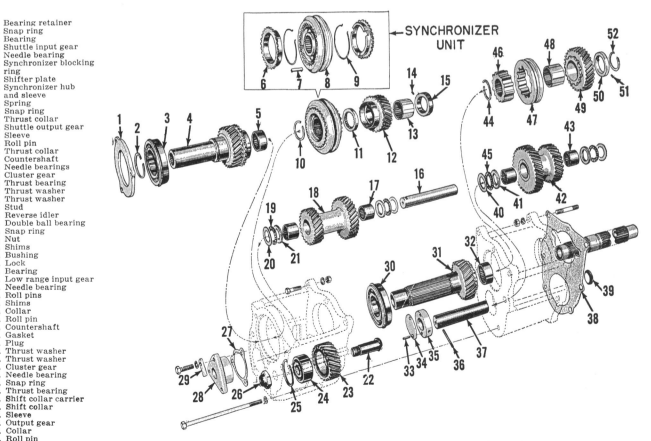

1. Bearing retainer
2. Snap ring
3. Bearing
4. Shuttle input gear
5. Needle bearing
6. Synchronizer blocking ring
7. Shifter plate
8. Synchronizer hub and sleeve
9. Spring
10. Snap ring
11. Thrust collar
12. Shuttle output gear
13. Sleeve
14. Roll pin
15. Thrust collar
16. Countershaft
17. Needle bearings
18. Cluster gear
19. Thrust bearing
20. Thrust washer
21. Thrust washer
22. Stud
23. Reverse idler
24. Double ball bearing
25. Snap ring
26. Nut
27. Shims
28. Bushing
29. Lock
30. Bearing
31. Low range input gear
32. Needle bearing
33. Roll pins
34. Shims
35. Collar
36. Roll pin
37. Countershaft
38. Gasket
39. Plug
40. Thrust washer
41. Thrust washer
42. Cluster gear
43. Needle bearing
44. Snap ring
45. Thrust bearing
46. Shift collar carrier
47. Shift collar
48. Sleeve
49. Output gear
50. Collar
51. Roll pin
52. Snap ring

Fig. C831—Exploded view of shuttle and low range gear units.

front bushing (28), countershaft (16) and cluster gear (18). The need and procedure for further disassembly is evident.

Thoroughly clean all parts and renew any which are damaged or worn. Tapered surfaces of input and output gears (4 & 12) as well as the mating tapered surfaces of the synchronizer blocking rings (6) must be free from nicks, burrs or foreign material. It will be impossible to shift the shuttle transmission if blocking rings do not turn freely on the tapered surfaces. New needle bearings can be installed by using a piloted driver against the numbered end of bearings.

Install cluster gear (18) in housing and position thrust washers (20 & 21) and thrust bearings (19) at each end of gear. NOTE: Inner thrust washers (21) are 0.125 thick and outer washers (20) are 0.031 thick. Install countershaft (16), shims (27), front bushing (28) and lock (29). Tighten the bushing retaining screws to a torque of 18 ft.-lbs. and check the cluster gear end play which should be 0.004-0.009. If end play is not as specified, vary the number of 0.005 thick shims (27). Install the reverse idler gear (shoulder toward front) and bearing assembly but do not install stud (22). Install low range input gear (31) into housing and position synchronizer, output gear (12), thrust collars (11 and 15) and sleeve (13) as gear (31) is being assembled. After selecting the proper thickness to give the synchronizer a minimum of end play, install snap ring (10). Available snap ring thicknesses are 0.076, 0.080, 0.083,

0.087 and 0.091. Notice that chamfered face of synchronizer is toward snap ring. Install the assembled input gear (4). Install the reverse idler gear stud (22), tighten nut (26) securely and stake in place.

Working on the low range gear box, remove the countershaft and collar (35) by grasping the roll pins with pliers. Lift cluster gear (42) with bearings and spacers from housing. The procedure for further disassembly is evident. To install the cluster gear needle bearings, use a piloted driver against numbered end of bearings. Shift collar (47) must slide freely on carrier (46), but keep in mind that excessive clearance may permit transmission to slip out of low range. If meshing teeth of shift collar and output gear are worn, renew the parts. Make sure that expansion plug (39) is properly seated. Leakage at this point will allow transfer of oil from torque tube to transmission.

Install cluster gear (42) in housing and position thrust washers (40 and 41) and thrust bearing (45) at each end of gear. Inner washers (41) are 0.125 thick and washers (40) are 0.031 thick. Install countershaft and collar assembly and tap end of shaft as shown in Fig. C832 to make certain that collar is seated and all end play is removed from the cluster gear. Install same number of shims (Fig. C833) as were originally removed; then, using a straight edge and feeler gage as shown, check the clearance between shims and edge of case. Vary the number of shims until clearance is 0.004-0.009.

Install the low range gear housing and shuttle gear housing as outlined in paragraph 98A.

DUAL RANGE GEAR BOX

The dual range gear box (Fig. C833B) is bolted to front of standard transmission and provides eight forward and two reverse gear ratios. This unit is available only on the series 400B and 600B.

NOTE: On series 400B and 600B with shuttle gears, refer to paragraphs beginning with paragraph 97.

CONTROL COVER

Series 400B-600B

99A. R&R AND OVERHAUL. Remove dip stick from cover, place the control lever in neutral and disconnect the regulator valve oil line (13—Fig. C833A). Remove the four retaining cap screws and lift cover from torque tube. Unwire and remove set screw retaining shifter fork to the rail. Using a punch and hammer, bump the rail forward and remove the rail, steel ball, spring and fork. The fork lever is keyed to the control lever and retained by a roll pin. If roll pin is removed, it will have to be bent back along surface of cover.

Inspect all parts and renew any which are damaged or worn. Mesh lock spring should have a minimum free length of 1-inch and require approximately 38 lbs. to compress it to a height of $\frac{13}{16}$-inch.

When reassembling, tighten the fork retaining set screw to a torque of 35 ft.-lbs. and secure with safety wire. Install the cover, making certain that fork is engaged and tighten the retaining screws to a torque of 35 ft.-lbs.

GEAR UNIT

Series 400B-600B

99B. REMOVE AND REINSTALL. To remove the dual range gear box, first detach (split) torque tube from transmission as outlined in paragraph

Fig. C832 — Installing countershaft and collar in low range gear box.

Fig. C833 — Checking end play of low range gear box cluster gear. End play should be 0.004-0.009.

94R. Unbolt and remove cover (54—Fig. C833B) from the low range gear housing and withdraw the input gear (31). Remove but do not lose shims (Fig. C828) which are located in front of the low range countershaft collar. Remove the snap ring located in front of the low range shift collar carrier then unbolt and remove the low range housing from transmission. Oil seal carrier can be removed from the transmission main shaft after removing its retaining snap ring; but do not lose the shims located in front of the main shaft front bearing cup. These shims control the transmission main shaft bearing adjustment. Refer to paragraph 102A. When installing the oil seal carrier over the main shaft splines, use a sleeve or shim stock to avoid damaging the seal lips. Be sure to install the seal carrier retaining snap ring.

99C. When reassembling, shellac both sides of the gasket and install gasket on transmission as shown in Fig. C829. Notice that gasket is installed so that reverse idler shaft hole is covered. If gasket is reversed, oil will transfer from torque tube to transmission. Install the low range gear housing and components and be sure that oil grooves (Fig. C830) in

low range output gear and shift collar carrier are toward rear. Tighten the retaining nuts to a torque of 35 ft.-lbs. Select the proper thickness of snap ring (Fig. C828) to give the parts a minimum of end play and install the snap ring, using care not to damage the main shaft splines. Available snap ring thicknesses are 0.076, 0.080, 0.083, 0.087 and 0.091.

Install shims (Fig. C828) in front of the low range countershaft collar, then install the front cover and tighten the retaining screws to a torque of 35 ft.-lbs.

99D. OVERHAUL. To overhaul the dual range gear box assembly, first remove the unit as outlined in paragraph 99B. Refer to Fig. C833B. Remove the countershaft and collar (35) by grasping the roll pins with pliers. Lift cluster gear (42) with bearings and spacers from housing. The procedure for further disassembly is evident. To install the cluster gear needle bearings, use a piloted driver against numbered end of bearings. Shift collar (47) must slide freely on carrier (46).

but keep in mind that excessive clearance may permit transmission to slip out of low range. If meshing teeth of shift collar and output gear are worn, renew the parts. Make sure that expansion plug (39) is properly seated. Leakage at this point will allow transfer of oil from torque tube to transmission.

Install cluster gear (42) in housing and position thrust washers (40 and 41) and thrust bearing (45) at each end of gear. Inner washers (41) are 0.125 thick and washers (40) are 0.031 thick. Install countershaft and collar assembly and tap end of shaft as shown in Fig. C832 to make certain that collar is seated and all end play is removed from the cluster gear. Install same number of shims (Fig. C833) as were originally removed; then, using a straight edge and feeler gage as shown, check the clearance between shims and edge of case. Vary the number of shims until clearance is 0.004-0.009.

Install the low range gear housing as outlined in paragraph 99C.

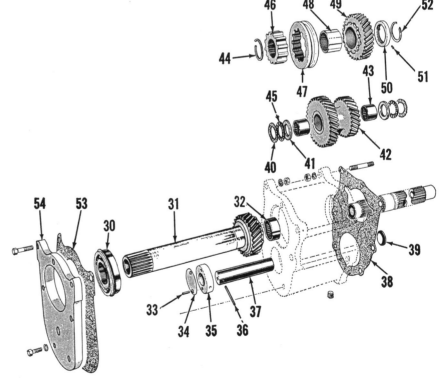

Fig. C833B — Exploded view of series 400B and 600B dual range gear box.

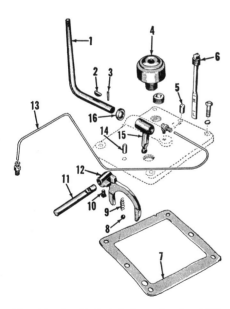

Fig. C833A—Exploded view of series 400B and 600B dual range control cover and related parts.

1. Control lever	9. Spring
2. Woodruff key	10. Lock screw
3. Roll pin	11. Rail
4. Breather	12. Fork
5. Dowel	13. Oil line
6. Dip stick	14. Dowel
7. Gasket	15. Fork lever
8. Detent ball	16. Oil seal

30. Bearing	38. Gasket	47. Shift collar
31. Input gear	39. Plug	48. Sleeve
32. Needle bearing	40. Thrust washer	49. Output gear
33. Roll pins	41. Thrust washer	50. Collar
34. Shims	42. Cluster gear	51. Roll pin
35. Collar	43. Needle bearings	52. Snap ring
36. Roll pin	44. Snap ring	53. Gasket
37. Countershaft	45. Thrust bearing	54. Front cover
	46. Shift collar carrier	

TRANSMISSION

TOP COVER

100. The transmission top cover can be removed after removing the dip stick and unscrewing the retaining cap screws.

OVERHAUL

101. **SHIFTER RAILS AND FORKS.** To overhaul the shift rails and forks, remove the transmission top cover, refer to Fig. C834 and proceed as follows: Unwire and remove set screws retaining forks to rails; then, using a punch, drive the rails forward and remove rails, forks, springs, balls and interlocking plugs. CAUTION: The steel balls are spring loaded and will fly out with force when rails are removed. The procedure for further disassembly is evident.

Length of interlock plugs should be 0.566-0.570. Mesh lock springs should have a minimum free length of one inch and require approximately 38 lbs. to compress them to a height of $\frac{13}{16}$-inch. Some early models of the 300 series had longer springs which should be discarded and the one-inch free length springs installed.

When reassembling, position the gear shift lever in control cover and turn the dog point retaining screw into cover until it bottoms or until a slight bind is felt when moving the shift lever. Back the screw off until lever moves freely and then tighten the jam nut securely. NOTE: Some early models were equipped with a shift lever pivot pin rather than a retaining screw.

Install support spring so that projecting end is on right side of cover and be sure that the two end coils have not doubled up thereby restricting the travel of the shift lever. Install the remaining parts, tighten the set screws securely and install the safety wire.

102. **MAIN SHAFT.** To remove the transmission main drive shaft (Fig. C835), drain transmission and torque tube and remove the transmission top cover. Detach (split) torque tube from transmission as outlined in paragraph 93 or 94R. On models with "Tripl-Range", remove the "Tripl-Range" gear box as outlined in paragraph 95A. On models with dual range shuttle gear assembly, remove the shuttle gear

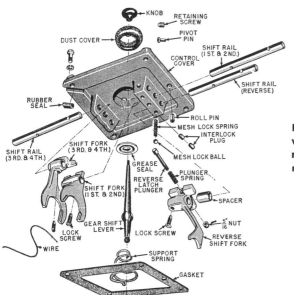

Fig. C834 — Exploded view of a typical transmission top cover, shifter rails and forks and associated parts.

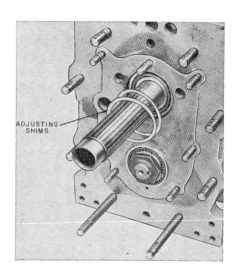

Fig. C836 — Location of shims used to adjust the end play of the transmission main shaft.

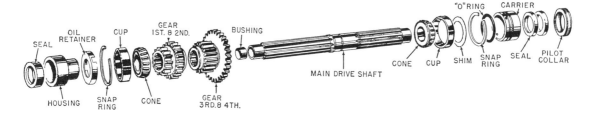

Fig. C835—Exploded view of the transmission main drive shaft, gears and associated parts. Bearings should be adjusted to provide an end play of 0.002-0.005.

and low range gear box assemblies as outlined in paragraph 98. On series 400B and 600B with dual range, remove the dual range gear box as in paragraph 99B. On models with four-speed transmission only, remove front cover from transmission.

On all models, do not damage or lose the adjusting shims (Fig. C836) located in front of the main shaft front bearing cup. Remove the complete power take-off assembly or the transmission case rear cover. Remove the seat, tool box and the top rear cover from transmission case. On models so equipped, disconnect the speedometer cable and remove the speedometer drive assembly.

Using Case spacer tool No. G15012 between cluster gears on main shaft as shown in Fig. C837, bump the main shaft forward out of transmission case and remove gears from above. NOTE: If tool No. G15012 is not available, make up a tool using the dimensions shown in Fig. C838.

Renew any damaged bearing cups and cones. Before installation of shaft, install the rear bearing cup and its positioning snap ring; also, press the front bearing cone on the shaft. Renew bushing in rear of main shaft if it is worn.

When installing the cluster gears, be sure that largest diameter gear on each cluster is toward front of tractor. Bump shaft into rear bearing cone and install the front bearing cup. Adjust the main shaft bearings as follows:

102A. Position the previously removed shims (Fig. C836) against the front bearing cup and on models with four-speed transmission, install the front cover pilot. Place dry gasket on transmission case, install front cover without seals and tighten the nuts to a torque of 35 ft.-lbs. On models with "Tripl-Range" install the front seal carrier without seals, place dry gasket on transmission case, temporarily install "Tripl-Range" gear box and tighten the nuts to a torque of 35 ft.-lbs. On models with dual range shuttle gears or on series 400B and 600B with dual range, install the front seal carrier without seals, place dry gasket on transmission case, temporarily install the range gear housing and tighten the nuts to 35 ft.-lbs.

On all models, tap both ends of main shaft with a soft faced mallet to be sure that bearings are seated, turn the main shaft several times and check the shaft end play which should be 0.002-0.005. If end play is not as

specified, vary the number of shims shown in Fig. C836. Adding shims reduces end play.

102B. With the mainshaft end play adjusted, remove the transmission front cover, "Tripl-Range" or low range gear box. Refer to Fig. C835. Install oil retainer at rear bearing and the double lipped seal in the seal housing. Start seal retainer over rear end of main shaft and drive same into position. NOTE: On models without power take-off, install expansion plug in rear of seal housing. It is advisable on early models to stake seal in housing as shown in Fig. C839.

On models with pto, it is advisable to mount the long pto drive shaft in lathe centers and check for run-out over entire length of shaft. If run-out exceeds 0.003 at any point, shaft must be straightened. NOTE: This applies to a new as well as a used shaft.

When reassembling, shellac both sides of the front gasket and install gasket on transmission as shown in Fig. C840. Notice that gasket is installed so that reverse idler shaft hole is covered. If gasket is reversed, oil will transfer from torque tube to transmission.

On models with four-speed transmission, install the opposed seal in front cover and position pilot on main shaft. On other models, install the opposed seals in seal carrier. On all models, use a seal sleeve or shim stock when installing seals over the main shaft splines.

Install the "Tripl-Range" unit as in paragraph 95B, the dual range shuttle unit as in paragraph 98A, or the dual range unit as in paragraph 99C.

Fig. C837 — Before driving the main shaft out of the transmission case, a spacer (Fig. C838) should be inserted to separate the cluster gears.

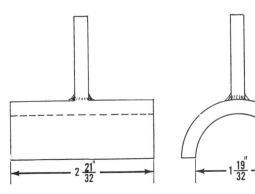

Fig. C838 — Necessary dimensions for making the spacer shown installed in Fig. C837.

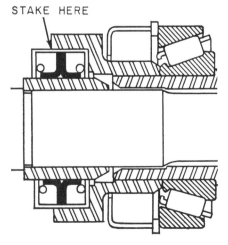

STAKE HERE

Fig. C839 — Transmission mainshaft rear oil seal and rear bearing installation.

103. **REVERSE IDLER.** To remove the reverse idler gear and shaft, remove the seat, tool box and both top covers from the transmission case. Loosen the jam nut and back-off the set screw shown in Fig. C841. Slide the gear shaft rearward into differential compartment and lift out the gear.

Inside diameter of the reverse idler bushing is 0.862-0.863. If bushing is worn, install a new one as shown in Fig. C842. Lead of the spiral oil groove in the bushing must be in the same direction as the gear operates; which is clockwise when viewed from rear of case. Notched end of bushing must project $\frac{1}{16}$-inch beyond face of gear. Roll ends of bushing into a chamfer and burnish bushing to an inside diameter of 0.862-0.863.

When reassembling, tighten set screw into the shaft groove and securely tighten the jam nut.

104. **BEVEL PINION (COUNTERSHAFT).** To remove the bevel pinion shaft, first remove the transmission main shaft as outlined in paragraph 102 and the differential and bevel ring gear assembly as outlined in paragraph 105. Refer to Fig. C839A and C839B. Using two large screw drivers, pry the second and third speed gears apart and use a feeler gage to determine the space between the gears as shown in Fig. C844 and C844A. The measured space should be 0.001-0.005. Record the exact measurement.

Unstake and remove adjusting nut from front end of shaft, bump shaft rearward and remove gears and spacers from above. Thoroughly examine all parts and renew any which are damaged or worn. Rear bearing cone can be heated in hot oil to facilitate installation.

NOTE: On series 200B, 300, 300B and 400B, the bevel pinion shaft is available as a separate replacement part and there is no fore and aft (mesh) position adjustment for the main drive bevel pinion. Proceed to install the pinion shaft as in the following paragraph 104A.

Series 350, 500B and 600B are fitted with Hypoid bevel gears and the gears are available as a matched set only. If the bevel gears, shaft bearings and/or transmission housing are renewed, the fore and aft (mesh) position of the bevel pinion must be adjusted as outlined in paragraph 104B before proceeding to install the pinion shaft as in the following paragraph 104A. Fore and aft position of the main drive bevel pinion is controlled by shims (Y—Fig. C839B or C844A).

104A. If none of the pinion shaft parts were renewed and the gap measurement in Fig. C844 or C844A was 0.001-0.005, install the same spacer (X). If none of the parts were renewed, but the gap measurement was not as specified, select the appropriate thicker or thinner spacer as required. If some of the parts were renewed, check their thickness as compared with the parts removed; then select the appropriate thicker or thinner spacer as required to maintain the gap of 0.001-0.005. Spacer (X) is available in increments of 0.005, ranging from 0.164 to 0.244.

Fig. C840 — Front view of transmission with gasket properly installed.

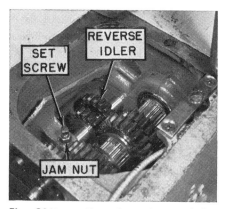

Fig. C841 — Top view of transmission showing the reverse idler gear installation.

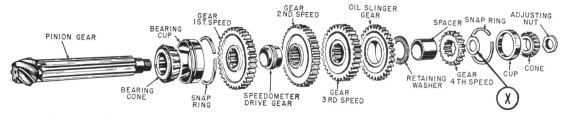

Fig. C839A—Exploded view of the transmission bevel pinion (countershaft), gears and associated parts used on series 200B, 300, 300B and 400B. A spacer is used instead of the speedometer drive gear on some models.

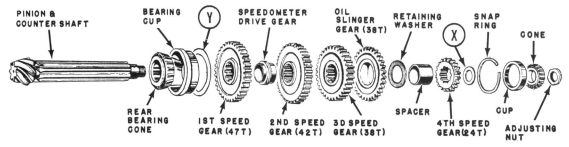

Fig. C839B—Exploded view of the transmission bevel pinion (countershaft), gears and associated parts used on series 350, 500B and 600B. Bevel gear mesh position is controlled by shims (Y).

After determining the proper thickness of spacer (X), install parts by reversing the removal procedure and tighten the adjusting nut until a rolling torque of 4-8 inch-lbs. (slight drag) is required to turn the shaft in its bearings. Bump both ends of shaft with a soft faced hammer to be sure bearings are seated and again check the rolling torque. Recheck gap between second and third speed gears to be sure it is as specified and stake the adjusting nut.

104B. **HYPOID GEAR ADJUSTMENT.** Series 350, 500B and 600B only. The main drive bevel pinion and ring gear are Hypoid gears and it is important that the pinion be properly located in the transmission case to insure proper tooth contact with the bevel ring gear. The fore and aft position of the pinion is controlled by shims (Y—Fig. C844B). To determine the proper shims to be used, use Case special cup holding fixture No. G1503 and pinion gage No. G1504 as shown in Fig. C844C and proceed as follows:

a. Note the etched reading (A—Fig. C844B) stamped on pinion.

b. Take feeler gage reading (B) as shown in Fig. C744C.

c. Make note of marking on transmission case.

d. Refer to the accompanying Hypoid Gear Chart and locate the pinion

reading on left hand side; then, follow this space until it intersects space from feeler gage reading at top.

e. Refer to notes at bottom right hand corner of chart to determine whether or not the markings on the transmission case indicate using the chart as is, or moving to the right or left in the chart reading.

Example: If etched pinion reading (A) is "4.030" and the feeler gage reading (B) is "0.025", determine shim size by checking intersection of horizontal and vertical spaces. In this case it is 0.031. If transmission is stamped (N), use a 0.031 thick shim, etc.

Shim sizes and the corresponding J. I. Case part numbers are as follows:
0.016 . . . . G17117
0.018 . . . . G17045
0.020 . . . . G17046
0.022 . . . . G17047
0.025 . . . . G17048
0.028 . . . . G17049
0.031 . . . . G17050
0.034 . . . . G17117 and G17045
Etc.

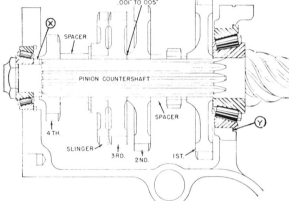

Fig. C844A — Sectional view of transmission case showing the countershaft (bevel pinion) installation on series 350, 500B and 600B. Thickness of spacer (X) determines clearance between second and third speed gears. Shims (Y) control the bevel gear mesh position.

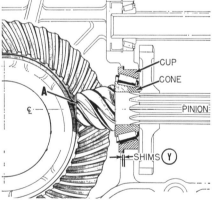

Fig. C844B—Sectional view of series 350, 500B and 600B transmission case showing bevel gear installation and mesh adjusting shims (Y).

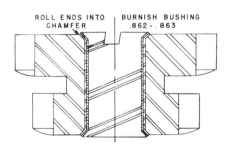

Fig. C842 — Sectional view of reverse idler gear showing proper installation of the bushing.

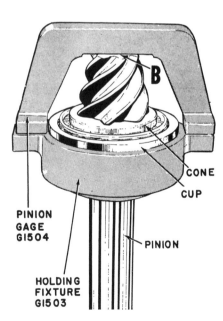

Fig. C844C—Series 350, 500B and 600B pinion and Case special gages to determine feeler gage reading (B). Before taking reading (B) be sure of the following.

1. Clean bearing cone must be completely pressed on pinion shaft.

2. Clean bearing cup must be seated in fixture G1503.

3. With cone seated in bearing cup (rotate several times) and all parts in vertical position, take feeler gage reading (B).

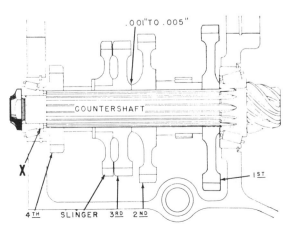

Fig. C844 — Sectional view of transmission case showing the countershaft (bevel pinion) installation on series 200B, 300, 300B and 400B. Thickness of spacer (X) determines clearance between second and third speed gears.

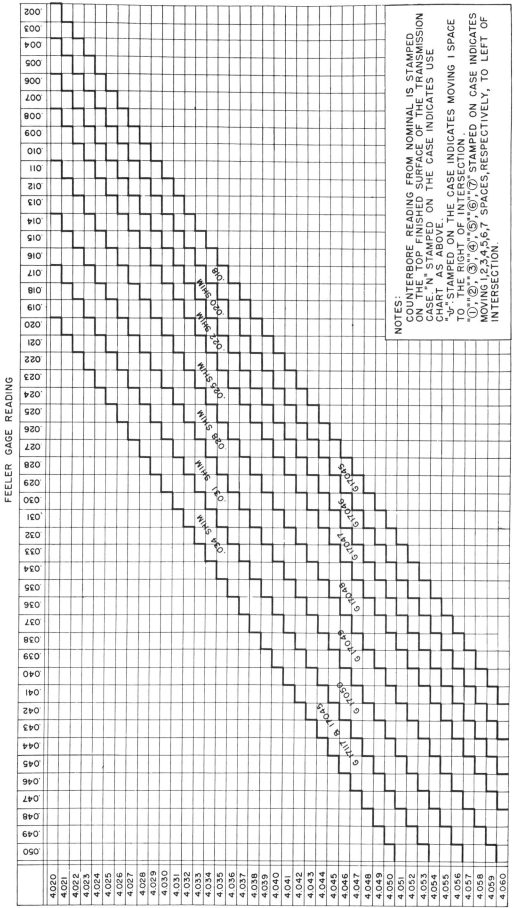

NOTES:
COUNTERBORE READING FROM NOMINAL IS STAMPED ON THE "TOP FINISHED SURFACE OF THE TRANSMISSION CASE."N" STAMPED ON THE CASE INDICATES USE CHART AS ABOVE.
"J" STAMPED ON THE CASE INDICATES MOVING 1 SPACE TO THE RIGHT OF INTERSECTION.
①"",②",③",④",⑤"",⑥"",⑦" STAMPED ON CASE INDICATES MOVING 1,2,3,4,5,6,7 SPACES,RESPECTIVELY, TO LEFT OF INTERSECTION.

FEELER GAGE READING

PINION READING

DIFFERENTIAL & MAIN DRIVE BEVEL GEARS

DIFFERENTIAL

105. **R&R AND OVERHAUL.** To remove the main drive bevel ring gear and differential assembly, first remove both of the final drive bull gears as outlined in paragraph 108. Remove both of the brake rod adjusting nuts and lock nuts. Unbolt and remove both brake covers, outer lined discs, actuating disc assemblies and inner lined discs. Support the differential assembly and remove both of the differential bearing carriers, using care not to mix, damage or lose any of the shims located between bearing carriers and transmission case. Refer to Fig. C845. Withdraw both of the combination differential side gear and bull pinion units and lift the bevel ring gear, differential and cross shaft assembly from the transmission case.

To disassemble the unit, bump out the three roll pins and extract the three pinion shafts and gears. Cross shaft can be pressed out of center wheel if renewal of either part is required. Bevel ring gear can be removed from center wheel by extracting the twelve rivets. When reassembling, make certain that rivets are tight.

The differential bearing cones can be pulled from side gears and the bearing cups can be pulled from bearing carriers at this time. New oil seals should not be installed until after the carrier bearings and bevel gear backlash are adjusted.

Reinstall pinion gears, shafts and roll pins. Pinions should have approximately 0.040 end play on shafts. Examine the differential side gears and renew the expansion plugs if there is any evidence of oil leakage. Use sealing compound around the plugs and be sure they bottom against shoulder in gears. Install the composition thrust washer and shims on one side of the center wheel, push side gear firmly against shims and check the backlash between the pinion gears and side gear in several positions. If backlash is not within the limits of 0.004-0.008, install a different thickness thrust washer and/or vary the number of shims. Shims are 0.010 thick and thrust washers are available in various thicknesses, ranging from 0.089 to 0.119. Set the side gear, shims and thrust washer aside for subsequent installation. Then, using the other side gear, select the proper thrust washer and shims to give it a backlash of 0.004-0.008 and set it aside for later installation. Do not interchange the side gears after making the backlash adjustment.

Install differential in transmission housing and position the respective side gears, thrust washers and shims. Install the originally removed shims (oil hole to bottom) against side of transmission housing, then install the differential bearing carriers (oil drain passage down) without seals and tighten the flat head machine screws. Temporarily install the brake covers

and tighten the stud nuts to a torque of 80-90 ft.-lbs. Make certain there is some backlash between the bevel pinion and ring gear, then mount a dial indicator with contact button resting on side of bevel ring gear and check the differential end play which should be 0.000-0.005. If end play is not as specified, vary the number of shims under one of the bearing carriers and recheck. NOTE: With end play adjusted, there must still be some backlash between the main drive bevel gears. Now, transfer shims from under one of the differential bearing carriers to the other until there is a backlash of 0.0055-0.0090 between the main drive bevel gear teeth. NOTE: Only transfer shims, do not add or remove shims or the previously determined bearing adjustment will be changed. Available shim thicknesses are 0.005 and 0.010.

With the bearings and backlash adjusted, remove the bearing carriers, but be careful not to mix or lose the shims. Install the carrier "O" rings and oil seals and be sure to lubricate the "O" rings. Reinstall the bearing carriers (oil drain hole toward bottom), using the previously determined number of shims. Use Case sleeve No. G13503 or shim stock to avoid damaging the seal lips as bearing carriers are installed. Tighten the brake cover nuts to a torque of 80-90 ft.-lbs.

MAIN DRIVE BEVEL GEARS
Series 200B-300-300B-400B

106. The main drive bevel pinion or ring gear are available separately. To renew the bevel pinion, refer to paragraph 104. The bevel ring gear is riveted to the differential center wheel and can be renewed as outlined in paragraph 105.

Series 350-500B-600B

106A. The main drive bevel pinion and ring gear are available as a matched set only. To renew the bevel pinion, refer to paragraph 104 for removal procedure and to paragraph 104B for Hypoid gear adjustment. The bevel ring gear is riveted to the differential center wheel and can be renewed as outlined in paragraph 105.

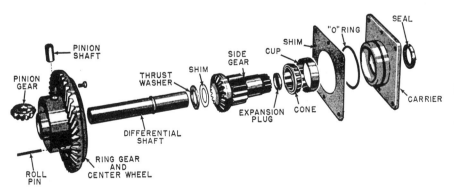

Fig. C845 — Exploded view of a typical differential, side gear, carrier bearings and associated parts.

FINAL DRIVE AND REAR AXLE

BRAKES

Brakes are of the double disc, self-energizing type located on outer ends of the combination bull pinion and differential side gear shafts.

BULL PINIONS

107. The final drive bull pinions are integral with the differential side gears and can be removed by following the procedure for R&R differential (paragraph 105).

BULL GEAR

108. To remove either bull gear, block-up tractor and remove the complete power take-off unit, seat, tool box and the top rear cover from transmission housing. Remove the fender, platform and the wheel and tire unit.

Refer to Fig. C846. Remove cotter pin (70) and nut (72) from inner end of axle shaft. Unbolt axle housing from transmission housing. Pull axle shaft and housing assembly away from transmission and lift out bull gear.

When reassembling, tighten the axle housing retaining nuts securely. Tighten nut (72) to provide a slight preload on the axle shaft bearings and install cotter pin (70).

AXLE SHAFTS AND HOUSINGS

109. The procedure for removing either rear wheel axle shaft and/or bearings is evident after removing the respective bull gear as outlined in paragraph 108.

Notice that the 200B, 300, 300B and 400B utility tractors have an inner and outer seal; whereas all general purpose tractors and the 350, 500B and 600B utility tractors have only the outer seal. The seal on general purpose tractors can be renewed without R&R of axle shaft.

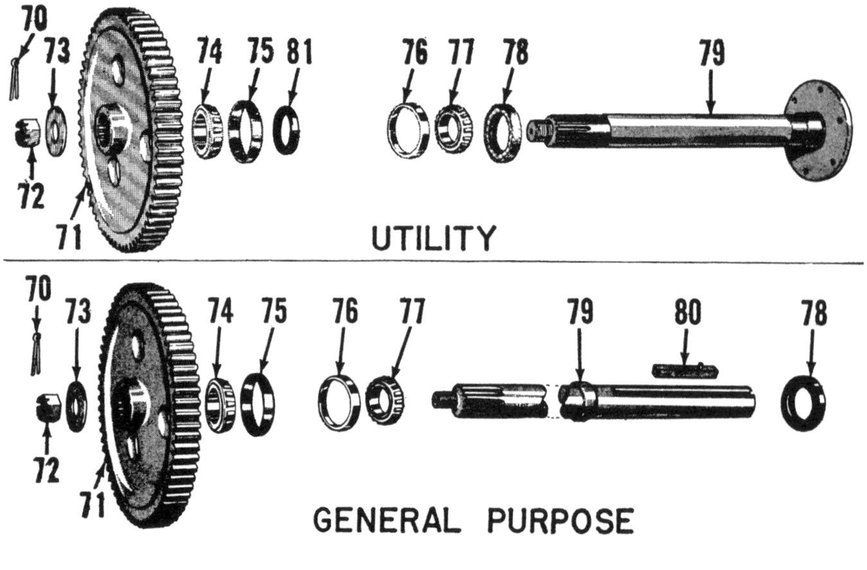

110. **ADJUSTMENT.** Jack-up rear of tractor, loosen jam nut and tighten each brake adjusting nut (Fig. C847) until a slight drag is felt when turning rear wheels by hand. Then loosen the adjusting nuts three turns and tighten the jam nuts. If brakes are not equalized, loosen the tight brake slightly.

111. **OVERHAUL.** Refer to Fig. C-847. Remove jam nuts, brake adjusting nuts and the cover retaining nuts. Remove cover, lined discs and actuating discs assembly. The need and procedure for further disassembly is evident. Linings are of the bonded type and are not available for field installation. If linings are worn, renew the complete lined disc.

Oil leakage on brake linings could be caused by a faulty seal in the differential bearing carrier or, by a leaking plug in outer end of side gear. When renewing plug, coat O.D. of same with sealer and be sure plug bottoms in gear. If carrier seal is to be renewed, remove the bearing carrier but do not mix, damage or lose any of the shims located between carrier and transmission housing. When installing the bearing carrier, use a sleeve or shim stock to avoid damaging the seal lip.

When reassembling, tighten the brake cover retaining nuts to a torque of 80-90 ft.-lbs. and adjust the brakes as in paragraph 110.

Fig. C846 — Exploded view showing bull gears, both type of wheel axle shafts and associated parts.

70. Cotter pin	75. Inner bearing cup	78. Outer seal
71. Bull gear	76. Outer bearing cup	79. Wheel axle shaft
72. Nut		80. Key (general purpose)
73. Washer	77. Outer bearing cone	81. Inner seal (200B, 300, 300B & 400B utility)
74. Inner bearing cone		

BELT PULLEY

Series 200B-300-300B-350-500B

The belt pulley unit is mounted on side of torque tube and is driven by a bevel gear located on the engine clutch shaft. To renew the driving bevel gear, refer to paragraph 89.

112. **R&R AND OVERHAUL.** Pulley unit can be removed from torque tube by unscrewing the four retaining cap screws. To disassemble the unit, remove the pulley retaining nut and lock washer and using a suitable puller, remove pulley from shaft. Refer to Fig. C848. Loosen set screw re-

taining sleeve and shaft assembly into housing and remove housing. Large oil seal in housing can be renewed at this time. Unstake and remove nut retaining bevel gear to inner end of shaft, then bump shaft and components out of sleeve. The need for further disassembly will be evident after an examination of the removed parts.

When reassembling, tighten nut on inner end of shaft to provide a very slight bearing drag; then lock the nut in place by staking a portion of the nut into shaft slot.

Fig. C849 — Sectional view of 200B, 300, 300B, 350 and 500B belt pulley attachment showing the unit in the inward or engaged position.

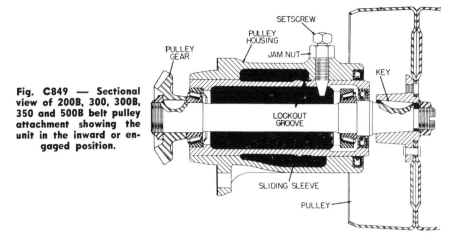

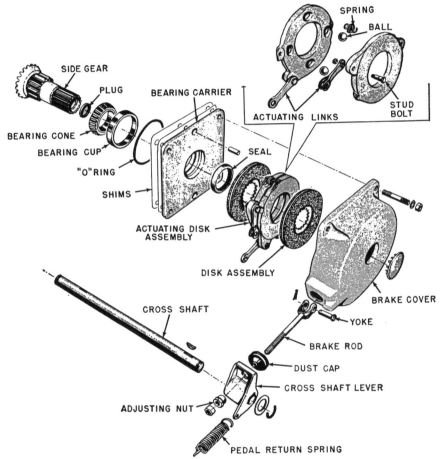

Fig. C847 — Exploded view of brakes, actuating linkage and associated parts.

Install the assembled sleeve in housing and with sleeve in the engaged position as shown in Fig. C-849, tighten the set screw and jam nut. Install pulley and tighten its retaining nut securely.

When installing pulley attachment on torque tube, vary the number of shims (Fig. C848) between pulley housing and torque tube to provide a bevel gear backlash of 0.004-0.008 or approximately $\frac{1}{16}$-inch movement at rim of pulley.

To disengage the pulley bevel gears when pulley is not in use, unscrew the set screw sufficiently to clear inner sleeve, pull outward on pulley until set screw can be seated in lockout groove of inner sleeve.

Series 400B-600B

The belt pulley attachment, when used, is a self-contained unit which is mounted on and receives its drive from the tractor pto output shaft.

112A. **OVERHAUL.** With unit removed from tractor, remove the pulley, then unbolt and remove the front bearing housing, rear bearing housing and output housing and disassemble the unit as shown in Fig. C849A. Thoroughly clean all parts and examine them for damage or wear. Note: Bevel gears are available as a matched set only.

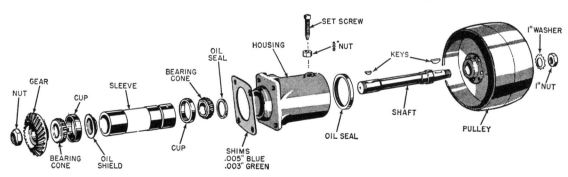

Fig. C848—Exploded view of 200B, 300, 300B, 350 and 500B belt pulley attachment. The unit is driven by a bevel gear on the engine clutch shaft.

When reassembling, refer to Fig. C849B, install the input gear and use the same front and rear bearing housing plastic shims as were originally removed. Tighten the bearing housing retaining screws securely. Then, insert the four ½x4½-inch bolts through the main housing and front bearing housing, secure with nuts and tighten same to a torque of 30 ft.-lbs. Check the end play of the input gear shaft. Gear shaft must turn freely with absolutely no binding tendency, but end play should never exceed 0.005. End play can be adjusted by varying the number of front and rear bearing housing shims. If only one shim is to be added or removed, do so at rear bearing housing; if two or more shims are to be added or removed, divide the amount equally between the front and rear bearing housing shim packs. Blue shims are 0.005 thick and brown shims are 0.010 thick.

Tighten nut on inner end of output shaft until a rolling torque of 10-15 inch-pounds is required to turn the shaft in its bearings and lock the nut in position by staking. Mount output shaft in a vertical position as shown in Fig. C849C and using a straight edge as shown, measure the distance (A) from machined surface of gear to machined surface of housing and use the proper number of shims when installing the output housing, as follows:

Dimension (A)	No. of Shims Required
2.175-2.176	1
2.177-2.179	2
2.180-2.183	3
2.184-2.187	4
2.188-2.191	5
2.192-2.194	6
2.195-2.198	7
2.199-2.201	8

Install output housing and the proper number of shims and tighten the housing retaining screws securely but make certain there is some bevel gear backlash as the screws are tightened. Install pulley and, while holding the input shaft stationary, check the bevel gear backlash which can be considered satisfactory if pulley rim has a back and forth movement of 1/64-inch. Pulley rim movement of 1/64-inch is equivalent to 0.008-0.010 backlash at bevel gears. If backlash is not as specified, it will be necessary to add or subtract shims at the front and rear bearing housings. It is important that proper input shaft end play be maintained; therefore, if any shims are added or removed at either location, the same thickness must be removed or added at the other location.

Fill pulley housing with one pint of S.A.E. No. 90 multipurpose lubricant.

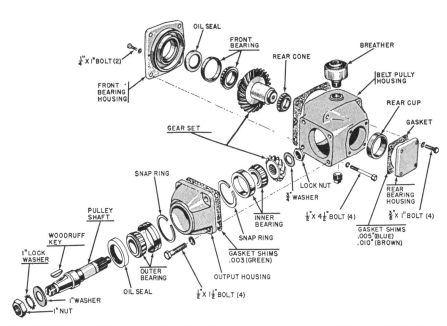

Fig. C849A—Exploded view of belt pulley unit used on series 400B and 600B. Bevel gears are available in a matched set only.

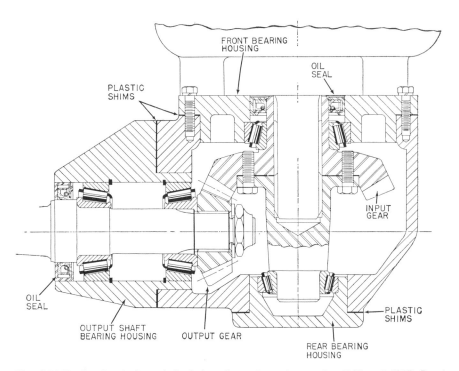

Fig. C849B—Sectional view of the belt pulley unit used on series 400B and 600B. Bevel gear mesh and backlash adjusting shims are shown.

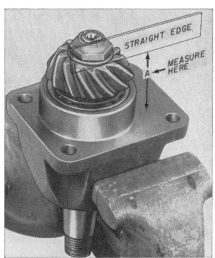

Fig. C849C — Measurement to determine the number of mesh adjusting shims required on series 400B and 600B belt pulley.

POWER TAKE-OFF

Series 200B-300-300B

113. **R&R AND OVERHAUL.** To remove the power take-off unit, drain transmission, remove the six retaining cap screws and remove complete assembly from transmission housing. Use extra care not to bump or bend the long drive shaft.

To disassemble the unit, remove oil seal retainer (35—Fig. C850), then remove pal nut (17) and adjusting nut (18). Buck up gear (24) and bump output shaft (32) rearward and out of housing. Rear bearing cone (29) will remain on shaft and both bearing cups (21 and 28) will remain in housing. Bump roll pin out of shift lever and remove washer (14), spring (13) and fork (12). Extract snap ring (10) and pull drive shaft assembly from housing. The procedure for further disassembly is evident. Mount the long drive shaft in lathe centers and check for run-out over entire length of shaft. If run-out exceeds 0.003 at any point, shaft must be straightened. NOTE: This applies to a new as well as a used shaft.

Use Fig. C850 as a guide during reassembly and when installing the output shaft (32), tighten nut (18) to provide the shaft with a very slight rotational drag. Bump shaft on each end and recheck the bearing adjustment. Install pal nut. Install pto unit on tractor, using care not to damage the double lipped seal at rear of transmission main shaft. Tighten the retaining screws to a torque of 35 ft.-lbs.

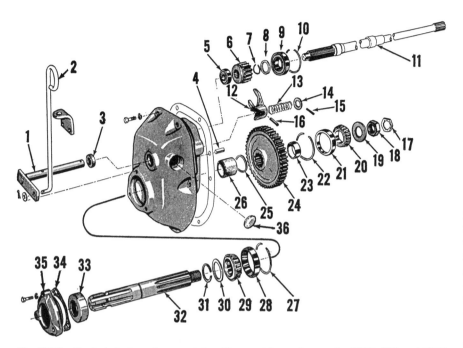

Fig. C850—Exploded view of power take-off assembly used on series 200B, 300 and 300B. Unit is mounted on rear of transmission housing.

1. Pto shift lever	10. Snap ring	18. Nut	27. Snap ring
2. Control rod	11. Pto drive shaft	19. Washer	28. Bearing cup
3. Oil seal	12. Drive gear shift	20. Bearing cone	29. Bearing cone
4. Dowel pin	fork	21. Bearing cup	30. Washer
5. Drive shaft rear	13. Shifter fork	22. Snap ring	31. Snap ring
bearing	throw-out spring	23. Spacer, front	32. Output shaft
6. Drive gear	14. Washer	24. Driven gear	33. Oil seal
7. Snap ring	15. Roll pin	25. Spring tension	34. Gasket
8. Washer	16. Roll pin	washer	35. Oil seal
9. Drive shaft front	17. Pal nut	26. Spacer, rear	retainer
bearing			36. Expansion plug

Series 350-500B

114. **R&R AND OVERHAUL.** To remove the power take-off unit, drain transmission, remove the six retaining cap screws and remove complete assembly from transmission housing. Use extra care not to bump or bend the long drive shaft.

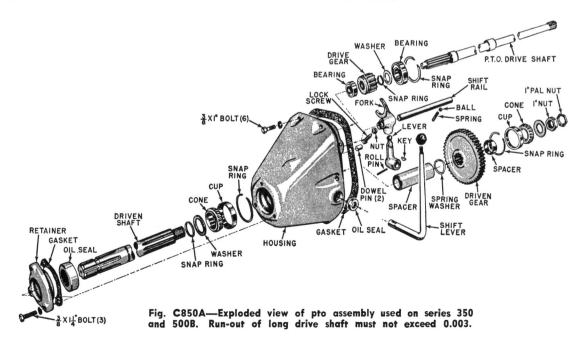

Fig. C850A—Exploded view of pto assembly used on series 350 and 500B. Run-out of long drive shaft must not exceed 0.003.

To disassemble the unit, refer to Fig. C850A and proceed as follows: Remove snap ring from front of housing, withdraw the long drive shaft and bearing assembly and lift out the drive gear. Remove oil seal retainer from rear of housing. Remove pal nut and 1-inch nut from front of driven shaft and bump driven shaft out through rear of housing. Lift driven gear, spacer and washer from housing. Remove roll pin retaining shift fork lever to hand shift lever and remove both levers. Using a ⅜-inch socket, universal and long extension, loosen shift fork set screw several turns and remove shift rail and fork. CAUTION: Be sure to cover opening in housing as shift rail is withdrawn to prevent personal injury from the flying detent ball. Refer to Fig. C850B.

Thoroughly clean all parts and renew any which are visibly damaged or worn. Place detent spring in housing bore. See inset, Fig. C850B. If end of spring is not flush with top of housing bore, add a washer under the spring or install a new spring of proper length. Mount the long drive shaft in lathe centers and check for run-out over entire length of shaft. If run-out exceeds 0.003 at any point, shaft must be straightened. Note: This applies to a new as well as a used shaft.

When reassembling, reverse the disassembly procedure and be sure the snap rings are properly seated. Tighten 1-inch nut on inner end of driven shaft to provide the shaft with a slight rotational drag. Bump shaft on each end and recheck the bearing adjustment. Install pal nut.

Install pto unit on tractor, using care not to damage the double lipped seal at rear of transmission main shaft. Tighten the retaining screws to a torque of 35 ft.-lbs.

CLUTCH AND OUTPUT (STUB) SHAFT

Series 400B-600B

114A. CLUTCH ADJUSTMENT. To adjust the pto clutch, first place the clutch hand lever in the disengaged (full downward) position, remove the pipe plug from side of housing and turn the pto stub shaft until the two spring loaded lockpins are visible through the pipe plug hole. Refer to Fig. C850C. Notice that one of the pins is in the engaged (IN) position, while the other pin is disengaged. With the adjusting tool inserted in the pipe plug hole, pull the engaged lock pin out with the forked end of the tool.

To tighten the clutch, turn the pto stub shaft counter-clockwise as shown until the disengaged lock pin drops into a notch. Check the adjustment by pulling the hand lever fully upward.

A properly adjusted clutch will require 60-65 pounds pull and will engage with a distinct snap of the lever.

114B. R&R AND OVERHAUL. To remove the pto clutch and output (stub) shaft, first drain oil from transmission and pto housing; then, unbolt and remove the complete pto unit from tractor. Use extra care not to bump or bend the long drive shaft. Release tension on the brake adjustment spring bolt (Fig. C850D), place hand lever in released position and remove the ⅜-inch bolt and spacer that connects hand lever to linkage. Refer to Fig. C850E, remove the ¼x½-inch screw from rear end of shift rail and using crocus cloth, polish all paint and foreign material from exposed end of rail. Withdraw the shift rail and fork. CAUTION: Detent ball under rail is spring loaded. To prevent personal injury and loss of ball, be sure to cover

opening in housing as shift rail is withdrawn. Unbolt base housing from main housing and separate the units.

114C. Refer to Fig. C850F, remove roll pin, extract shift yoke shaft and hand lever assembly and withdraw the spacer. Lift out the shift yoke. Remove snap ring from rear end of pto output shaft and bump output shaft and clutch assembly from housing. Extract snap ring and remove output shaft bearing from housing.

To disassemble the removed clutch, pull the clutch adjusting lock pin outward, unscrew the adjusting and throwout yoke assembly from the stub shaft and remove the pressure plate, bronze plates, steel plates, separator springs and back plate. Inspect or renew any parts which are damaged, cocked, or show wear.

Reassembly of the clutch plates and stub shaft will be simplified if a clutch drum is used as a jig. Place the stub shaft in the clutch drum and assemble the clutch plates in the order shown in Fig. C850G. Screw the ad-

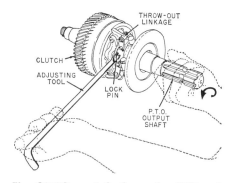

Fig. C850C — Adjusting series 400B and 600B pto clutch. A properly adjusted clutch will require 60-65 lbs. at end of control lever to engage same.

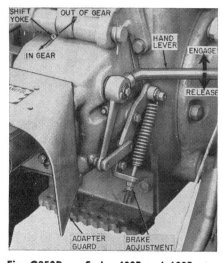

Fig. C850D — Series 400B and 600B pto unit installation, showing the clutch control linkage.

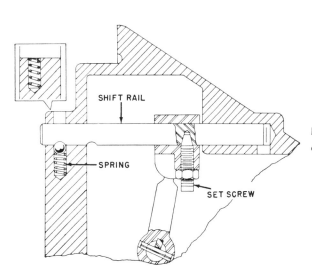

Fig. C850B — Series 350 and 500B pto shift rail sectional view.

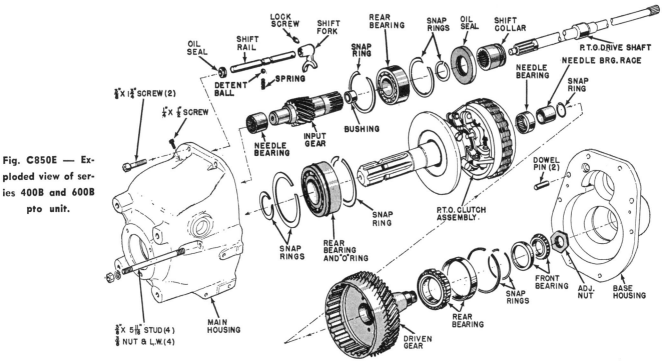

Fig. C850E — Exploded view of series 400B and 600B pto unit.

Fig. C850F—View of clutch installation in main housing when base housing is removed.

justing yoke assembly on the stub shaft until one of the lock pins can be engaged in the pressure plate. Engage the clutch and remove the assembly from the spare clutch drum.

When reassembling, reverse the disassembly precedure and adjust the clutch as in paragraph 114A.

PTO INPUT GEAR, DRIVEN GEAR (DRUM) & BASE HOUSING

Series 400B-600B

114D. **OVERHAUL.** Refer to Fig. C850E. With base housing separated from main housing as in paragraph 114B, proceed as follows:

Remove large Welch plug from base housing and, after unstaking, remove the adjusting nut from end of driven gear shaft. Bump shaft from housing. Remove oil seal and snap ring from input gear and remove input gear from housing.

If bushing in end of input gear is renewed, ream the bushing after installation to an inside diameter of 0.625. Be sure the $\frac{3}{16}$-inch breather hole in base housing is open and clean.

When reassembling, tighten the driven gear adjusting nut until the driven gear has an end play of 0.003-0.005 and lock the nut by staking.

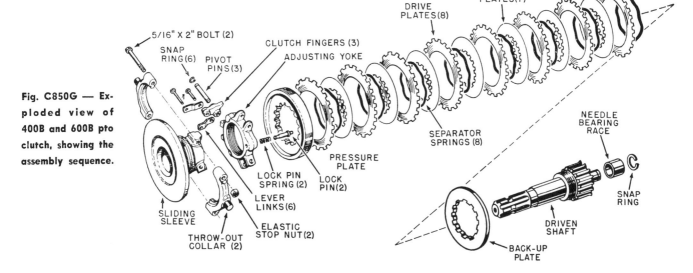

Fig. C850G — Exploded view of 400B and 600B pto clutch, showing the assembly sequence.

HYDRAULIC SYSTEM

Most of the troubles encountered with the hydraulic system on modern tractors are caused by dirt or gum deposits. The dirt may show up as the result of wear or partial failure of some part of the system. The presence of gummy deposits, however, usually results from inadequate fluids or from failure to drain and renew the fluid at the recommended intervals. These principles should be kept in mind when shooting trouble on the system and also when performing repair work on the pump, valves and work cylinders.

Thus, when disassembling a pump or valves unit, it is good practice generally to not remove any parts which can be thoroughly inspected while they are installed. Internal parts of the pump, valve and cylinders, when removed, should be handled with the same care as would be accorded the parts of a Diesel injection pump or nozzle unit, and should be soaked or cleaned with an approved solvent to remove gum deposits. Unless you practice good housekeeping (cleanliness) in your shop, do not undertake the repair of hydraulic equipment.

FLUID AND FILTER

Series 200B-300-300B-350-500B

115. The hydraulic fluid reservoir is located in the torque tube and it is recommended that fluid level be checked every 60 hours of operation. Fluid level should be maintained at

the full mark on dip stick. Never operate tractor with fluid level above the full mark.

Hydraulic system working fluid should be changed every 1000 hours of operation. Flush torque tube with tractor fuel and allow to drain completely. Remove cover cap from filter compartment of the hydraulic control housing, which on series 300 requires removal of hood. Refer to Fig. C851. Be careful not to damage the "O" ring. Remove the filter cartridge from housing and examine same for damage, especially at each end. NOTE: Fluid passes through the filter from inside to outside, therefore dirt and foreign material will be found on inside surface of cartridge. Thoroughly wash the filter with clean tractor fuel and blow dry with LOW PRESSURE air. When installing the element, be sure it seats properly on the spring loaded cap. Be sure "O" ring is in good condition, install cover cap and tighten the screws securely.

Remove and thoroughly clean the breather located on top of control housing.

Refill reservoir with multi-viscosity SAE 10W-30 oil, or as an alternate, Automatic Transmission Fluid Type "A" can be used. System capacity is 12 quarts for models with "Tripl-

Range" or shuttle gear boxes, 14 quarts for other models.

Series 400B-600B

115A. One common fluid in torque tube reservoir is used for the hydraulic power lift system and the "Case-O-Matic" torque convertor system. Service procedures for the fluid, filters and reservoir are covered in paragraph 94B.

TROUBLE-SHOOTING

116. The following trouble shooting paragraphs often save considerable time in pin pointing malfunctions in the hydraulic system.

The procedure for remedying many of the causes of trouble is evident. The following paragraphs will list the most likely causes of trouble, but only the remedies which are not evident.

116A. EAGLE HITCH LIFT ARMS LEAK DOWN. This trouble is caused by either internally or externally escaping oil. If there are no external leaks, check oil level in transmission and torque tube. If oil level in torque tube is getting lower and oil level in transmission is getting higher, there is probably an oil leak in the general area of the rockshaft operating cylinder. Refer to paragraph 127 and check for:

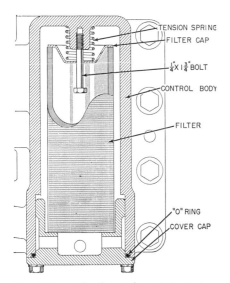

Fig. C851 — Sectional view of the hydraulic system filter compartment in the hydraulic control housing. Filter element should be cleaned at each fluid change.

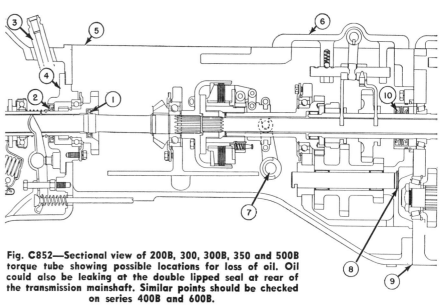

Fig. C852—Sectional view of 200B, 300, 300B, 350 and 500B torque tube showing possible locations for loss of oil. Oil could also be leaking at the double lipped seal at rear of the transmission mainshaft. Similar points should be checked on series 400B and 600B.

1. Pump drive gear inner seal
2. Pump drive gear outer seal
3. Tachometer drive sleeve
4. Hydraulic pump and tachometer drive gear housing
5. Control housing gasket
6. Rear top cover on torque tube
7. Hand clutch lever
8. "Tripl-Range" or shuttle-low range housing gasket or soft plug
9. Torque tube gasket
10. Transmission mainshaft front seals

a. Loose pipe plug in rockshaft cylinder
b. Faulty "O" ring on cylinder cap (some utility models only)
c. Faulty piston "O" rings
d. Pitted or scored cylinder walls
e. Scored piston

If oil level in transmission remains constant, there is probably no leakage at rockshaft cylinder and oil may be transferring back to the reservoir through the interlock. Refer to paragraphs 122 and 124 and check for:
f. Interlock end cap ball seats damaged
g. Interlock "O" rings leaking
h. Interlock plunger too long

116B. LOAD RAISES TOO SLOWLY. Could be caused by:
a. Load too heavy
b. Pressure relief valve setting too low, paragraphs 117 and 118
c. Insufficient supply of oil in reservoir
d. Binding rockshaft
e. Pump "O" rings leaking, paragraphs 120 and 121
f. Pump intake tube drawing air due to damaged "O" rings, paragraph 120
g. Faulty pump, paragraph 120 and 121
h. Over-heated oil (above 190° F.) due to too much oil in reservoir, paragraph 115
i. Control spool not moving to full raise position, paragraphs 125 and 126
j. Leaking control valve gaskets, paragraphs 125 and 126
k. Control spool oil bound, paragraph 125

l. Excessive clearance between control spool and body, paragraph 126

116C. THE HYDRAULIC PUMP IS NOISY OR PUMP OUTPUT IS LOW. Could be caused by:
a. Faulty pump "O" rings, paragraph 121
b. Faulty pump, paragraph 121
c. Relief valve pressure too low, paragraphs 117 and 118
d. Oil foaming—add defoament
e. Improper, dirty or contaminated oil, paragraph 115
f. Fractured pump intake tube, paragraph 120
g. Over-heating oil (above 190° F.) due to too much oil in reservoir, paragraph 115
h. Mis-alignment of drive gears, paragraph 120 (200B, 300, 300B, 350, 500B)
i. Insufficient gear backlash, paragraph 120 (200B, 300, 300B, 350, 500B)
j. Insufficient oil in reservoir, paragraph 115
k. Pump drive gear rubbing housing, paragraph 121 (200B, 300, 300B, 350, 500B)
l. Throwout fork loose or overshifting, paragraph 120 (200B, 300, 300B, 350, 500B)

116D. CHATTERING OR VIBRATION IN LINES. Could be caused by:
a. Excessive chamfer on control spool lands, paragraph 126
b. Interlock ball seats damaged, paragraph 123
c. Oil low in reservoir, paragraph 115

d. Oil flow restrictors not installed in remote cylinder couplers
e. Interlock plunger too short, paragraph 123

116E. OIL LEAKS IN GENERAL AREA OF CONTROL VALVE. Could be caused by:
a. Faulty oil seal in control valve body, paragraph 126
b. Faulty control valve body gasket, paragraph 125
c. Clogged breather
d. Faulty gasket between adapter base and torque tube
e. Faulty copper gasket at detent spring plug, paragraph 126
f. Porous casting, paragraph 126
g. Faulty control spool end cap gasket, paragraph 126
h. Faulty interlock gaskets, paragraph 122
i. Damaged "O" ring at filter cap, paragraph 115
j. Faulty seal around pump shift rod, paragraph 120
k. On series 400B and 600B, check for oil leaks at all "Case-O-Matic" fluid connections and valves.

116F. LOSS OF OIL FROM HYDRAULIC RESERVOIR could be caused by leaks at locations shown in Fig. C852. On series 400B and 600B, check for oil leaks at all "Case-O-Matic" fluid connections, valves, filter, radiator, etc.

SYSTEM OPERATING PRESSURE
Eagle Hitch System

117. To test the system operating pressure, connect Case test gage No. 4501AA or equivalent to the rock-

Fig. C852A — Pressure gage installation for checking the Eagle hitch system operating pressure.

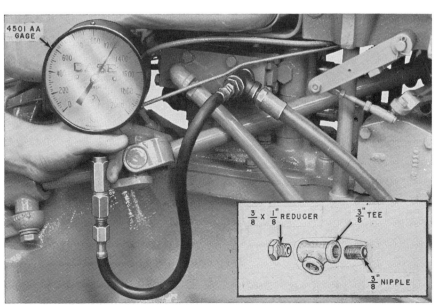

Fig. C852B—Pressure gage installation for checking the remote system operating pressure.

shaft housing as shown in Fig. C852A. Make certain that a snubber is used with the gage to prevent damage which might result from sudden surges in pressure. With hydraulic oil at operating temperature and engine running at rated speed, a pressure reading of 1200-1500 psi under steady load should be obtained.

If pressure is not as specified, vary the number of shims (41—Fig. C855) and recheck. The addition of one shim will change the pressure approximately 225 psi.

Remote System

118. The procedure for checking the remote system operating pressure is the same as outlined in paragraph 117, except the gage is connected as shown in Fig. C852B.

PUMP DRIVE GEAR
Series 200B-300-300B-350-500B

119. **R&R AND OVERHAUL.** To remove the pump drive gear, first detach (split) engine from torque tube as outlined in paragraph 90. Withdraw the clutch release bearing and disconnect the clutch release rod from the

pedal. Disconnect the flexible tachometer drive shaft, then unscrew and remove the tachometer drive gear and sleeve assembly from the drive housing. Remove the three cap screws retaining the drive housing to torque tube and remove the drive housing and attached parts as a unit as shown in Fig. C852C. The procedure for subsequent disassembly will be evident after an examination of the unit and reference to Fig. C852D.

When reinstalling the tachometer drive housing assembly over the clutch shaft, use shim stock as a sleeve to avoid damaging the clutch shaft seal. Be sure to use sealing compound on the housing gasket. Turn housing slightly to center release yoke and tighten the housing retaining screws to a torque of 25-30 ft.-lbs. Use new "O" ring and install the tachometer drive gear and sleeve.

Assemble the remaining parts by reversing the disassembly procedure.

Series 400B-600B

119A. The hydraulic pump drive gear is considered an integral part of the torque convertor unit and is not available as a separate replacement part. This same gear also drives the "Case-O-Matic" system oil pump. If gear is damaged, it will be necessary to renew the complete torque convertor unit as outlined in paragraph 94P.

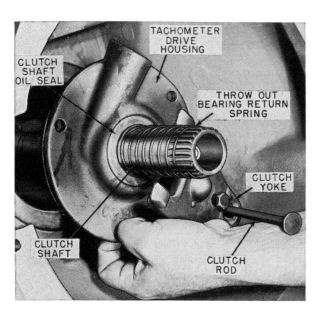

Fig. C852C — Removing the tachometer and hydraulic pump drive housing and attached parts from front wall in torque tube on series 200B, 300, 300B, 350 and 500B.

R&R PUMP

Series 200B-300-300B-350-500B

120. The hydraulic pump is located within the torque tube and is retained to the lower face of the adapter housing. To remove the pump, first remove the hood and proceed as follows:

Disconnect fuel line and remove throttle rod and fuel tank. Remove battery and battery tray. Thoroughly clean all dirt and foreign material from the hydraulic control valve (or valves) and remove the interlock assemblies. Disconnect throttle rod and on models so equipped, disconnect the speedometer cable. Remove the frame left side rail and the cap screw retaining the steering shaft bracket to torque tube. Remove cap screws retaining steering support mounting bracket to torque tube and lay steering support, dash and attached units forward on engine as shown in Fig. C853. Unbolt and remove adapter housing from torque

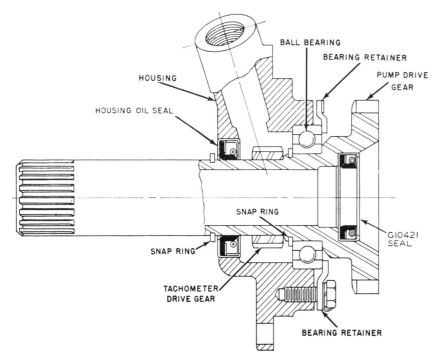

Fig. C852D — Sectional view of tachometer drive housing, showing the installation of the hydraulic pump drive gear on series 200B, 300, 300B, 350 and 500B.

tube. Remove intake pipe from pump housing, then unbolt and remove pump from adapter housing. Remove the ⅜-inch pipe plug (27—Fig. C854) from adapter housing and thoroughly clean any foreign particles from the housing passage.

Install pilot dowel (29) and "O" ring (28), then install pump and tighten the retaining screws to a torque of 45 ft.-lbs. Position "O" ring (34) in pump housing and install the intake pipe.

If oil is leaking around shift rod (21), it will be necessary to renew seal (24). To do so, loosen the fork retaining set screw (30), remove set screw (22) and shift button (23). Slide rod (21) out of position and remove fork (31). CAUTION: Hold hand over hole in housing as rod is removed because the detent ball is spring loaded and will fly out with force as the rod is withdrawn. Shift rod oil seal can be renewed at this time. When installing the shift rail, move same through seal from outside of housing. Install fork and shift button.

Install adapter housing, using the same number of gaskets as were originally removed. Remove belt pulley or pulley opening cover from side of torque tube and check the backlash between the hydraulic pump drive gears. If backlash is not between the limits of 0.007-0.010, vary the number of gaskets between the adapter housing and torque tube. Sealing compound should be used on the gaskets and the adapter housing screws should be tightened to a torque of 45 ft.-lbs.

Series 400B-600B

120A. To remove the hydraulic pump, first remove the hood and rear quarter panels unit and disconnect fuel line and gage wire from fuel tank. Then unbolt and remove fuel tank from tractor. Disconnect battery cables and remove battery and battery mounting tray. Uncouple steering shaft at one of the universal joints and unbolt the steering shaft intermediate support bracket. Disconnect the speedometer cable, tachometer cable at adapter housing, the "Case-O-Matic" heat indicator sending unit and oil line from the torque convertor pressure regulator valve housing and pressure gage oil line at "T" fitting. Disconnect the throttle rod and direct drive control linkage. Remove cap screws retaining the steering support mounting bracket to torque tube and lay steering support, instrument panel and attached units forward on engine.

Disconnect the "Case-O-Matic" oil lines, and the hydraulic lift system oil lines and control rods. Remove cap screws around perimeter and one Allen screw near the center of the adapter housing and, using a hoist and lifting eye, lift the complete assembly straight up and out of torque tube as shown in Fig. C818H. Use care during this operation not to distort the pump suction tubes.

When reassembling, reverse the disassembly procedure and tighten the pump retaining cap screws to a torque of 45 ft.-lbs.

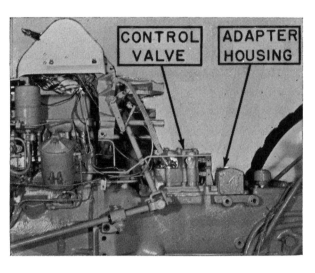

Fig. C853 — Instrument panel, steering support and attachment units laying forward on engine in preparation for removing hydraulic system adapter.

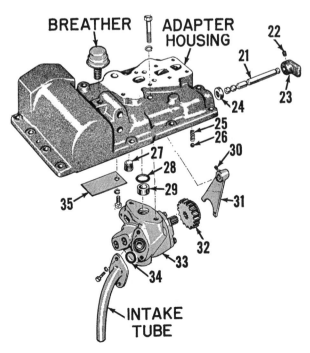

Fig. C854 — Hydraulic pump and associated parts exploded from the 200B, 300, 300B, 350 and 500B adapter housing.

21. Pump shift rod
22. Set screw
23. Shift button
24. Oil seal
25. Detent ball spring
26. Detent ball
27. ⅜-inch pipe plug
28. "O" ring
29. Pilot bushing
30. Set screw
31. Shift fork
32. Pump drive sliding gear
33. Pump assembly
34. "O" ring
35. Oil baffle

OVERHAUL PUMP

All Models

121. With the pump removed as outlined in paragraph 120 or 120A, refer to Fig. C855 and proceed as follows: Extract roll pin (40) and remove shims (41), spring (42), retainer (43) and ball (44). Remove cover (36) and disassemble the internal pump parts.

Thoroughly clean all parts and renew any which show any evidence of wear. If pumping gears are renewed, bearings (38) should also be renewed. Do not remove seat (45) unless the seat is visibly damaged. If seat is damaged, it can be removed by tapping hole in seat for a puller screw. Pull seat after heating bearing body in a dry oven to 225 degrees F.

When reassembling, renew all "O" rings and be sure that bearings are free to move parallel to gear journals. If any binding exists, dress the bearing flats with crocus cloth, but keep in mind that there should never be more than 0.001 clearance between the flats. Tighten cover bolts evenly to a torque of 35 ft.-lbs.

INTERLOCK

All Models

122. **REMOVE AND REINSTALL.** A safety interlock is located in the system circuit between the control valve (or valves) and the remote or rockshaft operating cylinder. An installation view of one interlock is shown in Fig. C856. To remove either one of the interlocks, proceed as follows:

Remove the interferring sheet metal, disconnect the hydraulic lines, then unbolt and remove the interlock from control valve.

When installing the interlock, be sure that gasket between interlock and control valve is positioned so that all oil ports are open. If gasket is reversed, it will cover the center part and restrict the flow of oil in the dump position.

123. **OVERHAUL.** With the interlock removed, thoroughly clean all dirt from outer surfaces and remove the cover assembly from both ends of the interlock body. (Some very early models of the 300 series have a cap on only one end.) The cover assemblies are identical except for a possible difference in the number of copper gaskets behind the end plugs. Do not interchange the cover assemblies. Tilt the block on end and remove plunger. Turn plugs out of caps and remove balls and springs.

Wash cap in solvent and blow dry. Inspect ball seat in cover where a fine line seat should be visible over the entire circumference. If there are nicks or breaks in the line, the cap must be reseated. Also check the ball for rough surfaces which might prevent good seating action. Secure the end cover in a vise and clean the seat. Drop ½-inch steel ball into position and tap the ball very lightly with a mild steel or brass rod. Tap until a fine line seat is obtained. NOTE: Excessive peening can do more harm than good. A broad seat will not prove to be as leak proof as a fine line seat. If satisfactory seat cannot be obtained, renew the end cover. Be sure to install a new ball after reseating. Inspect end surfaces of plunger and plugs for signs of peening which may result from vibration in the system. The distance from the ends of the plunger to the center section must not

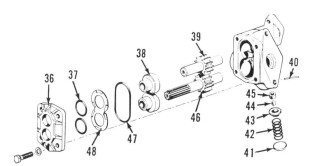

Fig. C855 — Exploded view of hydraulic pump.

36. Cover
37. "O" ring
38. Gear bearing
39. Driven gear & shaft
40. Roll pin
41. Shims
42. Detent valve spring
43. Ball retainer
44. Relief valve ball
45. Ball seat
46. Drive gear & shaft
47. "O" ring
48. Bearing spring plate

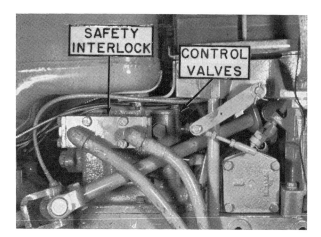

Fig. C856 — Typical installation view of one of the hydraulic system safety interlocks. Two are used on dual control valve system.

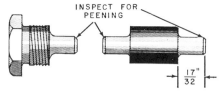

Fig. C858 — Inspecting the interlock plunger and end plug.

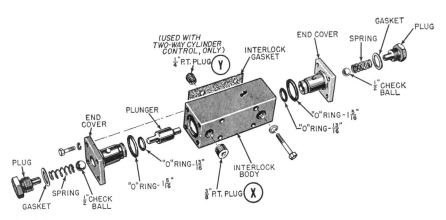

Fig. C857 — Exploded view of the hydraulic system safety interlock. The interlock is located in the system circuit between the control valve and cylinder.

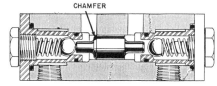

Fig. C859 — Sectional view of interlock showing proper installation of the plunger.

vary more than 1/64-inch from the dimension shown in Fig. C858. Shortness of the end of the plunger towards the pressure line end of the interlock block would cause vibration in the system during the lowering operation. This could only be corrected by renewing the plunger or by reversing it end for end.

When reassembling, insert ball and spring into cover, place copper gasket on end plug, making certain gasket is centered for a proper seal then tighten plug finger tight. NOTE: If there were two copper gaskets on the plug when it was removed, make certain two of them are reinstalled. Install new "O" rings on cover, then install cover assembly and tighten the retaining screws to a torque of 18 ft.-lbs.

Insert plunger through open end of block. Note: If the plunger has one end chamfered, it must be installed as shown in Fig. C859. This positions the chamfered end towards the port opening which is nearest to the mounting holes. Later models do not include a chamfer on the plunger and the plunger can be reversed end for end. Install cover assembly on other end of body and tighten screws to a torque of 18 ft.-lbs. Check for free movement of plunger within the interlock. Plunger must be free to slide under its own weight in the body bore when tilting the interlock assembly from end to end. A clearance of 0.020 inch to 0.050 inch between the seated balls and end of the plunger must be maintained. A lack of clearance may cause the balls to be forced off their seat resulting in a leak back of oil from cylinder to reservoir. Clearance between balls and plunger can be obtained by grinding a small amount of material from end of plunger.

When using the interlock assembly to actuate the Eagle Hitch attachment or single acting remote cylinder, a ⅜-inch countersunk head pipe plug (X—Fig. C857) must be installed in the exterior oil port. All interior oil ports remain open. Refer to Fig. C860. Notice that the port for line fitting is closer to mounting holes than port for pipe plug.

When the interlock assembly is used to actuate a double acting remote cylinder, a ¼-inch countersunk head pipe plug (Y—Fig. C857) is installed at the interior oil port.

After the unit is installed, test the system for back pressure as outlined in the following paragraph.

124. **TESTING FOR BACK PRESSURE.** When back pressure is present in the single acting system, the Eagle Hitch arms will lower slowly and the pump will labor even though the control lever is in dump position. Back pressure results when the plunger does not uncover the return port in the interlock block, thereby, restricting the flow of oil. A variation in hole location or length of plunger may cause this condition. Whenever a new plunger is installed, the system should be checked for signs of back pressure. This is especially true when changing from a plunger with a chamfer on one end to a plunger without chamfer.

If there is reason to believe that back pressure exists, proceed as follows: Move the hydraulic control lever to the full dump position, stop the tractor engine and install a gage in the tapped holes shown in Fig. C861. This is the opening normally plugged for a single acting system. Start engine and observe the pressure reading which should not exceed 75 psi at governed engine speed. A reading of 0-25 psi is desirable. If reading exceeds 75 psi, add another copper gasket behind the plug in the end cap at the point indicated in Fig. C860.

Fig. C861 — Pressure gage installation to check for back pressure in the safety interlock.

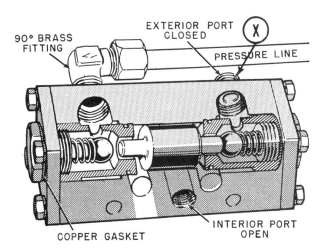

Fig. C860—Interlock connections for single acting system.

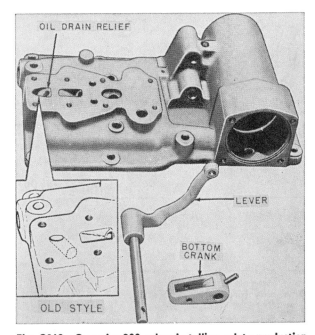

Fig. C862—On series 300, when installing a late production control valve assembly on a very early production adapter casting, it is necessary to chip out an oil drain relief. Refer to text.

CONTROL VALVE
All Models

125. **REMOVE AND REINSTALL.** To remove control valve (or valves), first remove the safety interlock (or interlocks) as outlined in paragraph 122 and proceed as follows: Disconnect fuel line and remove throttle rod and fuel tank. Drift out the roll pin connecting the valve spool to the bell crank and remove cover from top of valve. Lift off the control valve assembly. On dual valve systems, drift out the roll pin connecting the lower valve spool to bell crank and remove the valve assembly. Notice that roll pin slots in spools face down. Also notice that four of the eight attaching

screws are ¼-inch shorter than the other four. This is true for both dual and single valve systems. On series 300 only, refer to the following note.

Note: Series 300 only. During operation, any oil that is trapped behind the control spool drains off through the vertical hole in the control valve block, and then to the reservoir through a diagonal hole in the adapter casting. Late production adapter castings are relieved at the point where the two holes meet in order to provide for a free flow of oil. Refer to Fig. C862. The passage in the early style control valve blocks was drilled at an angle making it unnecessary to have a cored relief in the adapter casting. If, however, a late production control valve assembly (with vertical drain hole) is to be

installed on an early style base, a relief will have to be chipped out of the adapter casting. A gasket can be used to locate point of relief. Be absolutely certain that no chips or other foreign material fall into reservoir.

When installing the control valve (or valves), refer to the accompanying parts list as well as Figs. C862A, C862B and C862C to make certain that the correct parts are being installed. Tighten the retaining screws progressively and evenly to a torque of 18 ft.-lbs. Notice that four of the retaining screws are ¼-inch longer than the other four.

Make certain that the spool slots are down and that the control valve operating linkage is not worn. Test sliding action of spool in valve body to make

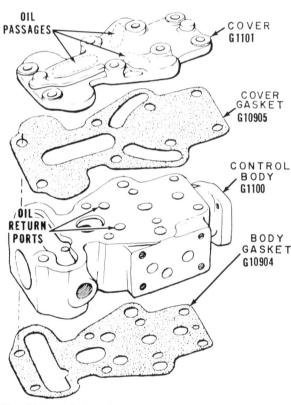

Fig. C862A—Series 300 and some models of the 350 series are equipped with the control valve body, cover and gaskets shown. Cover G1101 has three separate oil passages. Body G1100 can be identified by the two vertical oil return ports. Gaskets G10905 and G10904 cannot be used with any other control body or cover.

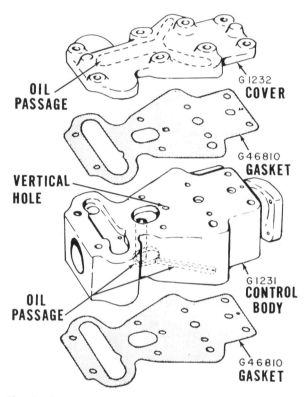

Fig. C862B—Some models of the 350 series are equipped with the control valve body, cover and gaskets shown. The same G1232 cover (three oil passages connected) and G-1231 control body (vertical hole and diagonal oil passage) are used on the 200B, 300B, 400B, 500B and 600B, but the gasket part number on these models is G46838. See Fig. C862C.

PARTS LIST
CONTROL VALVE COVER, BODY AND GASKETS

	200B	300	300B & 500B Eagle Hitch	300B & 500B Remote Control & Eagle Hitch	350 Some Models	350 Some Models	400B & 600B
Valve assembly	13983	13103	13983	13985	13703	13983	G14140
Valve body cover	G1232	G1101	G1232	G1232	G1101	G1232	G1232
Gasket between cover & body...........	G46838	G10905	G46838	G46838	G10905	G46810	G46838
Valve body	G1231	G1100	G1231	G1231	G1100	G1231	G1281
Gasket between valve body and valve body (dual system)...............	G46810	G10904	G46838	G46838	G10904	G46810	G46838
Gasket between valve body and adapter casting	G46810	G10904	G46838	G46838	G10904	G46810	G46838

certain that no binding exists. Binding could be caused by improper tightening of the mounting screws or the machined mounting surfaces of the cover, valve body and/or adapter casting are not smooth and flat. Surfaces must be smooth, flat and parallel within 0.001 total indicator reading.

Note: When installing one control valve on top of another valve to operate a remote cylinder, a ¼-inch pipe plug must be added to the body as shown in Fig. C862D.

126. **OVERHAUL.** With the control valve (or valves) removed as outlined in paragraph 125, refer to Fig. C863 and proceed as follows: Remove end cap (46) and gasket (45). Unscrew detent plug (52) and on models so equipped remove spring (50) and ball (53). Clean all dirt and paint from the exposed end of spool, then withdraw spool from cap end of control block. Bump roll pin from control spool and remove spring (48), sleeve and washers. Thoroughly clean all parts and renew any which are damaged or

worn. Check casting for porous condition and pay particular attention to polished surfaces of spool and spool bore. Spool lands should not have excess chamfer. A chamfer in excess of 0.0025 will result in undesirable port opening and may be manifested by chattering, vibration or uncontrolled drop.

On all except series 300, spool (No. G16670—Fig. C863A) is used. Originally the spool lands were machined with $\frac{7}{32}$-inch notches. When these spools are encountered, either modify the spool by grinding 0.010 deep flats as shown or install current production spool with ⅜-inch notches. Spool return spring should have a free length of 1½ inches and should require approximately 45 lbs. to compress it to a height of ¾-inch.

On control valve covers with part No. G1232, examine the oil passages. On early production of these covers, the oil passages were formed by a machining process and in some individual cases the center portion was not milled back far enough to uncover the small vertical hole in the control body. This condition can be corrected by removing the additional material shown in Fig. C864A. On later covers, the passages are cast in and no reworking is required. There should be only one washer (47-Fig. C863) and one washer (49) used. If more than one washer

is found at either end of the spring, install a new spool and one washer at each end of the spring.

When reassembling, make certain that all parts are perfectly clean. Coat O.D. of seal (55) with a suitable sealing compound and install seal flush with end of casting. Install spool in casting, using a suitable sleeve to avoid damaging the seal lip. Note: Sleeve (Fig. C864) can be made from a Case VT-5082 brake bushing. Install end cap and tighten the retaining screws to a torque of 10 ft.-lbs. If control valve is used to actuate the Eagle hitch, install detent ball (53-Fig. C-863) and spring (50). On all other applications, the ball and spring are omitted so that spool will return to neutral when lever is released.

ROCKSHAFT OPERATING CYLINDER

All Models

127. **R&R AND OVERHAUL.** Empty oil from cylinder by moving control lever to full dump position, disconnect links from lift arms, then disconnect control rod and oil lines. Remove the attaching cap screws and lift the entire rockshaft assembly from transmission housing. Refer to Fig. C865.

Bump the two roll pins down in the rocker arm and remove the connecting rod. On general purpose models and

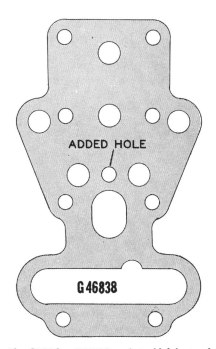

Fig. C862C—G46838 gasket which is standard on some models. This gasket is a universal type and can be used as a cover gasket or body gasket on all models.

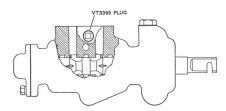

Fig. C862D — Location of plug to be installed in top control valve to operate a remote cylinder.

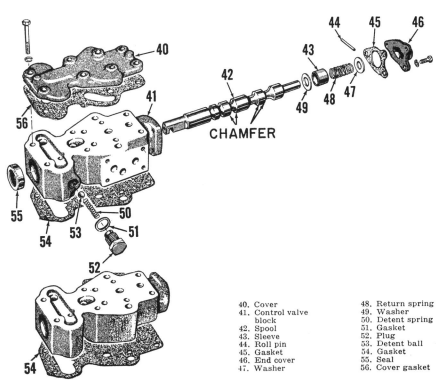

40. Cover	48. Return spring
41. Control valve block	49. Washer
42. Spool	50. Detent spring
43. Sleeve	51. Gasket
44. Roll pin	52. Plug
45. Gasket	53. Detent ball
46. End cover	54. Gasket
47. Washer	55. Seal
	56. Cover gasket

Fig. C863—Exploded view of a typical hydraulic system control valve. Detailed differences in the body (41), cover (40), gaskets (54 & 56) and spool (42) will be evident after an examination of the unit and reference to related illustrations.

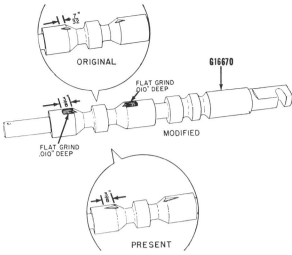

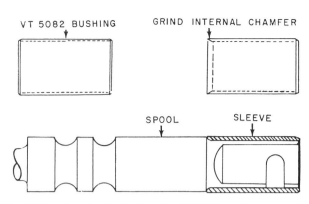

Fig. C864—Case brake bushing No. VT-5082 can be reworked and used as a sleeve to install control valve spool through oil seal.

Fig. C863A—View showing method of reworking spool No. G16670 on all models except series 300.

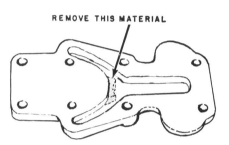

Fig. C864A—Some early production G1232 control covers require reworking as shown.

some utility models, remove plug from front end of cylinder and using a brass rod or wooden dowel, tap piston rearward out of cylinder. On other utility models, remove cap from front of cylinder and push piston forward out of cylinder.

Clean and inspect piston and cylinder in a conventional manner and be sure to renew all "O" rings. Thoroughly lubricate piston and "O" rings and reassemble. Be sure to use new roll pins in rocker arm. On models so equipped, use sealer on front pipe plug and tighten same securely. On other models, use a new, well lubricated "O" ring on cylinder cap and tighten the retaining screws to a torque of 60 ft.-lbs.

Install rockshaft housing by reversing the removal procedure.

ROCKSHAFT

All Models

128. CONTROL ROD ADJUSTMENT. Disconnect control rod, place control valve spool in neutral position, move hand control lever to neutral (vertical position) and adjust the length of the control rod so that attaching pins can be freely inserted without moving either the spool or lever.

129. QUADRANT ROD ADJUSTMENT. Place the quadrant lever stops in extreme raise and extreme drop position, start tractor engine and move hand control lever to raise position. When the draft arms reach the full raised position, the quadrant lever should contact the quadrant stop and throw the hand control lever back to neutral position. While the draft arms are still in full raised position, grasp the draft arms (Eagle claws) and pull upward. There should be approximately one inch free travel at end of

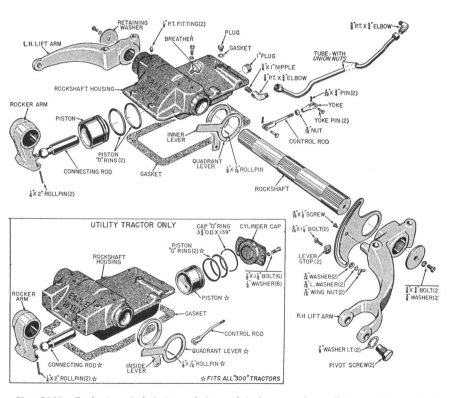

Fig. C865—Typical exploded view of the rockshaft, operating cylinders and associated parts. On some models, the inner and outer quadrant levers are one piece. Early model pistons are filled with two "O" rings as shown. Late models have one "O" ring and one cast iron ring.

arms before the rockshaft rocker arm strikes the internal stop on the housing. If sufficient free movement does not exist, shorten the quadrant rod to increase free travel or lengthen rod to decrease free travel. A lack of free movement may cause the rocker arm to bottom before the lift arms have reached full raised position. The control levers will not return to neutral and the pump will by-pass continuously.

130. **OVERHAUL.** The need and procedure for removing and/or overhauling the rockshaft and associated parts will be evident after an examination of the unit and reference to Fig. C865.

REMOTE CYLINDER
All Models

131. **OVERHAUL.** To disassemble the cylinder, refer to Fig. C866 and proceed as follows: Remove ½-inch street ells and remove the cap screws holding stop bar to cylinder head. With piston rod in extended position, move limit stop assembly away from cylinder head. Move cylinder head inward to allow removal of snap ring. Pull the piston rod and limit stop unit from the cylinder. After cylinder has been disassembled, all parts should be washed in gasoline or distillate, thoroughly inspected and those showing excessive wear renewed.

When reassembling, be sure to renew all "O" rings. The wiper dust seal is installed with lip to the outside. Use Case thimble 08668-AB or equivalent to install cylinder head with seals over piston rod. Lubricate the inside of cylinder and all packing rings with clean engine oil. Insert assembly into the cylinder, pushing cylinder head down to allow insertion of snap ring. Pull cylinder head up in place with piston rod and fasten stop-bar to cylinder head with four cap screws.

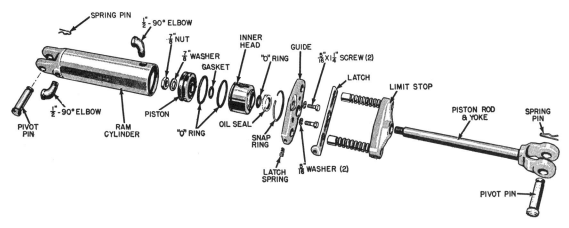

Fig. C866—Exploded view of remote control cylinder.

NOTES

NOTES

CASE

Models ■ 400   ■ 700B   ■ 800B

Previously contained in I & T Shop Service Manual No. C-9

SHOP MANUAL
J.I. CASE
SERIES 400-700B-800B

Serial number is stamped on the instrument panel name plate and on the left rear side of the engine block near the air cleaner.

BUILT IN THESE VERSIONS

	TRICYCLE		AXLE TYPE	
	Single Wheel	**Dual Wheel**	**Non-Adjustable**	**Adjustable**
DIESEL MODELS	401-701B-801B	401-701B-801B	400-402-405-420-700B-702B-705B-800B-802B-805B	401-403-701B-703B-801B-803B
NON-DIESEL MODELS	411-711B-811B	411-711B-811B	410-412-415-425-710B-712B-715B-810B-812B-815B	411-413-711B-713B-811B-813B

INDEX (By Starting Paragraph)

CONDENSED SERVICE DATA

GENERAL	Series 400 Non-Diesel	Series 700B Non-Diesel	Series 800B Non-Diesel	Series 400 Diesel	Series 700B Diesel	Series 800B Diesel
Engine Make	Own	Own	Own	Own	Own	Own
Engine Model	400	251S	251S	....	267D	267D
Number Cylinders	4	4	4	4	4	4
Bore—Inches	4	4	4	*	4⅛	4⅛
Stroke—Inches	5	5	5	5	5	5
Displacement—Cubic Inches	251	251	251	*	267	267
Compression Ratio, Gasoline and Diesel	6.5:1	6.5:1	6.5:1	15:1	15:1	15:1
Compression Ratio, Distillate	4.75:1	4.75:1	4.75:1	....	....	....
Compression Ratio, LP-Gas	7.46:1	7.46:1	7.46:1	....	....	....
Pistons Removed From?	Above	Above	Above	Above	Above	Above
Main Bearings, Number Of?	5	5	5	5	5	5
Main and Rod Bearings Adjustable?	No	No	No	No	No	No
Cylinder Sleeves	Wet	Wet	Wet	Wet	Wet	Wet
Electrical System Voltage	6 or 12	12	12	12	12	12
Generator and Starter Make?	A-L	A-L or D-R	A-L or D-R	A-L	A-L or D-R	A-L or D-R

TUNE-UP						
Firing Order	1-3-4-2	1-3-4-2	1-3-4-2	1-3-4-2	1-3-4-2	1-3-4-2
Valve Tappet Gap, Intake	0.012C	0.012C	0.012C	0.012C	0.012C	0.012C
Valve Tappet Gap, Exhaust	0.020C	0.020C	0.020C	0.012C	0.012C	0.012C
Valve Face Angle	45°	45°	45°	45°	45°	45°
Valve Seat Angle	45°	45°	45°	45°	45°	45°
Ignition Distributor Model (Wico)	X30169	X30169	X30169	....	....	....
Ignition Distributor Model (Auto-Lite)	IAD-6003-1B	IAD-6003-1B	IAD-5003-1B	....	....	....
Ignition Magneto Make	Own	Own	Own	....	....	....
Ignition Magneto Model	41	41	41	....	....	....
Ignition Timing, Retard	TDC	TDC	TDC	....	....	....
Ignition Timing, Advanced—Distributor	26°	26°	**	....	....	....
Ignition Timing, Advanced—Magneto	25° BTDC	25° BTDC	25° BTDC	....	....	....
Injection Timing	....	....	....	— See paragraph 119A, 120A or 121A —		
Timing Mark Location?	————— Crankshaft Pulley —————			——— Crankshaft Pulley ———		
Crankshaft Pulley Mark Indicating						
Retard Timing	Notch	Notch	Notch	....	....	....
Advanced Timing	Deg. Notch	Deg. Notch	Deg. Notch	....	....	....
Injection Timing	....	....	....	Deg. Notch	Deg. Notch	Deg. Notch
Breaker Contact Gap, Distributor	0.020	0.020	0.020	....	....	....
Breaker Contact Gap, Magneto	0.010	0.010	0.010	....	....	....
Spark Plug Make	Ch. of A-C	Ch. or A-C	Ch. or A-C	....	....	....
Model, Champion	8 Comm.	8 Comm.	8 Comm.	....	....	....
Model, AC	85S Comm.	85S Comm.	85S Comm.	....	....	....
Electrode Gap	0.025	0.025	0.025	....	....	....
Carburetor Make, Gasoline and Distillate	M-S	M-S	M-S	....	....	....
Carburetor Model, Gasoline and Distillate	TSX616	TSX616	TSX616	....	....	....
Carburetor Float Setting	¼-inch	¼-inch	¼-inch	....	....	....
Carburetor Make, LP-Gas	Ensign	Ensign	Ensign	....	....	....
Carburetor Model, LP-Gas	Xg8833	Xg3896	Xg3896	....	....	....
Engine Slow Idle rpm	600	600	600	600	600	600
Engine High Idle, No Load rpm	1600	1655	2000	††	1650	1950
Engine Full Load rpm	1450	1500	1800	††	1500	1800

SIZES—CAPACITIES—CLEARANCES						
(Clearances in Thousandths)						
Crankshaft Main Journal Diameter	2.9985	2.9985	2.9985	2.9985	2.9985	2.9985
Crankpin Diameter	2.7485	2.7485	2.7485	2.7485	2.7485	2.7485
Camshaft Journal Diameter	2.1215	2.1215	2.1215	2.1215	2.1215	2.1215
Piston Pin Diameter	1.35845	1.35845	1.35845	1.35845	1.35845	1.35845
Valve Stem Diameter, Intake	0.4025	0.4025	0.4025	0.4025	0.4025	0.4025
Valve Stem Diameter, Exhaust	0.4005	0.4005	0.4005	0.4005	0.4005	0.4005
Main Bearings, Running Clearance	2.0-4.6	2.0-4.6	2.0-4.6	1.6-4.6	1.6-4.6	1.6-4.6
Rod Bearings, Running Clearance	1.5-3.6	1.5-3.6	1.5-3.6	1.3-3.8	1.3-3.8	1.3-3.8
Piston Skirt Clearance	3-5	3-5	3-5	3.5-5.5	3.5-5.5	3.5-5.5
Crankshaft End Play	4-12	4-12	4-12	4-12	4-12	4-12
Camshaft Bearing, Diameter Clearance	1.4-5.4	1.4-5.4	1.4-5.4	1.4-5.4	1.4-5.4	1.4-5.4
Crankcase Oil, Including Filter—Quarts	9	9	9	9	9	9
Cooling System—Gallons	7.4	7.4	7.5	7.4	7.4	7.5
Transmission and Differential—Quarts	60	60	60	60	60	60
Hydraulic System, With Eagle Hitch—Quarts	10.5†	10.5†	10.5†	10.5†	10.5†	10.5†
Hydraulic System, Without Eagle Hitch—Quarts	9†	9†	9†	9†	9†	9†
Power Take-Off	—Uses same oil as hydraulic system—			—Uses same oil as hydraulic system—		
Power Steering—Quarts	***	***	***	***	***	***
Final Drive, High Clearance—Quarts	4.5	4.5	4.5	4.5	4.5	4.5
Main Fuel Tank—Gallons	21	22	22	21	22	22

*Series 400 Diesel prior to 8104711 same as 400 Non-Diesel. Series 400 Super Diesel same as 700B Diesel.

**When equipped with A-L dist. 26°; 28° with Wico.

***General purpose and high clearance, 3.75 quarts; others 2.0.

†Add one quart for each portable cylinder.

††Series 400 Diesel prior to 8104711 engine high idle no-load rpm is 1600, loaded rpm is 1450. Series 400 Diesel after 8104710 engine high idle no-load rpm is 1635, loaed rpm is 1500.

3

FRONT SYSTEM

TRICYCLE AXLE

Dual Wheel Tricycle Models

1. Dual front wheels are mounted on a horizontal axle which is retained to lower steering spindle by snap ring (2—Fig. C500). The procedure for removing axle from lower spindle or spindle from tractor is evident. When installing the wheels, tighten the adjusting nut to provide a noticeable bearing drag.

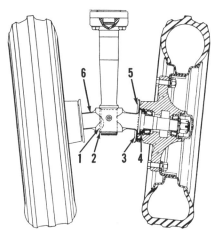

Fig. C500 — General purpose dual wheel tricycle front axle. Snap ring (2) retains the horizontal axle to the spindle.

1. Woodruff key	4. Seal
2. Snap ring	5. Retainer
3. "O" ring	6. Axle

Single Wheel Tricycle Models

2. The single front wheel fork is bolted to the upper steering spindle and the removal procedure is conventional. Shims (S—Fig. C501) at both ends of axle should be varied to obtain a noticeable bearing drag. Shims are available in thicknesses of 0.003 and 0.005.

STEERING KNUCKLES

Standard and Adjustable Axle Models

3. On adjustable axle models shown in Fig. C502, the front wheel spindle (10) is keyed and retained to the king pin (7) by a nut. On standard (non-adjustable) axle models, the wheel spindle is integral with the king pin as shown in Fig. C503. In either case, however, the king pin is carried in needle type roller bearings.

When installing the needle bearings, use a sizing arbor that is a close fit inside the bearing and has a wide enough shoulder to engage all of the end face of the bearing.

Note: Special arbors are available from J. I. Case for installing the Torrington needle bearings. Arbor No. A8346 is used for installing the upper bearing on adjustable axle models. Arbor No. A8354 is used for installing the upper bearing on standard non-adjustable axle models. Arbor No. A8351 is used for installing the lower needle bearing in all adjustable and non-adjustable axle models.

When installing thrust bearing (9) at lower end of king pin, make certain that the open flange end of same is down.

TIE-RODS AND DRAG LINK
Standard and Adjustable Axle Models

4. Tie-rod ends on all models are of the non-adjustable type. On adjustable axle models with two tie-rods, adjust the length of each rod an equal amount to provide a toe-in of $\frac{1}{4}$-$\frac{3}{8}$-inch. On standard non-adjustable axle models with one tie-rod, turn the tie-rod either way as required to obtain the recommended toe-in of $\frac{1}{4}$-$\frac{3}{8}$-inch.

The ball and socket type drag link ends on standard non-adjustble axle models should be adjusted to remove all free play without causing any binding tendency.

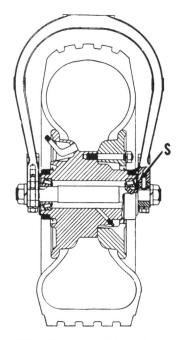

Fig. C501—One type of single front wheel available on general purpose tractors. The unit is equipped with a 9.00-10 tire. Another version with a one-piece wheel is available for use with a 6.50-16 tire. Shims (S) control the bearing adjustment.

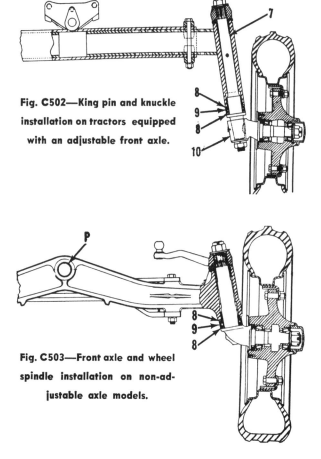

Fig. C502—King pin and knuckle installation on tractors equipped with an adjustable front axle.

Fig. C503—Front axle and wheel spindle installation on non-adjustable axle models.

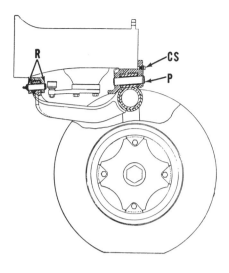

Fig. C504—Axle and radius rod pivot pins installation on models with an adjustable front axle.

AXLE PIVOT PINS AND BUSHINGS

Adjustable Axle Models

5. Radius rod is integral with the axle middle section on all adjustable axle models. To renew the axle and radius rod pivot pins and bushings, first place a floor jack under clutch housing and raise front of tractor enough to just take weight off of front wheels. Remove cap screws (CS—Fig. C504) retaining axle pivot pin (P) to pedestal and withdraw the axle pivot pin. Remove snap rings (R) from radius rod pivot pin and remove the pin. Raise front of tractor, allowing axle to drop down far enough to remove the pivot pin bushings from axle and integral radius rod. New bushings should be final sized after installation, if necessary, to provide a free running fit for the pivot pins.

Pivot pin diameters are as follows:
General Purpose Models
    Front 1.246-1.247
    Rear 0.997-0.998
High Clearance Models
    Front 1.745-7.747
    Rear 1.495-1.497

Standard Axle Models

6. Radius rod is bolted to axle main member. To renew the axle pivot pin and bushing, place a floor jack under clutch housing and raise front of tractor enough to just take weight off of front wheels. Unbolt the radius rod at rear end, remove the axle pivot pin retaining snap ring and remove the pivot pin (P—Fig. C503). Raise front of tractor, allowing axle to drop down far enough to remove the pivot pin bushing from axle main member. New bushing should be final sized after installation to provide a free fit for the pivot pin.

MANUAL STEERING SYSTEM

General Purpose and High Clearance Models

7. ADJUSTMENT. The steering gear unit is provided with two adjustments; one for the upper steering spindle end play, the other for the steering worm shaft end play. To make the adjustments, raise front of tractor to remove load from steering gear and remove the grille screen.

8. WORM SHAFT END PLAY. To adjust the steering worm shaft, remove cover (1—Fig. C505) and remove the worm shaft front bearing carrier (2). Vary the number of shims (3) located between the bearing carrier and the housing to remove all shaft end play without causing any binding tendency. Shims are available in thicknesses of 0.003, 0.005 and 0.010.

9. UPPER SPINDLE END PLAY. To adjust the upper spindle end play, remove the gear housing top cover as shown in Fig. C506 and tighten plug (19) to remove all spindle end play; then, back-off the plug until projection (P) on upper cover will engage slot in the adjusting plug. Before installing the upper cover, make certain that "O" ring (9) is in position.

10. R & R AND OVERHAUL. To remove the steering gear housing assembly, first drain cooling system and remove hood, radiator screen, grille and radiator. Raise front of tractor and unbolt lower spindle on dual wheel tricycle models, wheel fork on single wheel tricycle models or center arm on adjustable axle models from the upper steering spindle. Disconnect the steering worm shaft front universal joint, unbolt and remove the steering gear housing assembly from tractor.

Drain lubricant from housing and remove upper cover (20—Fig. C507) and adjusting plug (19). Unbolt and separate upper gear housing (17) from lower housing (15). Remove snap ring (18) and withdraw sector (21). Withdraw upper spindle from lower housing. Needle bearing (11) in upper housing, needle bearing (12) and seal (14) in lower housing can be renewed at this time.

Note: Special arbors are available from the J. I. Case Co., for installing

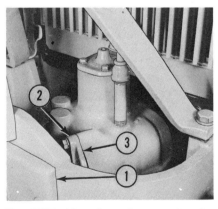

Fig. C505—Manual steering gear installation on general purpose and high clearance tractors. Steering worm shaft end play is adjusted with shims (3) between gear housing and front bearing carrier (2).

Fig. C506 — Manual steering gear for general purpose and high clearance models. Upper spindle end play is adjusted with plug (19), "O" ring (9) must be in good condition.

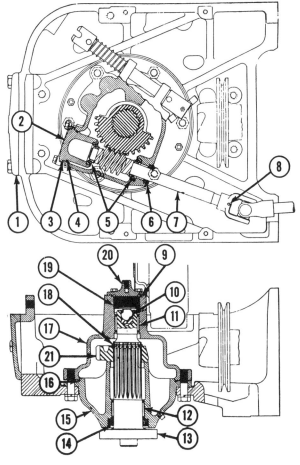

Fig. C507 — Components of the manual steering gear used on general purpose and high clearance tractors. Gear unit mesh and backlash is not adjustable.

1. Cover plate
2. Bearing carrier
3. Shims
4. "O" ring
5. Bearings
6. Oil seal
7. Worm shaft
8. Roll pin
9. "O" ring
10. Thrust ball
11. Upper needle bearing
12. Lower needle bearing
13. Upper steering spindle
14. Oil seal
15. Housing, lower half
16. "O" ring
17. Housing, upper half
18. Snap ring
19. Adjusting plug
20. Housing cover
21. Sector

the Torrington needle bearings. Arbor No. A8354 is used for installing the bearing in upper housing. Arbor No. A8352 is used for installing the bearing in lower housing.

Remove the wormshaft front bearing carrier (2) and withdraw the worm shaft from housing. The need and procedure for further disassembly is evident.

When reassembling, renew all "O" rings, gaskets and seals and when installing the sector gear, make certain that the single punch mark on steering spindle is located between the two punch marks on sector gear.

Adjust the gear unit as outlined in paragraphs 8 and 9.

Standard Non-Adjustable Axle Models

11. **ADJUSTMENT.** The steering gear unit is provided with three adjustments; one for the steering worm shaft end play, one for the sector shaft end play and the other for the front wheel turning angle. To make the adjustment, disconnect the drag link from the Pitman (drop) arm to relieve load from steering gear and slide the battery out and away from the gear housing.

12. **WORM SHAFT END PLAY.** To adjust the steering worm shaft bearings, remove the shaft front bearing carrier (1-Fig. C 508) and vary the

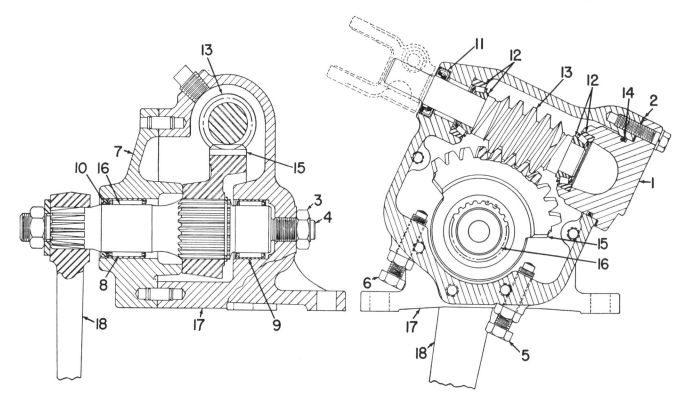

Fig. C508—Sectional views of manual steering gear used on standard non-adjustable axle models. Mesh position of the worm and sector gears is not adjustable.

1. Worm shaft bearing housing
2. Shims
3. Lock nut
4. Adjusting screw
5 & 6. Stop screws
7. Housing cover
8 & 9. Needle bearings
10. Oil seal
11. Oil seal
12. Worm shaft bearings
13. Worm shaft
14. "O" ring
15. Sector
16. Sector shaft
17. Gear housing
18. Pitman (drop) arm

number of shims (2) located between bearing carrier and housing to remove all shaft end play without causing any binding tendency when bearing carrier cap screws are tightened to a torque of 30 Ft.-Lbs. Shims are available in thicknesses of 0.003, 0.005 and 0.010.

13. SECTOR SHAFT END PLAY. To adjust the sector shaft end play, loosen locknut (3-Fig. C 508) and turn adjusting screw (4) in to remove all end play from the sector shaft (16) without causing any binding tendency.

14. WHEEL TURNING ANGLE. With the front wheels pointing straight ahead and drag link connected to the Pitman (drop) arm, turn the front wheels to the full right turn position and measure the turning angle (A-Fig. C 509); then, turn the wheels to the full left turn position and measure the turning angle (B). Right turn angle (A) must be the same as the left turn angle (B) and equal to 45 degrees for standard and industrial models; 40 degrees for orchard models. A protractor mounted as shown in Fig. C 510 will facilitate measuring the turning angle. Stop screw (5-Fig. C 508) controls the right turn angle and screw (6) controls the left.

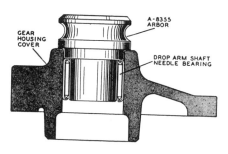

Fig. C511—Using case arbor A-8355 to install the sector shaft needle bearing in the gear housing cover on standard non-adjustable axle models.

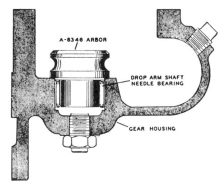

Fig. C512—Installing needle bearing in standard non-adjustable front axle steering gear housing using special Case arbor A-8346.

15. R&R AND OVERHAUL. To remove the steering gear assembly on orchard models it is first necessary to remove the interfering sheet metal; then, proceed as follows: On all models remove the batteries and tray unit. Remove the Pitman (drop) arm from sector shaft, disconnect the steering shaft universal joint, unbolt and remove the gear unit from tractor. Remove the gear housing cover (7—Fig. C 508), sector and sector shaft. Remove the worm shaft front bearing housing (1) and withdraw worm shaft forward and out of housing. The need and procedure for further disassembly is evident.

Using Case arbor A-8355 or equivalent, install needle bearing in gear housing cover, by pressing against end of bearing which has the stamped

identification, until locating shoulder on arbor seats against housing cover as shown in Fig. C511. Then install seal (10-Fig. C508) with lip facing inward until seal bottoms against the needle bearing.

Using Case arbor A-8346 or equivalent, install needle bearing in gear housing, by pressing against end of bearing which has the stamped identification, until locating shoulder on arbor seats against housing as shown in Fig. C512. Install oil seal (11—Fig. C508) with lip facing inside of gear housing.

Assemble the remaining parts by reversing the disassembly procedure, making certain that timing marks on sector shaft and gear are in register as shown in Fig. C513. Install gear unit on tractor and adjust same as outlined in paragraphs 11, 12, 13 and 14.

POWER STEERING SYSTEM

Note:—The maintenance of absolute cleanliness of all parts is of utmost importance in the operation and servicing of the hydraulic power steering system. Of equal importance is the avoidance of nicks and burrs on any of the working parts. On all models with power steering, the wormshaft end play is not adjustable; but, the gear unit sector spindle or shaft must be adjusted to zero end play. Refer to the last part of paragraph 25 and 26.

LUBRICATION AND BLEEDING
All Models

16. Fluid capacity of the power steering system is approximately 3¾ quarts for general purpose and high clearance models; 2 quarts for standard non-adjustable axle models. The J. I. Case Company specifies that only Automatic Transmission Fluid Type A should be used in the power steering system. Filter element in fluid reservoir should be renewed every 1,000 hours; or more often in severe dust conditions. Spring in reservoir cover must exert a downward seating pres-

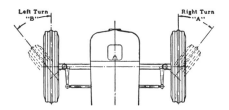

Fig. C509—Checking the turning angle of standard, industrial and orchard models. Angle (A) must be the same as angle (B).

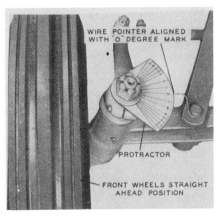

Fig. C510—A protractor can be installed on the spindles of standard, industrial and orchard models to check the front wheel turning angle.

Fig. C513—When assembling the steering gear on standard non-adjustable axle models, make certain that timing marks on sector and shaft are in register.

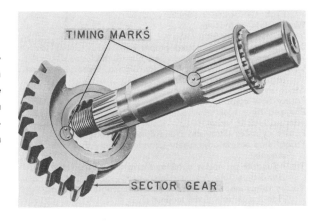

sure on filter cartridge to assure filtration of fluid. Note: A plugged filter element can be the cause of fluid bubbling out of the filter cap air vent.

After installing a new filter element, check the fluid level and if necessary, bleed the system as follows:

16A. BLEEDING, GENERAL PURPOSE AND HIGH CLEARANCE MODELS. To fill and bleed the power steering system, remove grille, filler cap from reservoir and filler plug and bleed plug from gear housing. Refer to Fig. C514. Fill the gear housing with Automatic Transmission Fluid, Type A, and install the gear housing filler plug. Screw the bleed plug in the gear housing one or two threads. Fill the system reservoir with Automatic Transmission Fluid, Type A, start engine and run at low idle speed. Turn front wheels **PARTLY** to the right and left three or four times to pump fluid into the torque motor and lines. Stop engine and refill the reservoir. Continue this procedure several times to pump a safe amount of

Fig. C514—Power steering components installed on general purpose and high clearance models. Notice the location of bleed plugs and filler plugs. The radiator is removed for illustrative purposes only.

oil into the torque motor before making full turns. Continue this process until the reservoir fluid level remains constant and fluid appears at the bleed plug in gear housing. Tighten the bleed plug.

16B. BLEEDING, STANDARD NON-ADJUSTABLE AXLE MODELS. To fill and bleed the power steering system, fill reservoir with Automatic Transmission Fluid, Type A, start engine and run at low idle speed. Disconnect the cylinder ram from the Pitman (drop) arm. Turn steering wheel to the right and left to pump fluid into the cylinder but stop turning the wheel before cylinder ram reaches the end of its stroke. Stop engine and refill reservoir. Continue this procedure several times to make certain there is oil in the cylinder before allowing the cylinder ram to make a full stroke. Continue this process, and allow the cylinder ram to make full extending and full retracting strokes until the reservoir fluid level remains constant. Reconnect the cylinder ram to the Pitman (drop) arm.

POWER STEERING SYSTEM TROUBLE SHOOTING CHART

	Loss of Power Assistance	Power Assistance in One Direction Only	Unequal Turning Radius	Erratic Steering Control	Fluid Foaming Out of Reservoir	Steering Cylinder Bottoming
Binding, worn or bent mechanical linkage	★		★	★		
Insufficient fluid in reservoir	★					
Faulty fan belt or pump drive belt	★			★		
Low pump pressure	★					
Internal leakage in torque motor (General Purpose and High Clearance models)	★					
Faulty or improperly installed control valve thrust bearing	★	★		★		
Valve thrust bearing nut improperly tightened	★			★		
Sticking or binding valve spool	★	★		★		
Faulty valve plungers and/or springs	★			★		
Improperly positioned steering shaft, restricting spool travel	★	★				
Check valve ball not seating	★				★	
Damaged or restricted hose or tubing	★	★			★	
Vane inlet and outlet connectors not seating on "O" rings, or "O" rings missing (General Purpose and High Clearance models)	★	★				
Pump to valve hose lines reversed	★					
Plugged balance passage at front of worm (General Purpose and High Clearance models)	★			★		
Wrong fluid in system	★			★	★	
Improperly adjusted tie rods			★	★		
Steering arms not positioned properly			★			
Timing marks on sector and shaft not aligned			★			
Vane and spindle timing marks not aligned (General Purpose and High Clearance models)			★			
Air in system				★	★	
Faulty flow control valve in pump				★		
Plugged filter element					★	
Internal leak in valve					★	
Faulty cylinder (Standard Non-Adjustable Axle models)	★			★		
Faulty stop screw adjustment (Standard Non-Adjustable Axle models)			★			★
Timing marks on sector shaft and arm not aligned (Standard Non-Adjustable Axle models)			★			
Faulty adjustment of cylinder retracted length (Standard Non-Adjustable Axle models)			★			★

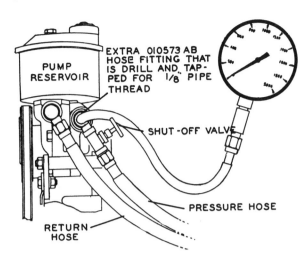

EXTRA 010573 AB HOSE FITTING THAT IS DRILL AND TAPPED FOR 1/8 PIPE THREAD

PUMP RESERVOIR

SHUT-OFF VALVE

PRESSURE HOSE

RETURN HOSE

Fig. C515—Pressure gage and shut-off valve installation for checking the power steering system operating pressure.

TROUBLE SHOOTING

All Models

17. The accompanying table lists troubles which may be encountered in the operation of the power steering system. The procedure for correcting most of the troubles is evident; for those not readily remedied, refer to the appropriate subsequent paragraphs.

SYSTEM OPERATING PRESSURE

All Models

18. A pressure test of the hydraulic circuit will disclose whether the pump or some other unit in the system is malfunctioning. To make such a test proceed as follows:

Connect a pressure test gage and a shut-off valve in a manner similar to that shown in Fig. C515. Note that shut-off valve is connected in the circuit between the gage and the control valve on the worm shaft. Open the shut-off valve and run the engine at idle speed until oil is warmed. Rotate the front wheels of the tractor to

extreme right or extreme left turn position and note gage reading which should be in the range of 725-825 psi on early production 400 series general purpose and high clearance tractors with an 800 psi relief valve; 800-900 psi on late production 400 series and all 700B and 800B general purpose and high clearance tractors with an 1100 psi relief valve; 1025-1125 psi on 400, 700B and 800B standard non-adjustable axle models with an 1100 psi relief valve. Note: To avoid heating the steering fluid excessively, don't hold wheels in this test position

Fig. C517—Exploded view of the combination flow control and pressure relief valve. Both 800 and 1100 psi relief valves are used.

for more than 30 seconds. Note also that the pressure reading for general purpose and high clearance models equipped with an 1100 psi relief valve is lower than might be expected. This is due to a normal pressure drop in the torque motor.

If the pressure is higher than specified, the relief valve is probably stuck in the closed position. If the pressure is less than specified, turn the wheels to straight ahead position; at which time, the gage reading should drop considerably. Now slowly close the shut-off valve and retain in closed position only long enough to observe the gage reading. Pump may be seriously damaged if valve is left in closed position for more than 10 seconds. If the gage reading increases to 725-825 psi on all models with an 800 psi relief valve or 1025-1125 psi on all models with an 1100 psi relief valve, with valve closed, the pump is O.K. and the trouble is located in the control valve, the power cylinder (or torque motor) or the connections.

While a low pressure reading with the shut-off valve closed indicates the need of pump repair, it does not necessarily mean the remainder of the system is in good condition. After overhauling the pump, recheck the pressure reading to make sure the control valve, cylinder (or torque motor)

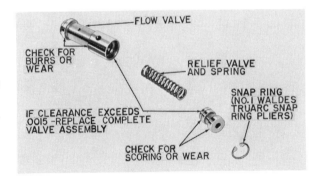

FLOW VALVE

CHECK FOR BURRS OR WEAR

RELIEF VALVE AND SPRING

SNAP RING (NO. I WALDES TRUARC SNAP RING PLIERS)

IF CLEARANCE EXCEEDS .0015 -REPLACE COMPLETE VALVE ASSEMBLY

CHECK FOR SCORING OR WEAR

and connections are in satisfactory operating condition.

Refer to paragraph 19 or 20A for data concerning the flow control and relief valve.

PUMP AND FLOW CONTROL VALVE

All Series 400-Early Series 700B-800B

19. **FLOW CONTROL AND RELIEF VALVE.** This combination valve can be removed from the pump by removing the outlet hose fitting bolt and valve cap fitting (Fig. C516). The spring and valve can then be lifted out without removing the pump from the engine. Valve can be disassembled by removing the snap ring (Fig.—C517) from flow valve. Remove any slight

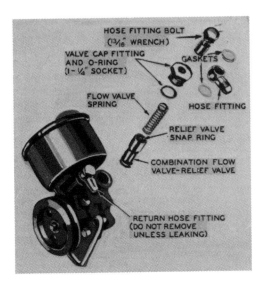

HOSE FITTING BOLT (13/16" WRENCH)

VALVE CAP FITTING AND O-RING (1-1/4" SOCKET)

GASKETS

FLOW VALVE SPRING

HOSE FITTING

RELIEF VALVE SNAP RING

COMBINATION FLOW VALVE-RELIEF VALVE

RETURN HOSE FITTING (DO NOT REMOVE UNLESS LEAKING)

Fig. C516—Details of combination flow control and pressure relief valve mounted in cover of power steering hydraulic pump. Flow valve limits output to a fixed quantity regardless of pump speed.

roughness or scratching from outside surface of flow valve and from its bore in pump cover, using crocus cloth. Check clearance of relief valve in flow control valve bore and if greater than 0.0015, install a new valve assembly as only the spring is available separately.

On all models with an 800 psi relief valve, the relief valve spring has a free length of approximately 1.66 inches and should require 25 - 27 pounds to compress it to a height of 1.18 inches. On all models with an 1100 psi relief valve, the relief valve spring has a free length of approximately 1.61 inches and should require 21-23 pounds to compress it to a height of 1.18 inches. The flow control valve spring has a free length of 2.23 inches and should require 16-18 pounds to compress it to a height of 1.12 inches.

When reassembling, tighten the valve cap fitting to a torque of 35 Ft.-Lbs., and the hose fitting bolt just enough to eliminate fluid leakage.

Note:—There is no external visual means of determining whether the system is equipped with an 800 or 1100 psi relief valve. The only method would be to test the pressure or remove the valve and test the spring. The 1100 psi valve is used to produce increased power assistance so that heavy, front mount-

Fig. C520—If a 0.009 feeler blade can be inserted as shown, install a new rotor set.

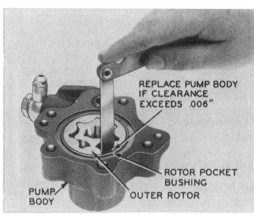

Fig. C521—If a 0.007 feeler blade can be inserted between driven rotor and pump body insert, install a new pump body and rotors assembly.

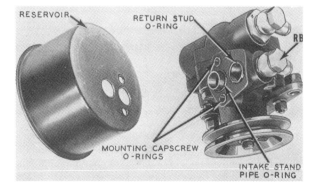

Fig. C518—First step in disassembly of power steering pump is to remove the reservoir. Fitting (RB) need not be removed unless it is leaking.

ed equipment will not cause the relief valve to open prematurely. The 1100 psi valve assembly (Case Part No. A-10121) can be installed on models which originally had the 800 psi valve. When making this change, the flow valve spring can be reused providing it meets the previously mentioned specifications.

20. R&R AND OVERHAUL PUMP. Procedure for removal and installation of the pump is self-evident. Order of disassembly is as follows: Remove reservoir cover, filter element and reservoir body (Fig.—C518). Do not remove the return hose fitting bolt (RB) unless same is leaking.

Remove pulley, separate the pump cover from the pump housing and lift the gasket out of the groove in the housing. Mark outer face of each rotor with chalk so that rotors will be reinstalled same way. Remove the outer (driven) rotor from the housing. Refer to Fig. C519.

To remove rotor shaft, extract the bearing snap ring (retainer) from pulley end of housing then press or bump shaft and bearing out of housing. Remove flow control and relief valve assembly from pump cover. Do not wash the sealed ball bearing.

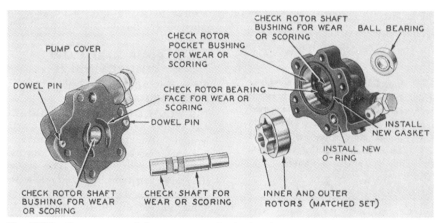

Fig. C519—Components of the power steering hydraulic pump. None of the bushings are available separately.

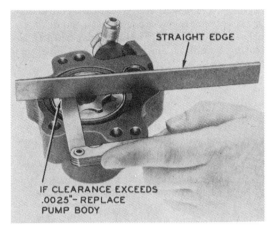

Fig. C522—If a 0.0035 feeler blade can be inserted between faces of rotors and undersize of straightedge, install a new pump body and rotors assembly.

Fig. C522C — Install the carrier on the pump shaft so that the carrier will rotate in the direction shown.

Check pump body and cover faces for wear; also check the bushings in the body and cover. If pump body or bushing in body is worn, install a new body and bushing assembly. If the cover or bushing in cover is worn install a new cover and bushing assembly. None of the bushings are available separately.

Partially reassemble the pump and check clearance between rotors at teeth, with feeler gages as shown in Fig. C520 and between outer rotor and body insert as shown in Fig. C521. If clearance between rotors exceeds 0.008; or if clearance between outer rotor and insert in pump body exceeds 0.006, install a new pump body and rotors assembly. Also check end clearance between end of rotors and body face as shown in Fig. C522. If end clearance exceeds 0.0025, install a new body and rotors assembly.

Always install new "O" rings and seals. Rotor shaft oil seal should be assembled with lip of same nearest the rotors.

Late Series 700B-800B

The belt driven power steering pump used on late production series 700B and 800B tractors is of the roller vane type and is built by Eaton Mfg. Co. Contained in the cover of the pump is a spring loaded flow control valve combined with a pressure relief valve (1, 2 & 3—Fig. C522A). The flow control valve limits the output of the pump to approximately 3.0 gallons per minute regardless of the pump speed.

Refer to paragraphs 19 and 20 for the early type flow control valve and power steering pump.

20A. **FLOW CONTROL AND RELIEF VALVE.** This combination valve can be removed from the pump by removing the valve cap (1—Fig. C522A). The spring (2) and valve (3) can then be lifted out without removing the pump from the engine. The valve is sold only as a complete unit. Remove any slight roughness or scratching from outside surface of flow valve and from its bore in pump cover, using crocus cloth.

20B. **PUMP OVERHAUL.** Procedure for removal and installation of the pump is self-evident. Order of

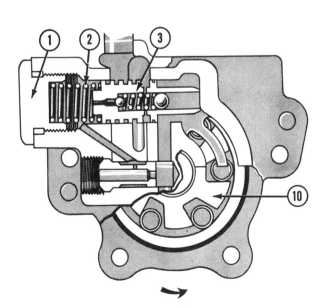

Fig. C522A — Sectional drawing of a typical roller type power steering pump. Valve cap is shown at (1), spring at (2), flow control and pressure relief valve at (3) and carrier at (10).

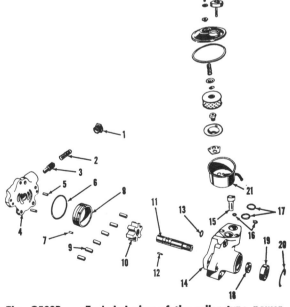

Fig. C522B — Exploded view of the roller type power steering pump used on late production series 700B and 800B tractors.

1. Valve cap	7. Cam positioning pin
2. Spring	8. Cam ring
3. Flow control and pressure relief valve	9. Rollers
	10. Carrier
4. Pump cover	11. Pump shaft
5. Dowel pins	12. Carrier drive pin
6. "O" ring	13. "O" ring
14. Pump body	
15. Venturi	
16. Gaskets	
17. Gaskets	
18. Oil seal	
19. Bearing	
20. Snap ring	
21. Reservoir	

disassembly is as follows: Remove reservoir cover, filter element and reservoir body (Fig. C522B). Remove pulley, separate the pump cover from the pump body (14) and lift the gasket (6) out of the groove in the body. Withdraw the rollers (9), carrier (10) and cam ring (8) from the pump body.

To remove rotor shaft, extract the bearing snap ring (20) from pulley end of housing, then press or bump shaft and bearing out of housing. Remove flow control and relief valve assembly from pump cover. Do not wash the sealed ball bearing (19).

Check pump body and cover faces for wear; also check the bushings in the body and cover. If pump body or bushing in body is worn, install a new body and bushing assembly as neither the bushing nor the body is available separately. If the cover or bushing in cover is worn, install a new cover and bushing assembly. None of the bushings are available separately.

Partially reassemble the pump and check clearance between carrier and

face of body and rollers and face of body as shown in Fig. C522D. If clearance between carrier and face of body or between rollers and face of body exceeds 0.0015, install a new pump body and/or carrier, rollers and cam ring.

Always install new "O" rings. Pump shaft oil seal should be installed with lip of same facing toward the pump carrier and rollers. Carrier (10—Fig. C522B) should be installed on the shaft as shown in Fig. C522C.

STEERING VALVES
All Models

21. R&R AND OVERHAUL. To remove the power steering valves, proceed as follows:

21A. On general purpose and high clearance models, remove cover from front of pedestal, remove grille and drain fluid from the torque motor. Disconnect lines from the valve body and remove the valve cap as shown in Fig. C523. Unlock and remove the thrust bearing nut from front of

wormshaft. Being careful not to lose or damage any of the valve parts, slide the valve and thrust bearing assembly from the wormshaft as shown in Fig. C524.

21B. To remove the power steering valves on orchard models, it is necessary to first remove the interfering sheet metal; then, proceed as follows: On all models with a standard non-adjustable axle, remove the batteries and tray unit and Pitman (drop) arm from sector shaft. Disconnect the lines from the valve body and disconnect the steering shaft universal joint. Unbolt and remove the gear housing assembly from tractor. Remove the valve cap, unlock and remove the thrust bearing nut from front of worm shaft. Being careful not to lose or damage any of the valve parts, slide the valve and thrust bearing assembly from the worm shaft as shown in Fig. C524.

22. With the valve assembly removed, disassemble the parts as shown in Fig. C525. Thoroughly clean and

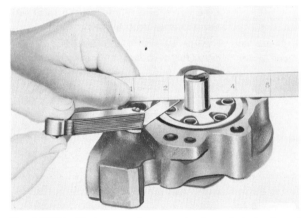

Fig. C522D—If clearance between the face of the body and the carrier and rollers exceeds 0.0015, renew the carrier, rollers and cam ring assembly and/or the pump body.

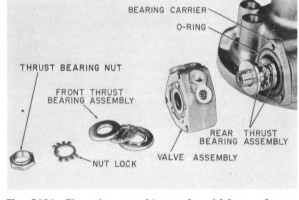

Fig. C524—The valve assembly can be withdrawn after removing the thrust bearing nut.

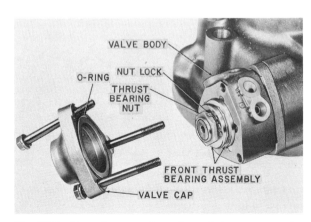

Fig. C523—Removing cover from front of the power steering valve body.

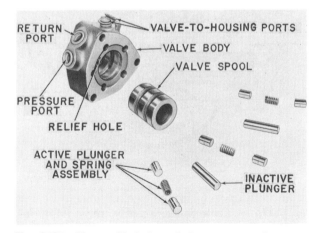

Fig. C525—Disassembled view of the power steering valve body.

examine all parts for being damaged or worn. The valve spool and body are mated parts and must be renewed as an assembly. The plungers and springs are available separately. The check valve in the return port of valve body can be removed with a screw driver and must be renewed as an assembly.

Inspect the brass tube seats in valve body. If they are damaged, tap the seats for a $\frac{5}{16}$-16 NF thread and remove the seats using a screw and nut puller arrangement. Press new seats into position, using the valve to housing tube as shown in Fig. C526.

When reassembling, be sure to lubricate all parts in Automatic Transmission Fluid, Type A. Install the valve spool so that the identification groove is located toward same side of body as the port identification symbols "PR" and "RT" are located. There is no particular order for installing the plungers and springs, but the plungers must have free movement and there must be one active plunger between the two inactive plungers as shown in Fig. C527.

23. To install the valve assembly, proceed as follows: Install the rear

Fig. C526—Using one of the tubes to press the brass tube seats in the valve body.

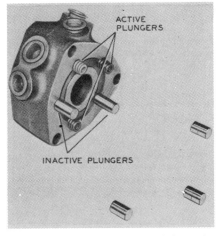

Fig. C527—There must be one of the active plunger sets between the two inactive plungers.

thrust bearing assembly on the wormshaft, then install the body with the port return and the pressure identification symbols toward front. Install the front thrust bearing assembly with the larger diameter washer toward the valve. Install the adjusting plate (Case Part No. A-8342) with the three original ⅜ x3 ¾ inch cap screws and washers and tighten the screws to a torque of 30 Ft.-Lbs. Refer to Fig. C528A. Install the nut lock washer and thrust bearing nut and tighten the nut to a torque of 20 Ft.-Lbs. Remove the A-8342 adjusting plate, install the valve cap and tighten the cap screws to a torque of 30 Ft.-Lbs.

23A. On standard non-adjustable axle models, install the complete gear unit on tractor and make the following adjustment:

Loosen the steering shaft universal joint clamp bolt and pull the steering wheel and shaft rearward as far as possible; then, allow the steering shaft to move forward until there is a clearance (x-Fig. C528B) between the universal joint and gear housing of $\frac{3}{32}$-inch for all standard non-adjustable axle models except special production tractors (402, 412, 702B, 712B, 802B and 812B). Clearance at (x—Fig. C528B) should be ⅛-¼-inch for special production models (model numbers 402, 412, 702B, 712B, 802B and 812B).

GEAR HOUSING, WORMSHAFT AND TORQUE MOTOR
General Purpose and High Clearance Models

24. **WORMSHAFT-R&R AND OVERHAUL.** To remove the steering wormshaft, first remove the valve unit as in paragraph 21A. Remove the radiator side panel and drive out the roll pin retaining the steering shaft front universal joint to the wormshaft. Re-

Fig. C528A — Tightening the power steering valve thrust bearing nut while using Case No. A-8342 adjusting plate to hold the valve in position.

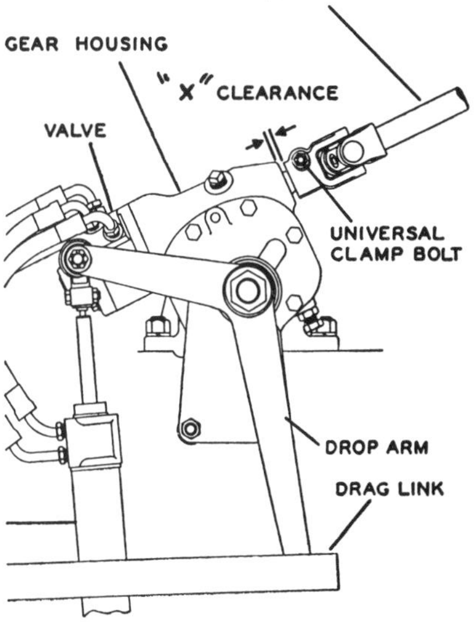

Fig. C528B—Checking the clearance between the steering shaft universal joint and gear housing on power steering equipped standard non-adjustable axle models.

move the wormshaft front bearing carrier and turn the wormshaft out of the housing. Refer to Fig. C529. The rear dust seal and quad-ring seal can be renewed at this time without removing the seal carrier. Note: Early series 400 general purpose models were fitted with a single lip type seal at the rear of the wormshaft. The late style seal carrier and seals can be installed on early models without modifying the housing. A good sealing compound should be applied to O. D. of carrier before installation. If the wormshaft rear needle bearing requires renewal, it will be necessary to remove the seal carrier. Install the rear bearing from rear of housing, using Case arbor No. A-9817 or equivalent, until shoulder on arbor seats against the seal carrier seating surface in housing. Press against end of bearing which has the stamped identification numbers. In-

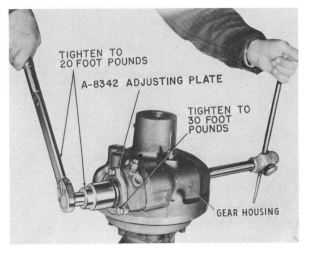

stall the front needle bearing in the bearing carrier, using Case arbor No. A-9818 or equivalent. Press the bearing in until shoulder on arbor seats against rear surface of carrier and be sure to press against end of bearing which has the stamped identification numbers.

Install the wormshaft, being careful not to damage the rear seal lip. Install the front bearing carrier and make certain the bleed hole in carrier is open and clean. Note: The similar carrier used on standard non-adjustable axle models does not have this bleed hole.

Install the steering valve unit as outlined in paragraph 23.

25. TORQUE MOTOR - R&R AND OVERHAUL. To remove the gear housing and torque motor assembly, first drain the power steering fluid and proceed as follows: Drain tractor cooling system and remove the hood, radiator screen, grille and radiator. Raise front of tractor and unbolt lower spindle on dual wheel tricycle models, wheel fork on single wheel tricycle models or center steering arm on adjustable axle models from the upper steering spindle. Disconnect the power steering hoses and the steering shaft front universal joint. Loosen the bolts retaining pedestal to engine and slide pedestal forward approximately ¾-inch. Unbolt and lift torque motor assembly from tractor.

Refer to Fig. C530. Back-off the vane inlet and outlet connectors two or three turns and remove the gear housing top cover, retainer plug, thrust ball and cover. Remove the sector retaining snap ring and lift off the sector. Remove the ten Allen head cap screws and lift off the vane upper cover. Use chalk or crayon and mark the two stationary vanes so they can be installed in their original position and with the same end up. Remove both stationary vanes, dowel pins and vane seals as shown in Fig. C531. Lift the movable vane, vane body and vane lower cover from spindle. Note: If either the movable vane or vane body are ordered as a replacement part, the following assembly will be furnished:

1. All seals and "O" rings.

2. Vane upper cover.

3. Vane body, stationary vanes and seals.

4. Movable vane.

5. Vane lower cover, complete with needle bearing and seal.

6. Cover bolts.

The combination oil and dust seal in the vane lower cover can be renewed without disturbing the lower needle bearing. Install the seal with lip facing the bearing. If bearing is damaged, press the bearing and seal out together. Install the lower bearing from bottom of cover, using Case arbor No. A-8352 or equivalent until shoulder on arbor seats against the oil seal seating surface in the cover. Press against end of bearing which has the stamped identification numbers.

The needle bearing in the gear housing should be installed from inside of housing, using Case arbor No.

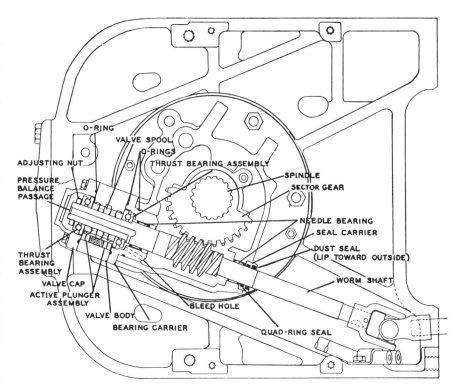

Fig. C529—Top sectional view of the power steering unit used on general purpose and high clearance models. Early series 400 general purpose and high clearance models were fitted with a single lip type seal at rear of worm shaft.

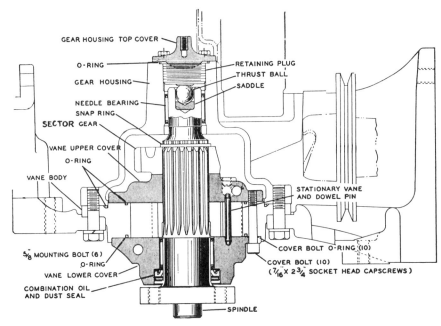

Fig. C530—Side sectional view of the power steering unit used on general purpose and high clearance models. The steering spindle should have zero end play without binding.

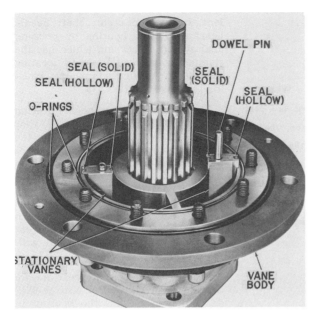

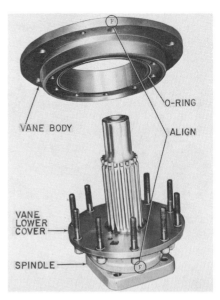

Fig. C531—Removing the stationary vanes, seals and dowel pins. Early series 400 general purpose models were not fitted with the hollow vane seal.

Fig. C533—When installing the vane body, make certain the "F" marks in the castings are aligned.

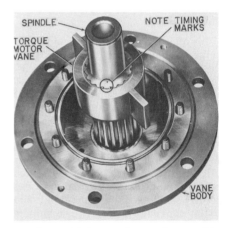

A-8354 or equivalent until shoulder on arbor seats against the inside of housing. Press against end of bearing which has the stamped identification number.

To reassemble the torque motor, proceed as follows: Using care not to damage the oil seal, position the vane lower cover on the spindle as shown in Fig. C532. Install the ten Allen head cap screws, then slip the "O" rings (well lubricated) over the cap screws and into the "O" ring grooves in cover. The "O" rings should hold the cap screws in position as shown. Install a new (well lubricated) "O" ring in underside of vane body and install the vane body on the lower cover so that the cast-in "F" marks

are aligned as shown in Fig. C533. The stationary vane grooves in body and vane dowel pin holes in lower cover can also be used as positioning guides. Install the movable vane so that timing marks on vane and spindle are in register as shown in Fig. C534. Install the stationary vanes, dowel pins and new, well lubricated seals in their original position with the same end up as in Fig. C532A. Installation will be simplified if the seals and vanes are installed together. It is important that the seals cover the entire length of the vanes and bottom against the lower cover. Any excess seal material should be trimmed off flush with top of vane. Note: Early production vanes were not fitted with

Fig. C534—Timing marks on the movable vane and spindle must be aligned during reassembly.

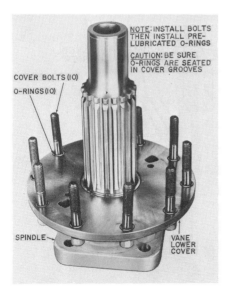

Fig. C532A—Installing the stationary vanes, seals and dowel pins. Early series 400 general purpose models were not equipped with the hollow vane seal.

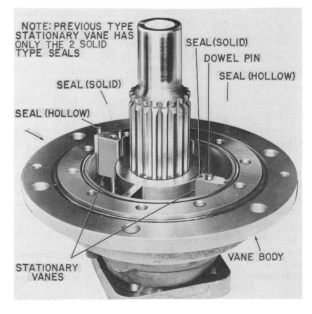

Fig. C532—Vane lower cover, Allen head cap screws and "O" rings installed on the steering spindle. The "O" rings must be seated in the grooves in the cover.

Fig. C532A—Installing the stationary vanes, seals and dowel pins. Early series 400 general purpose models were not equipped with the hollow vane seal.

15

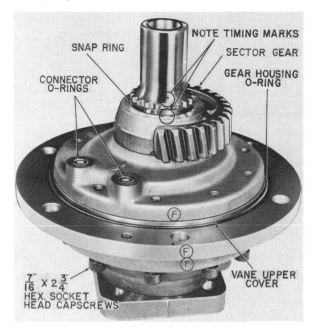

SNAP RING
CONNECTOR O-RINGS
NOTE TIMING MARKS
SECTOR GEAR
GEAR HOUSING O-RING
$\frac{7}{16}$ X $2\frac{3}{4}$ HEX. SOCKET HEAD CAPSCREWS
VANE UPPER COVER

Fig. C536 — When the vane housings are properly assembled, the "F" marks in the castings must be in register. Sector and spindle timing marks must also be in register.

Using Case arbor A-8346 or equivalent, install the sector shaft needle bearing in gear housing by pressing against end of bearing which has the stamped identification, until locating shoulder or arbor seats against housing as shown in Fig. C538.

The wormshaft front needle bearing should be installed with Case Arbor No. A-9818 or equivalent. Press the bearing in from inside end of carrier by pressing on end of bearing which has the stamped identification until shoulder on arbor seats against the carrier. Then install the two oil seals back to back in the carrier as shown in Fig. C539. These two seals prevent mixing of the SAE 90 gear oil with the power steering fluid.

Install the wormshaft rear needle bearing with Case arbor No. A-9817 or equivalent. Press the bearing in from outside of housing by pressing on end of bearing which has the stamped identification until shoulder on arbor seats against the oil seal counterbore in the housing. Install the wormshaft rear seal with lip facing inward. Refer to Fig. C540. Assemble the gear unit, making certain that the punched timing marks on sector and sector shaft are in register. Tighten the housing side cover cap screws to a torque of 30 Ft.-Lbs.; then, tighten the sector shaft end play adjusting screw (Fig. C541) to remove all end play without causing binding.

Install the steering valve unit as in paragraph 23, bleed the system and check and adjust if necessary the front wheel turning angle as in paragraph 14 and the steering shaft end clearance as in paragraph 23A.

the hollow seals. If there is any free play of these early vanes in the vane body, install the later type vanes and hollow seals.

Install new, well lubricated "O" rings on vane body, then install the vane cover so that the cast-in "F" markings are aligned as in Fig. C536 and tighten the ten Allen head cap screws securely. Align the timing marks as shown and install the sector and snap ring. Install new connector "O" rings as shown, then install the gear housing so that the connectors align with the "O" rings. Tighten both of the housing retaining cap screws. Tighten the vane inlet and outlet connectors enough to eliminate leakage, but do not over-tighten. Refer to Fig. C530. Install the thrust ball with flat side up. Install and tighten the retainer plug just enough to eliminate spindle end play; then install the top cover so that lug on cover engages slot in the retainer plug.

Install the power steering valve as outlined in paragraph 23, then install the complete unit on the tractor.

GEAR HOUSING
Standard Non-Adjustable Axle Models

26. **R&R AND OVERHAUL.** To overhaul the unit, first remove the gear unit and valves as in paragraphs 21B and proceed as follows: Remove the gear housing side cover and withdraw the sector and sector shaft. Bump the wormshaft front bearing carrier from the housing and remove the worm. The need and procedure for further disassembly is evident.

Using Case arbor No. A-8355 or equivalent, install needle bearing in the gear housing side cover, by pressing against end of bearing which has the stamped identification, until locating shoulder on arbor seats against housing cover as shown in Fig. C537. Then press the sector shaft seal in the housing side cover with lip facing inward until the seal bottoms against the needle bearing.

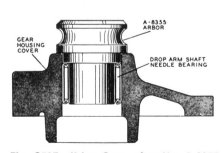

GEAR HOUSING COVER
A-8355 ARBOR
DROP ARM SHAFT NEEDLE BEARING

Fig. C537—Using Case arbor No. A-8355 to install the sector shaft needle bearing in the gear housing cover on standard non-adjustable axle models.

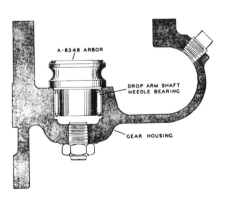

A-8346 ARBOR
DROP ARM SHAFT NEEDLE BEARING
GEAR HOUSING

Fig. C538—Installing needle bearing in the steering gear housing of a standard non-adjustable axle tractor, using special Case arbor A-8346.

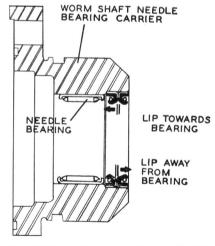

WORM SHAFT NEEDLE BEARING CARRIER
NEEDLE BEARING
LIP TOWARDS BEARING
LIP AWAY FROM BEARING

Fig. C539 — On standard non-adjustable axle models, the wormshaft front bearing housing seals should be installed back to back as shown.

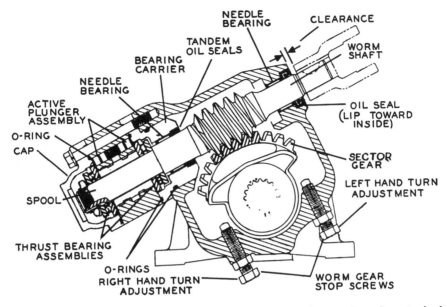

Fig.C540—Side sectional view of the power steering valve and gear unit used on standard non-adjustable axle models.

STEERING CYLINDER
Standard Non-Adjustable
Axle Models

27. **ADJUSTMENT.** With the cylinder disconnected from the drop arm and in the full retracted position, the center to center distance between the two ball studs should be 15.438 inches as shown in Fig. C542. If the retracted length is not as specified, loosen the clamp and make the necessary adjustment.

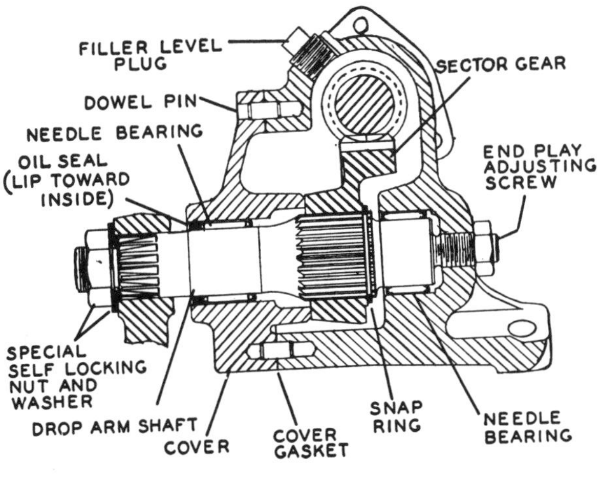

Fig. C541—End sectional view of the power steering gear unit used on standard non-adjustable axle models.

28. **OVERHAUL.** The only overhaul work which can be accomplished on the cylinder is to renew the seal and port seats. To renew the seal, remove the retaining snap ring, bronze washer, dust seal, plastic backing washer and "O" ring. See Fig. C542. Install a new, well lubricated "O" ring, then install a new plastic washer. Lubricate a new dust seal and install same with lip facing away from cylinder. Then install the bronze washer and snap ring.

To renew a damaged port seat, tap same with $\frac{5}{16}$-24 NF thread tap and pull the seat with a screw and nut type puller. A new seat can be pressed into position using the regular tube and fitting in a manner similar to that shown in Fig. C526.

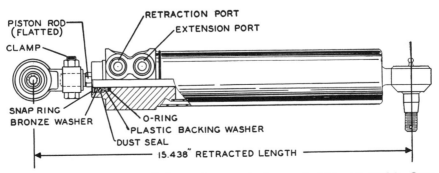

Fig. C542—Power steering cylinder used on standard non-adjustable axle models. Overhauling the unit consists of renewing the seals and port seats.

ENGINE AND COMPONENTS

R & R ENGINE

50. To remove the engine from series 400, 700B and 800B tractors, first drain cooling system and on series 800B the torque convertor compartment. On all models if the engine is to be disassembled, drain oil pan. Remove the hood and the radiator. On standard, non-adjustable front axle models, disconnect the drag link and rear end of radius rod. On tricycle and adjustable front axle models, disconnect the steering worm shaft front universal joint and on models with power steering, disconnect the pressure and return lines. On series 800B tractors, disconnect the oil cooler tubes from the oil manifold and control valve; then, remove or loosen the tube clamps and guards and lay the tubes out of the way. On all models, block-up tractor under clutch or torque convertor housing, unbolt front support (radiator bracket) from engine and roll the front support and wheels assembly away from tractor.

Disconnect the heat indicator sending unit from water manifold, ammeter wire from generator, battery cables and starter solenoid wires. Remove conduit that shields the wires along left side of engine block and lay the wires out of the way. Disconnect oil line from the hour meter pressure switch or engine block. On tractors equipped with a mechanical tachometer, disconnect the drive cable from the engine. On tricycle and adjustable axle models, loosen the clamp bolt on the steering shaft universal joint and unbolt the steering shaft support from clutch housing.

On Diesel models, shut-off fuel and disconnect fuel lines at top of tank and shut-off valve. Decompress the engine and disconnect the decompression linkage at lower right front end of fuel tank. Remove the rear nut on throttle rod. Disconnect the fuel pres-

sure gage line. On models so equipped, disconnect the fuel vacuum gage line at bleed valve connection above the first stage fuel filter. Disconnect the ether starting aid tube from intake manifold.

On non-Diesel models, shut off fuel and disconnect choke wire and fuel line at carburetor. Disconnect the throttle control rod at a point just under front end of fuel tank. Disconnect wire from coil, or on magneto equipped engines, d i s c o n n e c t the grounding wire.

On all models, swing engine in a hoist, unbolt engine from clutch or torque convertor housing and separate the tractor halves.

When connecting engine to clutch or torque convertor housing, tighten the flange bolts to a torque of 80 ft.-lbs.; flange cap screws and dust shield cap screws to a torque of 50 ft.-lbs. Be sure to bleed the power steering and Diesel fuel systems. On series 800B, refer to paragraph 156B when refilling torque convertor housing.

CYLINDER HEAD
Diesel Models

51. To remove any of the cylinder heads, first drain cooling system and remove hood. On some tractors it is necessary to remove the complete air cleaner assembly; however, in most cases only the air cleaner hose needs to be removed. Disconnect the heat indicator sending unit from water manifold and the ether starting aid tube from the air intake manifold. Remove the air intake, exhaust and water manifolds from cylinder heads. Disconnect the high pressure and return lines from injection nozzles and immediately cap the fuel line connections to eliminate the entrance of dirt. In some cases (depending upon which cylinder head is to be removed and the type of filtering system used) it

may be necessary to remove some fuel lines and/or a fuel filter assembly. Loosen the decompression shaft coupling (Fig. C550) between the cylinder heads and slide the coupling out of engagement. Remove the rocker arm covers, disconnect the decompression links and remove the rocker arms assemblies and push rods. Disconnect the fuel return lines from the fuel tank, remove the head retaining stud nuts and lift cylinder head from engine.

Install head gasket with copper side up and always use new "O" rings (Fig. C550). Place the cylinder heads in position, but do not tighten the stud nuts until after the air intake, exhaust and water manifolds have been permanently installed. Renew "O" rings on rocker arm oil tubes if rings are not in good condition.

Tightening torques are as follows:
Air intake and exhaust
  manifold stud nuts........25 ft.-lbs.
Water manifold cap screws..15 ft.-lbs.
Cylinder head stud nuts....100 ft.-lbs.
Rocker arm nuts and screws.40 ft.-lbs.

Cylinder head nut tightening sequence is shown in Fig. C551. Run engine at normal operating temperature for at least forty-five minutes and retorque the stud nuts, using an off-set box end wrench.

Non-Diesel Models

52. To remove any of the cylinder heads, first drain cooling system and remove hood. On some tractors, it may be necessary to remove the complete air cleaner assembly; however, in most cases only the air cleaner hose needs to be removed. Disconnect the heat indicator sending unit from water manifold. Disconnect throttle rod, choke wire and fuel line from carburetor. Remove the intake, exhaust and water manifolds from cylinder heads. Disconnect spark plug

INSTALL NEW O-RINGS

Fig. C550—Installing the Diesel engine cylinder head. Head gasket must be installed with the copper side up. Be sure the "O" rings are in good condition. The installation procedure for non-Diesel engines is similar.

TIGHTEN EVENLY TO 100 FOOT POUNDS

Fig. C551 — Cylinder head nut tightening sequence.

wires and remove rocker arm cover. Remove the rocker arms assembly and push rods. Remove the head retaining stud nuts and lift cylinder head from engine.

Install head gaskets with copper side up and always use new "O" rings. The "O" rings are in the same position as those on Diesel engines shown in Fig. C550. Place the cylinder heads in position, but do not tighten the stud nuts until after the intake, exhaust and water manifolds have been permanently installed. Renew "O" ring on rocker arm oil tubes if rings are not in good condition.

Tightening torques are as follows:
Intake and exhaust
manifold stud nuts........25 ft.-lbs.
Water manifold cap screws..15 ft.-lbs.
Cylinder head stud nuts....100 ft.-lbs.
Rocker arm nuts and screws .40 ft.-lbs.

Cylinder head nut tightening sequence is shown in Fig. C551. Run engine at normal operating temperature for at least forty-five minutes and retorque the stud nuts, using an off-set box end wrench.

VALVES AND SEATS

53. Intake and exhaust valves are not interchangeable and the intake valves seat directly in the cylinder head with a face and seat angle of 45 degrees and a seat width of $\frac{3}{32}$ inch for Diesels, 5/64 inch for non-Diesels. Exhaust valves seat on renewable seat inserts with a face and seat angle of 45 degrees and a seat width of 5/64 inch for Diesels, $\frac{3}{32}$ inch for non-Diesels. Seats can be narrowed, using 30 and 60 degree stones. The only recommended method for removing the exhaust valve seat inserts is by the use of a puller. Chill inserts in dry ice to facilitate installation with a suitable driver. Valve stem diameter is 0.400-0.401 for the exhaust, 0.402-0.403 for the intake.

On Diesel models, adjust both the

intake and exhaust tappet gap to 0.012 cold. On non-Diesels, adjust the intake tappet gap to 0.012 cold and the exhaust tappet gap to 0.020 cold.

VALVE GUIDES AND SPRINGS

54. The pre-sized intake and exhaust valve guides are interchangeable in non-Diesel engines, but are not interchangeable in Diesels. Press old guides out from bottom of head so they emerge at top of head.

Note: A stepped mandrel (0.4045 small diameter, 0.750 large diameter) is suitable for removing and reinstalling the valve guides.

Press new valve guides into top of cylinder head until top of guide protrudes $1\frac{1}{16}$ inches above top of head as shown in Fig. C553 or C554. Valve guides should not require reaming if properly installed.

Inside diameter of the installed valve guides should be 0.4045-0.4055. This provides a normal stem to guide clearance 0.0035-0.0055 for the exhaust, 0.0015-0.0035 for the intake.

55. Intake and exhaust valve springs are interchangeable. Renew any spring which does not meet the following test specifications:
Non-Diesels—Series 400, 700B & 800B
Pounds pressure
  @ 1 35/64 inches..........64-72
Pounds pressure
  @ 1 15/16 inches..........35-39
Free length, approx........2 41/64
Diesels—Series 400 Prior to 8104711
Pounds pressure
  @ 1½ inches...............66-76
Pounds pressure
  @ $1\frac{15}{16}$ inches...............35-39
Free length, approx........2 41/64
Diesels—Series 400 After 8104710,
  700B & 800B
Pounds pressure
  @ 1½ inches...............89-99
Pounds pressure
  @ $1\frac{15}{16}$ inches...............44-49
Free length approx.........2 19/32

VALVE ROTATORS
Non-Diesel Models

56. Normal servicing of the positive type exhaust valve rotators ("Rotocaps") consists of renewing the units. It is important, however, to observe the valve action after engine is started. The valve rotator action can be considered satisfactory if the valve rotates a slight amount each time the valve opens. A cut-away view of a typical "Rotocap" installation is shown in Fig. C555.

ROCKER ARMS

57. To remove the rocker arms from either cylinder head, first remove hood and rocker arm cover. Disconnect the decompression link on Diesel models; and on all models, unbolt and remove the rocker arm assembly from cylinder head.

Disassemble the unit, thoroughly clean all parts and check the following:
Rocker arm shaft
  diameter 0.872-0.873
Inside diameter of
  rocker arm
  bushing (installed) ..0.8745-0.8755

If valve stem contact end of rocker arms is worn, it can be refaced providing the original contour is maintained. Renew the oil wick in each exhaust valve rocker arm if wick is hard, carboned or damaged. Lower end of push rods can be resurfaced if they are rough. On Diesel models, examine trip pins in the decompressor shaft and renew them if they are bent or show wear.

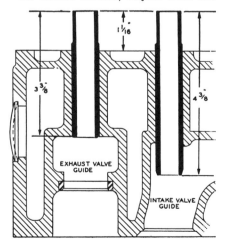

Fig. C553—Valve guide installation dimensions for Diesel engines. Notice that guides for the intake valves are longer than the exhaust.

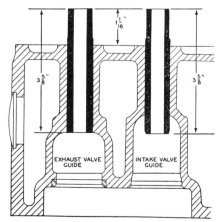

Fig. C554—Valve guide installation dimensions for non-Diesel engines. Intake and exhaust valve guides are interchangeable.

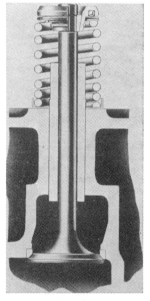

Fig. C555—Typical "Rotocap" installation on non-Diesel engines. Exhaust valves should rotate a slight amount each time the valves open.

New rocker arms are furnished complete with bushing. Replacement bushings, however, are available for installation in rocker arms that are otherwise in good condition. Install the bushings so that oil hole in bushing is in register with oil hole in rocker arm as shown in Fig. C556.

The bushings are pre-sized and if carefully installed, using a 0.873-0.8735 diameter arbor, will require no final sizing. It is infrequently necessary, however, to hone the bushings to remove any localized high spots.

When reassembling, oil holes in rocker arm shaft must face toward push rod side of engine. "O" ring on rocker arm shaft oil feed tube must be in good condition. After the rocker arm assembly is installed on cylinder head, vary the number of washers at each end of shaft until the end rocker arms (exhaust) are free on shaft without any noticeable end play. Refer to Fig. C557. Tighten the rocker arm retaining nuts and screws to a torque of 40 ft.-lbs.

On Diesel models, check the decompressor linkage as follows: With decompressor engaged, all four exhaust valves must be held open; conversely, when the decompressor is disengaged, all four exhaust valves must be completely released. Any malfunction is apparent after examining the linkage.

Valve tappets gap is listed in paragraph 53.

VALVE TAPPETS
(CAM FOLLOWERS)

58. The mushroom type cam followers ride directly in the unbushed cylinder block bores and can be remov-

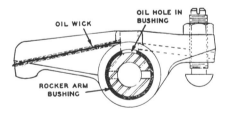

Fig. C556—Exhaust valve rocker arm, showing the oil wick installation. Oil hole in bushing must align with hole in all rocker arms.

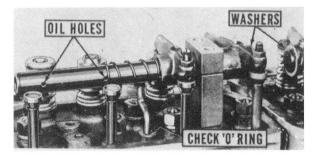

ed after first removing the camshaft as outlined in paragraph 66.

The standard size cam followers of 0.8095-0.8105 diameter have a normal clearance of 0.001-0.003 in the 0.8115-0.8125 diameter cylinder block bores. Followers of 0.010 oversize are available for service which requires that the bores be sized to 0.8215-0.8225. Before installing the followers, make certain that breather hole in each cam follower is open and clean.

TIMING GEAR COVER AND
CRANKSHAFT FRONT OIL SEAL

59. To remove the timing gear cover, first drain cooling system and remove the hood. On tractors equipped with a radiator shutter, remove the grille and disconnect the shutter control wire. On all models remove radiator. On standard, non-adjustable front axle models, disconnect the drag link and rear end of radius rod. On tricycle and adjustable front axle models, disconnect the steering worm shaft front universal joint and on general purpose and high clearance models, with power steering, disconnect hose lines from torque motor. On all models, unbolt front support (radiator bracket) from engine and roll the front support and wheels assembly away from tractor. On non-Diesel models, disconnect governor spring and throttle rod from governor lever.

On all models, remove the fan blades, belt and crankshaft pulley. Unbolt and remove timing gear cover from engine. The crankshaft front oil seal can be pressed out of cover and a new one pressed in with lip of same facing toward engine. Be sure to coat lip of a new seal with Lubriplate or equivalent before installing the timing gear cover.

Tighten the crankshaft pulley retaining cap screw to a torque of 100 ft.-lbs.

TIMING GEARS

60. Timing gear train on Diesel engines consists of the crankshaft gear, camshaft gear and the injection pump drive gear as shown in Fig. C558. On magneto equipped non-Diesel engines,

Fig. C557—Proper installation of rocker arms and shaft. Oil holes in the shaft must face toward push rods.

the timing gear train consists of the camshaft gear, crankshaft gear and the magneto drive gear as shown in Fig. C561. On non-Diesel engines equipped with a battery ignition distributor, the timing gear train consists of the crankshaft and camshaft gears only. Recommended timing gear backlash is 0.002-0.006.

61. **TIMING GEAR MARKS.** Gears are properly meshed when single punch marked tooth on crankshaft gear is meshed with the double punch marked tooth space on camshaft gear as shown in Figs. C558 and C561. On Diesel engines and magneto equipped non-Diesel engines, the single punch marked tooth on camshaft gear should be meshed with the double punch marked tooth space on the magneto or injection pump drive gear.

62. **CAMSHAFT GEAR.** The camshaft gear is keyed and press fitted to camshaft and is further positioned by a lock nut and lip type washer. Although the cam gear can be renewed without removing the camshaft, the J. I. Case Company recommended procedure for removing the gear is to first remove the camshaft and gear as a single unit as outlined in paragraph 66, then remove the lock nut and press the camshaft out of the gear. If the camshaft is not removed, however, the oil pan should be removed so that camshaft can be bucked up with a heavy bar while gear is being drifted on. This procedure will prevent possible loosening of the welch plug located in cylinder block at rear of camshaft.

When assembling the gear to the camshaft, tighten the lock nut to a

Fig. C558—View of a diesel engine timing gear train. Gears must be meshed with timing marks in register. Timing gear marks on non-Diesel engines are similar.

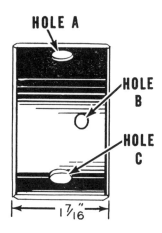

Fig. C560—Number two, three and four camshaft bushings are identical, but their installation procedure is different. Refer to text.

torque of 125 ft.-lbs.; then bend lip of washer against flat of nut. Timing marks are shown in Figs. C558 and C561.

63. **CRANKSHAFT GEAR.** Crankshaft gear is a press fit on the shaft and can be removed after timing gear cover is off by using a suitable puller with screws of same threaded into the two tapped holes in the gear.

To install the gear, heat same in hot oil (about 350° F) for approximately fifteen minutes, position the gear on shaft and drive it in position with a piece of steel pipe. Timing marks are shown in Figs. C558 and C561.

64. **INJECTION PUMP DRIVE GEAR.** The injection pump drive gear can be removed from the rear with timing gear cover in place after removing the injection pump and the shaft housing retaining screws; however, the J. I. Case Company recommends that the timing gear cover be removed and the backlash of the other timing gears be checked when renewing this gear. Make sure that thrust plunger, spring and thrust washer are in good condition and free. When reinstalling gear, mesh double mark on pump shaft gear with single dot mark on cam gear.

65. **MAGNETO DRIVE GEAR.** To remove the magneto drive gear, it is necessary to remove the magneto and drive housing. Refer to paragraph 145.

CAMSHAFT AND BEARINGS

The camshaft is carried in four steel backed, babbitt lined bushings. All camshaft journals are of the same diameter. Shaft end play is automatically maintained by a spring loaded thrust plunger (button) in front end of shaft and a bronze thrust washer interposed between the cam gear and the front bearing.

66. To remove the camshaft and gear unit, first remove the timing gear cover as outlined in paragraph 59 and the rocker arms assemblies and push rods as outlined in paragraph 57. Remove the oil pan and oil pump. Push the tappets (cam followers) up into their bores as far as they will go. If tappets will not stay in this position, make up eight long dowel pins slightly larger in diameter than the push rods and taper one end of the pins. Insert the long dowel pins down through the push rod openings in the cylinder head and force the tapered ends into the tappets. Lift the dowel pins up, thereby holding the tappets away from the camshaft and hold the dowels in the raised position with spring type clothes pins. Withdraw the camshaft and gear unit from cylinder block. After removing the lock nut, the camshaft can be pressed out of the cam gear. When reassembling, the lock nut should be tightened to a torque of 125 ft.-lbs.

Diameter of the camshaft bearing journals is 2.121-2.122. Recommended bearing journal clearance in bushings is 0.0014-0.0054.

Camshaft journal
   diameter 2.121-2.122
Bushing inside diameter.2.1234-2.1264
Running clearance, new..0.0014-0.0054
Rebush if clearance exceeds....0.0065
Thrust washer thickness.0.1225-0.1275

If the camshaft bushings require renewal, it will be necessary to perform the additional work of detaching (splitting) engine from clutch housing and removing clutch, flywheel and the welch plug which is located behind the camshaft rear bearing.

Bushings are pre-sized and will not require final sizing after installation if carefully installed, using a piloted arbor or mandrel of approximately 2.122 diameter. Bushings should be installed with joint at bottom and with oil holes in bushings in register with oil holes in cylinder block. The number one (front) bushing is 1$\frac{11}{32}$ inches long and both oil holes should align with oil holes in block. Number two, three and four bushings are identical and 1$\frac{7}{16}$ inches long as shown in Fig. C560. On number two and three bushings, holes (A & C) should align

Fig. C561—Timing gear train for magneto equipped non-Diesel engines. Models with battery ignition are similar except the camshaft and crankshaft gears only are used.

with oil holes in block and hole (B) will be against a blank surface. On number four bushing, hole (B) should align with oil hole in block and holes (A & C) will be against a blank surface.

When installing the camshaft and gear unit, mesh the gears so that timing marks are in register as shown in Figs. C558 and C561.

CONNECTING ROD AND PISTON UNITS

67. Connecting rod and piston units are removed from above after removing cylinder heads, oil pan and the connecting rod bearing caps.

When reinstalling, the numbers on the rod and the cap should be in register and face toward camshaft side of engine. Tighten the self-locking connecting rod nuts to a torque of 95-105 ft.-lbs. Some engines use rod bolts instead of nuts.

PISTONS, RINGS AND SLEEVES

Diesel Models

68. Each cast iron piston is fitted with three compression rings and one oil control ring. Check the pistons, sleeves and rings against the values which follow:

Series 400 Prior to 8104711
Compression ring
   end gap · · · · · · · · 0.013-0.023
Cylinder sleeve
   diameter · · · · · · · · 4.000-4.001
Renew sleeves if
   diameter exceeds · · · · · · · 0.002
Piston skirt
   diameter · · · · · · · 3.9955-3.9965
Piston skirt
   clearance · · · · · · · 0.0035-0.0055

Series 400 After 8104710, 700B and 800B

Compression ring end gap . . 0.013-0.023
Cylinder sleeve diameter . . 4.125-4.126
Renew sleeves if diameter
  exceeds 4.133
Renew sleeves if out-of-round
  exceeds 0.002
Piston skirt diameter 4.1205-4.1215
Piston skirt clearance 0.0035-0.0055

On all models, piston skirt clearance can be considered satisfactory when a ½-inch wide, 0.003 thick feeler ribbon is a loose fit and a ½-inch wide, 0.005 thick feeler ribbon is a tight, smooth fit. Check is made by withdrawing feeler ribbon from between piston skirt (90° to piston pin) and cylinder sleeve.

The procedure for renewing the wet type cylinder sleeves is outlined in paragraph 70.

Install the replacement type piston rings as shown in Fig. C562.

Non-Diesel Models

69. Each aluminum piston is fitted with three $\frac{3}{32}$ inch wide compression rings and one ¼ inch wide oil control ring. Check the pistons, sleeves and rings against the values which follow:
Compression ring
  end gap 0.013-0.023
Top compression ring
  side clearance 0.0025-0.004
Other compression rings
  side clearance 0.002-0.0035
Oil control ring
  end gap 0.013-0.023
Oil control ring
  side clearance 0.001-0.0025
Cylinder sleeve
  diameter 4.000-4.001
Renew sleeves if
  diameter exceeds 4.008
Renew sleeves if out-of-
  round exceeds 0.002
Piston skirt
  diameter 3.996-3.997
Piston skirt
  clearance 0.003-0.005

The J. I. Case company recommends measuring the piston skirt and the sleeve with micrometers to determine the skirt clearance.

The procedure for renewing the wet type cylinder sleeevs is outlined in paragraph 70.

Install the replacement type piston rings as shown in Fig. C562.

R & R CYLINDER SLEEVES

70. The wet type cylinder sleeves can be removed from cylinder block by using a suitable puller after connecting rod and piston units are out. Coolant leakage at bottom of sleeves is prevented by two packing rings.

Before installing new sleeves, thoroughly clean the mating and sealing surfaces of cylinder block and sleeves.

All sleeves should enter crankcase bores full depth and should be free to rotate by hand when tried in bores without the packing rings. After making a trial installation, remove the sleeves and install packing rings dry in grooves in sleeves, then lubricate only the outer face of the packing rings with a very light coat of petroleum jelly. Carefully lower the sleeves into cylinder block and press in place with hand pressure.

Fill the cooling system with cold water and check for leaks adjacent to the packing rings.

PISTON PINS

71. The 1.3583-1.3586 diameter full floating type piston pins are retained in the piston pin bosses by snap rings and are available in standard size only.

Piston pin fit in piston
  Diesels 0.0002-0.0009
  Non-Diesels 0.0001-0.0003
Piston pin fit in connecting rod bushing.
  Diesels 0.0004-0.0011
  Non-Diesels 0.0004-0.0011

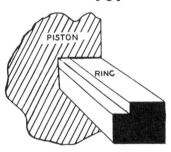

COMPRESSION RINGS

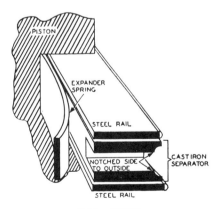

OIL RING

Fig. C562—Proper installation of replacement type piston rings.

CONNECTING RODS AND BEARINGS

72. Connecting rod bearings are of the steel-backed, non-adjustable, slip-in precision type, renewable from below after removing oil pan and bearing caps. Bearing material is copper-lead for Diesels, babbitt for non-Diesels.

When installing new bearing shells, make certain that the bearing shell projections engage the milled slot in connecting rod and bearing cap and that cylinder numbers on the rod and cap are in register and face toward camshaft side of engine. Tighten the self-locking connecting rod nuts to a torque of 95-105 ft.-lbs. Bearing inserts are available in undersizes of 0.002, 0.010, 0.020 and 0.030.

Check the connecting rods, bearing inserts and crankshaft crankpins against the values which follow:

Rod length C to C 10.499-10.501
Crankpin diameter 2.748-2.749
  Regrind if out-of-round 0.001
  Regrind if tapered 0.002
Rod bearing running clearance
  Diesel models 0.0013-0.0038
  Non-Diesel models 0.0015-0.0036
Renew bearing if running
  clearance exceeds 0.006
Rod side play 0.005-0.012
Rod nut torque . . . . 95-105 ft.-lbs.

CRANKSHAFT AND MAIN BEARINGS

73. Crankshaft is supported in five steel-backed, non-adjustable, slip-in precision type bearing inserts which can be renewed after removing oil pan and main bearing caps. Bearing material is copper-lead for Diesels, babbitt for non-Diesels. Crankshaft end play of 0.004-0.012 is controlled by one bronze half washer located on each side of the No. 3 main bearing cap. If the end play exceeds 0.020, renew the thrust washers. Thrust washers, 0.006 oversize, are available if standard thickness washers will not bring the end play within the limits of 0.004-0.012. If 0.006 oversize washers will not correct excessive end play, renew the crankshaft.

The oil grooved upper half of the main bearing inserts is not interchangeable with the non-grooved lower half. When installing new bearing shells, make certain that the bearing shell projections engage the milled slot in cylinder block and bearing cap and that numbered end (1, 2, 3, 4 & 5) of bearing caps face toward camshaft side of engine. Tighten the main bearing cap screws to a torque of 145-155 ft.-lbs. Bearing inserts are available in undersizes of 0.002, 0.010, 0.020 and 0.030.

74. To renew the crankshaft, first remove the engine as outlined in paragraph 50 and proceed as follows: Remove the flywheel and the crankshaft rear oil seal retainer. Remove the crankshaft pulley, timing gear cover, oil pan and oil pump. Remove the rod and main bearing caps and lift the crankshaft from engine.

Check the crankshaft and main bearing inserts against the values which follow:

Crankpin diameter (New)..2.748-2.749
Crankpin (0.010 under)....2.738-2.739
Crankpin (0.020 under)....2.728-2.729
Crankpin (0.030 under)....2.718-2.719
Main journal
   diameter (new) 2.998-2.999
Mains (0.010 under) 2.988-2.989
Mains (0.020 under) 2.978-2.979
Mains (0.030 under) 2.968-2.969
   Regrind if out-of-round..0.001
Main bearing running clearance
   Diesels 0.0016-0.0046
   Non-Diesels 0.002-0.0046
Renew bearing inserts if
   clearance exceeds 0.0065

Refer to paragraph 76 when installing the crankshaft rear oil seal; and on models with battery ignition, refer to paragraph 79 when installing the oil pump.

CRANKSHAFT OIL SEALS

75. FRONT SEAL. The crankshaft front oil seal is located in the timing gear cover and can be renewed after removing the complete front support and axle assembly and the crankshaft pulley.

76. REAR OIL SEAL. The crankshaft rear oil seal is contained in a retainer plate which is bolted to rear face of cylinder block. To renew the oil seal, first detach (split) engine from clutch or torque convertor housing as outlined in paragraph 153 or 156P and remove clutch and/or flywheel. Unbolt the seal retainer plate (Fig. C563) from the cylinder block and oil pan and remove the seal and retainer as a unit. Remove seal from retainer.

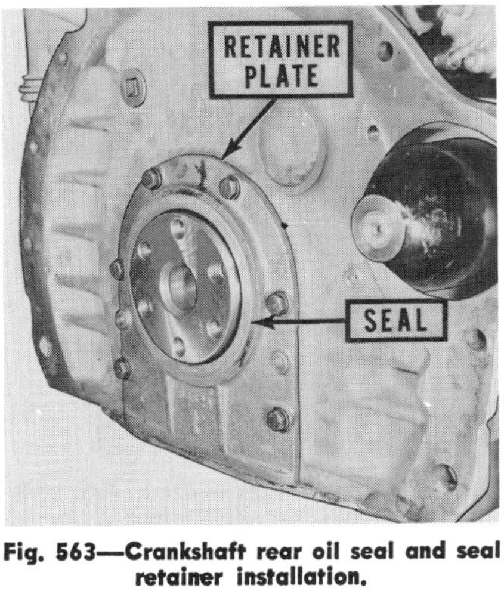

Using a new gasket, install the seal retainer plate (without seal) to cylinder block and tighten the retaining cap screws until they are snug. Do not install the two cap screws which secure the oil pan to the retainer plate.

Note: Gasket must extend full length of seal retainer. The strip of material across the lower part of a new gasket is to help retain its shape. Remove this strip before installation.

Mount a dial indicator as shown in Fig. C564 and check the concentricity of the seal retainer bore with respect to the crankshaft flange. Shift the seal retainer until the bore is concentric with crankshaft flange within 0.0025; then tighten the retaining cap screws to a torque of 25 ft.-lbs. Install and tighten the two cap screws securing oil pan to the seal retainer.

Fig. C564—Checking the concentricity of the crankshaft rear oil seal retainer with respect to the crankshaft flange. Concentricity should be within the limits of 0.0025.

Coat the metal outer circumference of the oil seal with Permatex and the neoprene lip with No. 110 Lubriplate. Carefully work lip of seal over crankshaft flange and tap the seal evenly into the seal retainer bore.

OIL PAN

77. On tricycle type tractors and models with an adjustable type front axle, the oil pan can be removed in the conventional manner. On models with a standard, non-adjustable front axle, it is necessary to remove radius rod before oil pan can be removed.

FLYWHEEL

78. To remove the flywheel from a 400, 700B or 800B series tractor, first detach (split) engine from clutch or torque convertor housing as outlined in paragraph 153 or 156P; then, on series 400 and 700B tractors, unbolt and remove the engine clutch from the flywheel. On all models, unbolt and remove flywheel. Renew the clutch shaft front oil seal and pilot bearing if they are damaged.

To facilitate installation of the flywheel ring gear, heat same to 350° F, then position ring gear on flywheel so that beveled end of flywheel teeth face toward front of tractor.

On series 400 tractors prior to tractor serial number 8063022, make certain that there are no burrs on mating surfaces of flywheel and crankshaft and install the flywheel as follows: Coat the mating surfaces of the crankshaft and the flywheel with Permatex and install flywheel before the Permatex hardens. Install the retainer "O" ring (Fig. C565). Coat the mating surfaces of the clutch shaft front oil seal and its retainer with Permatex and install seal in retainer so that lip of seal will face toward front

Fig. C563—Crankshaft rear oil seal and seal retainer installation.

Fig. C565 — Crankshaft rear oil seal, and clutch shaft installation. The flywheel "O" ring is not used on series 400 prior to Ser. No. 8063022.

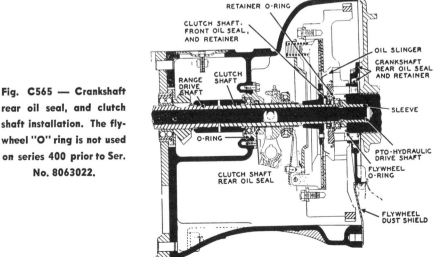

of tractor. Coat lip of seal with Lubri-plate. Install the seal retainer and tighten the flywheel retaining cap screws to a torque of 100 ft.-lbs. On series 400 tractors after serial number 8063021 and all series 700B tractors, a flywheel "O" ring is provided between crankshaft flange and flywheel. This eliminates the necessity of coating the mating surfaces of the crankshaft flange and the flywheel with Perma-tex. Otherwise, the installation procedure is the same as early 400 models.

On series 800B tractors there is an "O" ring seal between the outer diameter of the torque convertor and the flywheel.

On all models, the flywheel retaining cap screws should be tightened to 100 ft.-lbs. of torque.

OIL PUMP AND RELIEF VALVE

79. REMOVE AND REINSTALL. The gear type oil pump can be removed after removing the oil pan. When installing the oil pump, tighten the retaining cap screws to a torque of 25 ft.-lbs.

Note: When installing the oil pump on distributor equipped non-Diesel engines, time the pump as follows: Crank engine until number one piston is coming up on compression stroke and continue cranking until the DC mark on crankshaft pulley is in register with the pointer on the timing gear cover as shown in Fig. C566. Install oil pump so that slot in the dis-

tributor drive coupling is parallel with engine block and offset (narrowest side) of slot is toward engine block as shown in Fig. C567. Retime the distributor as outlined in paragraph 142.

80. OVERHAUL. With the pump removed from tractor, remove roll pin and drive gear (7—Fig. C568). Remove the pump cover, and withdraw the driven gear and shaft and the follower gear. Loosen the lock nut and remove the pressure relief valve adjusting screw, spring and relief valve. Thoroughly clean all parts and examine them for wear. The pump cover is a lapped fit to the pump body. The cover should be renewed if the gears have caused excessive wear on the cover. The pressure relief valve plunger (4) should have a free fit in the pump body. If the valve is scored or damaged, renew the valve. If valve bore is scored, renew the pump body. If pumping gears are worn or if the pump shaft is bent or worn, it is recommended that the complete pump be renewed.

Normal setting of the relief valve adjusting screw, on late series 400 and all series 700B and 800B, is when the distance from underside of the adjusting screw head to pump body is ½-inch as shown in Fig. C569. On early series 400 (see inset) the adjusting screw has a thinner head and the distance should be $\frac{13}{16}$ inch as shown. One complete turn of the adjusting screw will change the oil pressure approximately 1¾ psi. Turn screw in to increase pressure. Normal oil pressure is 35-40 psi., when oil is warm.

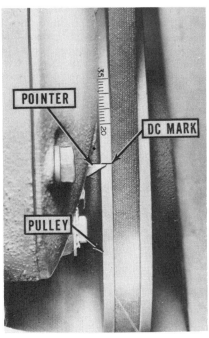

Fig. C566—When installing the oil pump on distributor equipped non-Diesel engines, position the pump shaft as in Fig. C567 when "DC" mark on pulley is in register with pointer.

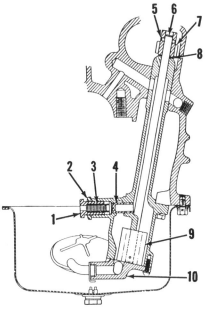

Fig. C568—Sectional view of the oil pump. Relief valve (4) must have a free fit in pump body.

1. Pressure adjusting screw	6. Pump shaft
2. Jam nut	7. Oil pump drive gear
3. Relief valve spring	8. Thrust washer
4. Relief valve plunger	9. Pumping gears
5. Roll pin	10. Cover

Fig. C567—Proper installation of the oil pump shaft on distributor equipped non-Diesel engines. Timing marks must be positioned as shown in Fig. C566.

Fig. C569—Initial adjustment of late style oil pressure relief valve is ½-inch as shown. Inset shows the early style used on early series 400 tractors.

CARBURETOR

(Except LP-Gas)

81. Gasoline and distillate models are equipped with a Marvel-Schebler TSX616 carburetor. Calibration data are as follows:

TSX616

Float setting...............¼ inch
Repair kit 286-1076
Gasket set 16-594
Inlet needle & seat........233-543
Idle jet 49-101L
Nozzle 47-267
Main adjusting needle seat...36-302
Idle adjusting needle........43-58
Main adjusting needle
 assembly complete 43-620
Venturi 46-438

LP-GAS SYSTEM

Series 400, 700B and 800B tractors are available with a factory installed LP-Gas system, using Ensign equipment. Like other LP-Gas systems, this system is designed to operate with the fuel supply tank not more than 80% filled. It is important when starting the engine to open the vapor or liquid valves on the supply tank SLOWLY; if opened too fast, the excess flow valves will shut off the fuel supply to the regulator.

An Ensign model Xg No. 8833 carburetor is used on series 400 tractors but a model Xg No. 3896 carburetor is used on series 700B and 800B tractors. Ensign model W No. 3490 regulator is used on all three series of tractors. There are three points of mixture adjustment on the LP-Gas system, plus an idle stop screw. Refer to Fig. C575.

85. **STARTING SCREW.** Immediately after the engine is started, bring the throttle to the **fully open** position and with the choke **closed**, rotate starting screw until the highest engine speed is obtained. A slightly richer adjustment (counter-clockwise until speed drops slightly) may be desirable for a particular fuel or operating condition. Average adjustment is 1½ turns open. Place the controls in operating position by completely opening the choke.

86. **IDLE STOP SCREW.** Idle speed stop screw on the carburetor throttle should be adjusted to provide a slow idle speed of 600 crankshaft rpm.

87. **IDLE MIXTURE SCREW.** With the choke **open**, engine warm and idle stop screw set, adjust idle mixture screw, located on regulator, until best idle is obtained. An average adjustment is 1½ turns open.

88. **LOAD SCREW.** In the Xg Ensign carburetor, the so-called load screw (D—Fig.C576) primarily controls the partial load mixture. The richer mixture needed for full power is supplied by a by-pass jet (L) which is opened and closed by the economizer diaphragm spring (Q) and the manifold vacuum. When the manifold vacuum drops below 4-6 inches Hg which will occur at full load (wide-open carburetor throttle) the diaphragm spring opens the non-adjustable power jet. When the vacuum is higher than 10-13 inches Hg which will occur at no load or light load, regardless of rpm, the diaphragm spring is overridden by the higher vacuum and the jet is closed off; at which time, the mixture is controlled by the load screw and idle screw to provide economizer action.

89. ADJUSTMENT WITHOUT LOAD. Average adjustment of the load screw, Fig. C575, as recommended by J. I. Case is 3 turns open. However to accurately set the load screw and regulator for the fuel and operating conditions as found in a certain locality, the J. I. Case recommendations are as follows:

The engine should be thoroughly warmed up to operating temperature of 180-195° F. The carburetor throttle valve and choke should be **fully open**. With the engine operating at its no load speed of 1600 rpm, rotate the load screw clockwise (lean) until the exhaust is uneven; then, rotate the load screw counter-clockwise (rich) until the engine loses speed or is sluggish; then, rotate the screw clockwise until the engine runs smoothly. Recheck the adjustment and lock the load screw in position. Recheck the idle mixture screw adjustment as per paragraph 87.

90. ENSIGN METHOD WITHOUT LOAD. First carefully adjust the idle mixture screw as in paragraph 87. Disconnect the economizer to carburetor flange vacuum line, and plug the carburetor flange connection. Run engine at high idle speed with hand throttle in position where governor does not regulate the rpm. Adjust load screw to obtain maximum rpm, note its position; then, carefully rotate the

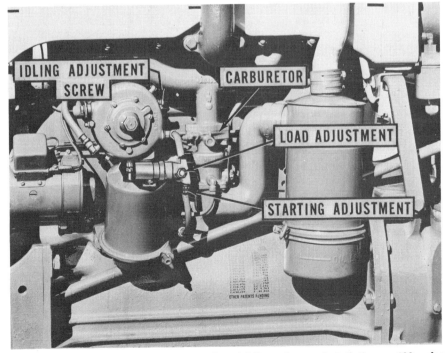

Fig. C575—Ensign model W regulator and model Xg carburetor installation on 400 series tractors. Installation is similar for series 700B and 800B tractors.

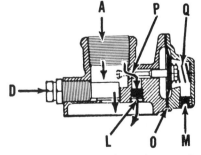

Fig. C576—Sectional view of economizer used on Ensign model Xg carburetor.

A. Fuel inlet
D. Load adjusting screw
L. Orifice
M. Vacuum connection
O. Diaphragm
P. Fuel passage
Q. Spring

screw in a lean direction (clockwise) until the rpm just begins to fall. Rotate the screw to the mid-point of these two positions and tighten the locknut. Unplug the carburetor flange connection and reconnect the econo-

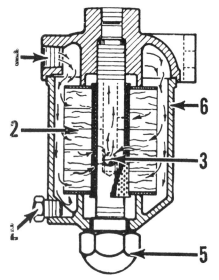

Fig. C577—LP-Gas filter used, filters liquid fuel.

L. Drain plug
1. Fuel inlet
2. Filter cartridge
3. Outlet passage
5. Stud nut
6. Filter bowl

mizer vacuum line. The power valve and jet are fully open when using this method.

91. ANALYZER AND VACUUM GAGE METHOD. In this method, the engine is operated with the carburetor throttle wide open and with sufficient load on the engine to hold the operating speed to 300 to 500 rpm slower than maximum operating rpm. One method of loading the engine is to disconnect or short out two or more spark plug wires. Do not disconnect any lines. Set the load screw to give a reading of 12.8 on the analyzer gasoline scale or 14.3 on an analyzer with a LPG scale.

92. Check the part throttle (partial load) mixture by reducing the opening of the throttle valve and the load on the engine until a manifold vacuum of 10-13 inches is obtained at the same rpm as used in paragraph 91. The power jet should be closed at this time and the analyzer should now read 13.8-14.5 on the gasoline scale or 14.9-15.5 on the LPG scale. If readings are lower than specified, fuel may be leaking past the power valve; if higher than specified, the power jet orifice may be too small.

LP-GAS FILTER

93. The Ensign No. 6257 filter used in this system is subjected to and should be able to stand high pressures without leakage. Unit should be drained periodically at the blow off cock (L—Fig. C577). When major engine work is being performed, it is advisable to remove the lower part of the filter, thoroughly clean the interior and renew the felt cartridge if same is not in good condition.

LP-GAS REGULATOR

94. **HOW IT OPERATES.** Fuel from the supply tank enters the regulating unit inlet (A—Fig. C578) at a tank pressure of 25 to 80 psi and is reduced from tank pressure to about 4 psi at the high pressure reduction valve (C) after passing through the strainer (B). Flow through high pressure reducing valve is controlled by the adjacent spring and diaphragm. When the liquid fuel enters the vaporizing chamber (D) via the valve (C), it expands rapidly and is converted from a liquid to a gas by heat from the water jacket (E) which is connected to the cooling system of the engine. The vaporized gas then passes, at a pressure slightly

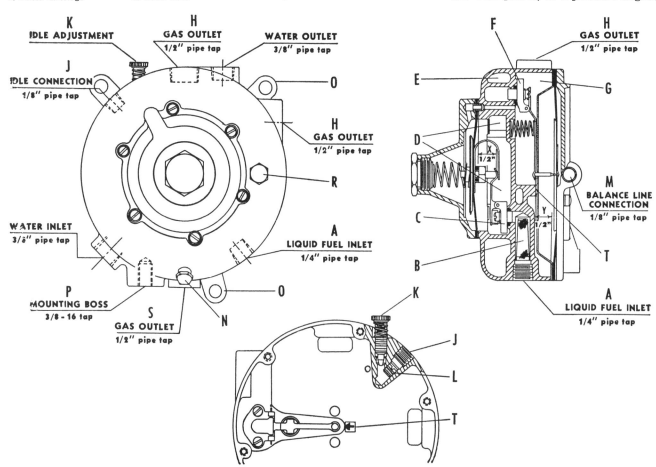

Fig. C578—Assembled and sectional views of Ensign model W LP-Gas regulator. Fuel is vaporized within the unit by heat from the engine cooling system.

B. Strainer   C. Inlet valve   F. Outlet valve   L. Orifice   T. Boss below atmospheric, via the low pressure reducing valve (F) into the low pressure chamber (G) where it is drawn off to the carburetor via outlet (H). The low pressure reducing valve is controlled by the larger diaphragm and small spring.

Fuel for the idling range of the engine is supplied from a separate outlet (J) which is connected by tubing to a separate idle fuel connection on the carburetor. Adjustment of the carburetor idle mixture is controlled by the idle fuel screw (K) and the calibrated orifice (L) in the regulator. The balance line (M) is connected to the air inlet horn of the carburetor so as to reduce the flow of fuel and thus prevent over-richening the mixture which would otherwise result when the air cleaner or air inlet system becomes restricted.

TROUBLE-SHOOTING

95. **SYMPTOM.** Engine will not idle with Idle Mixture Adjustment Screw in any position.

CAUSE AND CORRECTION. A leaking valve or gasket is the cause of the trouble. Look for leaking low pressure valve caused by deposits on valve or seat. To correct the trouble, wash the valve and seat in gasoline or other petroleum solvent.

If the foregoing remedy does not correct the trouble, check for a leak at high pressure valve by connecting a low reading (0 to 20 psi) pressure

gauge at point (R). If the pressure increases **after** a warm engine is stopped, it proves a leak in the high pressure valve. Normal pressure is 3½-5 psi.

96. **SYMPTOM.** Cold regulator shows moisture and frost after standing.

CAUSE AND CORRECTION. Trouble is due either to leaking valves as per paragraph 95 or the valve levers are not properly set. For information on setting of valve lever refer to paragraph 97.

REGULATOR OVERHAUL

If an approved station is not available the regulator can be overhauled as outlined in the following paragraph.

97. Remove the unit from the engine and completely disassemble, using Fig. C579 as a reference.

Thoroughly wash all parts and blow out all passages with compressed air.

Fig. C580—Using Ensign gauge 8276 to set the fuel inlet valve lever to the dimension as indicated at (X) in Fig. C578.

Inspect each part carefully and discard any that are worn.

Before reassembling the unit, note dimension (X—Fig. C578) which is measured from the face on the high pressure side of the casting to the inside of the groove in the valve lever when valve is held firmly shut as shown in Fig. C580. If dimension (X), which can be measured with Ensign gauge No. 8276 or with a depth rule, is more or less than ½-inch, bend the lever until this setting is obtained. A boss or post (T—Fig. C581) is machined and marked with an arrow to assist in setting the lever. Be sure to center the lever on the arrow before tightening the screws holding the valve block. The top of the lever should be flush with the top of the boss or post (T).

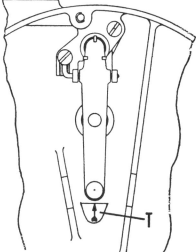

Fig. C581—Location of post or boss with stamped arrow for the purpose of setting the fuel inlet valve lever.

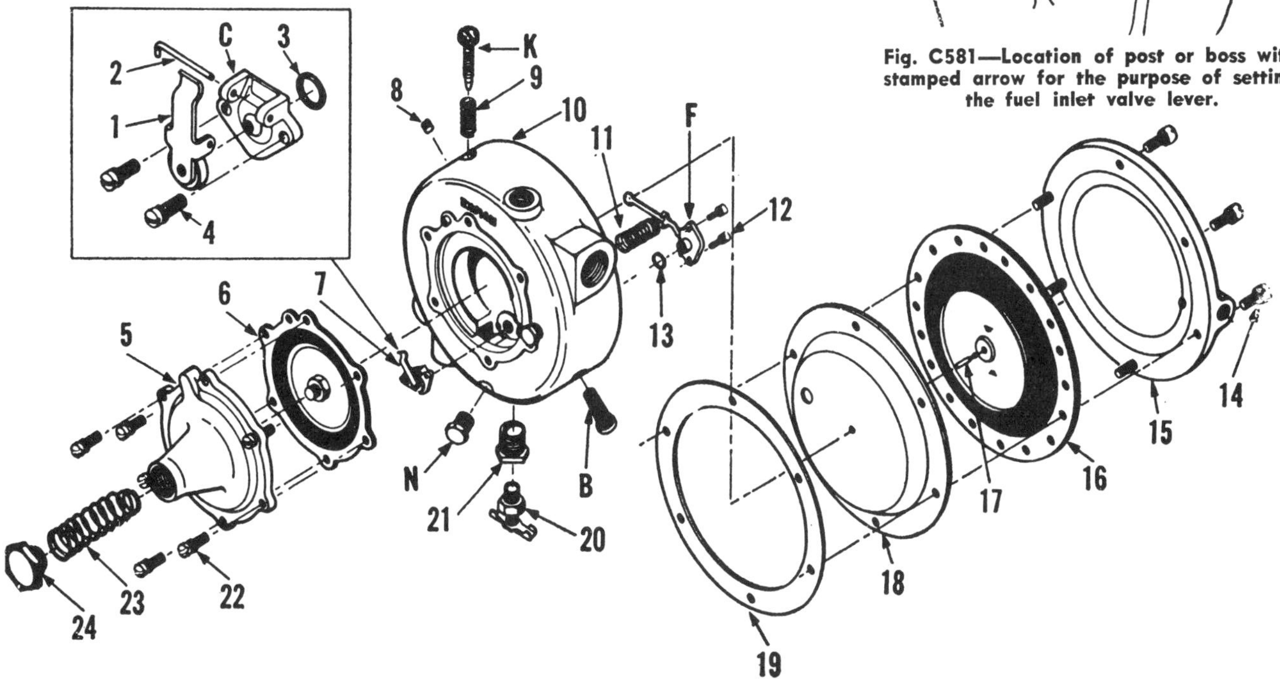

Fig. C579—Exploded view of the Ensign model W regulating unit used on LP-Gas engines.

B. Fuel inlet strainer	3. "O" ring	10. Regulator body	18. Partition plate
C. Valve seat	5. Regulator cover	11. Outlet diaphragm spring	19. Partition plate gasket
F. Outlet valve assembly	6. Inlet pressure diaphragm	13. "O" ring	20. Drain cock
K. Idle adjusting screw	7. Inlet valve assembly	15. Back cover plate	21. Reducing bushing
1. Inlet diaphragm lever	8. Bleed screw	16. Outlet pressure diaphragm	23. Inlet diaphragm spring
2. Pivot pin	9. Idle screw spring	17. Push pin	24. Spring retainer

DIESEL FUEL SYSTEM

American Bosch PSB; German (Robert) Bosch PES internal drive and German (Robert) Bosch PES external drive injection pumps are used on series 400, 700B and 800B tractors. Before servicing any diesel tractor, it should be determined which type of pump is used; then, refer to the appropriate paragraphs concerning the system using that type of pump. Refer to the accompanying identifying illustration.

The diesel fuel system consists of three basic units; the fuel filters, injection pump and injection nozzles. When servicing any unit associated with the fuel system, the maintenance of absolute cleanliness is of utmost importance. Of equal importance is the avoidance of nicks or burrs on any of the working parts.

Probably the most important precaution that service personnel can impart to owners of Diesel powered tractors is to urge them to use an approved fuel that is absolutely clean and free from foreign materials. Extra precaution should be taken to make certain that no water enters the fuel storage tanks. This last precaution is based on the fact that all Diesel fuels contain some sulphur. When water is mixed with sulphur, sulphuric acid is formed and the acid will quickly erode the closely fitting parts of the injection pump and nozzles.

BLEEDING THE SYSTEM

All Models

98. Refer to Fig. C585 or C585A. To completely bleed the system, fill the main fuel tank and open the tank shut-off valve.

Open the first stage filter bleed valve and retain in the open position until clear fuel (no bubbles) appears; then, close the valve.

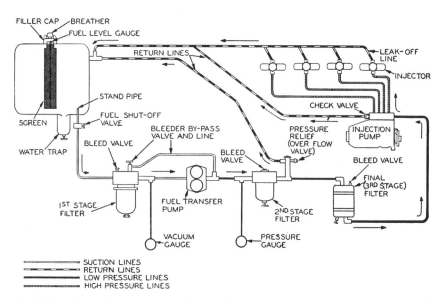

Fig. C585—Schematic view of the Diesel fuel system typical of that used on all PSB injection pump equipped tractors. The system is similar on PES internal drive equipped tractors, except that the fuel vacuum gage and the bleeder by-pass line are eliminated and the pressure gage is inserted between the final stage filter and the injection pump.

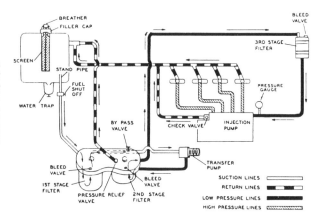

Fig. C585A — Schematic view of the fuel circuit used on PES external drive injection pump equipped tractors. Refer to Fig. C585 for tractors which are equipped with PSB injection pumps and PES internal drive pumps.

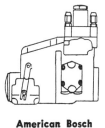

**American Bosch
PSB**

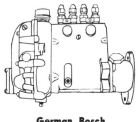

**German Bosch
PES (External Drive)**

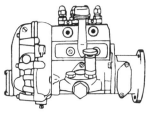

**German Bosch
PES (Internal Drive)**

Open the second stage filter bleed valve and the bleeder by-pass valve. Close the filter bleed valve when clear fuel appears but leave the by-pass valve open.

Open the final stage filter bleed valve and when clear fuel appears, close the bleed valve and the by-pass valve. Start engine.

NOTE: On tractors equipped with a PES injection pump, if fuel flows when bleeding the final stage filter but the engine won't start, bleed the fuel transfer pump as outlined in paragraph 98A.

98A. BLEED TRANSFER PUMP (PES INJECTION PUMPS). Make certain the fuel tank is full. Disconnect the injection pump excess fuel line at the check valve (CV—Fig. C585B), then remove the check valve. Loosen the cap nut (CN) on the inlet (rear) side of the fuel transfer pump. Decompress the engine and crank the engine with the starting motor until a steady stream of clean fuel (no bubbles) appears at the check valve hole and at the cap nut; then, reinstall the check valve, reconnect the excess fuel line, tighten the cap nut and start engine.

99C. FUEL RELIEF VALVE. A non-adjustable pressure relief valve is incorporated in the system to maintain a steady pressure in the final stage filter. Refer to Figs. C585 and C585A. To clean a sticky relief valve, remove same, disassemble and wash internal parts in a 50-50 mixture of denatured alcohol and benzol.

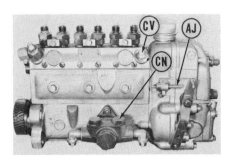

Fig. C585B — View of the left side of a removed PES external drive injection pump showing the high speed adjusting screw (AJ), check valve (CV) and cap nut (CN). Pumps used on four cylinder engines are similar.

SERVICING FUEL TANK, TANK FILTER, SCREENS AND RELIEF VALVE

All Models

99. FILTER IN TANK CAP. The fuel tank filler cap contains a vent (breather) and an edge wound paper filter element as shown in Fig. C586. A fully clogged filter will prevent the flow of fuel. Filter element should be removed and cleaned in an oil-free solvent, such as carbon tetrachloride, every 240 hours; or oftener, under severe dust conditions.

99A. WATER TRAP AND TANK STRAINER. The water trap, located at bottom of fuel tank, should be drained daily.

The strainer, located in the fuel tank, should be cleaned every 24 hours of operation. Strainer can be removed through the filler opening after extracting its retaining snap ring.

99B. FUEL TANK. Fuel tank of any diesel engine using "cracked" type fuels should be cleaned periodically to remove gum and varnish deposits. These deposits can be removed using a 50-50 mixture of denatured uncolored alcohol and benzol.

Fig. C586 — Fuel tank breather filter should be cleaned each 240 hours, or oftener, in severe dust conditions.

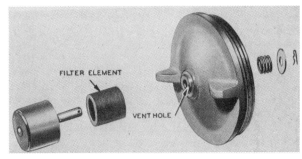

Fig. C586A — Right side view of a diesel engine fitted with a PES external drive injection pump.

SERVICING FIRST, SECOND AND THIRD STAGE FILTERS

Series 400, 700B and 800B (With Both Fuel Pressure and Vacuum Gages)

Refer to paragraph 100 for tractors, which have only a fuel pressure gage.

99D. FIRST STAGE FILTER. When the hand of the vacuum gage, located on the instrument panel, is in the red zone, the filter element should be cleaned.

Element can be washed in clean diesel fuel. If air under pressure is used to dry the element, introduce the air jet from inside the element. Never blow air against the outer surface.

99E. SECOND STAGE FILTER The cartridge or inner element of this filter must be discarded when clogged. A new element will normally project slightly beyond the top of the filter body so that element will be tightly compressed when the clamp nut is tightened. Since this filter is on the delivery side of the transfer pump, a partially plugged element results in an increase in pressure in the line between pump and filter.

When the hand on the fuel pressure gage, located on the instrument panel, is in the red zone, renewal of the second stage filter element is necessary.

NOTE: A restricted first stage filter element can affect the gage reading; therefore, it is recomended that the first stage filter be cleaned prior to checking the second stage filter.

99F. FINAL STAGE FILTER. This sealed type filter should be discarded and a new one installed every 2000 hours or whenever it cannot be bled by gravity head from a full fuel tank as in paragraph 98.

Series 700B and 800B (With Only Fuel Pressure Gage)

Refer to paragraphs 99D through 99F for tractors which have both fuel pressure and vacuum gages.

100. FUEL FILTERS. The fuel pressure gage on the instrument panel should be checked daily. As sediment gradually plugs a filter element, a pressure drop will occur. Any drop in pressure will be indicated by the hand of the pressure gage moving towards (or into) the RED zone on the gage. As soon as the hand reaches the RED zone, one or more of the filter elements needs servicing.

When servicing the fuel filters, always begin with the first stage filter and by a process of elimination, follow on through the second and final stage filters, if necessary, until the needle on the fuel pressure gage is again in the GREEN zone.

The first and second stage filters are of the renewable element type. The complete final stage filters should be renewed if it becomes contaminated as it is of the sealed type.

NOTE: It will be necessary to bleed the system as outlined in paragraph 98 after servicing any of the filters.

DIESEL (INJECTORS) NOZZLES

When testing or adjusting fuel injectors, do not place any part of your body in front of the injector nozzle. Fuel spray from an injector has sufficient penetrating power to puncture the flesh and destroy tissue. Should the fuel enter the blood stream it may cause blood poisoning.

In the event the skin is punctured from the discharge, immediately wash the injured part with boric acid solution, support the injured member with a splint and sling and get the patient to a physician as soon as possible.

108. LOCATING FAULTY INJECTOR. If one engine cylinder is misfiring, it is reasonable to suspect a malfunctioning injector in that cylinder. As in shorting out spark plugs in a spark ignition engine, the faulty injector is the one which, when its feed line is loosened (to deprive it of fuel), least affects the running of the engine. Remove the suspected injector from the engine as outlined in paragraph 109.

To check the diagnosis, install a new or rebuilt injector or one from a cylinder which is firing regularly. If the cylinder fires regularly with the other injector, the malfunctioning unit should be serviced as in paragraphs 110 through 113.

The J. I. Case Company recommends that whenever an injector is suspected of being faulty that the mating Powr-cel (pre-combustion chamber) be also removed, tested and serviced as outlined in paragraphs 118 through 118C.

109. R&R INJECTOR. Before loosening any lines, wash the connections and adjacent parts with fuel oil or kerosene. After disconnecting the high pressure and leak off lines, cap the exposed lines and fittings with plastic plugs to prevent the entrance of dirt. Remove injector clamp nuts and withdraw the injector using a pry bar with a right angle claw if necessary. Remove the solid copper gasket.

Thoroughly clean the injector recess in the cylinder head before re-inserting injector. The Case carbon reamer CD513 (shown in Fig. C587) is recommended for this job. Make certain that the old copper gasket has been removed (it may have stuck in the head recess seat). Install a new gasket and carefully insert the injector using care not to strike the nozzle tip against any hard surface. Torque the clamping nuts to 14-17 Ft.-Lbs.

110. CLEANING AND INSPECTING INJECTORS. Hard or sharp tools, emery cloth, crocus cloth, grinding compounds or abrasives of any kind should NEVER be used in the cleaning of injectors.

Injectors which do not check O.K. on the test stand or by visual testing, should be disassembled, inspected and cleaned as follows: Carefully clamp injector in vise and using a ¾-inch deep socket or box wrench, remove the nozzle cap nut as shown in Fig. C588. Remove the spray nozzle unit consisting of the valve (V—Fig. C589) and the tip (T). If valve cannot be readily withdrawn from tip with the fingers, soak the assembly in acetone, carbon tetrachloride or equivalent until parts are free.

Fig. C586B — Right side view of a diesel engine fitted with a PES internal drive injection pump. A PES external drive pump (Fig. C586A) is similar except that the transfer pump and check valve are on the left (engine) side of pump.

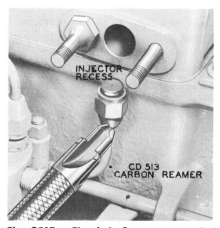

Fig. C587 — The J. I. Case recommended carbon reamer for the injector recesses in cylinder heads.

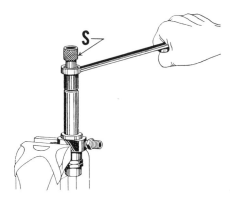

Fig. C588—Removing the injector cap nut. Use centering sleeve (S) when reassembling.

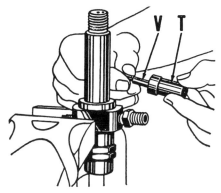

Fig. C589—After injector cap nut is removed, the nozzle tip or body (T) and valve (V) assembly can be lifted off.

The nozzle valve and body (tip) are a mated (non-interchangeable) unit and should be handled accordingly. Neither part is available separately.

All surfaces of the nozzle valve should be mirror bright except the contact line of the beveled seating surface. Polish the valve with mutton tallow used on a soft cloth or felt pad. The valve may be held by its stem in a revolving chuck during this operation. A piece of soft wood well soaked in oil, or a brass wire brush such as Case TSE7748 will be helpful

in removing carbon from the valve. If valve shows any dull spots on the sliding surfaces, or if a magnifying glass inspection shows any nicks, scratches or etchings, discard the valve and the nozzle body; or, send them to an authorized station for possible overhaul.

The inside of the nozzle tip (body) can be cleaned by forming a piece of soft oil soaked wood to a taper of the same angle as the valve seat in the body. A more efficient method is by using the soft metal special Case Nozzle Body Pressure Chamber Scraper TSE7747 and the TSE7735 Valve Seat Scraper. Some Bosch mechanics use an ignition distributor felt oiling wick instead of the soft wood for cleaning the pintle seat in the tip. In this procedure, the mechanic shapes the end of the wick to the seat contour and coats the formed end with tallow for polishing.

The orifice at the end of the tip can be cleaned with a wood splinter. Outer surfaces of the nozzle should be cleaned with a brass wire brush and a soft cloth soaked in carbon solvent. Any erosion or loss of metal at or adjacent to the tip orifice warrants discarding the nozzle body and valve.

The lapped surface of the nozzle body (tip), where it contacts the mating surface of the holder, must be flat and shiny. Any discoloration of these sealing surfaces should be removed by cleaning with tallow on a lapping plate using a figure 8 motion as shown in Fig. C590. If a magnifying glass inspection shows any nicks, scratches, etching or discoloration after cleaning, discard the injector assembly or send the holder and nozzle to an authorized station for re-lapping.

111. NOZZLE HOLDER. Clean the exterior of the holder (except the lapped sealing surface at the end of

same) using a brass wire brush and carbon solvent. The lapped sealing surface must be flat and shiny. Any discoloration should be removed by cleaning with tallow on a lapping plate using a figure 8 motion as shown in Fig. C591. If a magnifying glass inspection shows any etching, scratching or discoloration after cleaning, send the holder and nozzle assembly to an authorized station for re-lapping.

The cap nut for the holder should be cleaned of deposits using a brass wire brush and carbon solvent. Shallow irregularities of the gasket contacting lower surface, which would provide a possible leakage path across the sealing face, should be removed by reduction lapping using emery cloth laid on a flat surface. If the irregularities cannot be corrected in a reasonably short time, use a new nut.

112. SPINDLE AND SPRING. If the shop is not equipped with a nozzle tester, further disassembly of the injector should not be attempted. If a tester is available, remove the spindle and spring (Fig. C592) and wash parts in clean fuel. If spindle is bent, if end which contacts the valve stem is cracked or broken or if attached spring seat is worn, renew the spindle and seat. Reject the spring if same is rusted or shows any surface cracks. Reject upper seat if worn.

113. REASSEMBLY OF INJECTOR. After all of the component parts have been cleaned and checked, lay all of them in a clean container filled with clean diesel fuel. Withdraw the parts from the container as needed and assemble them while they are still wet. Observe the following points before and during assembly.

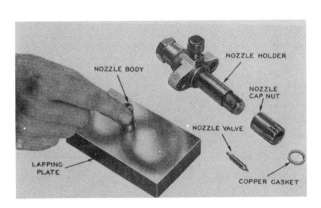

Fig. C590—Cleaning the lapped surface of the nozzle tip (body) using tallow on a lapping plate.

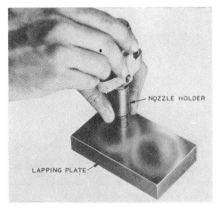

Fig. C591—Cleaning lapped end of nozzle holder, using figure 8 motion on tallow coated lapping plate.

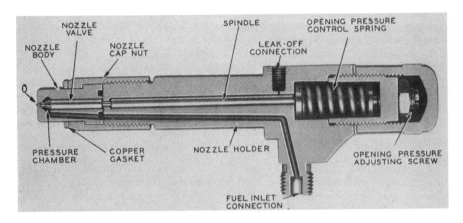

Fig. C592—Sectional view of Bosch throttling type pintle injector. Shops not equipped with a tester should refrain from disturbing the pressure adjusting screw.

Nozzle valve should have a minimum clearance (free fit) in the tip or body. If valve is raised approximately ⅜ of an inch off its seat, it should be free enough to slide down to its seat without aid when the assembly is held at a 45 degree angle.

If parts have been properly cleaned, all sliding and sealing surfaces of same will have a mirror finish. Any scratches, etching etc., on these surfaces warrants renewal; or, sending the mating pieces to an authorized Bosch service station for repairs.

It is highly desirable that the nozzle tip or body be exactly centered in the nozzle cap nut. A centering sleeve should be used for this purpose as shown in Fig. C588. Nozzle cap nut should be tightened to a torque of 50-55 Ft.-Lbs. and the sleeve should be kept free while tightening the nut.

After injector is assembled, set the opening pressure and check the spray pattern and leakage as outlined in paragraphs 114 through 117.

114. **TEST & ADJUST INJECTORS.** A complete job of testing and adjusting the injector requires the use of an approved nozzle tester. The nozzle should be tested for leakage, spray pattern and opening pressure.

115. OPENING PRESSURE. While operating the tester handle, observe the gage pressure at which the spray occurs. The gage pressure should be 1850-2000 psi. The drop in pressure which occurs when the valve opens should not exceed 300 pounds. A greater drop indicates a sticking valve. If the pressure is not as specified, remove the nozzle protecting cap, exposing the pressure adjusting screw and locknut. Loosen the locknut and turn the adjusting screw as shown in Fig. C593 either way as required, to obtain an opening pressure of 1850-2000 psi. Note: If a new pressure spring has been installed in a nozzle holder, adjust the opening pressure to 20 psi higher than others so as to compensate for subsequent fall off in pressure of the new spring.

Note: When testing and adjusting more than one injector of a set from the same engine, there must not be more than 100 pounds difference in opening pressure between any of the injectors.

116. LEAKAGE. The nozzle valve should not leak at any pressure below 1750 psi. To check for leakage, actuate the tester handle slowly and as the gage needle approaches 1750 psi, observe the nozzle tip for drops of fuel. If drops of fuel collect at nozzle valve orifice (O—Fig. C592) at pressures less than 1750 psi, the nozzle valve is not seating properly. Some slight over-flow at leak off port is normal but if greater than on a new nozzle, it indicates wear between the cylindrical surfaces of the nozzle valve and the body.

117. SPRAY PATTERN. Operate the tester handle slowly until nozzle valve opens; then, stroke the lever with light quick strokes at the rate of approximately 100 strokes per minute. If the nozzle has a ragged spray pattern as shown in the two right hand views of Fig. C594, the nozzle valve should be serviced. It is not necessary that the nozzle produce an audible chatter.

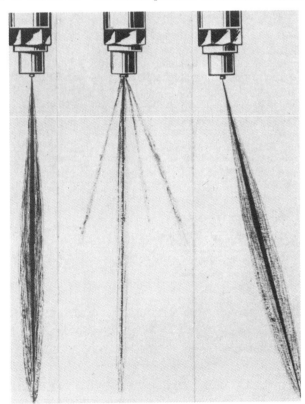

Fig. C594—Spray pattern shown at left is correct. Patterns shown at center and right are incorrect.

Fig. C593 — Adjusting the nozzle opening pressure.

POWRCELS
(Pre-combustion chambers)

These assemblies are mounted in the cylinder head directly opposite the injectors as shown in Fig. C595. The contacting surfaces of the cap and body as well as the stepped surfaces of the body where it fits the cylinder head recess are lapped fits to form a seal. The cap and body are mated parts and cannot be purchased individually.

118. The J. I. Case Company recommends that, whenever an injector is removed for servicing, the mating Powrcel be removed and checked or serviced. It is stated that in almost every instance where a carbon-fouled or burned Powrcel is encountered, the cause is traceable to either a malfunctioning injector, incorrect fuel or incorrect installation of the Powrcel. Manifestations of a fouled or burned unit are misfiring, exhaust smoke, loss of power and/or pronounced detonation (knock) at low engine speeds.

118A. **REMOVAL OF POWRCEL.** Removal of Nos. 1 and 4 Powrcels can be accomplished without removing the manifold. The air cleaner, however, must be removed to provide removal clearance for No. 4. The intake manifold must be removed to provide removal clearance for Nos. 2 and 3 Powrcels.

Using a 1⅛ inch socket, remove the clamp screw. If the cap cannot now be lifted out with a pair of pliers, use a thick washer and threaded screw or the Case CD511 cap puller as shown in Fig. C595A. To remove the body, it is sometimes necessary to use a puller (such as Case CD512 shown in Fig. C595B) which will engage the threads in the open end of the body. Refer to paragraph 118C for procedure to be followed when installing Powrcel.

118B. **CLEANING OF POWRCEL.** Clean all carbon from the front and rear crater of the Powrcel body using a brass carbon scraper or a shaped

piece of hard wood. Clean the first or front orifice (FO—Fig. C595) using a piece of hard wood soaked in Diesel fuel and shaped to enter the orifice. A soft brass tool may also be used.

After cleaning the exterior of the Powrcel with a brass wire brush, soak the parts in carbon solvent, then remove them and examine each piece for burning and "blow-by" or leakage. Reject any pieces that show signs of leakage, and reject the cap and body if there are any burned spots anywhere in the end recesses of the body. If cap and body are O.K. for leakage and burning, their lapped sealing faces should then be cleaned using a figure 8 motion on a tallow-coated lapping plate.

118C. **INSTALLATION OF POWRCEL.** Before installing a cleaned or new unit, the recess for same in the cylinder head must be thoroughly cleaned. Powrcel must fit freely into the recess and it is important that the sealing surfaces at inner step (S1—Fig. C595) be free of any irregularities which would provide a leakage path. Carbon deposits can be removed with a suitable brass wire brush if a special stepped reamer is not available.

Make sure that Powrcel body locating pin is in place, then install Powr-

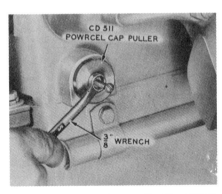

Fig. C595A—Using Case puller No. CD511 to extract the Powrcel cap.

cel with groove of same engaging the body locating pin. Insert the cap and tighten the clamp screw with a 1⅛ inch socket to a torque 100 ft.-lbs.

If the Powrcel locating pin is damaged, it will be necessary to remove the cylinder head to renew the pin.

PSB INJECTION PUMP

Refer to Fig. C596B which shows an American Bosch PSB single plunger fuel injection pump typical of that used on some series 400, 700B and 800B tractors.

The subsequent paragraphs will outline ONLY the injection pump service work which can be accomplished without the use of special, costly pump testing equipment. If additional service work is required, the pump should be turned over to an official Diesel service station for overhaul. Inexperienced service personnel should never attempt to overhaul a Diesel injection pump.

119. **LUBRICATION.** With the exception of the upper portion of the injection pump plunger, the pump and governor unit are dependent on the engine crankcase oil for lubrication. The J. I. Case Company recommends that engine crankcase be drained and refilled at least every 120 hours.

119A. **CHECK AND RESET TIMING.** Recommended injection timing is 26 degrees BTDC for series 400 prior to 8104711; 27 degrees BTDC for 400 after 8104710 and 700B series; 31 degrees BTDC for series 800B. To check the timing, proceed as follows: This check presumes that the engine will start and run but you wish to make sure that the timing is to specifications. Remove pump window cover as shown in Fig. C596. Operate decompression control so engine may be cranked by hand. Crank engine until the number one (front) piston is coming up on compression stroke and continue cranking engine until pointer on timing gear cover is exactly in regis-

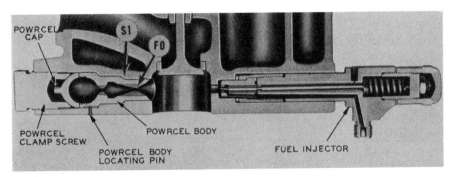

Fig. C595 — Sectional view of the Diesel engine combustion chamber. Nos. 1 and 4 Powrcels can be removed without removing the manifold.

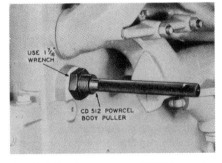

Fig. C595B—Case puller CD512 can be used to extract the Powrcel body.

ter with the correct degree mark on the crankshaft pulley. At this time, the beveled (red) tooth on the pump distributor gear should be visible in the timing window.

Remove plug from timing port in upper surface of drive shaft housing after first removing the pivot cap screw and governor arm attached to it. Sighting through the timing plug port, as shown in Fig. C596A, observe the position of the pointer on pump body with respect to the mark on pump shaft hub. If marks are not in register within $\frac{1}{32}$-inch as shown in left hand portion of Fig. C596A, the timing should be corrected as outlined in the next paragraph.

If the pointer is not in register with hub mark, remove the plate from side port and loosen each of the three cap screws which join the halves of the coupling shown in right hand portion of Fig. C596A. It will be necessary to rotate the engine crankshaft to gain access to all of the screws, which should be loosened only sufficiently to permit one coupling flange to be moved relative to its mating flange.

After coupling screws are loosened, rotate engine crankshaft to bring pulley mark back into register with gear cover pointer, and beveled (red) tooth of pump distributor gear into view in pump window. With engine thus restored to timing position, use a screw driver in the slot of driven coupling and rotate pump shaft until pointer on pump body is exactly in register with mark on pump shaft hub as shown in left hand portion of Fig. C596A. Tighten the three coupling cap screws.

119B. REMOVE PUMP. To remove the injection pump, close the fuel shut-off valve and drain the filters. Disconnect the governor linkage and remove fuel lines from pump. Unbolt and remove pump from drive housing. To install the pump, refer to the following paragraph.

119C. INSTALL AND TIME PSB PUMP. To install a new or repaired PSB injection pump, first assemble the drive coupling to same so that timing mark on coupling will be registered midway between the first and last of the eleven guide marks on the pump shaft hub as shown in Fig. C596B. Temporarily tighten the cap screws to the point where coupling can just be moved relative to the shaft hub flange. Remove the timing window cover from side of pump.

Crank engine until number one piston is coming up on compression stroke and continue cranking until the pointer on timing gear cover is exactly in register with the correct degree mark on

crankshaft pulley (26 degree mark for series 400, prior to 8104711; 27 degree mark for series 400 after 8104710 and 700B series; 31 degree mark for series 800B). At this time, the blank tooth on the pump driving shaft should be approximately in register with the upper outer hole in the mounting flange as shown in Fig. C596C. If the tooth is not so located, remove the drive shaft housing and remesh the drive gear.

Turn the pump cam shaft in the normal direction until the beveled (red) tooth is visible in the pump window and the pointer on pump housing is in register with mark on shaft hub as shown in the left half of Fig. C596A. At this time, the wide tooth on the coupling will be aligned with the outer pump mounting hole as shown in Fig. C596D. Install pump to engine. Timing pointer may go off register with line on coupling, as coupling splines go into mesh. After the pump mounting cap screws are tightened, recheck to make sure that pointer on pump body is exactly in

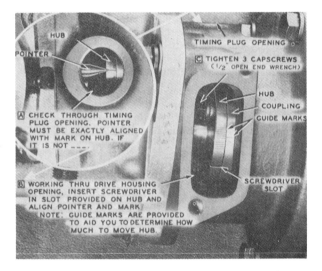

Fig. C596A — Injection pump is correctly timed if pointer on pump housing is registered with line mark on pump driving hub when crankshaft is positioned as described in text.

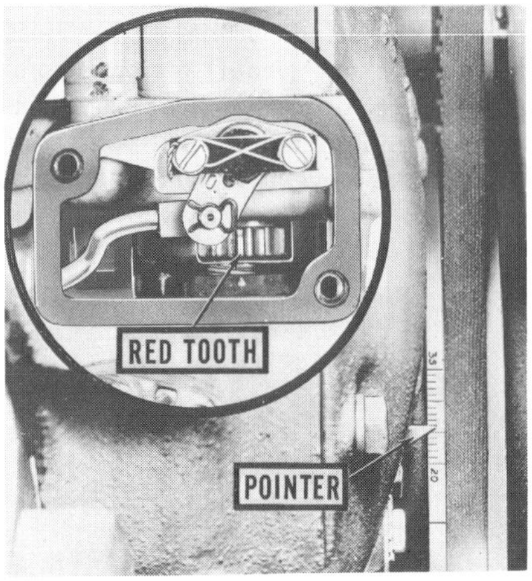

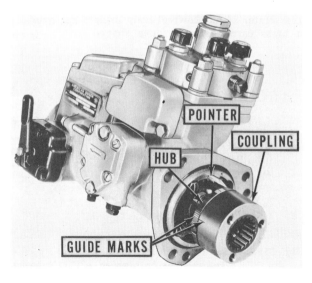

Fig. C596B — Assembled American Bosch PSB injection pump ready for installation. Notice that coupling guide mark must be registered midway between the first and last of the eleven guide marks on the hub.

Fig. C596 — With the number one cylinder on compression stroke and the correct number of degrees BTDC (see text), the beveled red tooth on pump distributor gear should be visible in the timing window.

register with mark on hub (Fig. C596A). If register is not exact turn hub with a screw driver until registration is obtained; then, tighten each of the three coupling cap screws securely. Reinstall the fuel lines and linkage.

119D. TRANSFER PUMP. The PSB pumps are equipped with a positive displacement, gear type transfer pump which is driven by a gear on the injection pump camshaft. The pump is designed to deliver fuel to the final stage filter under pressure.

If pump is not operating properly, it can be renewed as a complete unit or it can be disassembled and cleaned without removing the injection pump from the engine. A thorough cleaning job will often restore the performance of the transfer pump; but if parts are worn, it should be sent to an authorized repair station.

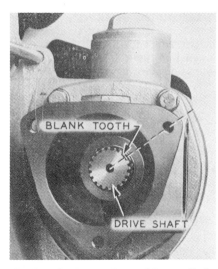

Fig. C596C—When piston of front cylinder is at injection point, the drive shaft attached to pump driving gear will be positioned as indicated.

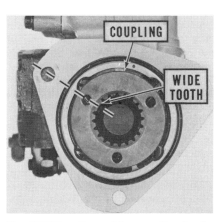

Fig. C596D — Pump drive coupling in the proper position for installing the PSB injection pump to Case Diesel engines.

PES (EXTERNAL DRIVE) INJECTION PUMP

Some series 700B and 800B tractors are equipped with a German (Robert) Bosch type PES, external drive, injection pump; which is a multiple plunger pump that rotates at camshaft speed. When the PES external drive pump is used, the fuel transfer pump is located on the left (engine) side of the injection pump. Refer to Fig. C597.

120. LUBRICATION. The camshaft and governor on the injection pump are lubricated by the oil contained in the pump camshaft compartment. Each time the engine crankcase oil is changed (at least every 120 hours of operation), the injection pump sump should be drained and refilled, with the same type oil as that used in the crankcase, until the oil just begins to drip from the overflow tube.

120A. CHECK AND RESET TIMING. Recommended injection pump timing is 31 degrees BTDC port closing for series 700B; 33 degrees BTDC port closing for series 800B. To check the timing, proceed as follows: Remove the front rocker arm cover and decompress the engine for hand cranking. Crank the engine until both valves of No. 1 cylinder are closed, then continue cranking until the pointer on the timing gear cover is exactly aligned with the correct degree mark on the crankshaft pulley. Remove the plug from the timing hole port on the pump drive housing. If the pump is properly timed, the mark on the pump gear will be exactly aligned with the mark on the pump flange as shown in Fig. C597A. If these marks aren't aligned loosen the pump to drive adapter cap screws and rotate the pump as required until the marks are aligned, then tighten the pump attaching cap screws.

120B. REMOVE AND REINSTALL. To remove the injection pump, first shut off the fuel, then disconnect all lines and control rod from the pump. Unbolt the pump and withdraw same rearward.

When reinstalling the pump first set the engine as follows: Remove the front rocker cover and decompress the engine for hand cranking. Crank the engine until both valves of the No. 1 cylinder are closed, then continue cranking until the pointer on the timing gear cover is exactly aligned with the correct degree mark on the crankshaft pulley (31 degree mark for series 700B; 33 degree mark for series 800B). Remove the plug from the timing hole port on the pump drive housing. Install the injection pump with the timing marks on the pump and gear hub

Fig. C597A—The PES external drive injection pump is properly timed when the timings marks are exactly in register as shown.

Fig. C597 — Right side view of a diesel engine fitted with a PES external drive injection pump.

in relatively close alignment. Install the four pump to drive housing cap screws, turn the pump either way as required to align the timing marks exactly and tighten the pump attaching cap screws. Reconnect the fuel lines and the control rod. After reinstallation is complete, bleed the filters as outlined in paragraphs 99D through 99F or paragraph 100. It may be necessary to loosen the high pressure fuel lines at the nozzles, crank the engine and retighten the lines after all air has escaped.

PES (INTERNAL DRIVE) INJECTION PUMP

Some 700B and 800B tractors are fitted with a German (Robert) Bosch PES internal drive injection pump. As shown in Fig. C598 the internal drive PES pump has the transfer pump mounted on the right side (outside) of the injection pump.

121. **LUBRICATION.** On the PES internal drive injection pump, the camshaft and governor are lubricated by the oil contained in the pump camshaft compartment. Each time the engine crankcase oil is changed (at least every 120 hours of operation), the injection pump sump should be drained and refilled, with the same type of oil as that used in the crankcase, until the oil just begins to drip from the overflow tube.

121A. **CHECK AND RESET TIMING.** Recommended injection timing is 31 degrees before top center, port closing for series 700B and 33 degrees before top dead center, port closing for series 800B. To check timing, remove front rocker lever cover and decompress and safety the engine for hand cranking. Crank engine until both valves of No. 1 cylinder are

closed, then slowly until the pointer on timing gear cover is in register with the correct timing degree mark on the crankshaft pulley. Remove the inspection plate from the side port (Fig. C598A). At this time, the **line mark** on end of pump body (viewed through side port) should be in register, within $\frac{1}{32}$-inch either way, with the line mark

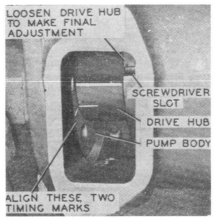

Fig. C598A—On PES internal drive injection pumps a line mark on the pump body is used as the index point for the line mark on the coupling hub. Final timing adjustment is by means of adjustable coupling.

on the injection pump hub as shown. If marks are in register, the timing is O.K. If marks are not in register, retime the pump as outlined in next paragraph.

Loosen each of the coupling screws which join the halves of the coupling. It will be necessary to rotate the crankshaft to reach all of the screws which should be loosened only sufficiently to permit coupling flange to be moved relative to its drive flange. After coupling screws are loosened, rotate the crankshaft to again bring the pulley timing mark into register with timing gear cover pointer when both valves of No. 1 cylinder are closed. With engine thus restored to timing position, use a screw driver in the slot in coupling and rotate pump shaft until line mark on pump body is exactly in register with the mark on pump shaft hub as shown in Fig. C598A. Tighten the coupling cap screws.

121B. **REMOVE AND REINSTALL.** Procedure for removal of pump from the engine is self-evident as is the procedure for installation. During installation, time the pump to the engine as outlined in paragraph 121A.

DIESEL GOVERNOR

The speed control governors built into the injection pumps seldom require adjustment and are provided with wire and lead seals which must be broken for access to the adjusting screws. If engine is difficult to start, lacks power, is sluggish, does not respond to load or is difficult to stop, inspect the system as outlined in paragraph 122 or 123.

Fig. C598 — Right side view of a diesel engine fitted with a PES internal drive injection pump. A PES external drive injection pump (Fig. C597) is similar except that the transfer pump is on the left (engine) side.

PSB INJECTION PUMPS

122. **SYSTEM INSPECTION.** Check the linkage from the hand throttle control lever to the governor and make sure the linkage is free and that full travel of the hand lever is obtained. Check to make certain that the roll pin at forward end of throttle rod is in a vertical position. If pin is not vertical, it may be disengaged from the throttle lever. To set the pin in a vertical position, loosen one of the adjusting nuts at rear of throttle rod and turn the rod as required. Make certain that the ball joints are threaded fully in so they will not bind in the steering column.

Excessive gum or varnish deposits on the governor control arm in the injection pump will cause sluggish governor action.

To check the control arm for freedom of movement with engine stopped, place the hand throttle lever in the stop position and remove the timing window cover from side of injection pump as shown in Fig. C599.

Move the control arm to the right as far as possible. When the arm is released, it must fall back of its own weight. If it does not, the arm is binding.

This binding can usually be eliminated by adding an approved fuel conditioner to the Diesel fuel. If the binding cannot be eliminated by this method, proceed as follows:

Refer to Fig. C599 and remove the hairpin cotter pin, safety wire and bridge. Move the control arm rearward and withdraw the control unit from the pump. Remove the hex nut and separate the control arm shaft from the bushing. To free-up the shaft in the bushing, use a fine lapping compound such as American Bosch No. BM10007 or equivalent. After the lapping operation, thoroughly wash the units and reassemble.

122A. **ADJUSTING ENGINE SPEED.** The recommended high idle no-load speeds are as follows:

Series 400 Prior to 8104711
Engine .... 1600 rpm
Pto .... 596 rpm
Belt pulley .... 1244 rpm

Series 400 After 8104710
Engine .... 1635 rpm
Pto .... 608 rpm
Belt pulley .... 1271 rpm

Series 700B
Engine .... 1650 rpm
Pto .... 615 rpm
Belt pulley .... 1295 rpm

Series 800B
Engine .... 1950 rpm
Pto .... 725 rpm
Belt pulley .... 1222 rpm

The recommended slow idle speeds are as follows:

Series 400 and 700B
Engine .... 600 rpm
Pto .... 223 rpm
Belt pulley .... 467 rpm

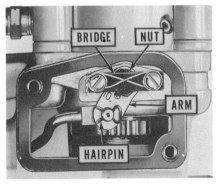

Fig. C599—Side view of PSB injection pump with the timing window cover removed. The control arm can be removed for cleaning purposes.

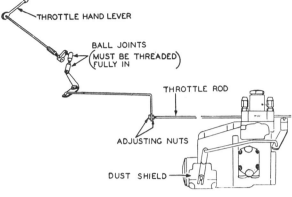

Fig. C600 — The high speed adjusting screw and speed control linkage adjustments for PSB injection pump equipped tractors. The high speed stop screw is accessible after removing the sealed dust shield.

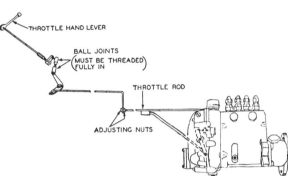

Fig. C601 — The speed control linkage used with PES injection pumps. Refer to text for speed adjusting procedure.

Series 800B
Engine .... 600 rpm
Pto .... 223 rpm
Belt pulley .... 376 rpm

Before attempting to adjust the engine speed, first make certain that the system linkage is in good condition and free of drag or friction as per paragraph 122.

122B. **LOW IDLE SPEED.** To check and adjust the engine low idle speed, which should be 600 rpm, start engine and allow to warm thoroughly. Move the hand control lever to the low idle position and, using an accurate tachometer, note the engine speed. If the engine speed is not 600-700 rpm, adjust the length of the throttle rod at adjusting nuts (Fig. C600) until 600 rpm is obtained.

122C. **HIGH IDLE NO-LOAD SPEED.** If the high idle no-load engine speed, is lower than specified in paragraph 122A when checked with an accurate tachometer, first check and adjust the engine low idle rpm as in paragraph 122B; then, recheck the high idle speed. If the high idle speed is still incorrect, disconnect the linkage rod from the injection pump control lever, hold the control lever in the wide open position and check the engine speed. If the speed is **correct** during the last check, the incorrect high idle speed with the linkage connected is due to bent or binding linkage preventing full travel (Refer to paragraph 122).

If after the above checks are made the engine speed is still low or if the engine speed is too high, remove the sealed dust shield (Fig. C600) from the side of the pump, rotate the high speed adjusting screw in or out as required until the correct engine rpm is obtained and lock adjusting screw with the jam nut.

PES INJECTION PUMPS

123. **LINKAGE INSPECTION.** Check all of the linkage from the hand throttle lever to the governor. Make sure that the hand throttle has full travel and the linkage is not bent, binding nor loose due to excessive wear.

123A. **ADJUSTING ENGINE SPEEDS.** The recommended high idle no-load engine speeds are as follows:
Series 700B
Engine .... 1650 rpm
Pto .... 615 rpm
Belt pulley .... 1286 rpm
Series 800B
Engine .... 1950 rpm
Pto .... 725 rpm
Belt pulley .... 1222 rpm

Recommended low idle speed is 600 rpm. Before adjusting the engine speeds, make certain that the linkage is in good condition (not bent, binding nor loose due to wear) and free with no excessive friction. The hand throttle lever should have complete travel.

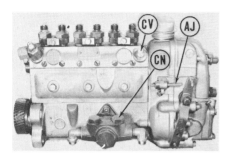

Fig. C601A—View of the left side of a removed PES external drive injection pump showing the high speed adjusting screw (AJ), check valve (CV) and cap nut (CN). Pumps for four cylinder engines are similar.

123B. LOW IDLE SPEED. To check and adjust the low idle speed, which should be 600 rpm, start the engine and allow to warm thoroughly. Move the hand control lever to the low idle position and, using an accurate tachometer, note the engine speed. If the engine speed is not 600-700 rpm, adjust the length of the throttle rod at the adjusting nuts (Fig. C601) until 600 rpm is obtained.

123C. HIGH IDLE NO-LOAD SPEED. If the high idle no-load engine speed is lower than specified in

paragraph 123A when checked with an accurate tachometer, first check and adjust the engine low idle rpm as in paragraph 123B; then, recheck the high idle speed. If the high idle speed is still incorrect, disconnect the linkage rod from the injection pump control lever, hold the control lever in the wide open position and check the engine speed. If the speed is **correct** during the last check, the incorrect high idle speed with the linkage connected is due to bent or binding linkage preventing full travel.

If after the above checks are made the engine speed is still low or if the engine speed is too high, remove the seal from the high speed adjusting screw (AJ — Fig. C601A), turn the screw in or out as required until the correct engine rpm is obtained and lock the adjustment with the jam nut.

123D. SURGE ADJUSTMENT. If the engine surges at engine low idle speed, remove the cap nut (Fig. C601B) and loosen the jam nut. Turn the supplementary idle adjusting screw in until surge is removed. NOTE: Do not turn the screw in far enough to increase the low idle speed.

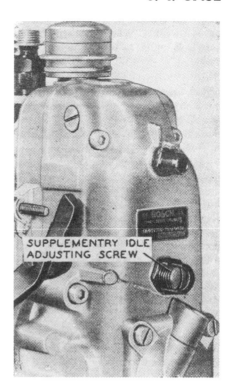

SUPPLEMENTRY IDLE ADJUSTING SCREW

Fig. C601B—View of the rear end of a PES external drive injection pump showing the supplementary idle (surge) adjusting screw.

NON-DIESEL GOVERNOR

Non-Diesel governors are of the centrifugal flyweight type with the weight unit mounted on the camshaft gear.

135. SPEED ADJUSTMENT. The specified governed speeds for non-diesel engines are as follows:

Series 400 — rpm
High idle no-load engine speed . 1600
Loaded engine speed 1450
Low idle engine speed 600
High idle no-load pto speed 596
Loaded pto speed 540
Low idle pto speed 223

Series 700B — rpm
High idle no-load engine speed . 1655
Loaded engine speed 1500
Low idle engine speed 600
High idle no-load pto speed 615
Loaded pto speed 557
Low idle pto speed 223

Series 800B — rpm
High idle no-load engine speed . 2000
Loaded engine speed 1800
Low idle engine speed 600
High idle no-load pto speed 1253
Loaded pto speed 1128
Low idle pto speed 376

Before attempting to adjust the governed speeds, make certain that the carburetor is properly adjusted. On same tractors there are one or two adjustable stop collars (S1 and S2—Fig. C608) which should be loosened before adjusting the engine speeds and repositioned, after the speeds are adjusted, to just contact boss on side of engine when the control lever is moved to the high and low speed positions.

A low idle speed of 600 rpm can be obtained as follows: Start engine and

Fig. C608—Some tractors are equipped with one or two governor control rod stop collars. Low idle stop collar is shown at (S2) high idle stop collar at (S1).

allow to run until it reaches operating temperature, move the hand throttle lever to the low idle position and turn the low idle speed stop screw on carburetor either way as required. In some cases it may be necessary to vary the length of the link (Fig. C609) by turning the adjusting nuts in-order-to obtain full travel. Reposition stop collar (S2—Fig. C608) if so equipped.

The recommended high idle speed can be obtained by turning the stop bolt on the hand control lever until

the correct speed is obtained. In some cases it may be necessary to vary the length of the link (Fig. C609) by turning the adjusting nuts in-order-to obtain full travel. Reposition stop collar (S1—Fig. C608) if so equipped.

NOTE: If the engine high idle no load rpm cannot be obtained, check the linkage synchronization as per paragraph 136. If, on series 800B tractors, the linkage is synchronized but 2000 rpm still cannot be obtained, refer to Fig. C610A and check to make certain the correct governor spring is installed.

136. LINKAGE. To synchronize the governor to carburetor linkage, dis-

If governor action is sluggish or if engine surges, check for worn or binding linkage or connect the carburetor rod and throttle control rod spring from governor lever. Hold the governor lever rearward as far as possible and measure the distance from the center of the ball joint hole in governor lever to the machined front face of cylinder block as shown in Fig. C610. Distance should be 2¾ inches as shown. If the measured distance is not as specified, bend the governor lever until the proper measurement is obtained.

NOTE: Lever should be removed before bending.

Hold the carburetor throttle lever in the wide open position and adjust the length of the governor lever to carburetor rod so that when rod is connected to governor lever, the center of the ball joint hole in lever will be 2⅞ inches from the machined front face of the cylinder block. Reconnect the governor spring and adjust the governed speed.

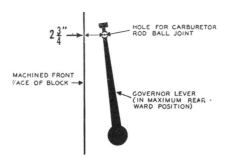

Fig. C610 — Checking the position of the non-diesel governor control lever. Refer to text if lever is not in the specified position.

137. OVERHAUL. The governor weight unit can be overhauled by removing only the timing gear cover cap as shown in Fig. C611. Withdraw the thrust bearing and sleeve from end of camshaft. Remove snap rings and pins which retain the weight units to the camshaft gear and withdraw the weight units. Disassemble the weight units, but be sure to keep the parts in order so they can be reassembled in their original position.

Check the weight pins and pin holes in camshaft gear and fingers for wear. Check the contacting surfaces of fingers and sleeves for wear. The sleeve

contacting surface of the fingers must be smooth and rounded. Weight fingers are sold only in sets of four. Finger contacting surface of weights must be smooth and free from burrs. Examine bolts and bolt holes for wear.

When assembling the weight units, install the bolts so that nuts will be on opposite sides when installed on camshaft gear as shown. Tighten the nuts to remove excessive end play from the weights, but make certain that weights and bolts turn freely. Weight units must pivot freely on camshaft gear. Thrust bearing must turn freely without drag.

Governor lever shaft must turn freely in its needle bearings. NOTE: The governor lever, lever shaft and finger are a balanced assembly and are available only as such for repairs. To renew the needle bearings, remove the finger, withdraw the lever and shaft and remove the worn bearings. Press new bearings in position, using Case arbor No. A8349 or equivalent. Always press against the end of the bearing that carries the identification. Install a new seal in the lever hub and be sure to install the bearing washer.

Recheck the linkage and adjust the governed speed.

Fig. C610A — Governor spring type "A" (Case part number A7294) has 18 active coils and is used with weights "W1" (Case part number A-7278) on series 400 and 700B non-diesel tractors. Governor spring type "B" (Case part number A-10617) has 17 active coils and is used with weights "W1" on series 800B tractors prior to serial number 8122783. Governor spring type "C" (Case part number A-8894) has 16 active coils and is used with weights "W2" (Case part number A7277) on series 800B tractors after serial number 8122782.

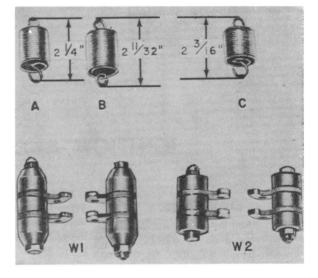

Fig. C609—In some cases it may be necessary to vary the length of the link in order to obtain full travel.

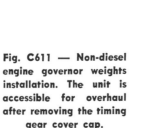

Fig. C611 — Non-diesel engine governor weights installation. The unit is accessible for overhaul after removing the timing gear cover cap.

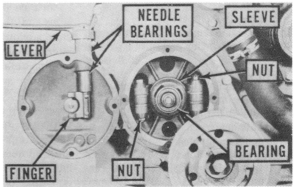

COOLING SYSTEM

RADIATOR

138. To remove the radiator, drain cooling system and remove hood and grille. On models so equipped, disconnect the shutter control wire. Disconnect the radiator hoses, then unbolt and remove radiator from tractor.

THERMOSTAT

139. Thermostat is located in the thermostat housing which is bolted to front of water manifold. The renewal procedure is evident. The thermostat starts to open at 175-180 degrees F. and is fully open at 195 degrees F.

WATER PUMP

140. **R&R AND OVERHAUL.** To remove the water pump, drain cooling system and remove hood and radiator. Disconnect hoses, remove fan blades and belt and unbolt pump from cylinder block. With the pump removed

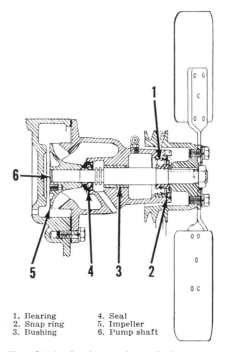

1. Bearing	4. Seal
2. Snap ring	5. Impeller
3. Bushing	6. Pump shaft

Fig. C612—Sectional view of the water pump. Shielded side of bearing (1) goes toward front of tractor.

from tractor, remove the impeller with a suitable puller attached to the two threaded holes. Remove pulley from pump shaft, extract snap ring (2 —Fig. C612) and bump shaft and bearing assembly out of pump body. Press seal out of pump body with a ¾ inch rod. Running clearance between pump shaft and bushing should be 0.0015-0.0035. Inside diameter of bushing (3) should be 0.874-0.875.

Renew the bushing and/or shaft if the running clearance exceeds 0.005. Bushing will not require final sizing if it is installed, using a closely fitting arbor.

Install bearing (1) so that shielded face is toward front of tractor. Pack the cavity between bushing and bearing with No. 2 gun grease. Do not use water pump grease. Press the impeller on the pump shaft until the distance from the rear face of impeller to gasket face of pump body is 1.130-1.140.

IGNITION AND ELECTRICAL SYSTEM

DISTRIBUTOR

141. Either an Auto-Lite No. IAD-6003-1B or a Wico No. X30169 battery ignition distributor may be used on Series 400, 700B and 800B tractors. Specification data on both distributors are as follows:

Auto-Lite IAD6003-1B
Breaker contact gap.......0.018-0.022
Breaker arm spring pressure.17-20 oz.
Cam angle ....42 degrees
Condenser capacity
 (microfarads) 0.023-0.026
 Advance data is in distributor degrees and distributor rpm.
Start advance .... 0° @ 275 rpm
Intermediate advance .. 1° @ 310 rpm
Intermediate advance .. 7° @ 490 rpm
Intermediate advance ..12° @ 645 rpm
Maximum advance ....13° @ 680 rpm

Wico X30169
Breaker contact gap.......0.019-0.022
Breaker arm spring pressure.20-28 oz.
Cam angle ....45 degrees
Condenser capacity
 (microfarads) 0.30-0.34
 Advance data is in distributor degrees and distributor rpm.
Start advance .... 0° @ 275 rpm
Intermediate advance.. 1° @ 300 rpm
Intermediate advance..10° @ 500 rpm
Maximum advance ...14° @ 1000 rpm

142. **INSTALLATION AND TIMING.** With the oil pump properly installed (Refer to paragraph 79), crank engine until number one piston is coming up on compression stroke and continue cranking until the "DC" mark on

crankshaft pulley is in register with pointer on timing gear cover as shown in Fig. C613. Install the distributor so that rotor arm is in the number one firing position and adjust the breaker contact gap to 0.020. Loosen the distributor clamp bolt, turn the distributor either way as required until breaker contacts are just beginning to open and tighten the clamp bolt. With the engine running at the high idle, no-load speed the spark should occur 26 degrees BTDC for all tractors equipped with the Auto-Lite distributor, 26 degrees BTDC for series 700B tractors equipped with the Wico distributor, 28 degrees for series 800B tractors equipped with the Wico distributor. Firing order is 1-3-4-2.

MAGNETO

143. J. I. Case model 41 magneto is used. Breaker contact gap is 0.010; condenser capacity 0.020 microfarads + or − 10%.

144. INSTALLATION AND TIMING. To install the magneto, crank engine until the number one piston is coming up on compression stroke and continue cranking until the "DC" mark on crankshaft pulley is in register with pointer on timing gear cover as shown

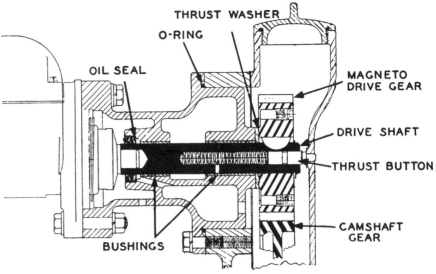

Fig. C615—Sectional view of the magneto drive housing.

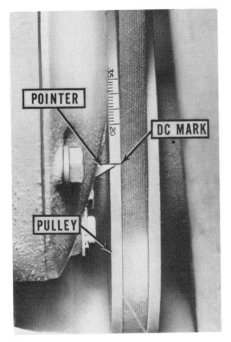

Fig. C613 — On non-Diesel engines, the static spark should occur when No. 1 piston is at top dead center of the compression stroke. Refer to text for running timing.

in Fig. C613. At this time, the slot in the magneto drive coupling (Fig. C614) will be 25 degrees beyond the horizontal center line in the direction of rotation as shown. If the coupling slot is not in this position, it will be necessary to remove and reinstall the magneto drive housing as outlined in paragraph 145.

Insert a spark plug wire in the upper right (No. 1) terminal of magneto cap and hold the other end of the wire about ⅛ inch from magneto frame. Turn the impulse coupling in normal direction of rotation until a spark occurs at the ⅛ inch gap. Without changing the setting of the magneto or the engine, install the magneto and tighten the cap screw and nut only finger tight. Rotate top of magneto toward engine as far as it will go, crank engine one complete revolution and again line up the "DC" mark with the pointer on timing gear cover. Rotate top of magneto away from engine until the impulse coupling just snaps; then, tighten the magneto mounting cap screw and nut securely. With the engine running at 1600 rpm, the spark should occur 25° before top center. Firing order is 1-3-4-2.

145. DRIVE UNIT—R & R AND OVERHAUL. To remove the drive unit, first crank engine until number one piston is coming up on compression stroke and continue cranking until the "DC" mark on crankshaft pulley is in register with the pointer on the timing gear cover. Remove the magneto, unbolt and withdraw the drive housing assembly being careful not to drop in the timing gear hous-

ing the thrust button and spring, located in forward end of drive shaft. Refer to Fig. C615. Extract the snap ring retaining the magneto drive gear to the drive shaft and using a suitable puller, remove the magneto drive gear. Withdraw the shaft and coupling from rear of housing. Remove and discard the oil seal located in the housing.

Inside diameter of the drive shaft bushings is 0.987-0.988. Drive shaft diameter is 0.985-0.9855. Renew the drive shaft and/or bushings if they are excessively worn. If the magneto drive gear is renewed, be sure to check the timing gear backlash which should be 0.002-0.006.

Install new seal in housing with lip of same facing toward front of tractor. Coat lip of seal with Lubriplate or

Fig. C614—With number one piston at top center of compression stroke, the magneto drive coupling should be in the position shown.

Fig. C616—Timing gear marks for magneto equipped non-Diesel engines.

petroleum jelly and install the drive shaft. Install the thrust washer and press the drive gear on the shaft just enough to install the retaining snap ring.

When installing the drive housing, be sure to renew the housing "O" ring. Remove the oil filler cap from timing gear housing in order to view the timing marks and install the drive housing assembly, meshing the single punch marked tooth on camshaft gear with the double punch marked tooth space on the magneto drive gear as shown in Fig. C616.

Note: When the drive housing is properly installed and the number one piston is in top dead center position, the slot in the magneto drive coupling (Fig. C614) will be 25 degrees beyond the horizontal center line in the direction of rotation as shown.

Install and time the magneto as outlined in paragraph 144.

ELECTRICAL UNITS

Diesel Models

146. Diesel tractors have a 12-volt system (two 6-volt batteries in series) with a negative ground. The following electrical units are used:

Generators
    Auto-Lite GHH6001J
    Delco-Remy 1100369
Regulators
    Auto-Lite VRT-4104A
    Delco-Remy 1118991
Starting Motors (Series 400 and 700B)
    Auto-Lite MCK-4005
    Delco-Remy 1113111
Starting Motors (Series 800B)
    Auto-Lite MCK-4007
    Delco-Remy 1113111

Test specifications are as follows:
Generators
Auto-Lite GHH6001J
  Brush spring tension.......24-26 oz.
  Field draw
    Volts 10.0
    Amperes 1.6-1.7
  Output (Cold)
    Maximum amperes 4.0
    Volts 14.0
  Maximum rpm 1150
Delco-Remy 1100369
  Brush spring tension........28 oz.
  Field draw
    Volts 12
    Amperes 1.58-1.67
  Output (Cold)
    Maximum amperes 20
    Volts 14.0
  Maximum rpm 2300

Regulators
Auto-Lite VRT-4104A
  Cut-out relay
    Air gap 0.031-0.034
    Point gap 0.015
    Closing voltage 12.6-13.6
    Opening amperes 3.0-5.0
  Voltage regulator
    Air gap 0.048-0.052
    Voltage range 14.3-14.7
    Voltage @ 80° F.............14.5
Delco-Remy 1118991
  Cut-out relay
    Air gap 0.020
    Point gap 0.020
    Closing range 11.8-14.0
    Voltage adjustment 12.8
  Voltage regulator
    Air gap 0.0750
    Voltage range 13.6-14.5
    Setting adjustment 14.0

Starting Motors
Auto-Lite MCK-4005 and MCK-4007
  Brush spring tension.......52-65 oz.
  No-load test
    Volts 10.0
    Amperes 130
    RPM 4700
  Lock test
    Volts 4
    Amperes 1300
    Torque, Ft.-Lbs. 33.0
Delco-Remy 1113111
  Brush spring tension........48 oz.
  No-load test
    Volts 11.5
    Maximum amperes 50
    Minimum rpm 6000
  Lock test
    Volts 3.3
    Maximum amperes 500
    Minimum torque, Ft.-Lbs......22

Non-Diesel Models

147. Early series 400 non-diesel tractors have a 6-volt electrical system; late series 400 and all series 700B and 800B tractors have a 12-volt system. Both 6 and 12-volt systems are negative grounded. Electrical unit application is as follows:

Early Series 400 (6-volt)
Generator
  Auto-Lite GHD6001AF
Regulator
  Auto-Lite VRR4104A
Starting Motor
  Auto-Lite MAX4095

Late Series 400 (12-volt)
Generator
  Auto-Lite GHH6001J
Regulator
  Auto-Lite VRT4104A
Starting Motor
  Auto-Lite MDL6015

Series 700B and 800B
Generators
  Auto-Lite GHH6001J
  Delco-Remy 1100369
Regulator
  Auto-Lite VRT4104A
  Delco-Remy 1118991
Starting Motor
  Auto-Lite MDL6015
  Delco-Remy 1107750

Test specifications are as follows:
Generators
Auto-Lite GHD6001AF
  Brush spring tension.......26-46 oz.
  Field draw
    Volts 5.0
    Amperes 2.0-2.2
  Output (Cold)
    Maximum amperes 8.0
    Volts 7.2
    Maximum rpm 1150
Auto-Lite GHH6001J
  Brush spring tension.......26-46 oz.
  Field draw
    Volts 10.0
    Amperes 1.6-1.7
  Output (Cold)
    Maximum amperes 14.7
    Volts 15.0
  Motoring draw
    Volts 10.0
    Amperes 2.9-3.3
Delco-Remy 1100369
  Brush spring tension........28 oz.
  Field draw
    Volts 12
    Amperes 1.58-1.67
  Output (Cold)
    Amperes 20
    Volts 14.0
    RPM 2300

Regulators
Auto-Lite VRR4104A
  Cut-out relay
    Air gap 0.031-0.034
    Point gap 0.015
    Closing voltage 6.3-6.8
    Opening amperes 4.0-6.0
  Voltage regulator
    Air gap 0.048-0.052
    Voltage range 7.0-7.4
    Voltage @ 80° F.............7.2
Auto-Lite VRT4104A
  Cut-out relay
    Air gap 0.031-0.034
    Point gap 0.015
    Closing voltage 12.6-13.6
    Opening amperes 3.0-5.0
  Voltage regulator
    Air gap 0.048-0.052
    Voltage range 14.3-14.7
    Voltage @ 80° F.............14.5

Delco-Remy 1118991
  Cut-out relay
    Air gap 0.020
    Point gap 0.020
    Closing range 11.8-14.0
    Voltage adjustment 12.8
  Voltage regulator
    Air gap 0.0750
    Voltage range 13.6-14.5
    Setting adjustment 14.0

Starting Motors
Auto-Lite MAX4095
  Brush spring tension......42-53 oz.

No-load test
  Volts 5.0
  Amperes 65
  RPM 4900
Lock test
  Volts 2.0
  Amperes 410
  Torque, Ft.-Lbs. 8.0
Auto-Lite MDL6015
  Brush spring tension......31-47 oz.
  No-load test
    Volts 10.0
    Amperes 60
    RPM 3200

Lock test
  Volts 4.0
  Amperes 225
  Torque, Ft.-Lbs. 6.0
Delco-Remy 1107750
  Brush spring tension........35 oz.
  No-load test
    Volts 10.3
    Maximum amperes 90
    Minimum rpm 3800
  Lock test
    Volts 5.8
    Maximum amperes 435
    Minimum torque, Ft.-Lbs......7.0

CLUTCH (SERIES 400 AND 700B)

Refer to paragraphs 156A through 156N concerning the torque convertor used on series 800B tractors.

CLUTCH

Some special tractors are equipped with a hand operated, over-center type clutch. Other models are equipped with a 12-inch spring loaded type.

ADJUSTMENT
Spring Loaded Type

148. Adjustment to compensate for lining wear is accomplished by adjusting the clutch pedal linkage, NOT by adjusting the position of the clutch release levers. The clutch linkage is properly adjusted when the pedal has a free travel of 1-inch when measured at platform. Adjustment is accomplished by varying the length of the throwout rod. Refer to Fig. C617.

Fig. C618 — Lock pin must be disengaged as shown before adjusting the over-center clutch used on some tractors.

Over-Center Type

149. Adjustment to compensate for lining wear is accomplished by adjusting the clutch unit as follows: Place the gear shift lever in neutral and the clutch hand lever in the disengaged position. Remove the hand hole cover from left side of clutch housing. Turn the engine flywheel until the clutch adjusting lock pin is aligned with the opening; then, pull the pin out and rotate it ¼ turn as shown in Fig. C618.

Working through the hand hole in clutch housing, turn the adjusting ring (AR—Fig. C622) counter-clockwise, one hole at a time until the operating lever requires a distinct pressure to engage the clutch. When adjustment is complete, release the lock pin, making certain that it engages a hole in the adjusting ring. Reinstall the clutch housing hand hole cover.

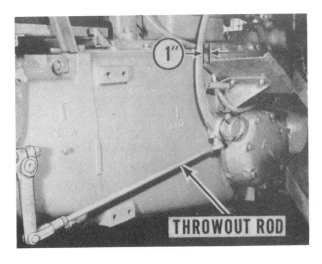

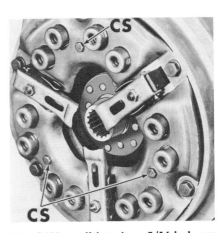

Fig. C617—Clutch throwout rod and foot pedal installation typical of all 400 and 700B series tractors equipped with a spring loaded clutch.

Fig. C620 — Using three 5/16-inch cap screws (CS) to compress the spring loaded clutch cover assembly prior to removing the assembly from flywheel.

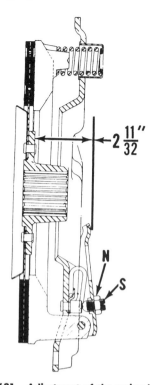

Fig. C621—Adjustment of the spring loaded clutch release levers is accomplished with screws (S). Notice that the lever height is measured from the machined face of the driven disc hub.

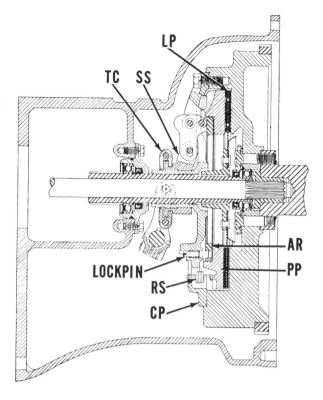

Fig. C622—Sectional view of clutch housing and over-center type clutch used on some tractors.

AR. Adjusting ring
CP. Cover plate
LP. Lined plate
PP. Pressure plate
RS. Release spring
SS. Sliding sleeve
TC. Throwout collar

R&R AND OVERHAUL
Spring Loaded Type

150. To remove the clutch, first detach (split) engine from clutch housing as outlined in paragraph 153, and proceed as follows:

Install three $\frac{5}{16}$ by 1¼ inches long cap screws (CS—Fig. C620) and tighten the screws evenly to relieve pressure from the lined plate. Unbolt and withdraw clutch from flywheel.

With the cover assembly still compressed with the $\frac{5}{16}$ inch cap screws, remove the cotter pin from each release lever pin, tap the pins out and remove the release levers. Place a bar across the clutch cover plate and use a press to maintain the spring compression while removing the three $\frac{5}{16}$ inch cap screws.

With the three cap screws removed, slowly release the compressing pressure and disassemble the clutch cover assembly.

Thoroughly clean all parts, inspect them and renew or recondition as required. Each of the twelve pressure springs should test at least 180 lbs. when compressed to $1\frac{13}{16}$ inches.

Reassemble the clutch by reversing the disassembly procedure and use a

spare clutch shaft to align the driven plate splines with the clutch shaft pilot bearing. Tighten the cover retaining cap screws to a torque of 50 ft.-lbs. and remove the three $\frac{5}{16}$ inch compressing cap screws.

Loosen lock nuts (N—Fig. C621) and turn the release lever adjusting screws (S) either way as required until the distance from the release bearing contacting surface of each release lever to the machined surface of the driven plate hub is $2\frac{11}{32}$ inches as shown. Tighten the lock nut when the adjustment is complete.

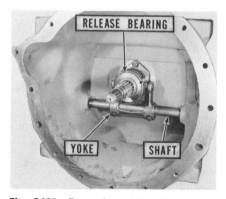

Fig. C625—Front view of the clutch housing, showing the spring loaded clutch release bearing installation.

Over-Center Type

151. To remove the over-center type clutch, first detach (split) engine from clutch housing as outlined in paragraph 153, and proceed as follows: Correlation mark the clutch cover and flywheel to assure correct reinstallation, remove the cap screws retaining cover to flywheel and withdraw the cover assembly and lined plate.

The procedure for disassembling and overhauling the clutch is conventional and evident after an examination of the unit. Pins should have a floating fit in the cover, levers and sleeve. Threads on adjusting ring must be in good condition and not burred. Lock pin spring must have sufficient tension to hold the pin in the engaged position.

When installing the clutch, align the previously affixed marks and adjust the unit as in paragraph 149.

TRACTOR SPLIT
Series 400-700B

153. To detach engine from clutch housing, drain cooling system and remove hood. On tractors equipped with a radiator shutter, remove the grille and disconnect the shutter control

cable and the head light wires. On standard, non-adjustable front axle models, disconnect the drag link and rear end of radius rod. On all models disconnect the heat indicator sending unit from water manifold, ammeter wire from generator, battery cables and starter solenoid wires.

On series 400 tractors, disconnect the hour meter bracket and oil line from hour meter pressure switch to engine block. If tractor is not equipped with an hour meter, disconnect oil line from engine block to oil pressure gage at connection above the starting motor. On series 700B detach the hour meter cable and oil pressure gage line from the engine.

On all models, remove the flywheel dust shield, and on tricycle and adjustable axle models, loosen the clamp bolt on the steering shaft universal joint and unbolt the steering shaft support from clutch housing.

On Diesel models, shut off fuel and disconnect fuel lines at top of tank and shut-off valve. Decompress the engine and disconnect the decompression linkage at lower right front end of fuel tank. Remove the rear nut on throttle rod and disconnect the fuel pressure gage line at inlet connection on second stage filter and vacuum gage line at bleed valve connection above the first stage fuel filter. Disconnect the ether starting aid tube from intake manifold.

On non-Diesel models, shut off fuel and disconnect choke wire and fuel line at carburetor. Disconnect the throttle control rod at a point just under front end of fuel tank. Disconnect wire from coil, or on magneto equipped engines, disconnect the grounding wire.

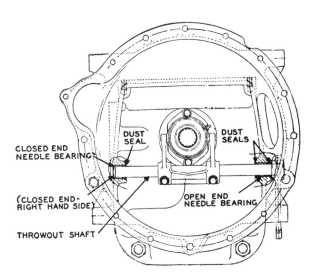

Fig. C627—Front view of clutch housing, showing the installation of the clutch throwout shaft needle bearings and seals.

On all models, swing engine in a hoist, unbolt engine from clutch housing and separate the tractor halves.

When connecting engine to clutch housing, tighten the flange bolts to a torque of 80 ft.-lbs., flange cap screws and dust shield cap screws to a torque of 50 ft.-lbs.

RELEASE BEARING OR COLLAR AND THROWOUT SHAFT
Series 400-700B

154. To remove the spring loaded clutch release bearing, detach (split) engine from clutch housing as outlined in paragraph 153 and disconnect the throwout rod from the lever. Withdraw the release bearing and sleeve from the release sleeve tube (Fig. C625).

154A. To remove the over-center clutch throwout collar, remove the hand hole cover from left side of clutch housing. Remove the bolts joining the collar halves and withdraw same from tractor.

154B. To remove the throwout shaft and yoke, first detach (split) engine from clutch housing as in paragraph 153 and remove the release bearing or collar.

Loosen the throwout yoke clamp screws. Pull the throwout shaft toward left until the Woodruff keys are ex-

posed (Fig. C626). Remove the Woodruff keys and withdraw the shaft and yoke.

If the throwout shaft needle bearings and seals are worn, drive them out and carefully install new ones. Bearings are pre-packed with grease. The right hand bearing is of the closed or shielded end type and should be installed so that closed end is flush with outside wall of clutch housing. Tap the right hand seal in place with lip toward inside of housing until the seal is $\frac{1}{32}$ inch beyond being flush with the inner edge of the bore. Install the left hand bearing so that bearing is centered in the bearing bore. Tap only on the end of the bearing that carries the identification, using a wooden dowel having the same outside diameter as the bearing. Seals should be installed so that they are about $\frac{1}{32}$ inch beyond being flush with the edge of the bore. Lip of inner seal should be toward inside of housing and lip of outer seal should be toward outside of housing. Refer to Fig. C627.

When reassembling, lubricate seal lips before installing the throwout shaft. Make certain that the yoke is centered on the release sleeve ears before tightening the yoke clamp screws.

CLUTCH FOOT PEDAL
Series 400-700B

155. The clutch pedal pivots on a needle bearing which rides on a sleeve surrounding the brake pedal cross shaft. The clutch pedal can be remov-

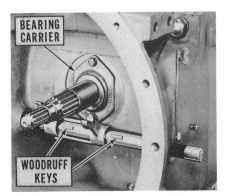

Fig. C626—Woodruff keys must be removed before the throwout shaft and yoke can be withdrawn.

ed after removing the left hand brake and brake lever. New needle bearing in clutch pedal should be packed with No. 1 gun grease. Lips of seals face away from the needle bearing. If the sleeve on which the needle bearing rides requires renewal, it will be necessary to remove the differential top cover, right brake, brake pedals and the brake cross shaft. Install new sleeve (end with "O" ring toward outside) so that the sleeve projects 1¾ inches from transmission case. Install new "O" rings in the cross shaft sleeves and coat the "O" rings with petroleum jelly before installing the brake cross shaft. Refer to Fig. C628.

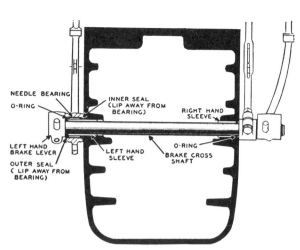

Fig. C628 — Brake pedal cross shaft installation showing the clutch pedal pivot bearing, seals and sleeve.

CLUTCH SHAFT AND SEALS

Series 400-700B

156. To remove the clutch shaft, first detach (split) engine from clutch housing and remove the clutch throw-out collar or release bearing. Unbolt the clutch shaft bearing carrier (Fig. C626) from clutch housing. Pull the clutch shaft and bearing carrier out, turn the unit 90 degrees to clear the throwout shaft and withdraw the assembly from tractor. Remove snap ring (S—Fig. C629) and press the clutch shaft and bearing out of carrier. The bearing can be removed from clutch shaft after removing snap ring (T), but be careful not to scratch shaft when snap ring is removed. Install the bearing with shielded side toward rear end of clutch shaft. Install a new seal in the bearing carrier with lip of seal toward rear. Note: If outer diameter of seal is not pre-coated (red or green), coat same with No. 1 Permatex before installation. Coat the rubber seal lip with No. 110 Lubriplate. When installing the clutch shaft, protect the seal lip with a Case No. A-8729 thimble or equivalent. Install the bearing retaining snap ring and the bearing housing "O" ring. Reinstall the assembled unit in the clutch housing.

To renew the clutch shaft front oil seal and pilot bearing, remove clutch from flywheel and unbolt the seal retainer. Renew the seal and the retainer "O" ring. Outer circumference of seal should be coated with Permatex and lip of same should be coated with Lubriplate prior to installation. Refer to Fig. C630.

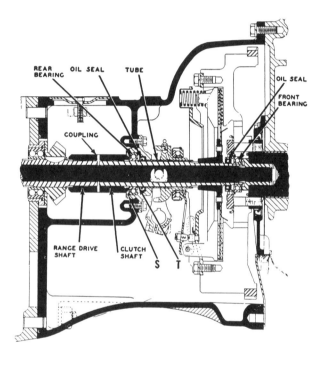

Fig. C629 — Sectional view of clutch housing, showing the clutch shaft and bearing installation.

Fig. C630 — Series 400 and 700B clutch shaft front oil seal installation. Clutch shaft pilot bearing is located just forward of the oil seal.

800B "CASE-O-MATIC" DRIVE

The "Case-O-Matic" drive, used in series 800B tractors, consists of the oil pump and manifold unit (Fig. C630K), the convertor control valve (Fig. C630J), and the torque convertor unit. The torque convertor unit includes the single stage torque convertor, the single disc direct drive clutch, the convertor regulator valve, the "Case-O-Matic" oil pump drive gear and the multiple disc "Main Powr Clutch" (Refer to Fig. C630A).

When disassembling any part of the "Case-O-Matic" system, it is a good practice generally to not remove any parts which can be thoroughly inspected while they are in-

stalled. All parts, when removed, should be handled with the same care as would be accorded the parts of a diesel injection pump or nozzle unit and should be soaked or cleaned with an approved solvent to remove gum deposits. Unless you practice good housekeeping (cleanliness) in your shop, do not undertake the repair of the "Case-O-Matic" equipped series 800B tractors.

An important precaution that service personnel should impart to the owners of Case series 800B tractors is to urge them to service the "Case-O-Matic" oil filter and the operating oil as described in paragraph 156B.

TROUBLE-SHOOTING

156A. When trouble-shooting the "Case-O-Matic" system first check the amount of oil in the system, the type and the condition of oil used and the condition of the filter element. Refer to paragraph 156B for information concerning the "Case-O-Matic" oil filter and the operating fluid.

If the "Case-O-Matic" oil temperature and pressure gages indicate some type of malfunction, first check the condition of the operating oil and the filter element; then, by using a master

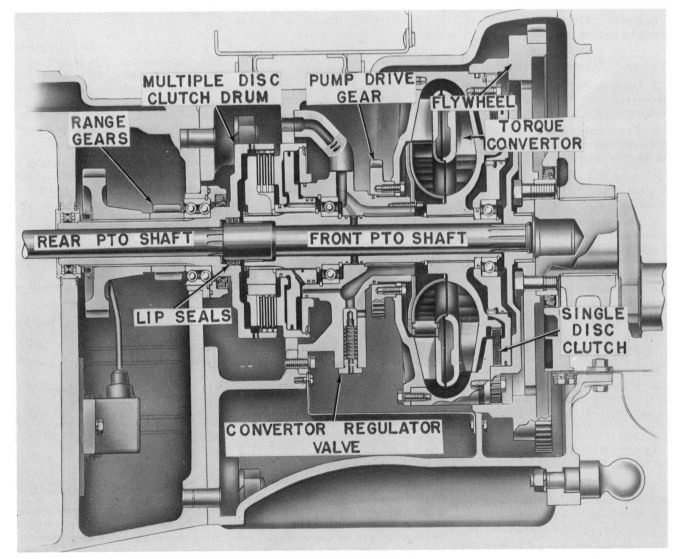

Fig. C630A—Cross section of the 800B "Case-O-Matic" torque converter unit. Refer to Figs. C630J and 630K for exploded views of the control valves and the oil pump assemblies.

gage check the condition of the instrument panel gages.

The following paragraphs list some of the troubles which may be encountered with the "Case-O-Matic" system. The procedure for correcting most of the troubles is evident; for those not readily remedied, refer to the appropriate subsequent paragraphs.

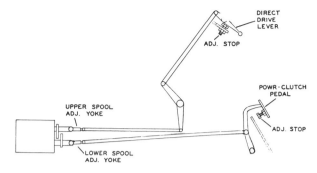

Fig. C630B — Drawing of the control valve operating linkage. To adjust the linkage refer to the text.

CONVERTOR OIL TEMPERATURE TOO HIGH. Could be caused by:

1. Operating the tractor too long in too high a work range (gear). Shift to a lower gear.

2. Restricted oil tube from pump to oil radiator or from radiator to control valve.

3. Dirty or plugged oil radiator and/or water radiator.

4. Engine overheated due to lack of water, incorrect timing, etc.

CLUTCH PRESSURE TOO LOW. Could be caused by:

1. Clutch operating linkage improperly adjusted. Refer to paragraph 156C.

2. Leaking, stuck or incorrectly adjusted regulator valve. Check as in paragraph 156E.

3. Restricted or leaking oil radiator or oil tubes from pump to radiator or radiator to control valves.

4. Pump and/or relief valve not operating correctly. Check as in paragraph 156D.

5. Leakage between the torque convertor unit base and the mating surface of the housing. Remove the torque convertor unit as outlined in paragraph 156L· and check surface (X — Fig. C630R) for nicks, burrs or foreign material which would prevent the torque convertor base from seating properly.

CLUTCH PRESSURE TOO HIGH. Could be caused by:

1. Pressure regulator valves stuck or improperly adjusted. Check as in paragraph 156E.

TRACTOR WILL NOT PULL A LOAD. Could be caused by:

1. Restricted or leaking oil radiator or external tubes.

2. Clutch operating linkage improperly adjusted. Refer to paragraph 156C.

3. Leaking, stuck or improperly adjusted regulator valve. Check as in paragraph 156E.

4. Faulty pump and/or relief valve. Check the pump pressure as in paragraph 156D.

5. Leakage between the torque convertor unit base and the mating surface of the housing. Remove the torque convertor unit as outlined in paragraph 156L and check surface (X— Fig. C630R) for nicks, burrs or foreign material which would prevent the torque convertor base from seating properly.

JERKING CLUTCH ACTION (will not start load out smoothly). Could be caused by:

1. An early type lower control valve spool. Refer to paragraph 156G.

2. Convertor oil pressure too low. Refer to paragraph 156F.

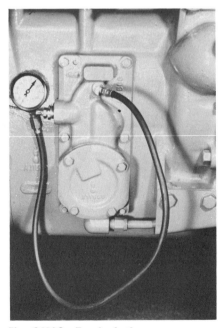

Fig. C630C—To check the pump pressure install a pressure gage as shown, run the engine at approximately 1800 rpm, shift the transmission to the neutral position and engage the multiple disc ("Main Powr") clutch. The gage should register 170-200 psi.

3. Clutch operating linkage improperly adjusted. Refer to paragraph 156C.

4. Sticking or leaking regulator valve. Check as in paragraph 156E.

5. Faulty pump or relief valve. Check as in paragraph 156D.

TRACTOR WILL NOT OPERATE IN DIRECT DRIVE. Could be caused by:

1. A worn bushing (B—Fig. C630Q) allowing operating oil to flow through the bushing instead of engaging the direct drive clutch. As this bushing is not available as an individual part, it will be necessary to renew the entire torque convertor unit, as per paragraph 156L.

2. Leakage between the torque convertor unit base and the mating surface of the housing. Remove the torque convertor unit as outlined in paragraph 156L and check the surface (X—Fig. C630R) for nicks, burrs or foreign material which would prevent the torque convertor base from seating properly.

Fig. C630D—A 300 psi capacity pressure gage can be installed as shown in order to check the multiple disc clutch operating pressure which should be 140-155 psi at approximately 1800 engine rpm with the clutch engaged.

WORKING FLUID, BREATHER AND FILTER

156B. It is recommended that the torque convertor fluid level (in convertor housing) be checked once each day immediately after stopping the tractor and the engine. Oil should be at the second (full) mark on the dip stick.

The torque convertor housing breather, which is at the top of the dip stick, should be cleaned in tractor fuel every 240 hours of operation or more often in extreme dust conditions.

Under normal operating conditions, the "Case-O-Matic" oil filter, located in the oil manifold on the left side of the tractor, should be serviced every 500 hours of operation and renewed every 1000 hours. If the filter element shows any visible damage, renew same. After engine is started and oil is warm, check for fluid leaks around filter.

Recommended working fluid for the "Case-O-Matic" system is 10W MS-DG (except when the temperature is below - 20 degrees F.; then, use 5W MS-DG). Capacity of the torque convertor housing is 4 gallons. The oil should be changed every 1000 hours or once a year which ever occurs first.

TESTS AND ADJUSTMENTS

156C. LINKAGE ADJUSTMENT. To adjust the control valve linkage proceed as follows: Disconnect the "Main Powr Clutch" adjustment yoke from the lower spool (Refer to Fig. C630B). Carefully move the lower spool either way until the detent ball is resting in the groove with the spool in the extended (rear) position; then, vary the length of the rod by turning the adjusting yoke until the holes in the spool and yoke are perfectly aligned

and install the pin and cotter pin. Turn the "Main Powr Clutch" pedal stop in; then, carefully push the pedal forward until the lower spool is in the neutral detent (forward) position. Hold the pedal in this position and adjust the stop on the pedal to just stop pedal movement at this time.

Move the direct drive lever to the engaged position (rearward), disconnect the direct drive lever adjusting yoke from the spool and carefully move the upper spool until the detent ball is resting in the groove with the spool in the extended (rear) position. Vary the length of the rod by turning the adjusting yoke either way as required so that the holes in the yoke and spool are properly aligned, then install the pin. Turn the direct drive clutch lever stop screw in; then, carefully push the direct drive lever forward until the detent ball has engaged the neutral detent in the upper spool. Hold the lever in this position and adjust the stop on the steering column to stop further forward movement of the lever.

156D. PUMP PRESSURE. To check the pump pressure, first run the engine until the torque convertor oil reaches operating temperature; then, proceed as follows: Install a test gage of at least 300 psi capacity as shown in Fig. C630C, shift the transmission into neutral and engage the "Main Powr Clutch". The gage reading should be 170-200 psi at approximately 1800 engine rpm.

If the pressure is low, remove the oil manifold and pump; then, check for leaking seals (2 and 13—Fig. C630K) in the oil manifold assembly, a malfunctioning relief valve and/or a faulty pump.

156E. CLUTCH PRESSURE. All oil used to operate the "Case-O-Matic" system must pass through the system oil filter; it is therefore essential that the filter element be clean before making any pressure check. Also, the control linkage must be properly adjusted as outlined as in paragraph 156C. To check the clutch pressure, first run the engine until the torque convertor oil is warm; then, proceed as follows: Install a test gage of at least 300 psi capacity as shown in Fig. C630D, shift the transmission to the neutral position and engage the "Main Powr Clutch". The gage reading should be 140-155 psi at an engine speed of approximately 1800 rpm.

If the pressure is higher than 155 psi or very erratic; remove regulator valve cap, withdraw the two plungers and two springs. Clean the plungers and the bore; then, reinstall the plungers and springs making certain that the plungers have free fits in their bores. Recheck the pressure and if the pressure is still high, remove a shim from between the large plunger and the springs or renew the springs.

If pressure is low, check the pressure regulator valve as follows: Install a Case A20275 test cap as shown in Fig. C630E in place of the original cap and turn the cap screw in until both the large and the small plungers (2 and 6 —Fig. C630J) bottom. Make the pressure check again.

If the pressure rises, remove the test cap and withdraw the two plungers (Refer to Fig. C630F) and two springs.

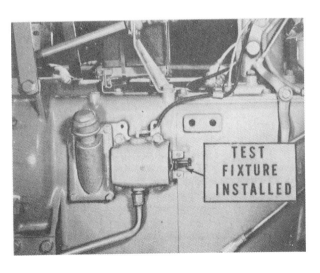

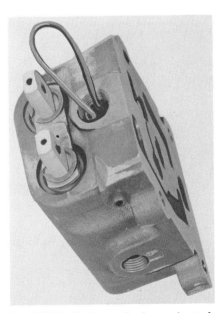

Fig. C630E—If the multiple disc clutch operating pressure is lower than 140 psi when checked with a gage installed as shown in Fig. C630D and as described in the text, install a Case test fixture A20275 as shown; then, proceed as outlined in the text.

Fig. C630F—If the small plunger is stuck, a wire can be bent and used to remove the plunger as shown.

Clean the plungers and the bore. Re-install the small plunger and the two springs, add a shim between the springs and the large plunger, then reinstall the large plunger and the plug (cap). NOTE: No more than two shim washers should be interposed between the springs and the large plunger and if after two shims are installed the pressure is still low, renew the plungers and/or springs. Recheck pressure.

NOTE: Never use a flat slug instead of a washer as it will not allow oil to pass through the small bleed hole in the plunger.

Possible causes for the pressure to remain low with the test cap installed are: A restricted oil filter element; restricted and/or leaking oil radiator or external oil tubes; faulty oil pump or relief valve; leakage between the torque convertor unit base and the mating surface of the housing. To check the oil pump and relief valve refer to paragraph 156D. To check for leakage between the convertor base and the housing, the torque convertor unit must first be removed as in paragraph 156L.

Inspection surface (X—Fig. C630R) for nicks, burrs or foreign material which would prevent the convertor base from seating properly and for oil leakage paths around the three oil passages.

156F. CONVERTOR PRESSURE. To check the convertor pressure, first run the engine until the torque convertor oil reaches operating temperature; then, proceed as follows: Install a test gage of at least 100 psi as shown in Fig. C630G, and with the engine running at approximately 1800 rpm, the gage reading should be 45-55 psi.

If the gage reading is not as specified, remove the oil manifold from the right side of the tractor and working through the opening, remove the regulator valve cap (Fig. C630H), spring and plunger. NOTE: The plunger is tapped to accommodate a 5/16-inch NC cap screw (3 or 4 inches long) which can be used as a puller if the plunger is stuck. Make certain the regulator valve plunger slides freely in its bore and that there is no foreign material which might keep it from seating properly. Reinstall the regulator valve assembly and tighten the cap to a torque of 50 Ft.-Lbs.

If the regulator valve is O.K., check the clutch pressure as outlined in paragraph 156E. If the clutch pressure is O.K., incorrect pressure is probably due to a faulty convertor unit. The torque convertor unit is available only as a complete assembly including the multiple disc clutch assembly.

CONTROL VALVE

156G. In cases where an objectionable "jerk" is encountered on early production 800B tractors, the clutch spool drain hole (Fig. C630I) can be drilled out to ½-inch diameter and a later chamfered lower control valve spool (10L—Fig. C630J) can be installed. Drilling the drain hole out to ½-inch diameter will not be beneficial unless the modified spool is also used. Make certain that no chips or dirt are allowed to enter the convertor housing when drilling the hole.

156H. R&R AND OVERHAUL. Disconnect the external oil lines and control rods from the control valve and remove the four retaining cap screws. Remove the valve housing, taking care not to loose or damage the two detent balls (11—Fig. C630J) and springs (12).

Withdraw the two detent balls and springs, remove the regulator valve cap (1), large plunger (2), springs (4 and 5) and the small plunger (6). NOTE: If the small plunger is stuck, a wire can be bent so as to engage the two holes in the plunger and used to withdraw same (Fig. C630F). Remove the upper and lower valve spools and check same for wear, scoring, nicks or burrs.

Remove by-pass valve Allen head plug (15—Fig. C630J) and check the ball (17) for seating. Check to make certain that all oil passages are open and clean. Renew the large and/or small plungers (2 and 6) if nicks, burrs or scoring is present on either. Spring (4) should exert 29 pounds at a length of $1\frac{21}{32}$ inches. The spring should be re-

Fig. C630G — To check the convertor oil presure, install a gage as shown; then, proceed as outlined in the text.

Fig. C630H—To remove the convertor regulator valve assembly, remove the oil manifold from the right side of the tractor and working through the opening, remove the cap (3) and withdraw the spring (2) and plunger (1). The plunger is tapped to accommodate a puller screw (see text).

Fig. C630I—When installing a late lower control valve spool on an early production 800B tractor, the drain hole shown should be enlarged to ½-inch diameter. Make certain that no chips or dirt enters the torque convertor housing during this operation. No benefit will be obtained from enlarging this hole if the late spool (10L—Fig. C630J) is not used.

newed if less than 27 pounds will compress it to $1\frac{31}{32}$ inches. Insert a ½-inch wood dowel in the small plunger and with the plunger on the end of the dowel, check to make certain that the plunger slides freely in its bore and seats properly.

If it is necessary to renew the two lip seals (13), pry same out and press new seals in, with the lip facing inward, until the top of the seal is flush with the lower part of the chamfer.

Wash all parts in a petroleum solvent and dry with compressed air; being sure to blow out all bores and passages in the valve body including the small pressure regulator orifice. Plug (14) through which the orifice is drilled will seldom need to be removed; however, a tag wire should be used to make certain that the orifice is open and clean. Check the orifice plug to be sure it is "bottomed" in its bore.

Reassemble in reverse of the disassembly procedure using Fig. C630J as a guide. After the control valve unit is reinstalled on the tractor, check and adjust the operating linkage and the clutch pressure as outlined in paragraph 156C and 156E.

OIL MANIFOLD

156J. The oil manifold assembly is located on the right side of the torque convertor housing and includes the oil pump, suction pipe and the oil filter assembly (Refer to Fig. C630K). Removal can be accomplished after removing the retaining cap screws and pulling the manifold from the dowel pins.

OIL PUMP

156K. R&R AND OVERHAUL. The pump can be removed after first unbolting and removing the oil manifold, then removing the two pump retaining cap screws. Remove snap ring (6) from end of drive shaft, remove Allen head screw (3A) from gear (3) and remove the gear. Remove the retaining cap screws, then separate the pump cover (8) from the pump body (5). Gears (10 and 10A) are pressed on shafts (9 and 11). If bushings (7) are renewed, they may require final sizing after they are pressed into the pump body and/or cover. Renew the two square section sealing rings (13) and all questionable parts. When reassembling, use a light coat of sealing

compound on the mating surfaces of the pump body and cover and tighten the cover to body cap screws to 6 Ft.-Lbs. of torque.

The relief valve assembly can be removed and reconditioned or renewed (Refer to Fig. C630K).

TORQUE CONVERTER UNIT

The torque convertor unit includes the single stage torque convertor, the single disc direct drive clutch, the converter regulator valve, the convertor oil pump drive gear and the multiple disc "Main-Powr Clutch". The

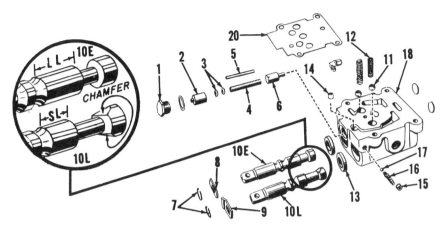

Fig. C630J—Exploded view of the 800B "Case-O-Matic" control valve. On late 800B control valves the upper spool is of the non-chamfered type (10E) and the lower spool is similar to the type (10L). Early control valve spools were both of the type (10E); however, some of the early valves may have the later lower spool (10L) installed to prevent an objectionable "Jerk" which is sometimes encountered.

LL. Long land	4. Spring (outer)	10E. Spool (Early, both; Late, upper)	14. Pressure regulator orifice plug
SL. Short land	5. Spring (inner)	10L. Spool (Late, lower)	15. Allen screw
1. Regulator valve cap	6. Small pressure regulator plunger	11. Detent balls	16. Spring
2. Large pressure regulator plunger	7. Snap rings	12. Detent springs	17. Check valve ball
3. Shims	8. Upper spool plate	13. Spool oil seals	18. Control valve body
	9. Lower spool plate		20. Gasket

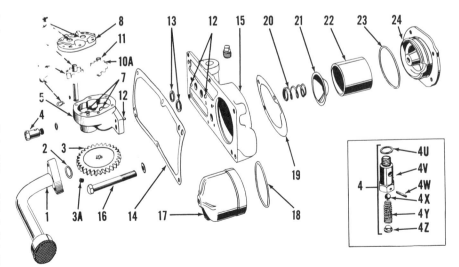

Fig. C630K—Exploded view of the "Case-O-Matic" oil pump, manifold and filter.

1. Intake pipe	4X. Valve ball	10A. Pump idler gear	17. Filter cover
2. Square section rubber	4Y. Spring	11. Pump idler shaft	18. "O" ring
3. Pump main drive gear	4Z. Cap	12. Dowel pins	19. Gasket
3A. Allen head set screw	5. Pump body	13. Square section rubber rings	20. Spring
4. Relief valve assembly	6. Snap ring	14. Gasket	21. Spring retainer
4U. Gasket	7. Bushings	15. Oil manifold	22. Filter element
4V. Body	8. Pump cover	16. Cap screw	23. "O" ring
4W. Roll pin	9. Pump drive shaft		24. Filter body
	10. Pump driven gear		

regulator valve assembly and the multiple disc clutch plates are the only sections of the convertor unit which have individual parts catalogued by the J. I. Case Co. If any section of the torque convertor unit other than the regulator valve or clutch plates is malfunctioning, the complete convertor unit must be renewed.

156L. REMOVE AND REINSTALL. To remove the torque convertor unit and multiple disc clutch as an assembly, first split the tractor as outlined in paragraph 156P. Disconnect all linkage wires and tubes which would hinder the removal of the fuel tank, bracket and instrument panel unit; then, remove the fuel tank, bracket and instrument panel assembly. Refer to Fig. C630L and install a Case A20165 loading fixture. Remove the filler opening from the left side and the torque convertor oil pump and manifold unit from the right side of housing. Install two $\frac{5}{16}$x1½-inch NC jack screws in the threaded holes on the torque convertor unit (Refer to Fig. C630M). Tighten the installed cap screws sufficiently to engage the multiple disc clutch; thus preventing the clutch plates from slipping out of alignment as the torque convertor unit is withdrawn. Remove the cap screws and stud nuts that retain the torque convertor unit to the rear of the housing and withdraw the convertor unit from housing (Refer to Fig. C630N).

Renew the double opposed lip seals (OS—Fig. C630P) installing same flush with the chamfer. Renew the convertor to flywheel gasket and check the bushing (Refer to Fig. C630Q) and the rear bearing surface of the front pto shaft. If the bushing is damaged, the complete torque convertor unit must be renewed as the individual bushing is not catalogued. NOTE: Failure of

Fig. C630M—Install two 5/16x1½-inch NC jack screws (S) in the threaded holes as shown; then, tighten same sufficiently to engage the multiple disc clutch, thus preventing the clutch plates from slipping out of alignment when the convertor unit is withdrawn.

this bushing will permit oil to flow through the bushing instead of engaging the direct drive clutch, resulting in difficulty operating the tractor in direct drive.

Before installing the torque convertor unit refer to Fig. C630R and make certain that the base of the unit and the mating surface (X) of the housing are free from nicks, burrs and foreign material, which would prevent sealing at this point. Install the convertor unit making certain that the unit is installed straight and not cocked in the housing; then, torque the retaining cap screws to 35 ft.-lbs. Remove the two jack screws. Reinstall the remainder of the parts by reversing the removal procedure.

NOTE: If it is necessary to withdraw the convertor unit, after it has been installed, check the front pto shaft to make certain it has not been pulled out and/or pulled lips of the seals over.

MAIN CLUTCH PLATES AND DRUM

Damage or wear of "Main Powr Clutch" parts is usually the direct result of improper linkage adjustment (Refer to paragraph 156C), low clutch pressure (Refer to paragraph 156E) or incorrect convertor pressure (Refer to paragraph 156F).

156M. CLUTCH PLATES. To renew the "Main Powr Clutch" (multiple disc) plates first remove the torque convertor unit as outlined in paragraph 156L. Back-off the two cap screws used to lock the plates during removal; then, remove the snap ring (Fig. C630S). Lift off the back plate and the alternate bronze and steel plates.

When reassembling, reverse the disassembly procedure, making certain the lugs on the back plate engage the

Fig. C630L—A Case A20165 loading fixture should be used as shown to remove the torque convertor unit.

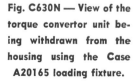

Fig. C630N — View of the torque convertor unit being withdrawn from the housing using the Case A20165 loading fixture.

squared notches in the steel plates and that reliefs in the piston are aligned with the "U" shaped notches in the steel plates. Use a front pto shaft and a multiple disc clutch drum (with seals) to align the teeth of the clutch discs; then, tighten the two jack screws maintaining the correct alignment of the discs. Reinstall the remainder of the parts, reversing the removal procedure.

After the tractor is reassembled, check and adjust the linkage as in paragraph 156C, the clutch pressure as outlined in paragraph 156E and the convertor pressure as in 156F.

156N. CLUTCH DRUM. The "Main Powr Clutch" drum is integral with the range drive shaft. To remove the drum and range drive shaft unit, refer to paragraph 161A.

TRACTOR SPLIT

156P. To detach the engine from the torque convertor housing, first drain cooling system, remove the hood and disconnect the head light wires. On models equipped with a radiator shutter, remove the grille and disconnect the shutter control wire. On standard, non-adjustable front axle models, disconnect the drag link and rear end of radius rod. On all models, disconnect heat indicator sending unit from water manifold, ammeter wire from generator, battery cables from batteries, starter solenoid wire from solenoid and oil cooler tubes from torque convertor control valve and oil manifold.

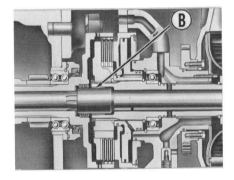

Fig. C630Q—Check the bronze bushing (B) for excessive wear or damage. Failure of this bushing will cause difficulty in operating the tractor in direct drive. Refer also to Fig. C630R.

Disconnect the hour meter cable and oil line from the engine block. On tricycle and adjustable axle models, loosen the clamp bolt on the steering shaft universal joint and unbolt the steering shaft support from the torque convertor housing.

On diesel models, shut off fuel and disconnect fuel lines at top of tank and shut-off valve. Decompress the engine and disconnect the decompression linkage at lower right front end of fuel tank. Remove the rear nut on throttle rod and disconnect the fuel pressure gage line at inlet connection on second stage filter and vacuum gage line at bleed valve connection above the first stage fuel filter. Disconnect the ether starting aid tube from intake manifold.

On LP-Gas models, shut off fuel and disconnect choke wire from carburetor. Remove the left side sheet under tank, then disconnect the fuel line from tank at filter. Disconnect the throttle control rod at a point just under front end of fuel tank. Disconnect wire from coil, or on magneto equipped engines, disconnect the grounding wire.

On all models, swing engine in a hoist, unbolt engine from convertor housing and separate the tractor halves.

When connecting engine to torque convertor housing, tighten the flange bolts to a torque of 80 ft.-lbs.

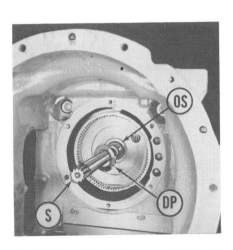

Fig. C630P — The double opposed oil seals (OS) can be easily pryed out after withdrawing the pto front shaft (S). When reassembling the tractor make certain that the oil deflector plate (DP) is against the clutch drum.

Fig. C630R—Examine the entire mating surface (X) for nicks, burrs and foreign material. Leakage around the oil passage holes may cause difficulty in operating the tractor. Oil leakage paths will usually be visible. The convertor base to housing cap screws and nuts should be tightened evenly to 35 Ft.-Lbs. of torque.

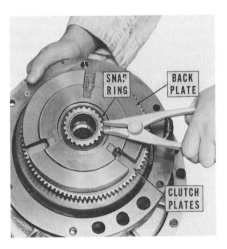

Fig. C630S — To remove the clutch plates, first loosen the jack screws used to engage the clutch during removal. Remove the snap ring and lift off the clutch plates.

TRANSMISSION

TOP COVER

All Models

157. **REMOVE AND REINSTALL.** To remove the transmission case top cover, first remove hood, batteries and battery tray. Disconnect wires from tail light and electrical outlet socket. Disconnect the instrument panel from the fuel tank, shut off the fuel and disconnect the fuel lines. Remove the fuel tank and lay the instrument panel on top of engine. Disconnect the throttle control rod and loosen the clamp bolt on the steering shaft universal joint. Place the gear shift lever in the neutral position, unbolt cover from transmission case and remove the transmission cover and steering column assembly. Note: Some mechanics prefer to remove the steering column assembly from cover before removing cover from transmission case.

Install the transmission case top cover by reversing the removal procedure.

SPLIT TRANSMISSION FROM CLUTCH HOUSING

Series 400-700B

158. To detach (split) the transmission from the clutch housing, first remove hood, batteries and battery tray. Disconnect wires from tail light and electrical outlet socket. Remove the front throttle rod and on diesel models, loosen the two bolts connecting the decompressor bracket to the fuel tank front support. On all models, disconnect the clutch throwout rod, unbolt the steering column from the transmission top cover and loosen the fuel tank front support. Raise and block-up the fuel tank and steering column assembly. Support both halves of tractor, unbolt transmission from clutch housing and separate the tractor halves.

When reassembling, renew the "O" ring seals (Fig. C631) and tighten the stud nuts securely.

SPLIT TRANSMISSION FROM TORQUE CONVERTER HOUSING

Series 800B

NOTE: Splitting the transmission from the torque convertor housing actually requires the complete removal of the torque convertor housing casting.

Fig. C631 — When the transmission housing has been separated from the clutch housing, it is good practice to always renew the "O" ring seals.

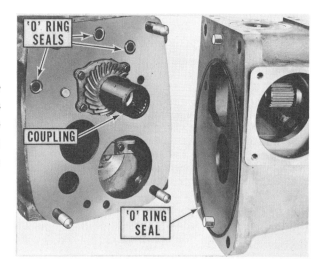

158A. To detach (split) the transmission from the torque convertor housing, first remove the torque convertor unit as described in paragraph 156L and the "Main Powr Clutch" drum using paragraph 161A as a guide; then, proceed as follows: Remove the oil deflector plate (DP—Fig. C631A),

Fig. C631A — On series 800B tractors, the range drive shaft and "Main Power Clutch" drive retainer is attached to the transmission housing with four Allen head screws. Two of the screws are shown at (AS). Rotation of the clutch drum 90 degrees will reveal the other two.

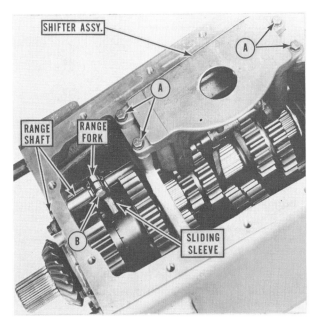

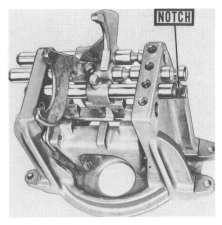

Fig. C632 — Top view of the 400 series transmission with the housing cover removed. The shifter assembly is retained by cap screws (A). Series 700B and 800B are similar.

Fig. C635—When reassembling the shifter unit, make certain that notch in the range shifter shaft is toward bottom.

then unbolt and separate the torque convertor housing from the transmission housing.

When reassembling, reverse the disassembly procedure, renewing all "O" rings and gaskets. Tighten the torque convertor housing to transmission stud nuts securely to prevent oil leakage.

OVERHAUL

Data on overhauling the various transmission components are outlined in the following paragraphs.

All Models

159. **SHIFTER SHAFTS AND SHIFTERS.** The transmission shift lever and associated parts can be removed and overhauled without removing the transmission top cover from tractor. The procedure for doing so is evident.

To remove the transmission shifter shafts and shifters, first remove the transmission top cover as outlined in paragraph 157 and proceed as follows: Remove cap screws (A—Fig. C632) and withdraw the shifter assembly from transmission case. Loosen cap screw (B), pull the range shifter shaft rearward and withdraw the range shifter fork.

To disassemble the removed unit, remove the set screws retaining the shifter forks to the shafts and bump the shifter shafts out of the bracket. Withdraw the detent balls and springs. Swing the shifter gate (3—Fig. C633) fully to the left or right and remove plunger and spring (14). Drive out roll pins (2) and remove gate. Guide (16) can be removed by driving out its retaining roll pin.

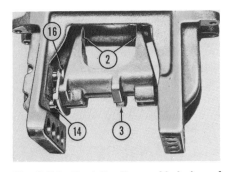

Fig. C633—Partially disassembled view of the transmission shifter assembly. Shifter gate (3) pivots on roll pins (2).

Fig. C634 — Exploded view of the transmission shifter rails and forks.

1. Bracket
2. Roll pins
3. Shifter interlock gate
4. Detent ball spring
5. Detent ball
6. Shifter shaft, 1st, 2nd, 5th and 6th speeds
7. Range rear shifter shaft
8. Shifter shaft, 3rd, 4th, 7th and 8th speeds
9. Shifter shaft, reverse
10. Reverse fork
11. Shifter fork, 3rd, 4th, 7th and 8th speeds
12. Range gear shifter block
13. Shifter fork, 1st, 2nd, 5th and 6th speeds
14. Interlock gate detent plunger and spring
15. Interlock gate detent plate
16. Interlock gate stop pin
17. Roll pin

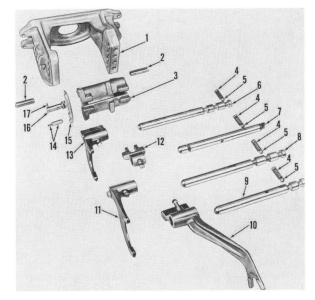

Thoroughly clean all parts and examine them for damage or excessive wear. Renew any parts which are questionable.

When reassembling, install guide (16) and its retaining roll pin. Install the shifter gate and its retaining roll pins. Swing the gate fully right or left and install plunger and spring (14). Install the detent springs and balls and insert the shifter shafts. Notch in range shaft should be toward bottom of assembly as shown in Fig. C635. Tighten the shifter fork retaining set screws securely and install the lock wire.

Install the range shifter shaft and fork, but do not tighten set screw (B—Fig. C632). Install the complete shifter assembly and tighten cap screws (A).

160. Make certain that the transmission is in belt pulley neutral, then center the range shifter sliding sleeve between the two range gears and tighten the fork set screw (B).

Series 400-700B

161. RANGE DRIVE SHAFT. To remove the range drive shaft from a series 400 or 700B tractor, first detach (split) the transmission from the clutch housing as outlined in paragraph 158 and remove the transmission case top cover. Withdraw the pto and hydraulic pump drive shaft.

Note: On early production series 400 tractors, there was a small snap ring located on the pto and hydraulic pump shaft behind the range drive shaft rear bearing. This snap ring should be removed and discarded. Refer to Fig. C637.

Unbolt the range drive shaft front bearing carrier from transmission case, pull the drive shaft assembly forward and tip front end of shaft down. Bend tab of lockwasher out of locknut and using a spanner wrench as shown in Fig. C638, remove the nut. If nut is difficult to loosen, slip the range drive shaft back into its rear bearing, engage the range sliding sleeve with one of the range gears, then engage both of the sliding gear clusters with the countershaft gears. With the lock nut removed, pull the drive shaft forward and out of case and withdraw the gear cluster from above. Be careful not to damage or lose the shims located behind the front bearing carrier, as they control the fore and aft position of the belt pulley drive pinion.

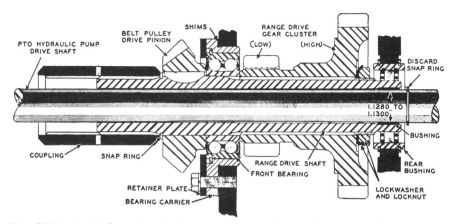

Fig. C637—Installation view of the range drive shaft and cluster gear. The snap ring located on the PTO and hydraulic pump drive shaft should be discarded. Although series 400 is shown, series 700B tractors are similar.

The range drive shaft can be pressed out of the belt pulley drive pinion, front bearing and carrier. The drive shaft rear bearing can be removed from the transmission case wall after removing one of the bearing retaining snap rings. The pto and hydraulic pump drive shaft bushing located in rear end of range drive shaft can be renewed if it is damaged. New bushing should be installed flush with end of shaft, using a 1.1285 diameter sizing arbor. If sizing arbor is not available, ream the bushing after installation to an inside diameter of 1.128-1.130. Note: Replacement range drive shafts are factory fitted with a reamed bushing.

Press the belt pulley drive pinion on the range drive shaft until gear seats against snap ring. Press the front bearing into the bearing carrier with bearing snap ring toward front of tractor. Install the retainer plate then press the bearing and carrier assembly on the range drive shaft until inner race of bearing seats against the belt pulley drive pinion. The next step in the assembly procedure is to adjust the fore and aft position of the belt pulley drive pinion. The position of the pinion gear (from machined front face of transmission case to front face of pinion) is controlled by shims between the

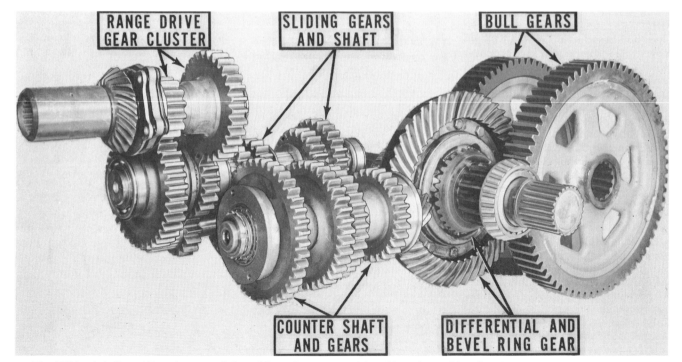

Fig. C636—J. I. Case 400 series transmission, differential and final drive. The unit provides eight forward speeds and two reverse. Series 700B and 800B tractors are equipped with a similar gear train.

transmission case and the range drive shaft front bearing carrier.

Install the range drive shaft and front bearing carrier assembly WITH-OUT shims or cluster gear and tighten the carrier retaining cap screws. Using J. I. Case gage No. A-8343 as shown in Fig. C639, measure and record the distance between the gear and the gage using feeler blades. Observe the front surface of the drive gear where the correct clearance is etched. Subtract the correct clearance from the measured clearance, thereby determining the proper thickness of shims to be inserted. For example, suppose the measured clearance is 0.020 and the etched clearance on front of gear is 0.006. By subtracting 0.006 from 0.020, you determine that 0.014 thickness of shims should be inserted. Shims are available in thicknesses of 0.003 and 0.005, so one 0.005 shim and three 0.003 shims should be inserted. You may encounter some gears that indicate no clearance; in which case, add shims until gear is a snug fit against gage No. A-8343.

Fig. C638—Using a spanner type wrench to remove the lock nut from rear end of the range drive shaft.

Fig. C639—Using Case gage No. A-8343 to determine the fore and aft position of the belt pulley driving bevel pinion gear.

Remove the range drive shaft and install the determined thickness of shims. Install the cluster gear, lock washer and lock nut. The remainder of the installation procedure is the reverse of the removal procedure.

Series 800B

161A. **RANGE DRIVE SHAFT.** To remove the range drive shaft and "Main Powr Clutch" drum from a series 800B tractor, first remove the torque convertor unit as outlined in paragraph 156L; then, proceed as follows: Shift the transmission into the neutral position, then unbolt and remove the transmission top (shifter) cover. Unbolt and withdraw the shifter rails and forks unit. Working through the two holes in the multiple disc clutch drum (Fig. C631A), remove two of the four Allen head screws (AS), which attach the clutch drum retainer to the transmission housing; then, rotate the clutch drum 90 degrees and remove the remaining two screws. Move the clutch drum forward and tip the front (clutch drum) end down. Bend tabs of the lock washer out of the lock nut and using a spanner wrench in a manner similar to that shown in Fig. C638, remove the locknut. Withdraw the clutch drum and range drive shaft from the front and the gears from the top.

The range drive shaft can be pressed out of the front bearing and carrier. The drive shaft rear bearing can be removed from the transmission case wall after removing one of the bearing retaining snap rings.

Reassemble in reverse of the disassembly procedure making certain that the locknut on the range drive shaft is

tightened securely, then bend a tab of the lockwasher into one of the notches in the nut.

All Models

162. **SLIDING GEAR SHAFT.** To remove the sliding gear shaft, first remove the range drive shaft as outlined in paragraph 161 or 161A. Remove the gear shifter assembly and the range shifter shaft and fork (Fig. C632). Disengage the snap ring (Fig. C640) from groove in the sliding gear shaft and work the snap ring rearward as far as possible. On early production series 400 tractors, insert a pry bar behind the snap ring on the front bearing and using a lead hammer, bump the low range gear rearward while prying the shaft forward as shown in Fig. C641. As soon as the shaft "breaks loose", it can be easily pulled forward by hand.

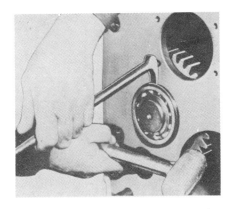

Fig. C641 — Pulling the early production series 400 sliding gear shaft forward. On later series 400 and all 700B and 800B models, the forward end of the sliding gear shaft is drilled and tapped for a puller screw.

Fig. C640 — Top view of 400 series transmission, highlighting the sliding gear shaft installation after the range drive shaft and shifter unit have been removed. Series 700B and 800B transmission are similar.

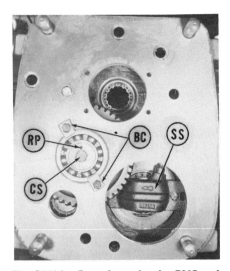

Fig. C641A—Front face of series 700B and 800B transmission case. Late series 400 is similar. Bearing clamps are shown at (BC), retainer plate cap screw at (CS), positioning roll pin at (RP) and speedometer drive support at (SS).

On series 700B and 800B tractors, it will be necessary to first remove the two bearing clamps (BC—Fig. C641A). Late series 400 and all 700B and 800B sliding gear shafts have a tapped hole in the forward end which will accommodate a puller screw to help move the shaft forward.

Pull the shaft forward, withdraw the sliding gears and work the snap ring off rear end of shaft. Pull the sliding gear shaft out of transmission case and withdraw the range gears.

The sliding gear shaft center and rear bearings can be removed from the transmission case walls by removing either one of their retaining snap rings. If the rear bearing is a tight fit in the case bore, it will be necessary to remove the differential case cover to tap the bearing out. To remove the front bearing from shaft, remove snap ring (on series 400) or cap screw and retainer plate (on series 700B and 800B) and press shaft out of bearing. Before installing the front bearing, make sure that the roll pin which positions the front washer is in place, install washer so that the roll pin is in the washer slot, then press the shaft into bearing until the bearing seats against the washer and the washer is against the shoulder on the shaft. Install the bearing retaining snap ring (on series 400) or cap screw and retainer plate (on series 700B and 800B).

Examine the range gear needle bearings. If they are damaged, press them out of the gears. The low range gear has two needle bearings. Place one of the bearings on J. I. Case arbor

No. A-8347 with the stamped identification marks on bearing against the arbor pressing shoulder. Press the needle bearing into gear until the stop shoulder on arbor seats against the gear. Turn the gear over and install the other needle bearing in the same manner. The high range gear has one long needle bearing. Install the bearing in a similar manner with Case arbor No. A8356. Oil hole in bearing must align with oil hole in gear as shown in Fig. C642.

To install the sliding gear shaft, start the shaft into the front bearing bore and slide the low range gear (1—Fig. C643 and 644) onto the shaft. Install the splined collar (12) onto the shaft and slip the sliding sleeve (2) onto the collar. Install the center bearing inner race (4) in the bearing. Install the high range gear onto shaft and push the sliding gear shaft through the center bearing, being careful not to damage the needle bearings in the range gears. Note: The small gear projections on the range gears must be toward the sliding sleeve. Install snap ring (5) on shaft and keep pushing the snap ring forward, as the shaft is pushed rearward. Be sure to "juggle" the range gears as the shaft is pushed rearward to prevent damaging the needle bearings.

Install the sliding gears and push the sliding gear shaft rearward until rear end of shaft starts in the rear bearing. With inner splines of collar (12) aligned with splines on shaft, drive the shaft in place. Move snap ring (5) forward and into the shaft

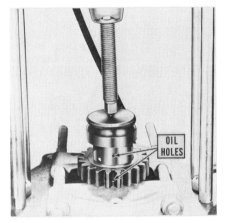

Fig. C642—Installing needle bearing in the high range gear. Oil holes must align.

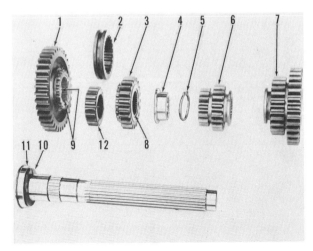

Fig. C643 — Exploded view of the 400 series sliding gear shaft.

1. Low range gear	3. High range gear	5. Snap ring
2. Sliding sleeve	4. Bearing inner race	6. Sliding gears

Fig. C644—Proper installation of the sliding gear shaft and gears.

7. Sliding gears	9. Needle bearings	11. Front bearings
8. Needle bearings	10. Washer	12. Collar

Fig. C645—Removing the reverse idler and oil slinger gear which cannot be accomplished until after the sliding gear shaft has been removed.

Fig. C648—Using home made puller to remove the countershaft front bearing carrier.

groove. Install the range drive shaft and shifters as outlined in previous paragraphs.

All Models

163. **REVERSE IDLER AND OIL SLINGER GEARS.** To remove the reverse idler and oil slinger gears, first remove the sliding gear shaft and gears as outlined in paragraph 162. Using a $\frac{5}{16}$ inch Allen wrench, remove the idler shaft lock screw as shown in Fig. C645, slide the idler shaft out of the front of the transmission case and remove the gears.

On all models the 1.435-1.436 diameter reverse idler shaft should have a clearance of 0.0035-0.0065 in the bronze reverse gear bushing. If the clearance is excessive the reverse idler gear and factory installed bushing unit and/or reverse idler shaft should be renewed. The bushing is not available separately.

On series 400 tractors renew the oil slinger gear bushing and/or reverse idler shaft if clearance between bushing and shaft is excessive.

On series 700B and 800B tractors the nylon oil slinger gear should be renewed if clearance between it and reverse idler shaft is excessive.

Install the shaft and gears by reversing the removal procedure, but make certain that the counterbored end of the lockscrew hole in the shaft is up. The Allen head set screw must be tightened fully into the casting. If the counterbored end of the hole is not up, the lock screw cannot be turned fully in.

All Models

164. **COUNTERSHAFT (BEVEL PINION SHAFT).** To remove the countershaft and integral main drive bevel pinion, first remove the sliding gear shaft as outlined in paragraph

162 and the differential and main drive bevel ring gear assembly as outlined in paragraph 165. Working through the front opening in the transmission case, remove the countershaft front bearing carrier retainer clip (Fig. C646) or speedometer drive support bracket (SS — Fig. C641A). Bend tab of locking washer (21—Fig. C647) out of notch in adjusting nut (20). Using a spanner type wrench, remove adjusting nut (20) and withdraw

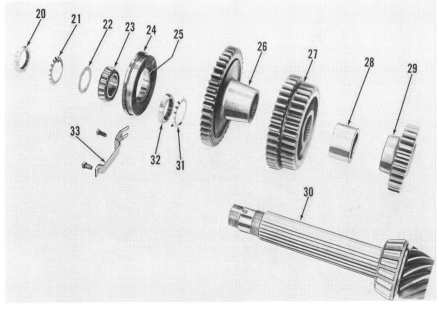

Fig. C646 — Front view of the 400 series transmission showing the location of the countershaft front bearing carrier retaining clip.

Fig. C647—Exploded view of the 400 series countershaft and related parts. Countershaft must be withdrawn rearward from transmission case which necessitates removal of the differential assembly. Series 700B and 800B countershaft parts are similar.

20. Adjusting nut	25. Bearing cup	29. 4th and 8th speed gear
21. Locking washer	26. 1st and 5th speed gear	30. Countershaft
22. Lip washer	27. 2nd and 6th, 3rd and 7th speed gear	31. Lockwasher
23. Bearing cone	28. Spacer	32. Locknut
24. Front bearing carrier		33. Retainer clip

59

lock washer (21) and lip washer (22). Using a puller with screws of same threaded into the two tapped holes in the countershaft front bearing carrier as shown in Fig. C648, pull the bearing carrier from bore in transmission case.

Bend tab of locking washer (31—Fig. C647) out of notch in locknut (32) and using a spanner type wrench, remove nut (32). Pull the countershaft from rear of the transmission case and withdraw the gears from above.

The rear bearing cone can be pulled from the shaft and the bearing cup can be tapped from the case bore. Front bearing cup can be pressed out of its carrier.

Note: The countershaft and integral main drive bevel pinion is not available as a separate replacement part. If the shaft or

pinion gear is damaged, it will be necessary to install the bevel pinion and ring gear matched set. Refer to paragraph 169.

Install the countershaft from the rear and install the gears and spacer. Slide lockwasher (31—Fig. C647) on the shaft so that both tabs of washer enter the holes provided in the front gear (26). Install and tighten nut (32) securely while tapping rear end of shaft with a lead hammer.

Note: There must be no end play between any of the gears, spacer or rear bearing.

Bend tab of lockwasher into one of the notches in nut (32). Position the front bearing carrier in the case wall so that the two threaded holes are in a horizontal plane. Install the bearing cone, lip washer (22), lockwasher (21)

and nut (20). Tighten nut (20), thereby pulling the bearing carrier back in place. When the bearing carrier is seated, continue to tighten the nut until 20-30 inch-pounds of torque is required to turn the countershaft in its bearings.

Turning torque can be checked by installing a ⅝-inch NF cap screw in the tapped hole (in place of speedometer drive gear on series 700B and 800B tractors) in forward end of countershaft and using a torque wrench.

On series 400 tractors, install the bearing carrier retaining clip (33). On series 700B and 800B tractors, install the speedometer drive support (SS—Fig. C641A). Tighten the two support retaining cap screws securely before tightening the Allen head set screw.

DIFFERENTIAL AND MAIN DRIVE BEVEL GEARS

DIFFERENTIAL

All Models

165. **REMOVE AND REINSTALL.** To remove the differential and main drive bevel ring gear assembly, remove the final drive bull gears as outlined in paragraph 171 and both brake assemblies. Unbolt and remove the differential case top cover. Attach a chain hoist to the differential assem-

bly as shown in Fig. C650 and remove both of the differential bearing carriers, being careful not to lose the shims located between the bearing carriers and the differential case. Withdraw the bull pinion shaft and integral differential side gears and hoist the differential assembly from case. Be careful not to lose or damage the thrust washer located between the right hand side gear and the differential.

The differential bearing cones can be pulled from side gears and the bearing cups can be pulled from bearing carriers at this time. New oil seals should not be installed until after the carrier bearings and bevel gear backlash are adjusted.

166. To install the differential, lower same into the case, using a hoist, and install the left hand pinion and side gear unit.

167. Install the thrust washer on right side of the differential, then install the right hand pinion and side gear unit. With the "O" rings and oil seals removed from the differential bearing carriers, install the bearing

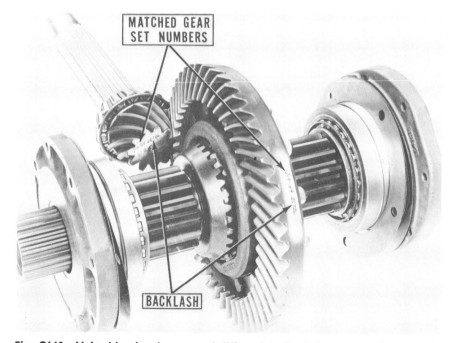

Fig. C649—Main drive bevel gears and differential. Bevel ring gear and transmission countershaft are available as a matched set only.

Fig. C650—Using hoist to lift the differential and bevel ring gear unit from transmission case.

carriers, using the same number of shims as were originally removed and tighten the retaining cap screws to a torque of 50 ft.-lbs. Make certain there is some backlash between the bevel pinion and ring gear, then add or remove shims from either bearing carrier until the carrier bearings have a slight drag. Then, remove one 0.003 shim from either bearing carrier to obtain the desired bearing pre-load. Note: There must still be some backlash between the bevel gears. Now, observe rear end of pinion and the edge of the ring gear where the matched gear set number and the proper backlash for this set of gears is etched. Refer to Fig. C652. The desired backlash for this matched set of gears is 0.007. Turn the bevel ring gear until the "X" marked tooth on pinion is located between the two "X" marked teeth on ring gear as shown and check the backlash with a dial indicator. If the backlash is not as specified, transfer shims from under one bearing carrier to the other until the desired backlash is obtained.

CAUTION: Only transfer shims, do not add or remove shims or the carrier bearing adjustment will be changed. Shims are available in thicknesses of 0.003, 0.005 and 0.010.

With the bearings and backlash adjusted, remove the bearing carriers, but be careful not to mix or lose the shims. Install the bearing carrier oil seals as follows: Soak the leather outer seal in oil until lip is soft and pliable. Coat the outer circumference of the seal with No. 1 Permatex and install seal in carrier so that lip of seal faces

differential. Press the outer seal in just enough so that inner seal can be started in carrier bore. Coat the outer circumference of the inner seal with No. 1 Permatex and install seal in carrier so that the synthetic rubber lip faces the differential. Press both seals in until the synthetic rubber inner seal is just flush with the bearing carrier bore. Install new "O" rings on bearing carriers and reinstall the carriers, using the previously determined number of shims. Use J. I. Case thimble No. A-8341 as shown in Fig. C653 to protect the seal lips when

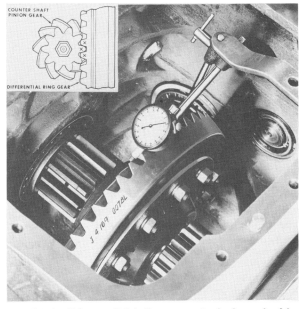

Fig. C652—Using a dial indicator to check the main drive bevel gear backlash. Backlash should be the value stamped on the gears.

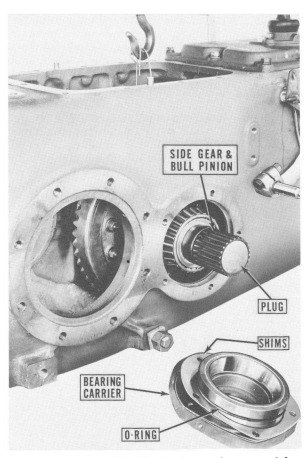

Fig. C651—Differential right bearing carrier removed from transmission case. Soft plug in end of side gear must be in good condition.

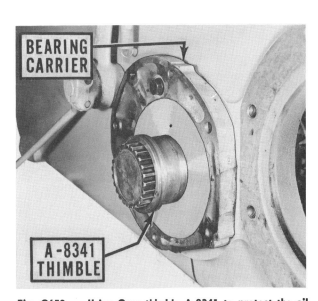

Fig. C653 — Using Case thimble A-8341 to protect the oil seals when installing the differential bearing carriers.

installing the bearing carriers. Tighten the carrier cap screws to a torque of 50 ft.-lbs.

168. **OVERHAUL.** With the differential removed from tractor, unbolt and remove the bevel ring gear and withdraw the pinion pin and pinions. The pinion pin has loose fit in the differential case (center wheel) and should fall out when ring gear is removed. If the pin is frozen, it can be drilled, tapped and pulled out.

The differential cross shaft has a very tight press fit in the differential case. Do not attempt to renew the shaft unless a press of at least six ton capacity is available. If a new shaft is installed, the hub on left hand side of differential case must be flush with the shoulder on the shaft. This will leave a small shoulder on the right hand side for the thrust washer.

MAIN DRIVE BEVEL GEARS
All Models

169. The main drive bevel pinion and ring gears are available in matched sets only. To renew the bevel pinion gear which is integral with the transmission countershaft, refer to paragraph 164. To renew the bevel ring gear, it is necessary to remove the differential as in paragraph 165. Refer also to paragraphs 166 and 167.

FINAL DRIVE AND REAR AXLE

BULL PINIONS
All Models

170. The final drive bull pinions are integral with the differential side gears and can be removed by following the procedure for R & R differential (paragraph 165).

BULL GEARS
All Models

171. To remove either of the final drive bull gears, remove the pto and hydraulic sump housing as outlined in paragraph 183, block-up tractor and remove the rear wheels. Remove the fenders and platform. Remove two of the cap screws retaining axle sleeve to transmission housing and install guide bolts as shown in Fig. C654, then remove the remaining cap screws. Bend lip of lockwasher away from nut and remove the nut retaining bull gear to axle shaft. Pull the axle and sleeve assembly away from transmission housing and withdraw the bull gear, being careful not to lose the thrust washer between bull gear and inner bearing cone.

When reassembling, tighten the axle sleeve retaining cap screws to a torque of 100 ft.-lbs. Tighten the bull gear retaining nut to provide a heavy drag on the axle shaft bearings and bend lip of lockwasher down over the nut.

AXLE SHAFTS AND HOUSINGS
All Models Except High Clearance

172. **R & R AND OVERHAUL.** To remove either axle shaft and/or housing, first remove the differential housing top cover, fender, platform and wheel. Remove two of the cap screws retaining axle sleeve to transmission and install guide bolts, then remove the remaining cap screws. Working through top opening in transmission housing, bend lip of lockwasher away from nut and remove the nut retaining bull gear to axle shaft. Pull the axle and sleeve assembly away from transmission housing, being careful not to lose the thrust washer between bull gear and inner bearing cone.

173. Remove dust guard from outer end of axle sleeve and push the axle shaft out of the sleeve. The outer bearing cone and collar can be driven from shaft and bearing cups can be pulled from housing. Refer to Fig. C655.

When reassembling, install the collar with chamfered end against shoulder on axle shaft and press the bearing cone on until it seats against the collar. Install axle shaft in the sleeve and drive the bearing cone on inner end of shaft.

174. Install the axle and sleeve assembly, making certain that the thrust washer is between bull gear and inner bearing cone. Tighten the axle sleeve retaining cap screws to a torque of 100 ft.-lbs. Tighten the bull gear retaining nut to provide a heavy drag on the axle shaft bearings and bend lip of lockwasher down over the nut.

175. Soak the combination outer oil seal and dust seal in oil until the leather lip is soft and pliable. Press the seal (lip toward transmission) in the dust guard until seal is flush with

Fig. C654—Removing bull gears. Notice that axle sleeve is supported on guide bolts.

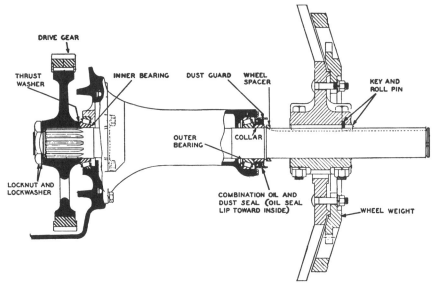

Fig. C655—Rear wheel axle shaft and sleeve assembly used on general purpose tractors. Orchard and standard type tractors are similarly constructed.

inner surface of the guard. Install the guard and seal assembly. Note: On general purpose tractors, it will be necessary to drive the wheel spacer, which is a light press fit, off the shaft before installing the seal and guard. Be sure to reinstall the spacer with the chamfer on inner bore of same toward transmission.

MAIN AXLE SHAFT
High Clearance Models

176. R & R AND OVERHAUL. To remove the main (upper) axle shaft (Fig. C656), first remove the differential housing top cover, fender, platform and wheel. Unbolt and remove the axle drive (drop) housing from main axle sleeve as shown in Fig.

C657. Remove two of the cap screws retaining the main axle sleeve to transmission and install guide bolts, then remove the remaining cap screws. Working through top opening in transmission housing, bend lip of lockwasher away from nut and remove the nut retaining bull gear to the main axle shaft. Pull the main axle and sleeve assembly away from transmission housing, being careful not to lose the thrust washer between bull gear and inner bearing cone.

Remove locknut and washer retaining pinion gear (Fig. C657) to outer end of the main axle shaft and push shaft from the axle sleeve. The procedure for removing and installing the shaft bearings is evident. Lip of oil seal goes toward the transmission.

When reassembling, install the pinion gear on the main axle shaft, tighten the retaining nut securely and bend the lock washer lip down over the nut.

Install the axle and sleeve assembly, making certain that the thrust washer is between bull gear and inner bearing cone. Tighten the main axle sleeve retaining cap screws to a torque of 100 ft. lbs. Tighten the bull gear retaining nut to provide a heavy drag on the axle shaft bearings and bend lip of lockwasher down over the nut.

Install the plastic gasket and the drive (drop) housing. Tighten the drive housing retaining cap screws to a torque of 100 ft.-lbs.

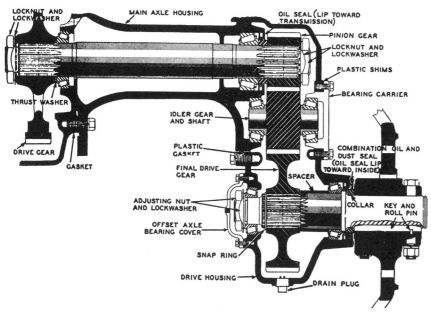

Fig. C656—Sectional view of the final drive housing assembly used on high-clearance tractors.

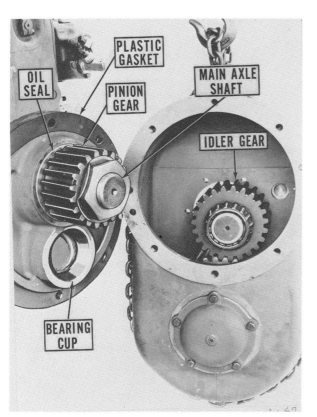

Fig. C657 — High-clearance drop housing removed from the main axle sleeve. Gasket between the two housings is made of plastic.

Fig. C658—Plastic shims used to adjust the high-clearance drop housing idler gear bearings. When bearings are properly adjusted, a noticeable effort will be required to turn the idler gear by hand.

DRIVE HOUSING IDLER AND OFFSET AXLE

High Clearance Models

177. **R & R AND OVERHAUL.** To remove either one of the drive housings, support tractor, remove wheel and unbolt the drive housing from the main axle sleeve. Refer to Fig. C657.

Withdraw the idler gear and bearings from the housing. Remove the offset axle bearing cover and remove the adjusting nut from inner end of the offset axle. Refer to Fig. C656. Remove the snap ring from shaft and bump the offset axle out of housing. The need and procedure for further disassembly is evident.

Before reassembling and adjusting the shaft bearings, remove the drive pinion gear (Fig. C657) from outer end of the main axle shaft. With the offset axle and gear removed from housing, install the idler gear and bearings in the drive housing and bolt the drive housing with plastic gasket installed to the main axle sleeve. Tighten the retaining cap screws to a torque of 100 ft.-lbs. Add or remove plastic shims (Fig. C658) between bearing carrier and housing until a very noticeable effort is required to turn the idler gear by hand. Shims are available in thicknesses of 0.003 (green), 0.005 (blue) and 0.010 (brown).

Unbolt and remove the drive housing from the main axle sleeve, lift out the idler gear but do not remove the idler gear bearing carrier. Install the pinion gear on outer end of the main axle shaft, tighten the nut securely and bend lip of lockwasher down over the nut.

Drive the outer bearing cone on the offset axle until the cone seats against the shoulder. Slip the spacer on the shaft and slide the shaft into the

housing as you install the final drive gear. Install the gear retaining snap ring. Drive the inner bearing cone onto shaft and install the lockwasher and nut. Tighten the nut until a very noticeable effort is required to turn the offset axle shaft by hand, then bend lip of lockwasher down over the nut. Install the inner bearing cover. Drive the collar onto shaft until the collar seats against shoulder on shaft. Lubricate the combination oil seal and dust seal and install the seal with lip facing the transmission. Install the idler gear and bolt the drive housing to the main axle sleeve. Tighten the retaining cap screws to a torque of 100 ft.-lbs.

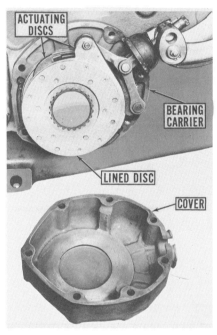

Fig. C660—The brake lined discs can be removed after removing the outer cover as shown.

BRAKES

Brakes are of the double disc, self-energizing type located on outer ends of the combination bull pinion and differential side gear shafts.

All Models

178. **ADJUSTMENT.** To adjust the brakes, tighten the adjusting nut (Fig. C659) on each brake until both brake pedals have a free travel of ¾-1 inch. Equalize by loosening the adjustment on the tight brake.

179. **OVERHAUL.** The lined discs can be removed and relined after removing the brake cover as shown in Fig. C660. The need and procedure for disassembling and overhauling the actuating disc assemblies is evident. If the brake linings are oil soaked, the oil seals in the differential bearing carriers are leaking or the Welch plug in outer end of the combination side gear and bull pinion shaft is leaking. The procedure for renewing the Welch plug is evident. To renew the seals, proceed as follows:

Remove the differential cover and support differential with a hoist.

Unbolt and remove the differential bearing carrier, being careful not to damage or lose the shims located between the bearing carrier and transmission housing. Press the old seals out of the carrier. Soak the leather outer seal in oil until lip is soft and pliable. Coat the outer circumference of the seal with No. 1 Permatex and install seal in carrier so that lip of seal faces the differential. Press the outer seal in just enough so that inner seal can be started in carrier bore. Coat the outer circumference of the inner seal with No. 1 Permatex and install seal in carrier so that the synthetic rubber lip faces the differential. Press both seals in until the synthetic rubber inner seal is just flush with

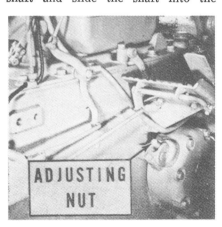

Fig. C659—The brake adjusting nut on the left side of an 800B series tractor. Pedals should have ¾-1 inch free travel. The right side is provided with a similar adjustment. Series 400 and 700B tractors are similar.

Fig. C661 — Using Case thimble A-8341 to protect oil seals when installing the differential bearing carrier.

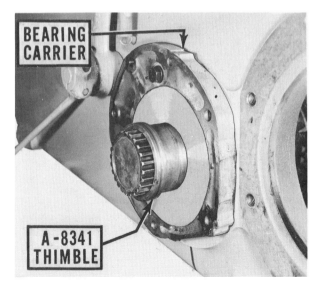

the bearing carrier bore. Install new "O" ring on bearing carrier and reinstall the bearing carrier, using the same shims between carrier and transmission housing as were removed. Use J. I. Case thimble No. A-8341 as shown in Fig. C661 to protect the seal lips when installing the bearing carrier. Tighten the carrier cap screws to a torque of 50 ft.-lbs.

BELT PULLEY

Series 400 and 700B

The belt pulley unit is mounted on right side of clutch housing and is driven by a bevel gear located on the transmission range drive shaft. To renew the driving bevel gear, refer to paragraph 161.

180. **R & R AND OVERHAUL.** To remove the belt pulley unit, first drain oil from the pulley compartment of the clutch housing and pour the oil back into the transmission. Drain plug is located on left side of the clutch housing. Disconect the pulley operating rod and unbolt pulley from the clutch housing. Withdraw the pulley unit, being careful not to damage or lose the shims located between pulley housing and clutch housing.

Remove pulley and clutch fork assembly (C — Fig. C662). Working through the clutch opening in side of pulley housing, slide the sliding sleeve (16) onto the outer shaft (17). Remove

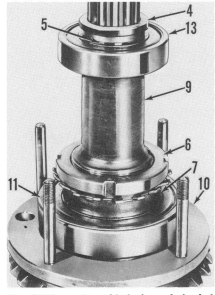

Fig. C663 — Assembled view of the belt pulley inner shaft. Refer to Fig. C662 for legend.

Fig. C664 — Disassembled view of the pulley clutch housing. Clutch fork is retained to shaft with a roll pin.

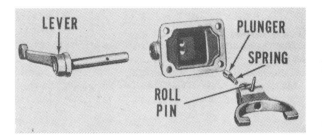

the four nuts (12) and using a pry bar between bearing retainer (10) and pulley housing, pry the inner shaft assembly out of pulley housing. Withdraw the sliding sleeve (16). Remove bearing cap (2) and remove outer shaft (17) by bumping on its inner end with a wooden dowel. If oil seal (1) is damaged, press it out of bearing cap (2). Press sleeve (18) off the outer shaft, remove snap ring and press the shaft out of bearing (3). Press the clutch stop collar (14) off the inner shaft, remove snap ring and press the inner shaft out of middle bearing (13). Unlock and remove nut (6) from inner shaft. Press the inner shaft out of bearing (11). If pilot bushing (15) in the inner shaft is damaged, it can be removed by using a ¾ inch tap as a puller.

Install the bushing with J. I. Case No. A-8348 sizing arbor and press the bushing in until shoulder on arbor seats against the shaft. Using the arbor puller included in Case A-8358 service tool set, remove the arbor from the pilot bushing.

Install the pulley shafts and bearings by reversing the removal procedure. Oil seal (1) should be installed with lip facing the pulley housing. If the outer circumference of the seal is not pre-coated with a sealer, coat it with No. 1 Permatex before installation. When installing bearing cap (2) tighten the cap screws to a torque of 20-25 ft.-lbs. Tighten the pulley nut to a torque of 160 ft.-lbs., then bend lip of lockwasher over the nut.

To disassemble the pulley clutch housing, swing the clutch fork fully to one side and remove the roll pin which secures the fork to the lever shaft. Withdraw the lever shaft, fork, plunger and spring. Check and renew any parts which are damaged. Plunger or detents in housing must not be worn. Refer to Fig. C664.

Note: Install the upper rear pulley housing cap screw before installing the pulley clutch assembly.

With the pulley housing "O" ring (X—Fig. C662) removed, install the belt pulley assembly, using the same number of shims as were removed.

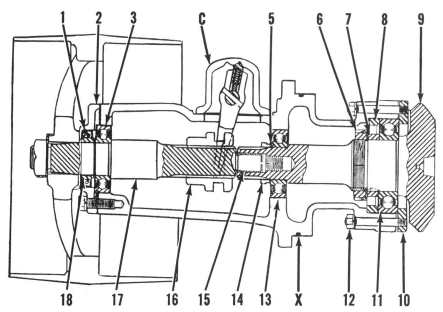

Fig. C662—Sectional view of series 400 and 700B belt pulley unit. The unit is driven by a bevel pinion gear on the range drive shaft.

1. Oil seal	8. Large spacer	13. Center bearing
2. Bearing cap	9. Inner shaft and gear	14. Stop collar
3. Outer bearing	10. Inner bearing	15. Pilot bushing
5. Snap ring	retainer plate	16. Sliding sleeve
6. Lock nut	11. Inner bearing	17. Outer shaft
7. Small spacer	12. Nuts	18. Sleeve

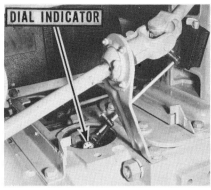

Fig. C665—Using a dial indicator to check the backlash between the pulley driving bevel gears. Backlash should be 0.006.

Slide the batteries out and remove the clutch housing top cover. Mount a dial indicator as shown in Fig. C665 and check the backlash between the driving bevel gears. Add or remove shims between pulley housing and clutch housing until the backlash is 0.006. Shims are available in thicknesses of 0.003 and 0.005. With the bevel gear backlash adjusted, remove the pulley unit and install the housing "O" ring.

Series 800B

The belt pulley unit is mounted on the pto housing at the rear of the tractor and is driven by the pto shaft.

180A. **R&R AND OVERHAUL.** To remove the belt pulley assembly it is

necessary to just unbolt and withdraw the unit from the pto housing.

Disassembly procedure will be evident after an examination of the unit and reference to Fig. C665A. Adjust the bearings (19) to a pre-load of 15-

20 inch-pounds with nut (23). Mesh position and backlash of the bevel gears (8 and 18) is controlled by shims (3 and 25). The recommended backlash of the matched gears is marked on the gears.

POWER TAKE-OFF

The pto clutch is of the multiple disc, over-center, wet type mounted in the pto and hydraulic pump (sump) housing. The hydraulic (sump) housing should be filled to the proper level with Automatic Transmission Fluid Type A. Clutch slippage and chatter can result if ordinary crankcase oil is used in the hydraulic system. If the pto stub (output) shaft creeps after clutch has been disengaged, check for cocked clutch plates, worn stub shaft splines and/or broken clutch plate separator springs. Failure to maintain the proper clutch adjustment will result in excessively worn and scored clutch plates.

CLUTCH AND OUTPUT (STUB) SHAFT

181. **CLUTCH ADJUSTMENT.** To adjust the pto clutch, first place the clutch hand lever in the disengaged (full forward) position, remove the pipe plug from side of sump housing

and turn the pto stub shaft until the two spring loaded lockpins are visible through the pipe plug hole. Refer to Fig. C666. Notice that one of the pins is in the engaged (IN) position, while the other pin is disengaged. With the adjusting tool inserted in the pipe plug hole, pull the engaged lock pin out with the forked end of the tool.

To tighten the clutch, turn the pto stub shaft counter clockwise as shown until the disengaged lock pin drops into a notch. Check the adjustment by pulling the hand lever fully rearward.

A properly adjusted clutch will require 50 pounds pull and will engage with a distinct snap of the lever.

182. **R & R AND OVERHAUL.** To remove the pto clutch and stub shaft, engage the clutch by pulling the hand lever fully rearward, and disconnect the hose lines and control rods from the hydraulic system valve housing. Remove the upper two cap screws retaining valve housing to sump housing and install ½ by 3 inch studs in their place. Remove the stud nuts securing bottom of valve housing and remove the pump and valves unit as shown in Fig. C667. Be careful not to lose the rubber seals located in the inlet and outlet pipes. Disconnect the clutch throwout linkage at side of housing and working through the hydraulic pump opening in rear of housing, loosen the two cap screws retaining the throwout fork to the shaft. Pull the throwout shaft from side of housing and withdraw the clutch throwout fork. Unbolt the pto bearing carrier from housing and withdraw the pto clutch and stub shaft assembly.

Note: The clutch throwout upper linkage and cross shaft can be removed from housing and the "O" rings and felt washers can be renewed if they are damaged. The procedure is evident.

To disassemble the removed clutch and stub shaft unit, remove the retaining snap ring and press stub shaft out of bearing carrier. To remove bearing, collar and oil seal from bear-

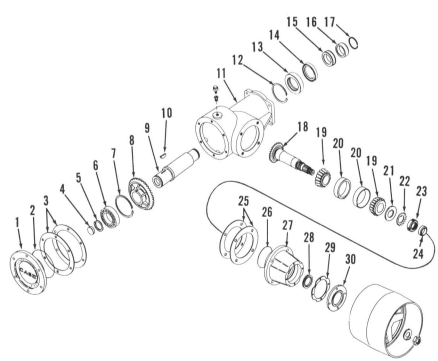

Fig. C665A—Exploded view of the 800B belt pulley unit. Drive bevel gears (8 & 18) are available only in a matched set.

1. Drive shaft rear bearing carrier	8. Drive bevel gear	16. Collar
2. "O" ring	9. Drive shaft	17. Snap ring
3. Shims (0.003 & 0.005)	10. Woodruff key	18. Drive bevel gear
4. Plug	11. Housing	19. Bearing cone
5. Snap ring	12. Snap ring	20. Bearing cup
6. **Bearing**	13. Bearing	21. Washer
7. Snap ring	14. Oil seal	22. Lock washer
	15. Inner bearing race	23. Spanner nut

24. **Collar**	27. **Housing**
25. Shims (0.003 & 0.005)	28. Oil seal
26. "O" ring	29. Gasket
	30. Oil seal **carrier**

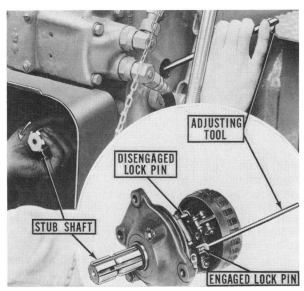

Fig. C666—Adjusting the pto clutch. To engage a properly adjusted clutch, approximately 50 lbs. pull on the lever will be required.

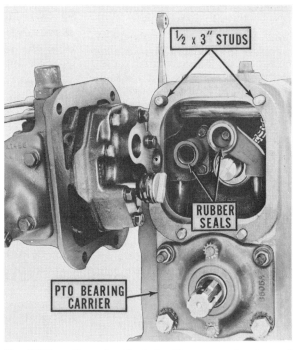

Fig. C667 — Removing the hydraulic pump and valves unit from sump housing. Rubber seals in the pump inlet and outlet pipes must be in good condition.

ing carrier, remove snap ring (Fig. C669) from front of bearing carrier and using a suitable step plate against the collar, press the collar, oil seal and bearing forward and out of carrier. Pull the clutch adjusting lock pins outward, unscrew the adjusting and throwout collar assembly from the stub shaft and remove the pressure plate, bronze plates, steel plates, separator springs and back plate. Inspect or renew any parts which are damaged, cocked, or show wear.

If the needle pilot bearings (Fig. C670) are damaged, they can be removed by using a suitable puller. Install the rear pilot bearing, using J. I. Case No. A-9031 arbor. Press on the end of the bearing which is stamped with the identification and press the bearing in until stop shoulder on arbor is against the stub shaft. Install the front pilot bearing in a similar manner, using J. I. Case arbor No. A-9032.

Reassembly of the clutch plates and stub shaft will be simplified if an extra clutch drum is used as a jig. Place the stub shaft in spare clutch drum and assemble the clutch plates in the order shown in Fig. C670. Smooth side of pressure plate goes toward the clutch plates. Screw the adjusting and throwout collar assembly on the stub shaft until one of the lock pins can be engaged in the pressure plate. Engage the clutch and remove the assembly from the spare clutch drum.

Press the bearing into carrier with shielded side of bearing toward rear and install the snap ring. Coat the seal

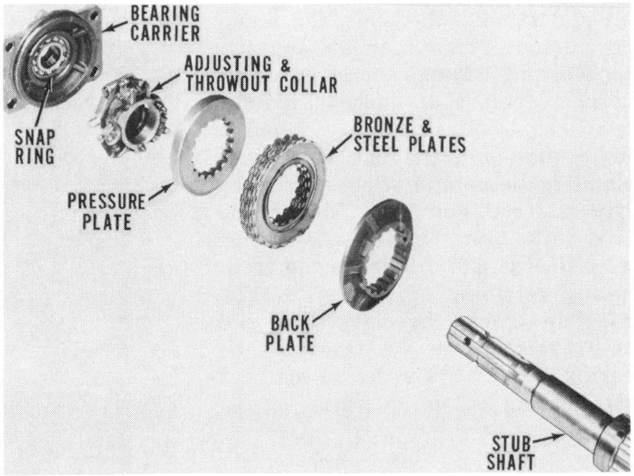

Fig. C669—Partially disassembled view of the pto clutch and stub shaft. Seven bronze and six steel clutch plates are used as shown in Fig. C670.

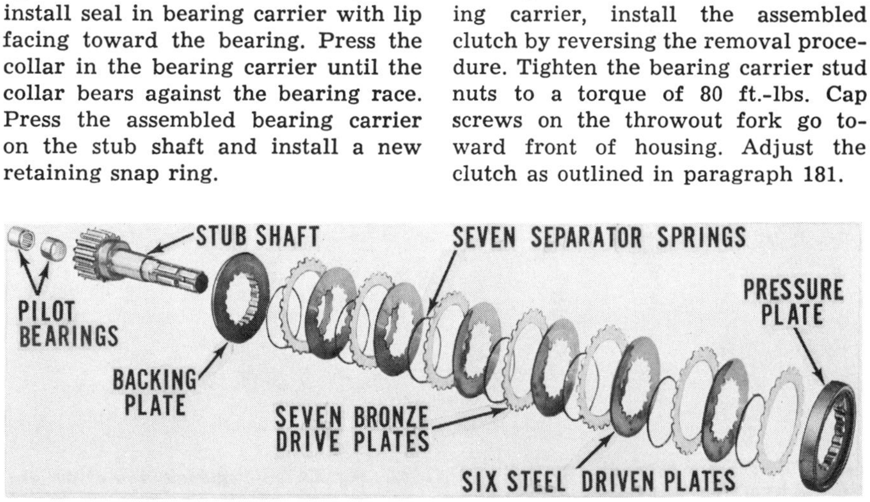

lip with White Petroleum Jelly and install seal in bearing carrier with lip facing toward the bearing. Press the collar in the bearing carrier until the collar bears against the bearing race. Press the assembled bearing carrier on the stub shaft and install a new retaining snap ring.

Using a new "O" ring on the bearing carrier, install the assembled clutch by reversing the removal procedure. Tighten the bearing carrier stud nuts to a torque of 80 ft.-lbs. Cap screws on the throwout fork go toward front of housing. Adjust the clutch as outlined in paragraph 181.

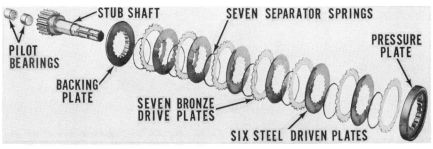

Fig. C670—Exploded view of the pto clutch plates, separator springs and stub shaft. Make sure that clutch plates are not cocked.

PTO & HYDRAULIC PUMP (SUMP) HOUSING

183. REMOVE AND REINSTALL. To remove the pto and hydraulic pump housing as an assembly, first disconnect the control rods and hydraulic lines from the hydraulic valve housing. Remove the hose bracket from top of housing. Attach a chain hoist to housing as shown in Fig. C671, unbolt and remove housing from tractor. If the driving shaft does not come rearward with the housing, it can easily be withdrawn from the transmission case.

NOTE: Early production series 400 tractors were equipped with a snap ring on the drive shaft. This snap ring is not necessary and should be discarded. Refer to Fig. C672.

Housing can be installed by reversing the removal procedure.

PTO & HYDRAULIC PUMP DRIVE UNIT

184. R & R AND OVERHAUL. To remove the drive unit, first remove the complete pto and hydraulic pump housing as in paragraph 183, then unbolt and remove the hydraulic pump and valve unit from rear face of housing. Working through the pump opening in rear face of housing, remove the two cap screws retaining the pump intake tube and the one cap screw retaining the pump outlet tube to the drive unit. Unbolt and remove the drive unit from front face housing. Refer to Fig. C673. Remove the front bearing retaining snap ring (8—Fig. C674) and press the drive shaft and front bearing out the front of the bearing carrier. Press the oil seal and rear bearing out of carrier.

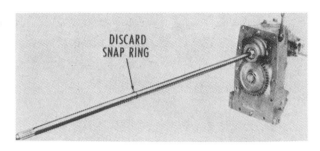

Fig. C672—Pto drive shaft on some early production series 400 tractors was equipped with an unnecessary snap ring. Discard the snap ring if shaft is so equipped.

Reassemble the drive unit, making certain that beveled edge of spacer (4) is toward rear and that lip of oil seal (1) faces rearward. Install the assembled drive unit by reversing the removal procedure and make certain that the breather hole (Fig. C673) is open.

CLUTCH DRUM

185. To remove the clutch drum, first remove the clutch plates and stub shaft unit as outlined in paragraph 182, then remove the pto and hydraulic pump housing as in paragraph 183. Unlock and remove the nut retaining the pto drive gear to the shaft. Withdraw the drive shaft from rear of housing and remove the gear. Press the clutch drum out of housing. Pry the clutch drum oil seal out of housing, remove the snap ring behind the drum bearing and press the bearing rearward and out of housing.

When reassembling, install the drum bearing and its retaining snap ring. Install J. I. Case No. A-8904 spacer plates on the clutch drum; then, using shim stock or J. I. Case No. A-8909 thimble, install the oil seal on drum with lip toward the drum. See Fig. C 675. Carefully press the drum and bearing into the housing until the shoulder on the drum is against the bearing. Then remove the spacer

Fig. C673 — Front view of the pto sump housing, showing the driving gears. Breather hole in drive unit housing must be open and clean.

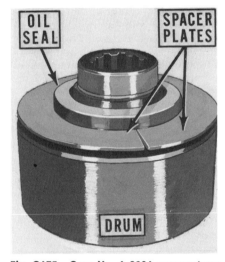

Fig. C675—Case No. A-8904 spacer plates are used when installing the drum so that oil seal will be properly pressed into the housing.

Fig. C671—Pto and hydraulic sump housing removed from transmission housing. The unit is extremely heavy and should always be handled with a hoist.

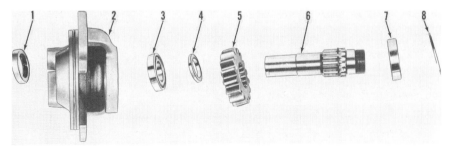

Fig. C674—Exploded view of the pto drive unit shown installed in Fig. C673.

1. Oil seal
2. Bearing carrier and housing
3. Rear bearing
4. Spacer
5. Drive gear
6. Hydraulic pump drive shaft
7. Front bearing
8. Snap ring

plates. Refer to Fig. C676. Install a new "O" ring, spacer washer and snap ring on the pto drive shaft and install the drive shaft and drive gear. Do not overtighten the nut which retains the drive gear to the shaft. The nut should be tightened only enough to remove all end play plus ¼ turn, Bend lip of lockwasher down over the nut.

Install the remaining parts by reversing the removal procedure and adjust the clutch as in paragraph 181.

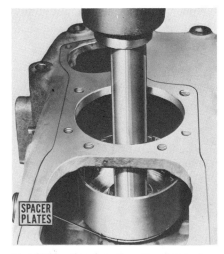

Fig. C676—Pressing the pto clutch drum into the sump housing. When drum is seated, be sure to remove the spacer plates.

HYDRAULIC SYSTEM

Most of the troubles encountered with the hydraulic system on modern tractors are caused by dirt or gum deposits. The dirt may enter from the outside, or it may show up as the result of wear or partial failure of some part of the system. The presence of gummy deposits, however, usually results from inadequate fluids or from failure to drain and renew the fluid at the recommended intervals. These principles should be kept in mind when shooting trouble on any of the existing systems and also when performing repair work on pumps, valves and work cylinders.

Thus, when disassembling a pump or valves unit, it is good practice generally to not remove any parts which can be thoroughly inspected while they are installed. Internal parts of the pump, valve and cylinders, when removed, should be handled with the same care as would be accorded the parts of a Diesel injection pump or nozzle unit, and should be soaked or cleaned with an approved solvent to remove gum deposits Unless you practice good housekeeping (cleanliness) in your shop, do not undertake the repair of hydraulic equipment.

The combined pump and valves unit is mounted on the rear face of the combined pto and hydraulic sump housing which is attached to the rear face of the transmission housing. The gear type pump is driven by a fixed drive from the engine and operates independently of the main (traction) clutch, transmission or pto clutch.

The hydraulic control on Eagle Hitch equipped tractors is of the dual valve float type that provides implement float as well as pressure for both raising and lowering the hitch. On models without Eagle Hitch, the hydraulic control is of the dual valve type without the float feature.

FLUID AND FILTER

190. It is recommended that the hydraulic system working fluid (Automatic Transmission Fluid, Type A) be changed after every 1000 hours of operation.

Remove the magnetic drain plug from the reservoir housing and allow system to drain. Remove the filter cover as shown in Fig. C680, withdraw the filter element and wash same in clean fuel.

Reinstall the filter element and cover and refill the reservoir to the level plug with Automatic Transmission Fluid, Type A.

Approximate fluid capacity is 10.5 quarts for models with Eagle Hitch, or 9 quarts for models without Eagle Hitch. Add 1 quart for each portable cylinder connected to the system.

On Eagle Hitch equipped tractors, the rockshaft housing is filled with approximately 1 pint of SAE 90 transmission oil.

TROUBLE SHOOTING

191. The following paragraphs should facilitate locating troubles in the hydraulic lift system of Eagle Hitch equipped tractors. On models without Eagle Hitch, only the appropriate items will apply.

UNIT WILL NOT LIFT OR PORTABLE CYLINDERS WILL NOT EXTEND OR RETRACT. Could be caused by:

a. Low system operating pressure, paragraph 195.

b. Insufficient fluid in reservoir, paragraph 190.

c. Worn pump or damaged pump drive coupling, paragraph 193.

d. Relief valve not seating, paragraph 197.

e. Return ball in head not seating, paragraph 202 or 207.

f. Dislocated seal at upper end or clogged screen at lower end of inlet pipe, paragraph 194.

g. Faulty work cylinder (mounted or portable), paragraph 211.

h. Missing or faulty auxiliary spool, (Relief valve may chatter, but unit will not lift).

i. Loose or faulty spool shifter arm, paragraph 198.

j. Relief valve ball in left hand head not seating, paragraph 200 or 205.

k. System overloaded.

l. Hydraulic pressure not equalized at couplings.

SLUGGISH LIFTING ACTION. Could be caused by:

a. Low system operating pressure, paragraph 195.

b. Insufficient fluid in reservoir, paragraph 190.

c. Worn or damaged pump, paragraph 193.

d. Weak relief valve spring, paragraph 197.

e. Partially plugged screen on inlet pipe, paragraph 194.

f. Collapsed or plugged hose line.

UNIT WILL NOT HOLD LOAD IN RAISED POSITION. Could be caused by:

a. Damaged "O" ring on work cylinder piston, paragraph 211.

b. Outlet ball seat in left hand side of valve body is leaking, or ball is not seating, paragraph 201 or 206.

c. Return ball seat in left hand head is leaking, or ball is not seating, paragraph 202 or 207.

d. Relief valve ball seat in left hand head is leaking, or ball is not seating, paragraph 200 or 205.

e. Auxiliary spool in left hand head is too long and return ball is prevented from seating.

Fig. C680—Removing the oil filter element from the hydraulic and pto sump housing. Filter element can be cleaned in clean fuel.

HAND LEVER WILL NOT STAY IN RAISED OR LOWERED POSITION DURING THE RAISING OR LOWERING CYCLE. Could be caused by:

a. Knife edge on spool holding poppet is damaged, or poppet is incorrectly installed, **paragraph 203 or 210.**

b. Spring for spool holding poppet is damaged or weak, **paragraph 203 or 210.**

c. Low system operating pressure, **paragraph 195.**

d. Relief valve plunger stuck in raised position, **paragraph 197A.**

e. System overloaded.

f. Wrong fluid used in system, **paragraph 190.**

g. Restricted hose.

h. Damaged linkage or wrong cylinder.

HAND LEVER WILL NOT RETURN TO NEUTRAL FROM RAISED POSITION. Could be caused by:

a. Low system operating pressure, **paragraph 195.**

b. Return ball in head is not seating, **paragraph 202 or 207.**

c. Relief ball in left hand head is not seating, **paragraph 200 or 205.**

d. Worn or damaged pump, **paragraph 193.**

e. Relief valve plunger stuck in lowered position, **paragraph 197A.**

f. Distorted hand control lever linkage.

NO DOWN PRESSURE ON LOWERING CYCLE. Could be caused by:

a. Lower return ball in right hand head is not seating, **paragraph 202 or 207.**

b. Return ball check spring improperly assembled. See illustrations.

c. Lower auxiliary spool in right hand head is too long and ball is prevented from seating.

d. Down pressure relief valve ball in right hand head is not seating, **paragraph 208.**

e. Float ball in right hand head is not seating, **paragraph 209.**

HAND LEVER WILL NOT STAY IN FLOAT POSITION. Could be caused by:

a. Float position retaining plunger tip is damaged, or plunger spring is missing, **paragraph 210.**

HAND LEVER WILL NOT OPERATE. Could be caused by:

a. Damaged linkage.

b. Damaged or missing valve spool shifter arm, **paragraph 198.**

c. Missing thrust ball under lower spool shifter arm shaft, **paragraph 198.**

d. Damaged spool, **paragraph 203 or 210.**

HOSES RUPTURING. Could be caused by:

a. System relief pressure to high, **paragraph 195.**

b. Relief valve sleeve installed upside down, **paragraph 197A.**

c. Relief valve plunger seized in lowered position, **paragraph 197A.**

MIXING OF TRANSMISSION OIL AND HYDRAULIC FLUID. Could be caused by:

a. Plugged hydraulic sump housing breather.

b. Damaged or incorrectly installed seal in hydraulic housing.

R & R PUMP AND VALVES UNIT

192. To remove the hydraulic pump and valves unit, disconnect the hose lines and control rods from unit and remove the upper two cap screws retaining the unit to reservoir housing. Install ½ x 3 inch studs in place of the upper two cap screws and remove the lower two screws. Withdraw the

pump and valves unit as shown in Fig. C681.

When reinstalling the unit, reverse the removal procedure, make certain that the rubber seals in the pump inlet and outlet pipes are in good condition and be sure to align the pump drive coupling slots.

OVERHAUL PUMP

193. With the pump and valve unit removed as outlined in paragraph 192, place pump in a soft jawed vise, remove the eight cap screws retaining the pump cover to the valve body and lift off the cover as shown in Fig. C682. Remove the idler gear and shaft

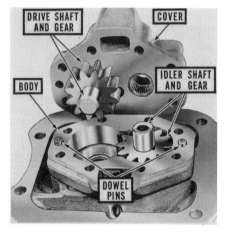

Fig. C682—Hydraulic pump cover removed from front face of valve housing. Dowel pins position the pump cover with respect to the pump body and valve housing.

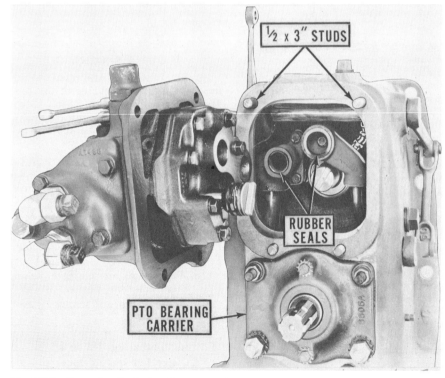

Fig. C681—Removing the hydraulic pump and valves unit from sump housing. Rubber seals in the pump inlet and outlet pipes must be in good condition.

and withdraw the pump body as shown in Fig. C683. To remove the pump drive shaft coupling, place the pump cover, gear and shaft assembly in a vise and remove the coupling as shown in Fig. C684. Coupling has a right hand thread.

Lift the valve spool holding poppets out of the valve body as shown in Fig. C685 and renew the poppets if knife edge of same is excessively worn. Poppet springs have a free length of $1\frac{3}{8}$ inches and should require 18 pounds to compress them to a height of $1\frac{3}{16}$ inches.

Check the valve body, pump cover faces and gears for scoring or wear. The total clearance between the pump gears and the body and cover faces should be 0.0015-0.0025. Check the pump gear shafts needle bearings. Needle bearings can be removed, as shown in Fig. C686, using J. I. Case tool No. 3913-AA. If tool No. 3913-AA or equivalent is not available, the bearings can sometimes be removed hydraulically by filling the bearing cavity with grease and ramming downward on same using a close fitting shaft struck sharply with a hammer. Bearings should be installed with

the flat (lettered) end facing the pump gears and can be installed with a piloted pressing tool (J. I. Case No. 08665-AB) or equivalent.

When reassembling, install the spool holding poppets as shown in Fig. C685. Knife edge on end of poppet must fit into the groove in the valve spool. Install the pump body and dowels as shown in Fig. C683 and insert the idler shaft and gear. Assemble the pump drive shaft, gear and coupling in the cover and install the cover as shown in Fig. C682. Tighten the cover cap screws in the sequence shown in Fig. C687 and to a torque of 35 ft.-lbs.

PUMP INLET AND OUTLET PIPES

194. To remove the pump inlet and outlet pipes, first remove the pump and valves unit as outlined in paragraph 192 and remove the rubber seals from the pipes. Refer to Fig. C681. Remove the mounting cap screws and withdraw the pipes from housing.

If screen at lower end of inlet pipe is dirty, clean the screen and flush the complete reservoir housing with kerosene.

When installing the pipes, make certain that lower end of outlet pipe is located in the filter base. A small amount of wheel bearing grease will facilitate holding the rubber seals in position when installing the pump and valves unit.

CHECK AND ADJUST RELIEF VALVE PRESSURE

195. To check and/or adjust the relief valve unloading pressure, remove the $\frac{1}{8}$ inch pipe plug from top of valve housing and install a suitable pressure gage (at least 1600 psi capacity) as shown in Fig. C688. Start engine and allow pump to operate until the hydraulic fluid is at the normal operating temperature of 95-110 de-

grees F; then, set the engine throttle lever to provide a pto speed of 250 rpm. Move the outer hand control lever forward to the full raise position. The relief valve should unload

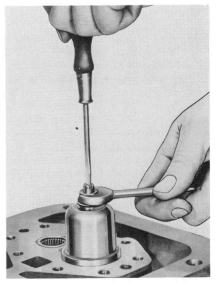

Fig. C686—Using J. I. Case special tool number 3913-AA to remove the pump shaft needle bearings.

Fig. C687—When assembling the hydraulic pump, the cover cap screws should be tightened in the sequence shown. Tightening torque is 35 Ft.-Lbs.

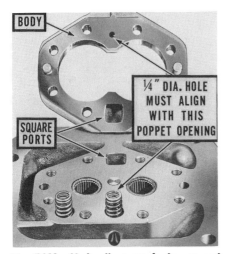

BODY

¼" DIA. HOLE MUST ALIGN WITH THIS POPPET OPENING

SQUARE PORTS

Fig. C683—Hydraulic pump body removed from valve housing. When installing the body, make sure that the proper ports align.

Fig. C684—Remove the right hand threaded coupling from the pump drive shaft.

Fig. C685 — The valve spool holding poppets are accessible after removing the hydraulic pump body. Notice the position of the knife edges when reassembling.

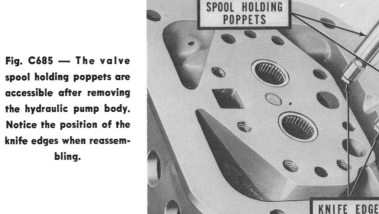

SPOOL HOLDING POPPETS

KNIFE EDGE

when the pressure is within the range of 1175-1225 psi for models with Eagle Hitch or 980-1130 psi on models without Eagle Hitch.

If the unloading pressure is not as specified, refer to the following paragraphs:

On models without Eagle Hitch and early 400 models with Eagle Hitch where relief valve has one spring, the unloading pressure can be changed by varying the number of shims between the relief valve cap nut and the spring. Refer to Fig. C689. If spring is being renewed on models with Eagle Hitch, refer to paragraph 197.

On late series 400 and all 700B and 800B models with Eagle Hitch where the relief valve has two springs, pressure adjusting shims are not normally used; however, if the unloading pressure is less than 1175 psi, shims can be added to increase the pressure.

OVERHAUL VALVE BODY AND HEADS

196. To overhaul the valve body and heads, first remove the pump and valves unit as outlined in paragraph 192. Procedure for bench disassembly and overhaul of the various components are as follows:

197. RELIEF VALVE. On models without Eagle Hitch, the relief valve is equipped with one 08405AB relief valve spring and shims are provided for adjusting the unloading pressure of 980-1130 psi.

On early series 400 tractors with Eagle Hitch, relief valve is equipped with one A7778 relief valve spring and shims are provided for adjusting the unload-

ing pressure of 1175-1225 psi. The relief valve cap nut A7779 on these early models was $2\frac{7}{32}$ inches long as shown in Fig. C690. The A7778 spring is no longer available for service and when it becomes necessary to renew the spring, No. A-9354 relief valve spring kit (Fig. C691) is provided and should be used with the original $2\frac{7}{32}$ inches long spring cap nut A7779. The two springs (A9303 & 08405AB) are shorter than the original A7778 spring. To compensate for the shorter springs, spacer (A9304) must be installed in cap nut (A7779) with chamfer on spacer toward top of nut as shown in Fig. C692 and the original pressure adjusting shims are not normally required. If, after checking the operating pressure as in paragraph 195, the presure is less than 1175 psi, shims can be added to increase the pressure.

Late model series 400 and all 700B and 800B Eagle Hitch tractors are factory equipped with two relief valve springs (A-9303 and 08405AB) and the relief valve cap nut is $1\frac{13}{32}$ inches long as shown in Fig. C693. CAUTION:

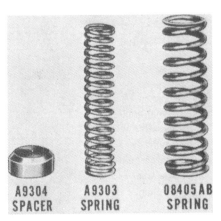

Fig. C688—Checking the hydraulic system pressure. PTO shaft must be turning at 250 rpm when making this check.

Fig. C690—Relief valve spring and cap nut used on early production 400 series tractors. For service, spring No. A-7778 is no longer available and must be replaced with kit No. A-9354 shown in Fig. C691.

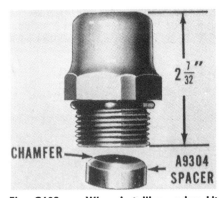

Fig. C692 — When installing spring kit A-9354, be sure that chamfer on spacer is toward top of cap nut.

Fig C689—Location of the relief valve adjusting shims for models without Eagle hitch and early series 400 tractors with Eagle hitch where the relief valve has one spring.

Fig. C691—No. A-9354 relief valve spring kit used to replace the early relief valve spring No. A-7778. This kit is used in conjunction with spring cap No. A-7779.

Fig. C693—Relief valve springs and cap nut used on late production 400 series and 700B and 800B tractors equipped with Eagle hitch.

Never install spacer (A-9304) in the late model 1$\frac{13}{32}$ inches long nut. To do so, would damage the pump.

On these late models, pressure adjusting shims are not normally used; however, if the unloading pressure is less than 1175 psi, shims can be added to increase the pressure.

197A. To service the relief valve, remove the cap nut and spring (or springs) and on models so equipped, save the relief valve unloading pressure adjusting shims located between the cap nut and spring. Using a pair of long nose pliers, check the fit of the relief valve plunger in the sleeve.

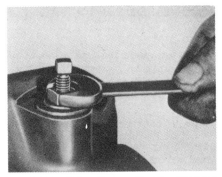

Fig. C694—Using Case puller No. 3914-AA to remove the hydraulic system pressure relief valve seat.

Fig. C695 — Installing the pressure relief valve sleeve (seat). Seat should be installed with the drilled ports downward.

NOTCH MUST REGISTER WITH HOLE

RETAINER

Fig. C696—Proper installation of the relief valve sleeve (seat) retainer. Notch in the retainer must register with the return hole.

The plunger should have a smooth sliding fit with no noticeable side play. Excessive clearance between plunger and sleeve will cause loss of pressure at low pump speeds. Check the bleed hole at flange end of plunger for any obstruction. This hole must be open to allow oil to fill the valve chamber.

To remove the relief valve sleeve (seat), press grease in the plunger hole in sleeve so as to catch any metal cuttings; then tap the sleeve hole with a $\frac{3}{8}$ x 16 thread tap. Using a bolt and sleeve puller (Case No. 3914-AA puller is recommended), pull the sleeve (seat) from the valve body as shown in Fig. C694. The sleeve retainer will come out with the seat. Thoroughly wash and clean the valve body to remove the metal cuttings.

Install a new relief valve seat with drilled ports downward as shown in Fig. C695. Press the sleeve completely into the valve body using a piloted pressing tool (Case No. 08669AB). Top of sleeve must be 1$\frac{7}{16}$ inches below top of sleeve bore. Using the same piloted pressing tool, press the sleeve retainer into position against the sleeve so that notch in retainer will register with return hole in valve body as shown in Fig. C696.

Install the relief valve plunger, spring (or springs), shims (if used) and the relief valve cap nut.

When the pump and valve unit is installed on tractor, check the system unloading pressure as in paragraph 195.

198. **VALVE CONTROL LEVERS.** To service the control levers only, it is not necessary to remove the pump and valve unit from tractor.

Remove the rear cover plate from the valve housing and on models with Eagle Hitch, use long nose pliers and remove the retaining E-ring as shown

in Fig. C697. On all models, remove the machine screw retaining the lower spool shifter arm to the control shaft and withdraw the shaft and arm. On models with Eagle Hitch, be careful not to lose the small thrust ball located at lower end of shaft. Refer to Fig. C698. On all models, loosen the machine screw retaining the upper spool shifter arm to the upper control shaft and withdraw the shaft and arm. If there is evidence of fluid leakage around the control lever shafts, renew the felt seals and "O" rings. Assemble the seals and "O" rings in the sequence shown in Fig. C699.

On models without Eagle Hitch, install the control levers so that when the valve spools are in neutral position, the control levers will be parallel to the mounting face of valve housing as shown in Fig. C700.

THRUST BALL

Fig. C698—When removing and reinstalling the control arms on models with Eagle hitch, be careful not to lose the thrust ball at lower end of shaft.

Fig. C697 — On tractors equipped with Eagle hitch, the "E"-ring must be removed before withdrawing the valve spool control arms.

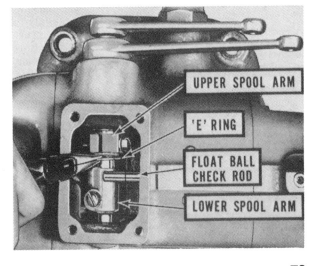

UPPER SPOOL ARM

'E' RING

FLOAT BALL CHECK ROD

LOWER SPOOL ARM

On models with Eagle Hitch, install the control levers so that when the valve spools are in neutral position, both levers will align and the center line of the pin hole in the long lever will be approximately 5½ inches from the mounting face of the valve housing as shown in Fig. C701.

Procedure for installing the levers is as follows: Using care not to damage the seals, insert the control lever and shaft for the upper valve spool and as the shaft is being pushed downward, install the spool lever and its retaining machine screw. Install the control lever and shaft for the lower valve spool and as the shaft is being pushed downward, install the spool lever and its retaining machine screw. On models with Eagle Hitch, be sure to install the small thrust ball under the shaft. Install the E-ring.

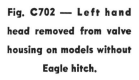

Fig. C702 — Left hand head removed from valve housing on models without Eagle hitch.

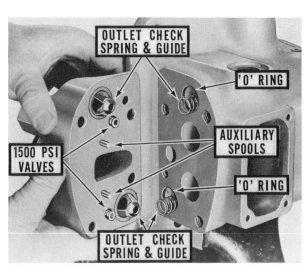

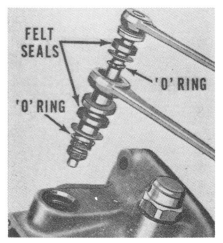

Fig. C699—J. I. Case hydraulic system control arm seals and "O" rings.

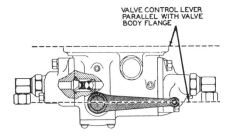

Fig. C700—On models without Eagle hitch, control levers should be installed parallel to mounting face of valve housing.

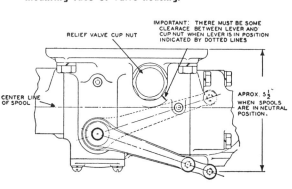

199. VALVES, SPOOLS AND VALVE HEADS (MODELS WITHOUT EAGLE HITCH). The left and right valve heads are similar except the left head contains two 1500 psi auxiliary relief valves. Both heads contain two outlet check valves and two return check valves. Refer to Figs. C702 and C704.

200. AUXILIARY RELIEF VALVES. These non-adjustable (high pressure safety) valves are located in the left hand valve housing head and are preset to unload at 1500 psi. Refer to Fig. C702. These valves can be serviced without removing the pump and valves unit from tractor as follows:

Disconnect the hose lines, unbolt and remove the left head from housing as shown. Withdraw the relief springs and balls and check the ball seats, which are integral with the head, for leakage, roughness, etc. If seats are not badly worn or scored, a new ball (¼ inch diameter) can be reseated to same by tapping lightly with a hammer and soft drift as shown in Fig. C703.

If the seat is damaged beyond the reseating stage, install a new head and seat the new valve ball by tapping it lightly with a hammer and soft drift. The auxiliary relief valve springs have a free length of 1¼ inches and should require 23 pounds pressure to compress them to a height of 1 1/16 inches. When reassembling, tighten the head retaining cap screws to a torque of 35 ft.-lbs.

201. OUTLET CHECK VALVES. Each end of the valve body contains two outlet check valves. To service these valves, which are illustrated in Figs. C702 and C704, first remove the pump and valves unit from reservoir housing as outlined in paragraph 192; then, unbolt and remove both heads from the valve housing as shown. Lift out the springs and balls.

Leakage at the outlet valves can be checked by placing the valve body in a vise so that one of the head ends is facing up and in a level position. Place the outlet balls on their seat and fill the valve port with gasoline. If the seat leaks, the valve can be reseated by tapping the ball with a

Fig. C701 — Correct installation of the valve control levers on tractors equipped with Eagle hitch.

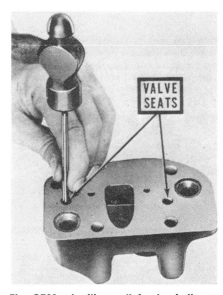

Fig. C703—Auxiliary relief valve balls can usually be reseated by tapping the ball lightly with a soft drift.

hammer and soft drift. If the seats are damaged beyond the reseating stage, press grease in the seat hole to catch any metal cuttings, tap the seat with a $\frac{9}{16}$ x 18 tap and remove the seat with a screw type puller.

Install the new seat with chamfered end up and press or drive the seat firmly into place, using a piece of ¾ inch round bar stock.

202. RETURN CHECK VALVES. Each of the valve body heads contain two ball type return check valves as shown in Fig. C705. Since the valves and the seats are wholly contained in the removable valve heads, they can be serviced without removing the pump and valves unit from the reservoir housing. To remove valves from either head, disconnect hoses, unscrew elbows and extract valve springs and balls. If valve seats need servicing, remove the valve head or heads from the valve body.

Leakage at the return valves can be checked by placing the head in a vise and in a level position. Place the return valve ball on the seat and fill the port with gasoline. If the seat leaks, the valve can be reseated by tapping the ball with a hammer and soft drift.

If the valve seat is damaged beyond the reseating stage, it will be necessary to renew the valve body head.

Reinstall the return valve ball and spring (small diameter end contacting the valve ball) by reversing the removal procedure; then reinstall the valve head and tighten the retaining cap screws to a torque of 35 ft.-lbs.

203. VALVE CONTROL SPOOLS. To remove the valve control spools, first remove the pump and valve unit from reservoir housing as outlined in paragraph 192. Remove the control valve levers as in paragraph 198 and the pump body as in paragraph 193. From pump side of valve body, withdraw the spool holding poppets as shown in Fig. C706. Remove both heads from valve body and withdraw the valve control spools.

Renew the spool holding poppets if knife edge of same is excessively worn. Poppet springs have a free length of 1⅜ inches and should require 18 pounds to compress them to a height of 1⅜ inches. If spools or body are excessively scored, renew the damaged part.

When reassembling, install the valve control spools as shown in Fig. C707 by reversing the removal procedure.

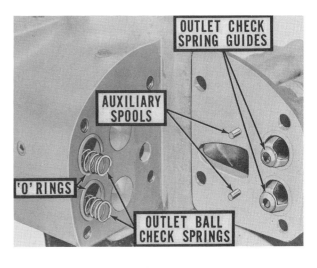

Fig. C704 — Right hand head removed from valve housing on models without Eagle hitch.

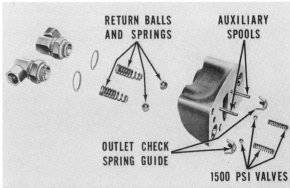

Fig. C705—Exploded view of left hand valve housing head for models without Eagle hitch.

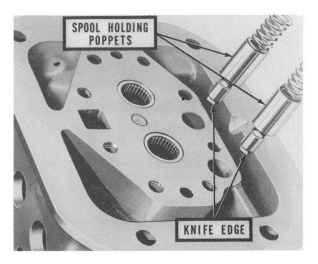

Fig. C706 — The valve spool holding poppets are accessible after removing the hydraulic pump body. Notice the position of the knife edges when reassembling.

204. VALVES, SPOOLS AND VALVE HEADS (MODELS WITH EAGLE HITCH). The left and right valve heads are similar, except: The left head contains two 1500 psi auxiliary relief valves. The right head contains the down pressure relief valve ball and spring as well as the float ball and check spring. Both heads contain two outlet check valves and two return check valves. Refer to Figs. C708, C709, C710 and C711.

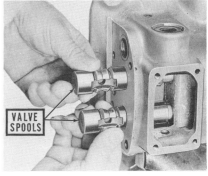

Fig. C707—Correct installation of the valve control spools on models without Eagle hitch.

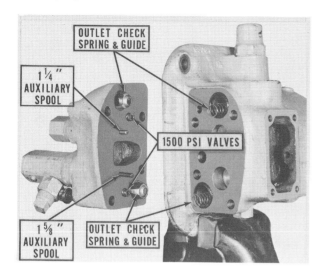

Fig. C708—Left hand cylinder head removed from valve body on models with Eagle hitch.

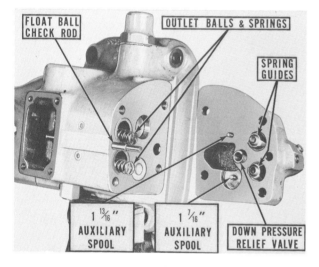

Fig. C709 — Right hand head removed from valve housing on models with Eagle hitch. Notice the installation of the float ball check rod and the down pressure relief valve.

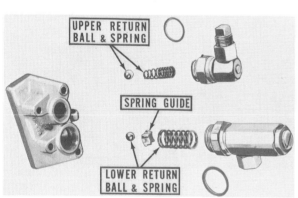

Fig. C710—Exploded view of left hand valve housing head on models with Eagle hitch.

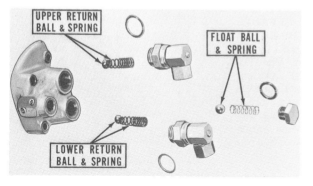

Fig. C711—Exploded view of right hand valve housing head on models with Eagle hitch. Notice the float ball and spring.

205. AUXILIARY RELIEF VALVES. These non-adjustable (high pressure safety) valves are located in the left hand valve housing head and are preset to unload at 1500 psi. Refer to Fig. C708. These valves can be serviced without removing the pump and valves unit from tractor as follows:

Disconnect the hose lines, unbolt and remove the left head from housing as shown. Withdraw the relief springs and balls and check the ball seats, which are integral with the head, for leakage, roughness, etc. If seats are not badly worn or scored, a new ball (¼ inch diameter) can be reseated to same by tapping lightly with a hammer and soft drift as shown in Fig. C703.

If the seat is damaged beyond the reseating stage, install a new head and seat the new valve ball by tapping it lightly with a hammer and soft drift. The auxiliary relief valve springs should require 28 pounds to compress them to a height of 1.00 inch.

When reassembling, make certain that the 1¼ inch auxiliary spool is at top and the 1⅝ inch auxiliary spool is at bottom of head as shown in Fig. C708. Tighten the head retaining cap screws to a torque of 35 ft.-lbs.

206. OUTLET CHECK VALVES. Each end of the valve body contains two outlet check valves. To service these valves, which are illustrated in Figs. C708 and C709, first remove the pump and valves unit from reservoir housing as outlined in paragraph 192; then, unbolt and remove both heads from the valve housing as shown. Lift out the springs and balls.

Leakage at the outlet valves can be checked by placing the valve body in a vise so that one of the head ends is facing up and in a level position. Place the outlet balls on their seat and fill the valve port with gasoline. If the seat leaks, the valve can be reseated by tapping the ball with a hammer and soft drift. If the seats are damaged beyond the reseating stage, press grease in the seat hole to catch any metal cuttings, tap the seat with a $\frac{9}{16}$ x 18 tap and remove the seat with a screw type puller.

Install the new seat with chamfered end up and press or drive the seat firmly into place, using a piece of ¾ inch round bar stock.

207. RETURN CHECK VALVES. Each of the valve body heads contain two ball type return check valves as shown in Figs. C710 and C711. Since the valves and the seats are wholly contained in the removable valve heads, they can be serviced without removing the pump and valve housing unit from the reservoir housing. To remove the valves, from either head, disconnect hoses, unscrew elbows and extract valve springs and balls. If the valve seats need servicing, remove the valve head or heads from the valve body.

Leakage at the return valves can be checked by placing the head in a vise and in a level position. Place the return valve ball on the seat and fill the port with gasoline. If the seat leaks, the valve can be reseated by tapping the ball with a hammer and soft drift.

If the seats are damaged beyond the reseating stage, it will be necessary to renew the valve body head.

Reinstall the return valve ball and spring (small diameter end contacting the valve ball) by reversing the removal procedure; then reinstall the valve head and tighten the retaining cap screws to a torque of 35 ft.-lbs.

208. DOWN PRESSURE RELIEF VALVE. The down pressure relief valve is located in the right hand head as shown in Fig. C709. To service the valve, remove the head as shown and withdraw the valve spring and ball.

Leakage at the down pressure valve can be checked by placing the head in a vise and in a level position. Place the ball on the seat and fill the port with gasoline. If the seat leaks, the valve can be reseated by tapping the ball with a hammer and soft drift. If the seat is damaged beyond the reseating stage, press grease in the seat hole to catch any metal cuttings, tap the seat with a ⅜-16 tap and remove the seat with a screw type puller.

Install the new seat with chamfered end up and press or drive the seat firmly into place, using a piece of ½ inch round bar stock.

When reassembling, make certain that the 1 3/16 inch auxiliary spool is at top and the 1 7/16 inch auxiliary spool is at bottom of head as shown in Fig. C709. Tighten the head retaining cap screws to a torque of 35 ft.-lbs.

209. FLOAT VALVE. The hydraulic float valve is located in the right hand head as shown in Fig. C711. To service the valve, remove the right hand head from valve body; then, remove the valve retaining screw plug and withdraw the spring and ball as shown.

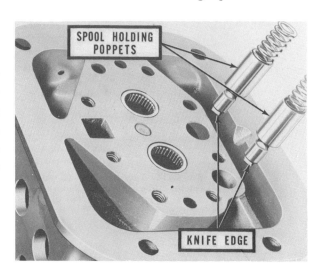

Fig. C712 — The valve spool holding poppets are accessible after removing the hydraulic pump body. Notice the position of the knife edges when reassembling.

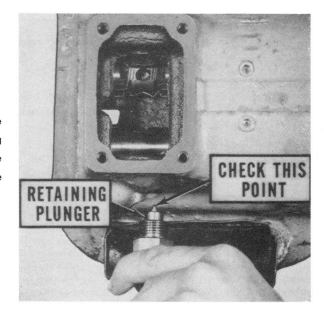

Fig. C713—Removing the float position retaining plunger from the valve body on models with Eagle hitch.

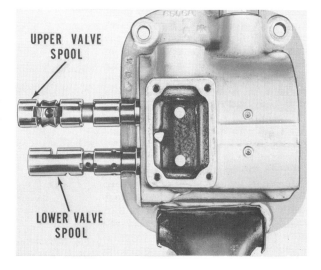

Fig. C714 — Correct installation of the valve control spools on models with Eagle hitch. Note the position of the shifter arm notches.

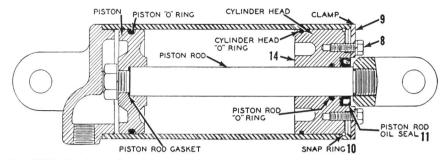

Fig. C715—Sectional view of typical Eagle hitch mounted cylinder used on general purpose tractors. The mounted cylinder used on standard four wheel tractors is the same except a spacer is positioned on the piston rod between the piston and the cylinder head (14).

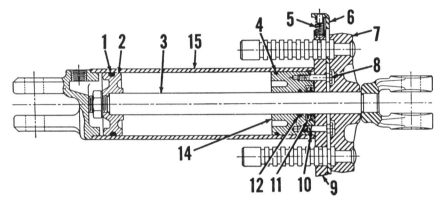

Fig. C16—Sectional view of the portable work cylinder.

1. Piston "O" ring	5. Detent	10. Snap ring
2. Piston	6. Stop bar latch	11. Lip seal
3. Piston rod	7. Stop bar	12. "O" ring
4. Head "O" ring	9. Head clamp	14. Cylinder head

Leakage at the ball seat can be checked by placing the valve head in a level position. Place the ball on the seat and fill the valve port with gasoline. If the seat leaks, the valve can be reseated by tapping the ball with a hammer and soft drift. If the seat is damaged beyond the reseating stage, it will be necessary to renew the head.

210. VALVE CONTROL SPOOL. To remove the valve control spools, first remove the pump and valve unit from reservoir housing as outlined in paragraph 192. Remove the control valve levers as in paragraph 198 and the pump body as in paragraph 193. From pump side of valve body, withdraw the spool holding poppets as shown in Fig. C712. Remove both heads from valve body and remove the float position retaining plunger as shown in Fig. C713. Withdraw the valve control spools.

Renew the spool holding poppets if knife edge of same is excessively worn. Poppet springs (except float position plunger spring) have a free length of $1\frac{13}{16}$ inches and should require 18 pounds to compress them to a height of $1\frac{3}{16}$ inches. If spools or

body are excessively scored, renew the damaged part.

When reassembling, install the valve control spools exactly as shown in Fig. C714 by reversing the removal procedure. Notice the different positions of the shifter arm notches in the upper and lower spools.

WORK CYLINDERS

211. Method of removal of the portable or mounted work cylinders is self-evident. To disassemble work

Fig. C717—Speed regulating valve used to control the lowering speed of the implement.

cylinders, refer to Figs. C715 and C716 and proceed as follows: Remove the four cap screws (8) from the cylinder head clamp (9). Push the cylinder head (14) inward slightly, and remove snap ring (10); then withdraw piston rod assembly from cylinder. Wash all parts and thoroughly inspect same for wear, scoring and scratch marks.

Reassemble the work cylinder by reversing the disassembly procedure. Renew all "O" rings and install the lip type seal (11) with the lip facing toward the rod end of the cylinder.

212. SPEED REGULATOR. A speed regulating valve, Fig. C717 is inserted in the outer left hand hose line between the work cylinder and pump to control the lowering speed of the cylinder.

To retard the lowering speed, loosen the adjusting screw stop screw, then rotate the adjusting screw in a clockwise direction.

Method of removal and disassembly of the speed regulating valve is self-evident. Piston should have a smooth sliding fit of minimum clearance in body. Rubbing surfaces must be smooth. Don't forget to reinstall the stop screw.

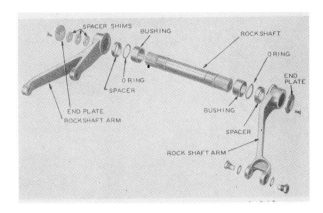

Fig. C718—Exploded view of the rockshaft, arms and associated parts.

ROCKSHAFT AND EAGLE HITCH

213. ROCKSHAFT. To remove the rockshaft, disconnect the lift links and hydraulic cylinder from rockshaft arms. Remove end plates and bump arms from rockshaft. Remove the "O" ring spacers and withdraw the rockshaft. Refer to Fig. C718.

If the shaft bushings are worn, new pre-sized bushings can be installed, with a sizing arbor. The sizing arbor (Fig. C719) can be made, using the dimensions shown.

When reassembling install the rockshaft, "O" rings and spacers. Install arms on the shaft splines so that arms are in the same plane, then install the end plate on right end of shaft. When installing the left hand end plate, vary the number of spacer shims (Fig. 718) to remove end play from the rockshaft without causing binding. Reconnect the lift links and cylinder. Install approximately one pint of SAE 90 transmission oil in the rockshaft housing.

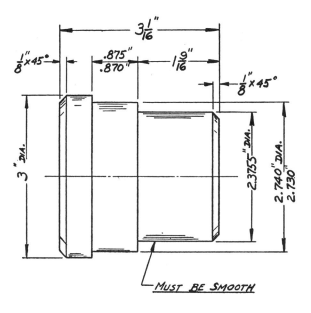

Fig. C719 — Dimensions for making an arber to install the rockshaft bushings.

214. EAGLE HITCH. Overhauling the Eagle hitch consists of disassembling the units and renewing worn parts. Refer to Figs. C720 and 721.

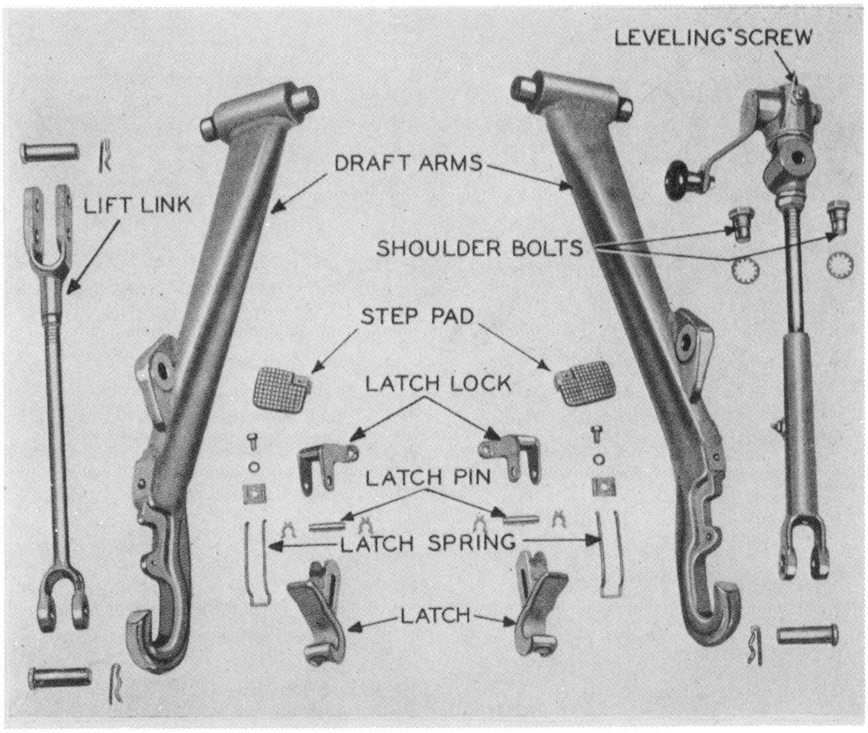

Fig. C720—Exploded view of Eagle hitch draft arm and latch assembly. Refer to Fig. C721 for exploded view of leveling screw.

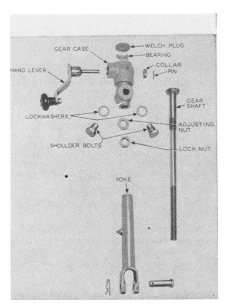

Fig. C721—Exploded view of the Eagle hitch leveling screw.

NOTES